Small Millets and Pseudo Cereals for Nutritional and Health Security

NIPA GENX ELECTRONIC RESOURCES & SOLUTIONS P. LTD.
New Delhi-110 034

Small Millets and Pseudo Cereals for Nutritional and Health Security

Editor in Chief

Gopal Kumar Sharma, FAFST(I)
Former Additional Director & Head
Grain Science and Technology Division
Defence Food Research Laboratory (DFRL)
Defence Research & Development Organisation (DRDO)
Ministry of Defence (MOD), Government of India (GOI)
Siddhartha Nagar, Mysuru, Karnataka 570 011, India

Editors

Anil Dutt Semwal, FAFST(I)
Director
Defence Food Research Laboratory (DFRL)
Defence Research & Development Organisation (DRDO)
Ministry of Defence (MOD), Government of India (GOI)
Siddhartha Nagar, Mysuru, Karnataka 570 011, India

Pal Murugan Muthaiah
Head, Technology Transfer and Exhibition Division
Defence Food Research Laboratory (DFRL)
Defence Research & Development Organisation (DRDO)
Ministry of Defence (MOD), Government of India (GOI)
Siddhartha Nagar, Mysuru, Karnataka 570 011, India

Janifer Raj Xavier
Scientist
Fruits and Vegetable Science Division
Defence Food Research Laboratory (DFRL)
Defence Research & Development Organisation (DRDO)
Ministry of Defence (MOD), Government of India (GOI)
Siddhartha Nagar, Mysuru, Karnataka -570011, India

nipa

NIPA GENX ELECTRONIC RESOURCES & SOLUTIONS P. LTD.
New Delhi-110 034

NIPA GENX ELECTRONIC
RESOURCES & SOLUTIONS P. LTD.
101,103, Vikas Surya Plaza, CU Block
L.S.C.Market, Pitam Pura, New Delhi-110 034
Ph. +91 11 27341616, 27341717, 27341718
E-mail: newindiapublishingagency@gmail.com
www: www.nipabooks.com
For customer assistance, please contact
Phone: + 91-11-27 34 17 17
Fax: + 91-11-27 34 16 16
E-Mail: feedbacks@nipabooks.com

ISBN: 978-93-90591-63-3

Composed and Designed by NIPA.

Preface

Millets and pseudo cereals are one of the oldest cultivated crops known to humans and staple food for millions across the globe. Since, centuries, they are the major source of fodder and feed material for livestock. Millets are small-grained cereal grasses widely grown in African and Asian countries in semiarid tropical regions. Millets were classified into two main groups i.e. major millets (sorghum and pearl millet) and small millets based on the grain size. The small millets were represented by eight species, namely finger millet (*Eleusine coracana (L.)*), little millet (*Panicum sumatrance*), kodo millet (*Paspalum scrobiculatum* (L.)), foxtail millet (*Setaria italica (L.)*), barnyard millet (*Echinochloa frumentacea* (L.)), proso millet (*Panicum miliaceum* (L.)), teff (*Eragrostis tef*) and fonio millet (*Digitarial exilis*)

Pseudo cereals are the non-grass seeds that are used in much the same way as cereals (true cereals are grasses). Their seed can be ground into flour and otherwise used as cereals. Amaranth (*Amarnthus sp.*), buckwheat (*Fagopyrum esculentum*), chia (*Salvia hispanica*), quinoa (*Chinapodium quinoa*) and cockscomb (*Celosia sp.*) are commonly used pseudo cereals across globe. Pseudo cereals are also known as super foods due to presence of high content of vitamins, minerals and anti-oxidants.

Health benefits of millets and pseudo cereals are ample as they have been implicated to take care of coronary artery disorder, helps in weight loss, reduce risk of colon cancer, help to reduce high blood pressure, control diabetics, helps in slowing down muscle degradation, aids in sleep, help in reducing menstrual cramps and preventing gallstones, aids breast milk production and also improve skin elasticity. They are loaded with anti-oxidants and vitamins which remove free radicals and balance enzymatic reaction in human body and known as natural blood detoxifier. They promote digestion due to high fibre content and take control of bowel urine. They can be grown in marginal and poor soil under varied climatic condition. They have been neglected over centuries due to advancement of research and development in major cereals.

Though millets are good for health but their excess use has got adverse effect on human health. Millet possesses goitrogens which interfere production of thyroid hormones and know to inhibit iodine uptake. Goitre is also caused by skin, anxiety, depression and slows thinking.

Emergence of industrial society i.e. metropolitan cities, more women folk taking job, increase in income, standard of living and educational level, changing life style and increase in awareness about the health related issues, etc., brought revolutionary change in dietary habit across the globe. As a result, there is continuous and steady increase in demand for nutrient rich cereals, pseudo cereal and millets. When compared to major millets and cereals, availability of scientific information data on traditional food based on small millets and pseudo cereals are scanty for small millets and pseudo cereals.

The prime focus of this book titled 'Small Millets and Pseudo cereals for Nutritional and Health Security' is to present systematically the importance as well as critical research carried out across the world on these climate complaint crops, their origin, cultivation methods, pre and post-harvest information, processing, chemical composition nutritional composition, physico-chemical characteristics, processing technology, health promoting attributes, traditional and modern foods to scientific fraternity, student community and general public. The book will be highly useful to common consumer also for making nutritious food of choice.

The 20 chapters in this book have been authored by reputed contributors having in-depth knowledge of their specialization from government, industry and academia. We, as editorial team whole heartedly place a record of gratitude to all renowned authors who came forward and devoted their valuable time and treasure of knowledge in the form of chapters in their domain of work. We, the editors strongly believe that this book will be an essential reference for researchers, academicians, students for research and development activities as well as government officials and policy makers to address the issue of malnutition in rural areas. This book will certainly drive future research in unexplored areas of traditional foods based on nutricereals and pseudocereals.

Editors

Contents

1

Small Millets and Pseudo Cereals An Overview

Pal Murugan Muthaiah, Janifer Raj Xavier, Gopal Kumar Sharma **and** ***Anil Dutt Semwal***

Defence Food Research Laboratory, Siddarthanagar, Mysore-570011, India

Introduction

Small millets and Pseudo cereals are staple food for millions of small and marginal farming community and household in developing and under developed countries. They are major source of fodder and feed material for livestock. They are nutritionally alike or superior to major cereal grains with additional advantage of proteins (gluten-free), enhanced fiber, lower glycemic index and bioactive phyto chemicals made them an appropriate balanced health food. They perform well in marginal and poor soil under varied climatic condition. These crops are complaint to challenges faced due to climate change and are suitable substitute to major cereals to meet food grain requirement and eradication of malnutrition of increasing world population.

Small Millets

Millets are small annual cereal grasses other than wheat, oat, barley and rice belonging to family Poaceae. Millets are the first crops to be cultivated prior to Stone Age. The millets are classified as major and minor millets on basis of size of grains and area under farming. Sorghum (*Sorghum bicolour*), maize (*Zea mays*) and pearl millet (*Pennisetum glaucum*) are considered as major millets. Finger millet (*Eleusine coracana*), barnyard millet (*Echinochloa frumantacea*), foxtail millet (*Panicum italicum* L.), kodo millet (*Paspalum scrobiculatum*), proso millet (*Panicum miliaceum*), fonio millet (*Digitaria exilis*), teff (*Eragrostis tef*) and little millet (*Panicum miliare*) are called minor millets due to their tiny sized seeds. They are also called as bird's food or miscellaneous cereals. The minor millets are also known as nutri-cereals as they contain all the essential nutrients (Sharma *et al.*, 2006).The word millet has its origin from the French word "Mille" meaning thousands which explains that a handful of millet contains thousands of seed (Taylor and Mohammad, 2008). Seeds of small millets are mostly round to oval in shape and 1000 seed weight is in the range of 1.9-5.5 g while volume ranges from and 1.3 - 3.8 ml. (Vikaspedia, 2019). More than 6000 millet varieties were grown throughout different nations with grain colour varying from pale yellow to grey white and red (sahajasamrudha.org/images/Millet percent20Booklet.pdf).

The small millet grains or caryopsis resemble dehusked rice or paddy in the morphological features. The caryopsis has a discrete husk, bran, endosperm and an embryo which contributes to 2 percent of total weight of kernel. The husks of small millets are non-edible, as that of paddy and bran has to be removed to obtain milled millets for food application. As that of cereals grains, husk accounts for 15 to 20 percent of the kernel, endosperm occupies 75 percent of kernel where as bran contributes to about 5 percent respectively. Foxtail, little and proso millets have glossy seed coat made of single layer and husk while seed coat and husk of kodo and barnyard millet are multi layered. Most of the small millets seed coat is greyish colored except foxtail millet and kodo millet which are of brown color and yellow color respectively. As that of paddy, the husks of small millets are non-edible and unusually difficult to digest. Milling of millets is necessary to prepare edible grains as that of paddy and other cereals. Small millets can be cultivated all type of soil *i.e* marginal saline, low fertile under varying climatic conditions, rain fed, irrigated conditions, montane ecosystem from mean sea level to 2000 m of MSL and also as late seasonal crops. The crop duration ranges from 2-4 months. Small millets may occupy area between 18 to 20 million hectare with production ranges between 15-18 million tonnes of grain. India is a major producer of millet in the world.

Pseudo Cereals

Pseudo cereals are an ill-defined group of crops species that are not closely related to one another and are less closely related to cereals. But, like cereals, they have starchy, dry seeds. Cereals such as wheat, oat, barley, rice and millets belong to monocotyledonous species, classified by taxonomists, based on important characteristics shared in the grass family, Poaceae, where as pseudo cereals *i.e* buckwheat (*Fagopyron esculentum*) belong to Polygonaceae family, quinoa (*Chenopodium quinoa*) belongs to Chenopodiaceae family, amaranth and cockscomb belong to family Amaranthaceae and Chia belongs to Lamiaceae family are of dicotyledonous plants. Their seeds can be consumed as that of cereals grain or can be milled into flour for different food preparations. Botanically, pseudocereals are neither grasses nor true cereal grains. Pseudocereals are typically high in protein and other nutrients and are gluten-free.

History of Millets

Lake inhabitants of Switzerland grew the millets during Stone Age. Cultivation of millets in arid climates of Africa and Northern part of China was reported during Neolithic era. It is staple food for India, China and Korean civilization during Neolithic era. Asia and Africa are the major centres of origin for the millets (sahajasamrudha.org/images / Millet percent 20 Booklet.pdf). Tropical West Africa is the centre of origin for pearl millet and Uganda for finger millet. From Africa, finger millet was introduced to India and Europe before 3000 and 2000 years respectively. Since ancient times, millets are part of human food system.

Paleoethnobotanical studies at archeological sites Harappa and Mohenjo-Daro showed the presence of millet as part of Indian food system. Millets were also used by earlier civilization *i.e* Classical Africans, Mayans, Incas, and Aztech in their traditional dishes. The king of Shang Dynasty/Yin dynasty ruled in lower yellow river valley of China during 2000 BC was called as Hou Chi meaning 'The ruler of millet'. References about millets were also found in Indian Vedic Literature Sathapatha Brahmana and Kalidasa's Shakuntala. Millets were considered as sacred in China since ancient times. Earliest record related to millet crop cultivation dates back to 2800 BC. Hebrew bible had mention about millets. Middle East, Northern Africa and Sumerian di et also cultivated millets as staple food during 2500 BC. Hanging gardens of Babylon had millets as precious crops and it is believed that Egyptians learned cultivation practices of millets from Africans during trade practices with Eritrea and Somalia circa during 3000 BC.

Finger millet was developed in Africa from *Eleusine coracana* subsp. *africana*, probably in the Ethiopian region. It was introduced to India perhaps more than 3,000 years ago (Hugh Dogget, 1986). De Wet *et al.* (1984) recognized five races in finger millet. Proso millet an ancient crop was well known to the Romans, and became the 'common millet'. This is essentially a crop of the temperate regions, but is also grown in the sub-tropics and on high ground in tropical winters. Foxtail millet is yet another ancient crop, probably domesticated in eastern Asia, and known to the Chinese as early as 2,700 BC (Purseglove, 1985). Cobley (1976) noted that more than 12 rather variable groups of cultivars have been recognized. Little millet is found in Northern India and Southeastern Asia in wild form. It will yield some grain and useful fodder under very poor conditions. Some forms mature in as little as two-and-a-half months (Purseglove, 1985). Around 4,000 years ago Japanese domesticated Japanese barnyard millet *Echinochloa crusgali.* It belongs essentially to the temperate zone. Barnyard (Sawa) millet, *E.colona,* was domesticated in India, Central African Republic, Tanzania and Malawi. Chromosome numbers of two species are 2n=54 and 2n=36 while hybrids between them are sterile. The wild form is wide spread as a tropical weed. De Wet *et al.* (1983a) distinguished four races. Kodo is a wide spread tropical weed. The crop has been grown for at least 3,000 years, yet scientists could not find any clear racial differentiation. Wild, weed and cultivated types merged in all the characters studied. Teff is a major cereal crop in Ethiopia, particularly least porous, heavy soils that predominate in the Central Plateau. Nevertheless, the crop has not become important outside Ethiopia.

History of Pseudo Cereals

Buckwheat, amaranth, quinoa, chia and cocks comb are four major groups of Pseudocereals documented worldwide. Amaranth and quinoa were widely grown in Mesoamerica during Hispanic period. They were grown throughout Latin America for thousands of year and expected to be cultivated as early as 4000 BC (Palomino *et al.*, 2008). Amarnth crop is originated from Peru of South American continent. Amaranth grain features a long and colorful history in Mexico. It had been a serious food crop of the Aztecs of Mexico, and a few have estimated amaranth was domesticated before 6,000 and 8,000 years. The Aztecs used the grains in ceremonies for making religious idols using a mixture of amaranth grains and honey. Once formed, the idols were worshipped before being broken into pieces and distributed for people as religious sacred things. Quinoa, an Andean crop is originated near Lake Titicaca in Peru and Bolivia. It was major crop cultivated by pre-Columbian civilizations and was replaced by other cereals on the arrival of the Spanish, despite being an area staple food at the time. Historical evidences showed that its domestication of quinoa by the peoples of America may have occurred between 3,000 and 5,000 years BC (FAO and CIRAD, 2015). Archeological studies also reported about the presence of quinoa in tombs of Tarapacá, Calama and Arica in Chile and in several regions of Peru (Cardenas, 1944). During Spanish invasion, quinoa was well developed technologically and was cosmopolitan within and beyond Inca territory. The primary Spaniard to notice the cultivation of quinoa was Pedro de Valdivia who, on noticing the planted crops around Concepción, recorded that, for food, the native Indians also sowed quinoa among other plants. In his royal commentaries, Garcilaso de la Vega, describes quinoa cultivation along with other and resembling millet or short-grain rice. He also mentions the primary shipment of seeds to Europe, which were unfortunately dead on arrival and unable to germinate, perhaps due to the high humidity of the ocean voyage.

Buckwheat, *Fagopyrum sps* is believed to originate in Southeast Asia around 5 or 6 thousand years ago. From there it spread to Central Asia, the center East, then Europe. Buckwheat was documented in Finland by a minimum of 5300 BCE. It had been finally brought to North America within the 1600s. Two buckwheat species, those are majorly cultivated for food consumption in Asia, common buckwheat-*Fagopyrum esculentum* and tartary buckwheat-*Fagopyrum tartaricum* (Oomah and Mazza, 1996). *Celosia cristata*, belongs to Amaranthaceae and is usually referred to as cockscomb. It is also known as crested *celosia*, foxtail amaranth, fire-flame bush and red cockscomb.

Chia is obtained from seeds of desert plant *Salvia hispanic* belonging to the mint family Lamiaceae. The Aztecs kingdom of Mexico possesses the first record of Chia since 3500 BC. It was used as important ingredients in the Aztec diet. Teotihuacan and Toltec people of Mexico cultivated chia between 1500 and 900 B.C. These Ancient people used chia seeds for many purposes and trusted chia provided some magic power. In Mayan, "chia" means "strength" because of large amounts of energy provided by chia seeds. Ancient warriors of Mexico attributed their stamina to this tiny seed. Runners of Mexican Tarahumara tribe drink Iskiate prepared using mixture of chia seeds, lemon, and water for endurance. Like quinoa, chia also lost its importance on Spanish invading Mexican territory. To promote Spanish religion, they banned all religious use of chia seeds. Despite of hurdles, cultivation of chia has been carried out certain region of Mexico by tribal farmers and reemerged for our modern-day use. Latin countries and Australia are major producer of chia seeds. Chia seed can be included in many foods as it does not have seamy flavor. It can substitute eggs as it forms thickened gel when mixed with liquid (Neville and Kerry, 2015)

Brief about Small Millets and Pseudo Cereals

Millets and pseudo cereals play an essential role in meeting livelihood of one third of world population who depend dry land or rain fed agriculture for their sustainability. They perform well in adverse climatic condition in different countries. The vernacular name prevailed in different regions of India for small millets and pseudo cereals in described (Table 1).

Finger Millet

Finger millet, *Eleusine corocona* derived its name from finger like terminal inflorescences. (de Wet, 2006; Quattrocchi, 2006). Seeds of finger millet bear reddish brown to red color (Fig 1). It is a short plant with abundant tillers and shallow root system matures within 3-6 months depending upon the climatic condition and varieties. It is a tropical crop, grown from sea level to 3,000 mean sea level (MSL) (FAO, 2012) and most widely cultivated in India and Africa. It performs well in rain fed conditions and provide better yield in response to fertilizer application and irrigation management. It is cultivated extensively in southern part of India especially in Karnataka and part of Tamil Nadu. Finger millet also known as ragi contains high quantity of calcium, protein with essential amino acids namely thiamine, riboflavin and niacin composition along with vitamin A, B and phosphorous. It is also good for constipation, high blood cholesterol and intestinal cancer due to high fibre content. Ragi flour is a major ingredient in ragi balls, popularly known as ragi mudde, made into flatbreads, leavened dosa and thinner, unleavened rotis.

Fig. 1: Finger millet

Table 1: Vernacular names of millets and pseudo cereals in different Indian languages

English	Others	Botanical	Hindi	Kannada	Tamil	Telegu	Malayalam	Marathi	Gujrathi	Bengali	Oriya	Kashmiri
Finger Millet	Rajika	*Eleusine coracana*	Mandua /Madua	Ragi	Kelvargu / Kezh-varagu	Ragulu	Moothari	Nachni	Bhav / Nagali / Bavto	Mandua / Madua	Mandia	-
Foxtail Millet	Moha Millet / Italian Millet	*Setarai italica*	Kangni	Navane / Priyangu Thene	Tenai	Korra / Korralu	Thina	Rala	Ral Kang	Syama Dhan	-	Shol
Kodo Millet	Pakodi / Manakodra	*Paspalum scrobiculatum*	Kodra	Harka	Varagu	Arikelu	Varagu	Harik	Kodra	Kodoad-han	Kodus	
Proso Millet	French Millet / Common Millet	*Panicum miliaceum*	Barri	Baragu	Panivar-agu	Varigulu / Varagalu	Panivaragu	Vari	Vari	-	Chinna	Pingu
Barnyard Millet	Japanese Mil-let / Sawank	*Echinochloa frumantacea*	Jhangora / Shama	Samai	Kuthi-ravaali	Odalu / Bonta Chamula	-	Shamul	Sama	Shamula	-	-
Little Millet	Goudli / Gondola	*Panicum miliare*	Kutki	Same	Samai	Sama	Chama	Sava	Gadro	Kangani	Suan	Ganuhaar
Teff	William lovegrass	*Eragrostis tef*	-	-	-	-	-	-	-	-	-	-
Fonio Millet	Hungry rice	*Digitaria exilis,*										
Amaranth	Rajakira	*Amaranthus Sps /*	NA		Keerai Vidai	Thotakoora Ginjalu	Cheera Vitu					
Quinoa	Quinoa	*Chenopodium quinoa.*		-	-	-	-	-	-	-	-	-
Buck Wheat	buckwheat	*Fagopyrum esculentum*	Kuttu									Trumba
Cocks-comb	Cockscomb	*Celosia sps cristata*	Lalmur-ga anaaj	-	Kozipoo vidai	-	-	-	-	-	-	-
Chia	Salvia	*Salvia his-panica*		Ciya bijakalu	-	-	-	-	-	-		-

Source: Dayakar *et al.*, 2017; https://www.millets.res.in/pub/2018/The_Story_of_Millets.pdf

Proso Millet

Proso millet, *Panicum miliaceum* known as broomcorn due to its compact panicle droops like an old broom (Fig 2). It is also known as temperate millet or common millet and grain possess high nutritional value. Proso millet grain varies in color from brown to black, red to orange, white cream, yellow and is spherical to oval in shape. Seeds also have fibre content and hull have to be removed by milling before consumption. The plant thrives in all soil types and climate conditions. It is a short season crop, grows three to four feet height and mature within 60 to 75 days after planting and usually grown as late seeded summer crop in case of main crop failure (Taylor *et al.*, 2008).

Fig. 2: Proso millet

Kodo Millet

Foxtail millet (*Setaria italica*) derived its name from foxtail like panicle. The panicles vary in color and size accordingly the plants are called yellow foxtail, green foxtail, or giant foxtail (Fig 3). Semi-arid regions of world widely cultivate this millet while subtropical and temperate zones also supports cultivation. The height of the plant varies from1-5 feet and has shallow root system. Although, foxtail millet perform in low level of irrigation or water content it not recover from drought condition. It matures in 65-70 days. Foxtail millet can be planted when it is too late to plant most other crops. The grain structure shows resemblance to paddy grain and the outer husk is removed before cooking. The main production areas are in Japan, China, India and Eastern Europe (Chenjiaju, 1986).

Fig. 3: Foxtail millet

Kodomilet

Kodo Millet (*Paspalum scrobiculatum* var. *scrobiculatum*) is a wide spread tropical weed also known as cow grass, rice grass, ditch millet, native paspalum, or Indian crown grass. Centre of origin is tropical Africa and domesticated in India 3000 years ago (de Wet *et al.*, 1983b). It is annual grass which grows up to 3 to 4 feet and has inflorescence with 40-6 racemes. The grain is enclosed in hard, corneous, persistent husks that are difficult to remove. The grain colour varies from light red to dark grey (Fig 4). The fibre content of the whole grain is very high. As with other food grains, the nutritive value of Kodo millet protein could be improved by supplementation with legume protein (Deshpande *et al.*, 2015). It is a minor grain crop in India, and an important crop in the Deccan plateau. It is cultivated in Gujarat, Karnataka and Tamil Nadu. Kodo millet is said to be poisonous after rain duc to a fungal infection. Winnowed, clean healthy grain seems to pose no health problem. The fungal disease Paspalum ergot causes nervousness; lack of muscular coordination, staggering gait, spasms depression and in extreme cases causes death in animals (Antony *et al.*, 2003).

Fig. 4: Kodomilet

Little Millet

Little millet, (*Panicum sumatrense*) is important catch crop in terms of its early maturity and resistance to adverse agro-climatic conditions of high drought as well as water logging. The plant grows up to 3-4 feet height and bears oblong panicle (Rao *et al.*, 2017). It has received less attention of breeders because of their tiny seeds. It is cultivated throughout India up to altitudes of 2,100 m in different cropping system. It is generally consumed as rice and any recipe that demands staple rice can be prepared using little millet (Venkatesh Bhat *et al.*, 2018).

Fig. 5: Little millet

Barnyard Millet

Barnyard millet, (*Echinochloa frumentacea*) is a grain crop of lesser importance. It is very drought resistant but is also capable of withstanding waterlogging conditions. It is generally grown as a rain fed crop. The grains of Barnyard millet are consumed as that of rice. They are also used in making rice pudding (kheer) (Fig 6). The digestibility of protein is 40 percent. Mostly grain is used for consumption by poor, but sometimes it is brewed the beer and bird feed. The straw makes good fodder for cattle. Its green fodder is very much relished by cattle. Barnyard millet is grown in India, China, Japan, Malaysia and East Indies. In China and Japan it is said to be grown as a substitute crop when the rice crop fails (Bhat *et al.*, 2018). It is an herbaceous annual, which tillers sparsely. Plant grows up to a height of 60 to 120 centimeter. The inflorescence is a panicle with densely crowded unawned spikelet. Colour of the grain is generally yellow or white.

Fig. 6: Barnyard millet

Teff

Teff, (*Eragrostis tef*) is the most significant cereal crop of Ethiopia, particularly in the poorly drained, heavy soils that predominate in the Central Plateau. Nevertheless, the crop has not become important outside Ethiopia (Fig 7). It is cultivated on over one million hectares annually while maize and wheat is grown less than a million hectares (Arendt and Zannini, 2013; Bultosa, 2016). Teff is a sexually propagated self-pollinated annual (Tareke, 1975; Tadese, 1975). It is tufted or bunching grass with thin narrow stems and a broad crown and colour of seed varies from white to deep red-brown. Teff seeds are among the smallest of all cereal grains, usually measuring less than 1 mm (0.04 inch) diameter. (Herak, 2016)

Fig. 7: Teff

Fonio Millet

Fonio, *Digitaria exilis* also called as hungry rice, is cultivated as a cereal crop throughout Savannas, West Africa as well as Guinea and Nigeria as a staple food (Fig 8). West Africa cultivates about 300,000 hectares annually and 3-4 million people are fed (Vietmeyer, 1996). It is the oldest West African cereal and its cultivation is thought to date back to 5000 BC. Fonio is rich in amino acids such as cysteine and methionine, which are devoid in other common cereals (Kura *et al.*, 2003). Two kinds of fonio *i.e.* white and black are being cultivated. White fonio is cultivated in Senegal, Chad and Nigeria and in raised plateaus of central Nigeria. The black fonio cultivation is limited to Bauchi and Jos of Nigeria as well as northern Benin and Togo (Idris *et al.*, 2015). The crop can grown in poor, shallow and rocky soils, but is not grown outside Africa. *D. iburua* is a white seeded form (Purseglove, 1985).

Fig. 8: Fonio Millet

Pseudocereals

Amaranth

Amaranth originated in America and is one of the oldest food crops in the world, with evidence of its cultivation reaching back as far as 6700 BC. Amaranth, name derived from the Greek word Amaranthus that means eternal flower and it is considered as an important nutritional crop (Fig 9).The genus Amaranth consists of nearly 60 species, several of which are cultivated as leaf vegetables, grains or ornamental plants, while others are weeds.. Three species - *Amaranthus cruenus*, *A. hypochondriacus*, and *A. caudatus* — are commonly grown for their edible seeds. Amaranth is also called as a super food as it contain high quality of carbohydrates, dietary fiber, lipids as omega-3 and omega-6, essential amino acids and phytochemicals such as squalene, tocopherols, phenolic compounds, flavonoids, phytates, vitamins and mineral (Manuel *et al.*, 2018).

Fig. 9: Amarnath

Quinoa

Quinoa (*Chenopodium quinoa* Willd.) is a pseudocereal belonging to Andean regions of South America (Matiacevich *et al.*, 2006). Quinoa belongs to the Chenopodiaceae family, genus Chenopodium. The classification of quinoa was first made from morphological traits such as plant and fruit colour (Fig. 10). Despite the wide variation observed, quinoa is considered to be one single species (Michala *et al.*, 2009). For practical reasons, quinoa, like maize, has been classified as a race. Quinoa sps collected from Ecuador, Peru, and Bolivia, has been classified into 17 races, however, more races may exist (FAO and CIRAD, 2015).

Fig. 10: Quinoa

Buckwheat

Buckwheat, *Fagopyrum esculentum Moench* is one of the traditional crops cultivated in Asia, Central and Eastern Europe (Wijngaard and Arendt, 2006). It is categorized as a pseudocereal, so it shows both differences and similarities with cereals. Common buckwheat (*Fagopyrum esculentum Moench*) is the most common grown species. It is an annual, dicotyledonic plant from the family of Polygonaceae (Skrabanja *et al.*, 2003). Its seeds are edible and have triangular shape (Fig 11). However, the shape, size, and color of the seed may vary depending upon the species and variety. The hull may be a glossy or dull brown, black or grey. The dehulled buckwheat seed, called the groat, resembles the cereal kernel in its gross chemical composition and structure (Misra and Jain, 2019).

Fig. 11. Buckwheat

Chia

Chia means strength and Mexican tribes used the tiny black and white seeds as an energy booster. Chia, "*Salvia Hispanica*" belonging to the mint family Lamiaceae. Chia seeds are consumed as whole grain food and can be absorbed during digestion (Fig 12). The word chia is derived from a Spanish word chian-means oily and it is oilseed. Chia seeds are nutrient "power house" loaded with healthy omega-3 fatty acids, fiber, poly unsaturated fatty acids and protein-including all essential amino acids, calcium and several other minerals (Cahill, 2003). The presence of higher content a of fibre help in diabetes mellitus by slowing down the digestion process and release of glucose, it also improves the peristaltic movement of intestine and reducing plasma cholesterol. The biological value of chia is superior to cereals due to enhanced level of calcium, magnesium and potassium in comparison to milk (De Tucci, 2006)

Fig. 12. Chia

Cockscomb

Cockscomb (*Celosia sps*) is a minor group among the edible and ornamental plants belonging to Amaranthacea. It is also known as crested celosia, fox-tail amaranth, fire-flame bush and red cockscomb. The tight, velvety texture of the cockscomb flowers appear as if brain tissue to some people(Fig 13). The fluffy, light, airy texture of the plume types blow freely during a breeze and are planted more often. (Esmail and Snafi, 2015) The precise origins of cockscomb within the wild are unknown, though some assign the geographical origins to the dry slopes of Africa and India also because the dry rocky regions of both North and South America. Celosia comes from the Greek word 'kclos', meaning burned; apparently pertaining to the design of the brightly colored flowers in some species. Cockscomb is widely believed to possess developed from red fox, which is listed as native to India but is common within the wilds in China.

Fig. 13. Cockscomb

Agriculture Importance of Small Millets and Pseudo Cereals

According to UN report, the world population was estimated to have reached 7.8 billion populations as of March 2020 with growth rate of around 1.05 percent per year. The demand of food will increase proportionately with growth in world population. At the moment about 50 percent of world's total calorie intake springs directly from cereals (Awika, 2011). Rice, wheat and maize being major staple food contribute maximum followed by sorghum and millets to certain extend. Small millets and pseudo cereals are suitable for cultivating in arid, semi-arid or mountain zones as rain fed crops, under marginal and sub marginal conditions of soil fertility and moisture. As that of small millets, pseudo cereals also mature within 120-150 days and perform well in rain fed and conditions. The average grain yield is almost a half to three tonnes per hectare. Presently, small millets and pseudo cereals are cultivated in areas where they produce a more dependable harvest compared with any other crop. This has been largely responsible for their continued presence and cultivation in many parts of the world. There is now an increasing realization of this fact, and a greater awareness that these crops merit more research and development (Rao, 1986)

Scientific development in the field of genetic engineering, genomic mapping, mutational and hybrid breeding by agricultural universities and international crop research institute resulted in the development of high yielding varieties of millets throughout the world (Kumar *et al.*, 2018). Newly developed hybrids of millets and pseudo cereals were resistant to diseases, pests and drought and increased per hectare yield compared to their parental varieties (Joel, 2005). Millets and pseudo cereals have abundant natural diversity, and the release of new varieties and hybrids increases this variation by several folds. In addition to above, the advancements in the post-harvest operations of millets have eased their processing.

The traditional methods of processing such as drying, winnowing, hand pounding, etc. used for the decortication of grains were labor intensive and time consuming. Development of crop specific threshers, decorticator, destoner, milling machinery and polishers improved the processing and led utilization of millets and pseudo cereals in development of nutritious food products. The increase in cultivation area of millets and pseudo cereals than traditional staple crops pave way to the existing agrarian challenges and can prove a milestone in achieving United Nations commitment to end malnutrition in all its forms by 2030 (UNICEF, 2016). The nutritional value of small millets and pseudo cereals in comparison with major staple food crops is shown (Table 2).

With rapid urbanization, changing life styles and technological advancement, world population is facing several health disorders and chronic diseases. As per 2018 Global Nutrition report, 150.8 million are stunted, 50.5 million are wasted and 38.3 million are overweight. Obesity among adults are at record levels with 38.9 percent of adults overweight or obese, stretching from Africa to North America and increasing among adolescents. Women have a higher burden than men when it comes to certain forms of malnutrition: one third of all women of reproductive age have anaemia and women have a higher prevalence of obesity than men. Millions of women are still underweight. Nutritionally imbalanced diet is responsible for most of these diseases (Global Nutrition Report, 2018).

Millets for Nutritional Security

Macronutrients

Average carbohydrate content of small millets and pseudo cereals varies from 56.88 to 82.00 g/100 g and 31.41 to 74.1 g/100 g respectively. Least carbohydrate content has been reported in barnyard millet and cockscomb. The protein content of small millet is almost similar to the major millet with range of protein content from 7.44 to 11.44 g/100. Finger millet, which has been reported to contain low protein in the range of 4.76 to 11.70 g/100 g in different studies (Panghal *et al.*, 2006; Singh and Raghuvanshi, 2012). The protein content of pseudo cereals varies from 11.3 to 30.75 g/100g with highest protein content in cockscomb. The perusal (Table 2) clearly revealed that among small millets, finger millet contain low level of fat followed by teff while highest fat content has been reported little millet (NIN, 2017). Among the pseudo cereals chia has been reported to have higher fat content of 24.8 g/100 g followed by quinoa *i.e.* 7.0 g/100 g. Chia and quinoa is called as super foods as they contain higher content of protein and fat compared to wheat and rice (Mariana *et al.*, 2019). Among pseudo cereals, cockscomb exhibited highest level of crude fibre content 10.22 g/100g followed by buckwheat 9.35 g/100g and quinoa 8.0 g/100g, which is 78.5 percent higher than rice and wheat; this makes millets to control the post- prandial blood glucose level. The calorific value varies from 300-473 kcal /100g of grain consumed for small millets and pseudo cereals.

Micronutrients

The minerals and vitamins are known as micronutrients and required in very small quantities to maintain the body functions. The mineral content in small millets and Pseudo cereals ranges from 1.72 to 4.5 g/100 g and 0.22 to 4.8 g/100 g respectively which is several folds higher than the staple cereals like

Table. 2: Proximate composition of millets and pseudocereals in comparison with wheat, rice and major millets

Crops	Moisture (g)	Protein (g)	Fat (g)	Ash (g)	Dietary Fibre (g)			Carbohydrate (g)	Energy (Kcal)
					Total	Insoluble	Soluble		
Finger millet1	10.89±0.61	7.16±0.63	1.92±0.14	2.04±0.34	11.18±1.14	9.51±0.65	1.67±0.55	66.19±1.19	334.0
Foxtail millet	14.23±0.45	08.92±1.09	2.55±0.13	1.72±0.27	06.39±0.60	4.29±0.82	2.11±0.34	67.30±5.70	353.4
Barnyard Millet	8.7	10.5	3.6	2.0	12.6	8.4	4.2	62.2	368.0
Proso millet	15.35	11.74±0.86	3.09±1.18	2.73±0.72	-	-	-	67.09 ± 4.79	352.5±1.62
Kodo Millet	12.0	8.3	1.4	2.6	9.0	-	-	65.9	349.
Little Millet	12.0	7.7	4..7	4.5	7.6	-	-	67.0	352
Teff	7.37	12.77	3.19	2.2	10.6	-	-	63.87	383
Fonio Millet	4.68	4.44	1.11	3.1	2.2			84.47	378
Amarnth	9.89	14.59	5.74	2.78	7.02	5.76	1.26	59.98	356
Quinoa	5.04	14.29	7.14	2.7	7.1	-	-	64.33	375
Buck Wheat	9.75	13.25	3.4	2.1	10	-	-	71.5	343
Cockscomb		30.79±2.72	19.98±1.04	0.22±0.06	-			31.41±1.92	248±9.2
Chia	7.14±0.26	18.8 ±1.20	32.16 ±0.29	4.8±0.8	33.37±0.26	30.4±0.35	2.89±0.09	4.59±0.34	473.11±
Rice	09.93±0.75	07.94±0.58	0.52±0.05	0.99±0.42	02.81±0.42	1.99±0.39	0.82±0.22	78.24±0.68	369.0
Wheat	10.58±1.11	10.59±0.60	1.47±0.05	1.42±0.19	11.23±0.77	9.63±0.19	1.60±0.75	64.72±1.74	438.0
Sorghum	09.01±0.77	09.97±2.45	1.73±1.60	1.39±0.34	10.22±0.49	8.49±0.40	1.73±0.40	67.68±1.03	329.0
Pearl millet	08.97±0.60	10.96±0.26	5.43±0.64	1.37±0.17	11.49±0.62	9.14±0.58	2.34±0.42	61.78±0.85	363.0

Source: Longvah *et al*., 2017; Ugare *et al*., 2011; Himanshu *et al*., 2018; Deshpande *et al*., 2015; Kundgol, 2014; Akansha *et al*., 2018 Steadman *et al*., 2001; Adegbaju *et al*., 2019; Abdellatif ASA, 2018; https://fdc.nal.usda.gov/fdc-app.html#/food- details/361081/nutrients.

wheat (1.5%) and rice (0.6%). The highest level of minerals was reported in little millet and chia seeds. Among pseudo cereals, cockscomb has very low level of minerals content. The details of micronutrient content of millets have been depicted (Table 3 and 4). National Nutrition Monitoring Bureau (NNMB) reported wide spread deficiency of calcium and iron among working population, lactating women and growing children. Studies on calcium and vitamin D over period of 50 years revealed that high prevalence of twin nutrient (Ca and vitamin D) deficiency among Indian population (Harinarayan *et al.*, 2019).

NNMB data (1981–2011), also reported about decline in dietary intake of calcium both in rural and tribal population in the past four and a half decades, which resulted osteoporosis among large chunk of adult population. It is observed from the Table 3 that calcium content of small millets and pseudo cereals is in the range 10-364 mg/100 g with lowest amount in fonio millets. Finger millet owned highest level of calcium (364 mg/100g) followed by barnyard millet (190 mg/100g), amaranth (181 mg/100), cockscomb and teff. Calcium content is about eight to twelve times higher than wheat, rice and other major millets. The dietary intervention for inclusion minor millet and pseudo cereals will play major role in reduction of osteoporosis among Indian population. Iron is an important micronutrient, which is essential for various functions in human body. Iron Deficiency Anemia (IDA) remains to be important public health problem in India. It is estimated that about 20 percent of maternal deaths are directly related to anemia (Anand *et al.*, 2014). Small millets and pseudo cereals are rich source of iron. Little millet (20 mg/100g), barnyard millet (18.6 mg/100g), teff (15.7 mg/100g) and cocks comb (15.5 mg/100g) contain higher quantity of iron which is only 8-10 g less than required daily value.

Magnesium is required for more than 300 biochemical reactions in the body. It aids to sustain normal nerve and muscle function, maintain a healthy immune system, keeps the heartbeat steady, and helps bones remain strong. It also helps adjust blood glucose levels. It also helps in the production of energy and protein (Magnesium in diet: MedlinePlus Medical Encyclopedia). 30-420 mg of magnesium required depending upon the age and sex. The magnesium content of small millets ranges from 171-181 mg/100g except fonio millet. Nearly half of the recommended daily allowance of magnesium can be obtained by consuming small millets. As that of small millets, pseudo cereals are also rich source of magnesium with content ranges from 10.69 to 502 mg/100g. Quinoa reported to have high level of magnesium followed by chia.

Manganese is an important trace mineral that is naturally present in many foods and available as a dietary supplement. It is involved in amino acid, cholesterol, glucose and carbohydrate metabolism, reactive oxygen species scavenging, bone formation, reproduction and immune response blood clotting and hemostasis in conjunction with vitamin K (https://ods.od.nih.gov/factsheets/Manganese-HealthProfessional/). Food Nutrition Board (FNB), USA recommended adequate intake of 0.003 mg-2.3 mg per day for leading healthy life . Small millets and pseudo cereals are good source of magnesium and they possess more than required level of magnesium compared to rice and wheat. Among small millets teff (3.8 mg/100g) contains high level of manganese followed by finger millet (3.1 mg/100g). The manganese content of pseudo cereals ranges from 1-64-5.29 mg/100 mg. Finely J W (1999) reported that dietary iron intakes and iron status (measured by serum ferritin concentration) appear to be inversely associated with manganese absorption.

Phosphorus is one of the mineral contained in each cell in our body. Phosphorus is available in higher quantity in bones and teeth and in our genes. Our body requires phosphorus to make energy and to carry out many important chemical processes. Depending upon the age and sex, average daily-recommended amount of phosphorous varies from 100 mg to 1250 mg. The phosphorus deficiency can cause loss of appetite, anemia (low red blood cell counts), muscle weakness, coordination problems, bone pain, soft and deformed bones, a higher risk of infection, a feeling of burning or prickling in the skin and confusion. Chia, teff, quinoa and amaranth contain higher quantities of phosphorous when compared to other small millets and pseudo cereals. It is almost higher or equal to wheat, rice and major millets.

Potassium is a mineral highly required for proper kidney and heart function, muscle contraction and nerve transmission. Depending upon the age and sex, average daily-recommended amounts phosphorous varies from 400 mg to 3400 mg. Minimum potassium intake results in increased blood pressure, deplete calcium in bones and increase the risk of kidney stones. Small millets and pseudo cereals relatively contain higher amount of potassium compared to rice and wheat. Highest level of potassium is found in quinoa (732 mg/100g) followed by finger millet (400 mg/100g) and teff (380 mg/100g).

Besides above minerals, small millets and pseudo cereals are also rich in sodium, selenium and zinc. Amaranth possesses higher amount of selenium, which 16 times higher than rice. Fonio millet contains higher quantities of zinc. Zinc along with iron play important role in enhancing the immunity. Small millets and pseudo cereals are also good source of both water and fat-soluble vitamins except vitamin C. The thiamine, riboflavin and niacin content are comparable to that of rice and wheat. The highest thiamine content, *i.e.* 1.644 mg/100 g, is found in chia followed by kodo millets *i.e.* 1.644 mg/100 g kodo millet. Buckwheat, quinoa, proso millet and foxtail millets are found to have higher amount of vitamin riboflavin, which is an essential for element of two major coenzymes, flavin mononucleotide (FMN; also known as riboflavin-5'-phosphate) and flavin adenine dinucleotide (FAD). These co-enzymes play major roles in energy production, cellular function, growth and development and metabolism of fats, drugs and steroids. Buckwheat, barnyard millet and proso millet contain high amount of niacin-vitamin B_3, which is essential for turning food to energy. Small millets and pseudo cereals are excellent source of vitamin B_6 and B_7. Total folate content of small millets especially foxtail millet, finger millet and little almost equal to folate content in wheat. Folate is highly require for making genetic material *i.e.* DNA and cell division in our body. Folate content found to be low among the pseudo cereals when compared to staple cereals. Quinoa, finger millet and little millet are rich in ergocalciferol (vitamin D_2) that is highly required for making strong bones and also for absorption of calcium from food and supplements. Small millets and pseudo cereals are also rich in tocopherols and phyloquinones. The incorporation and bio-fortification of small millets and pseudo cereals in the diet may help in eradicating nutritional deficiencies.

Health Benefits of Small Millets and Pseudo Cereals

Epidemiological studies conducted on consumption of nutritionally balanced whole grain cereals and their products reduced risk of cardiovascular diseases, type II diabetes, gastrointestinal cancers and a range of other disorders among population (McKeown, 2002). Millets and pseudo cereals, which are predominantly consumed as whole grain products, have rich treasure of important health-promoting phytochemicals and considered as functional foods. Study conducted by Kumari and Sumathi (2002) on the influence of consumption of finger millet on hyperglycemia in non-insulin-dependent diabetes mellitus (NIDDM) showed that glycemic index of finger millet was found to be lower than rice and wheat. They reported that presence of higher quantities of polyphenols in finger millet flour which is responsible for low glycemic index. Polyphenols are popularly known for its ability to reduce starch digestibility and absorption. Millets are also rich in antioxidants such as naringenin, kaempferol, luteolin, apigenin, (+)-catechin/(–)-epicatechin, daidzein and epigallocatechin gallate which play significant role in inhibition of starch digestive enzymes (Tadera *et al.*, 2006).

A study conducted by Chethan *et al.* (2008) on the effect of finger millet polyphenols (FMP) on aldose reductase (AR) enzyme inhibition activity clearly showed the positive influence. Aldose reductase (AR) enzyme plays a significant role in cataractogenesis, which lead to retinopathy and subsequently night blindness in human beings. Phenolic constituent in FMP such as gallic, protocatechuic, p- hydroxybenzoic, p-coumaric, vanillic, syringic, ferulic, trans-cinnamic acids and quercetin are attributed for inhibition of cataract eye lens.

Millets are rich in antioxidant molecules such as phenolic acids (daidzein, gallic, coumaric, syringic and vanillic acids) and were reported for their wound healing properties, by reduction of oxidative stress triggered during the inflammatory responses in diabetic rats (Rajasekaran *et al.*, 2004). Regular consumption of small millets help in fighting cancer, cardiovascular disease and aging, as they are rich in antioxidants, fiber and complex carbohydrates.

Asharani *et al.* (2010) reported high level of total antioxidant capacity in finger millet flour (15.3 + 3.5 mM TE/g) followed by proso millets (5.1 + 1.0 mM TE/g) foxtail (5.0 + 0.09 mM TE/g) and little millet (4.7 + 1.8 mM TE/g). Study conducted by Chandrasekara and Shahidi (2010) also showed that soluble as well as bound fractions of kodo, finger, proso and other millets grains are rich in phenolic compounds with antioxidant, metal chelating and reducing power (Saleh *et al.*, 2013). Study conducted by Nishizawa and Fudamo (1995) reported that increased consumption of proso millet is linked with reduced risk of chronic diseases, such as elevated serum cholesterol Rodriguez *et al.* (2014) reported that regular consumption of amaranth in diet helps in the reduction of diseases caused by inflammation as amaranth protein hydrolyzates activates functional peptides. Regular intake of amaranth grain in food helps in bone development and prevent osteoporosis as amaranth grain contain more calcium than other seed (Laparra and Haros, 2016). Cherkas *et al.* (2006) reported that duodenal peptic ulcer and chronic gastritis caused by *Helicobacter pylori* can be effectively treated with amaranth oil application. As seed of amaranth rich in manganese, it is good for maintaining sugar levels during gluconeogenesis and helps in preventing diabetes (Lee *et al.*, 2013). Quinoa grains popularly recognized for high quality protein (particularly rich in essential amino acids) and for their carbohydrate, oil, mineral and vitamin content. It is also considered as a rich source of dietary fibre and other bioactive compounds, such as polyphenols and triterpenoids. Studies adopting different biological approaches have confirmed that the bioactive components present quinoa have hypocholesterolemic, anti-oxidant, anti-inflammatory and anti-cancer effects (Fuentes and Gonzalez, 2015) Fagopyritols (Mono-, di-, and trigalactosyl derivatives of D-chiro-inositol) are specific carbohydrate compounds first identified in buckwheat and named after the latin name of this crop. Ortmeyer *et al.* (1993) and Fonteles *et al.* (2000) reported that D-chiro-inositol could positively affect the blood glucose level and insulin activity. The study conducted by Kawa *et al.* (2003), at University of Manitobahad shown that buckwheat extract equally efficient in lowering blood glucose level and activating insulin as synthetic D-chiro-inositol. Nestler *et al.* (1999) reported that D-chiro- inositol could help to control development of polycystic ovary.

Table 3. Mineral profile of millets and pseudo cereals in comparison among wheat, rice and major

Crops	Calcium	Iron	Magnesium	Manganese	Phosphorous	Potassium	Selinium	Sodium	Zinc
Foxtail millet	15.2±1.28	1.26±0.44	122. ±5.9	0.33±0.05	102±5.2	94±10.7	14.12±2.2	3.35±0.4	1.65±0.18
Finger millet	364±58	4.62±0.36	146±10.7	3.19±0.88	210±58.4	443±59.6	15.30±6.23	4.75±0.14	2.53
Barnyard Millet	22-190	7.6-18.6	190	280	-	-	-	-	0.7
Proso millet	20-23	4.0-5.2	117-153	0.6-1.81	230-281	250-320	0.1	8.2-10	1.4-2.4
Kodo Millet	10.0-31.0	0.7-3.6	130-166	1.10-2.9	215-310	144-170	-	4.6-10	0.7-1.5
Little Millet	12-15.3	7.6-20	120-266	-	220-455	129	-	8.1	1.47-11.0
Teff	165.2	15.7	181.0	3.8	425.4	380.0	-	15.9	4.8
Fonio Millet	0.018	8.48	0.46	25.8	0.17	0.14	-	0.017	36.15
Amarnth	181	9.33	325	5.29	374	433	16.46	2.70	2.66
Quinoa	86.3	10.5	502	-	411	732	-	2.4	4.1
Buck Wheat	19.7 ± 0.3	3.03 ± 0.11	267 ± 0.9	1.64 ± 0.02	490 ± 5.8	565 ± 2.9	-	-	2.92 ± 0.01
Cockscomb	178.08	15.25	39.64	1.73	38.01	128.33	-	71.32	7.25
Chia	63.10	7.7	335.0	2.7	860.0	407.0	-	16.0	4.6
Rice	7.9±1.26	0.65±0.11	19.30±6.99	0.73±0.21	96±16.30	108±10.9	1.01±0.13	2.34±0.28	1.21±0.12
Wheat	20.40±2.46	1.77±0.38	30.69±2.77	0.63±0.09	110±9.8	148±7.0	-	1.54±0.48	0.88±0.07
Sorghum	27.60±3.7	3.9±0.9	133±14	1.19±0.1	274±35.7	328±25.1	26.29±11.08	5.42±0.2	1.96±0.3
Pearl millet	27.35±3.3	6.42± 1.0	124±19.5	1.12±0.1	289±25.3	3.65±1.8	30.40±5.22	4.11±0.09	2.76±0.36

Source: Indian food composition Tables NIN-2017; Ugare *et al*., 2011; Himanshu *et al*., 2018; Akansha, *et al*., 2018; Abdellatif ASA, 2018; Steadman *et al*., 2001; Adegbaju *et al*., 2019; Kulczyński *et al*., 2019.

Table 4: Vitamin profile of millets and pseudo cereals in comparison among wheat, rice and major millets

C	Thiamine B_1	Riboflavin B_2	Niacin B_3	Panto-thenic acid B_5	Total B_6	Biotin B_7	Total Folates	Ascor-bic Acid	Ergocalcif-erol D_2	Tecoph-erols E	Phyloqui-nones
	Mg					ug		mg	ug	mg	ug
Finger millet	0.37±0.041	0.17±0.008	1.34±0.02	0.29±0.19	0.05±0.007	0.88±0.05	34.66±4.97		41.46±3.12	0.16±0.01	3.0±0.44
Foxtail millet	0.29±0.054	0.20±0.018	1.49±0.08	0.63±0.07	0.07±0.017	1.49±0.18	39.49±4.52	-	-	0.07±0.02	3.75±0.63
Barnyard Millet	0.33	0.10	4.20	-	-	-	-	-	-	-	-
Proso millet	0.41	0.28	4.50	1.20	-	-	-	-	-	-	-
Kodo Millet	0.59	0.11	3.20	0.82	-	-	-	-	-	-	-
Little Millet	0.26±0.042	0.05±0,018	1.29±0.02	0.60±0.07	0.04±0.005	6.03±0.57	36.20±7.04		3.75±0.80	0.55±0.16	4.47±0.38
Teff	0.183	0.033	0.909	-	0.097	-	18	-	-	-	-
Fonio Millet	0.39	0.07	3						-	-	-
Amarnth	0.04±0.007	0.04±0.07	0.52±0.05	0.28±0.03	0.33±0.023	1.87±0.24	24.65±3.21		53.96±3.38	0.15±0.03	2.50±0.87
Quinoa	0.83	0.22	1.70	0.62	0.21	0.62	1.73		-	2.08	2.0
Buck Wheat	0.101	0.425	7.02	1.233	0.21	-	30	0.473	-	0.31	4.3
Cockscomb											
Chia	1.644	0.161	3.08	-	0.473	87					
Rice	0.05±0.019	0.05±0.006	1.69±0.13	0.57±0.05	0.12±0.012	0.60±0.12	9.32±1.93	-	-	0.09±0.04	1.50±0.50
Wheat	0.46±0.067	0.15±0.0041	2.68±0.19	1.06±0.21	0.26±0.036	1.03±0.58	30.09±3.79		17.49±3.51	0.26±0.09	1.50±0.47
Sorghum	0.35±0.039	0.14±0.014	2.10±0.09	0.27±0.02	0.28±0.023	0.70±0.06	39.42±3.13		3.96±0.30	0.06±0.01	43.82±4.84
Pearl millet	0.25±0.04	0.20±0.03	0.86±0.10	0.50±0.05	0.27±0.09	0.64±0.05	36.11±5.05		5.65±0.27	0.24±0.02	2.85±0.63

Source. Indian food composition Tables NIN-2017; Ugare *et al.*, 2011; Himanshu *et al.*, 2018; Akansha, *et al.*, 2018; Abdellatif ASA, 2018; Steadman *et al.*, 2001; Adegbaju *et al.*, 2019; Kulczyński *et al.*, 2019.

Da Silva *et al.* (2017) reported that the consumption of chia seed in regular diet reduced levels of serum cholesterol as it has high concentrations of dietary fiber and unsaturated omega-3 fatty acids. Chia proteins and chia bioactive peptides can block key markers of cholesterol synthesis, such as 3-hydroxy-3-methylglutaryl coenzyme A reductase (HMG-CoA reductase) (Coelho *et al.*, 2018). Vuksan *et al.* (2017) demonstrated that the consumption of 30 g/1000 kcal of chia seed during six months by overweight and diabetic adults resulted in weight loss and C-reactive protein and enhanced level of adiponectin. Due to the composition of chia, high in dietary fiber and low in carbohydrates, this seed has demonstrated the ability to increase satiety and reduce the desire to eat. Hence, it can be inferred that small millets and pseudo cereals have potential to protect against age-onset degenerative diseases. This area needs to be explored as these diseases are engulfing Indian population. India is the largest producer of millets so we should try to capture the world market by supplying appropriate validated functional food.

Anti-Nutritional Factor in Small Millets and Pseudo Cereals

Anti-nutritional factors such as saponins, tannins, phytic acid, gossypol, lectins, protease inhibitors, amylase inhibitor and goitrogens present in food reduce nutritional value of food, interfering with digestibility and absorption of nutrients. The presence of antinutritional factors such as phytates and tannins reduces bioavailability of nutrient present in millets. The presences of higher quantities of fibre in small millets also lower the bioavailability. Study conducted by Rao *et al.* (1983) on absorption pattern of iron in humans from millet. Compared to quinoa, buckwheat contains negligible amount of tannin and ranging from 0.5 to 4.5 percent (Franchischi *et al.*, 1994). Verma (2003) studied 7 different cultivars of buckwheat and showed that phytate content in whole flours ranged from 623.9 to 995.9 mg/100g while in sieve flour it was reduced by 68.45 percent with values ranging from 184 to 339.4 mg/100g in different cultivars. The presence of some anti-nutritive compounds such as oxalates, alkaloids, phytates and saponins have been reported in minimal amounts in Celosia species (Gupta *et al.*, 2005).

Many traditional processing methods such as roasting/popping, soaking, germination and fermentation have been reported to have a significant impact on decreasing antinutritional content of millets (Jaybhaye and Srivastav, 2015). Studies conducted by Platel *et al.* (2010) showed that malting of millets increased bio- accessibility of iron by 300 percent and of manganese by 17 percent. Germination and soaking also showed positive and significant effect on reduction of phytic acid and tannins. The other traditional methods such as boiling, pressure cooking and modern processing technologies such as extrusion cooking showed reduction in anti-nutrients and increased bioavailability of minerals and nutrients (Nirmala *et al.*, 2000).

Conclusion

Small millets and pseudo cereals can be grown in extreme climatic and weather condition in poor to marginal fertile soil with short growing season for obtaining better yield when major cereals failed to provide desired yield. Major quantity of small millet and pseudo cereals produced in the world is consumed as food in developing countries. Millets and pseudo cereals are generally low glycemic food, rich source of gluten free protein with abundant quantities of vitamin, minerals, and phytochemicals such as polyphenols, tocopherols, phytosterols, and dietary fiber. They are generally recognized as climate change complaint crop due to their adoptability and resistance to pest, disease and drought condition. They are traditionally consumed in the regular diet among the world population due to their health benefits. With alarming increase in world population, shrinkage of cultivated area due to alkalinity and salinity with yield plateau of major cereals it inescapable necessity on the part of farmers, scientific community and policy makers to use small millets and pseudo cereals as future crops for sustainable feeding of world population. The existing limitation of small millets and pseudo cereals such as presence of antinutritional factors, low sensory acceptability and low yielding characteristic requires scientific intervention such as post harvest processing, development convenience and ready to eat food products, biotechnological intervention such as sequencing, genomic mapping and mutation breeding for developing high yielding varieties and hybrids. Development of composite flour food using flour of small millets and pseudo cereals with attractive packaging will definitely motivate the farming community to large-scale cultivation. Dietary intervention by policy makers in terms of including millet and pseudo cereals in midday meal programme, food for pregnant women and adolescent person will improve the nutrient deficiencies among them. In addition to above promoting utilization of millet and pseudo cereals in urban areas will open new avenues for farming community, entrepreneurs and traders to improve their income and standard of living.

References

Abdellatif, A. S. A. (2018). Chemical and technological evaluation of quinoa (Chenopodium Quinoa Willd) cultivated in Egypt. Acta Scientific Nutritional Health, 2(7), 42-53.

Adegbaju, O. D., Otunola, G. A., andAfolayan, A. J. (2019). Potential of Celosia species in alleviating micronutrient deficiencies and prevention of diet-related chronic diseases: a review [J]. AIMS Agriculture and Food, 4(2), 458-484.

Al-Snafi, A. E. (2015). Chemical constituents and pharmacological importance of Agropyron repens–A review. Research Journal of Pharmacology and Toxicology, 1(2), 37-41.

Alvarez-Jubete, L., Arendt, E. K., andGallagher, E. (2009). Nutritive value and chemical composition of pseudocereals as gluten-free ingredients. International Journal of Food Sciences and Nutrition, 60(sup4), 240-257.

Anand, T., Rahi, M., Sharma, P., andIngle, G. K. (2014). Issues in prevention of iron deficiency anemia in India. Nutrition, 30(7-8), 764-770.

Antony, M., Shukla, Y., andJanardhanan, K. K. (2003). Potential risk of acute hepatotoxicity of kodo poisoning due to exposure to cyclopiazonic acid. Journal of Ethnopharmacology, 87(2-3), 211-214.

Arendt, E. K., andDal Bello, F. (2008). Gluten-free cereal products and beverages. Cork.

Asharani, V. T., Jayadeep, A., and Malleshi, N. G. (2010). Natural antioxidants in edible flours of selected small millets. International Journal of Food Properties, 13(1), 41-50.

Awika, J. M. (2011). Major cereal grains production and use around the world. In Advances in cereal science: implications to food processing and health promotion (pp. 1-13). American Chemical Society

Ballogou Venerande Y., Soumanou Mohamed, Toukourou Fatiou and Hounhouigan Joseph D.2013. Structure and Nutritional Composition of Fonio (Digitaria exilis) Grains : A Review, International Research Journal of Biological Sciences; 2(1):73-79.

Bazile, D., Bertero, H. D., and Nieto, C. (2015). State of the art report on quinoa around the world in 2013.

Bhatt, R., Kukal, S. S., Busari, M. A., Arora, S., andYadav, M. (2016). Sustainability issues on rice–wheat cropping system. International Soil and Water Conservation Research, 4(1), 64-74

Blando, F., andOomah, B. D. (2019). Sweet and sour cherries: Origin, distribution, nutritional composition and health benefits. Trends in Food Science and Technology, 86, 517-529.

Britto, P. R. (2017). Early Moments Matter for Every Child. UNICEF. 3 United Nations Plaza, New York, NY 10017.

Chandrasekara A, Shahidi F. 2010. Content of insoluble bound phenolics in millets and their contribution to antioxidant capacity. J Agric Food Chem ;58:6706-14

Chauhan, M., Sonawane, S. K., andArya, S. S. (2018). Nutritional and nutraceutical properties of millets: A review. Clinical Journal of Nutrition and Dietetics, 1(1), 1-10.

Cherkas, A., Zarkovic, K., Cipak Gasparovic, A., Jaganjac, M., Milkovic, L., Abrahamovych, O., and Zarkovic, N. (2018). Amaranth oil reduces accumulation of 4-hydroxynonenal-histidine adducts in gastric mucosa and improves heart rate variability in duodenal peptic ulcer patients undergoing Helicobacter pylori eradication. Free Radical Research, 52(2), 135-149.

Chethan, S., Dharmesh, S. M., and Malleshi, N. G. (2008). Inhibition of aldose reductase from cataracted eye lenses by finger millet (*Eleusine coracana*) polyphenols. Bioorganic and Medicinal Chemistry, 16(23), 10085-10090.

Cobley, L. S. (1976). An introduction to the botany of tropical crops (No. 2. ed.). Longman.

Coelho, M. S., Soares-Freitas, R. A. M., Arêas, J. A. G., Gandra, E. A., andde las Mercedes Salas-Mellado, M. (2018). Peptides from chia present antibacterial activity and inhibit cholesterol synthesis. Plant Foods for Human Nutrition, 73(2), 101-107.

da Silva, B. P., Anunciação, P. C., da Silva Matyelka, J. C., Della Lucia, C. M., Martino, H. S. D., and Pinheiro-Sant'Ana, H. M. (2017). Chemical composition of Brazilian chia seeds grown in different places. Food Chemistry, 221, 1709-1716.

Dayakar Rao, B., Bhaskarachary, K., Arlene Christina, G. D., Sudha Devi, G., Vilas, A. T., and Tonapi, A. (2017). Nutritional and health benefits of millets. ICAR Indian Institute of Millets Research (IIMR), Rajendranagar, Hyderabad, 112.

De Francischi, M. L. P., Salgado, J. M., and Leitao, R. F. F. (1994). Chemical, nutritional and technological characteristics of buck wheat and non-prolamine buckwheat flours in comparison of wheat flour. Plant Foods for Human Nutrition, 46(4), 323-329.

De Tucci, J. (2006). Chia, la semilla que reduce el colesterol. Magazine Bayres Today, 5.

De Wet JMJ, Brink DE, Rao KP, and Mengesha MH. 1983b. Diversity in kodo millet, *Paspalum scrobiculatum*. Economic Botany, 37(2): 159-163.

De Wet, J. M. J., Rao, K. P., Brink, D. E., and Mengesha, M. H. (1984). Systematics and evolution of Eleusine coracana (Gramineae). American Journal of Botany, 71(4), 550-557.

Deshpande, S. S., Mohapatra, D., Tripathi, M. K., and Sadvatha, R. H. (2015). Kodo millet nutritional value and utilization in Indian foods. Journal of Grain Processing and Storage, 2(2), 16-23.

Drewnowski, A. (2020). Impact of nutrition interventions and dietary nutrient density on productivity in the workplace. Nutrition Reviews, 78(3), 215-224.

Finley, J. W. (1999). Manganese absorption and retention by young women is associated with serum ferritin concentration. The American Journal of Clinical Nutrition, 70(1), 37-43.

Fonteles, M. C., Almeida, M. Q., andLarner, J. (2000). Antihyperglycemic effects of 3-O-methyl-D-chiro-inositol and D-chiro-inositol associated with manganese in streptozotocin diabetic rats. Hormone and Metabolic Research, 32(04), 129-132.

Franchischi MLPde, Salgado JM, Leitao RFF. 1994. Chemical, nutritional and technological characteristics of buckwheat and non-prolamine buckwheat flour in comparison to wheat flours. Plants Foods Hum. Nutr 46(4): 233-239.

Francisco Fuentes and Ximena Paredes-Gonzalez, 2015 Nutraceutical perspectives of quinoa: Biological properties and functional applications Chapter 3.5 In FAO andCIRAD. State of the Art Report of Quinoa in the World in 2013:286-299.

Gupta, S., Lakshmi, A. J., Manjunath, M. N., and Prakash, J. (2005). Analysis of nutrient and antinutrient content of underutilized green leafy vegetables. LWT-Food Science and Technology, 38(4), 339-345.

Harinarayan, C. V., andAkhila, H. (2019). Modern India and the tale of twin nutrient deficiency–calcium and vitamin D–nutrition trend data 50 Years-retrospect, introspect, and prospect. Frontiers in Endocrinology, 10, 493.

Herak, D. (2016). Image analysis of the shapes and dimensions of Teff seeds (Eragrostis tef). Agronomy Research, 14(S2), 1299-1305.

Himanshu, Manish Chauhan, Sachin K. Sonawane, S.S. Arya. 2018. Nutritional and Nutraceutical Properties of Millets: A Review, Clinical Journal of Nutrition and Dietetics ;1(1);1-10.

https://fdc.nal.usda.gov/fdc-app.html#/food-details/168918/nutrients.Accessed 15 March 2020

https://fdc.nal.usda.gov/fdc-app.html#/food-details/361081/nutrients.Accessed 15 March 2020

https://ods.od.nih.gov/factsheets/Manganese-HealthProfessional/) Accessed 20 March 2020

Hulse, J. H., Laing, E. M., andPearson, O. E. (1980). Sorghum and the Millets: Their composition and nutritive value. Academic Press.

Index, F. G. (2010). A searchable catalogue of grass and forage legumes. Rome, Italy.

Jaybhaye, R. V., andSrivastav, P. P. (2015). Development of barnyard millet ready-to-eat snack food: Part II. Food Science Research Journal, 6(2), 285-291.

Joel, A., Kumaravadivel, N., Nirmalakumari, A., Senthil, N., Mohanasundaram, K., Raveendran, T., and Mallikavangamudi, V. (2005). A high yielding Finger millet variety CO (Ra) 14. Madras Agric J, 92, 375-80.

Kannan, S. M., Thooyavathy, R. A., Kariyapa, R. T., Subramanian, K., andVijayalakshmi, K. (2013). Seed production techniques for cereals and millets. Seed node of the revitalizing rainfed agriculture network Centre for Indian knowledge systems (CIICS), 1-39.

Kawa, J. M., Taylor, C. G., andPrzybylski, R. (2003). Buckwheat concentrate reduces serum glucose in streptozotocin-diabetic rats. Journal of Agricultural and Food Chemistry, 51(25), 7287-7291.

Kulczyński, B., Kobus-Cisowska, J., Taczanowski, M., Kmiecik, D., andGramza-Michałowska, A. (2019). The chemical composition and nutritional value of chia seeds—Current state of knowledge. Nutrients, 11(6), 1242.

Kumar, A., Tomer, V., Kaur, A., Kumar, V., and Gupta, K. (2018). Millets: a solution to agrarian and nutritional challenges. Agriculture and Food Security, 7(1), 1-15.

Kumari, P. L., and Sumathi, S. (2002). Effect of consumption of finger millet on hyperglycemia in non-insulin dependent diabetes mellitus (NIDDM) subjects. Plant Foods for Human Nutrition, 57(3), 205-213.

Kundgol, N. G., Kasturiba, B., Math, K. K., and Kamatar, M. Y. (2014). Screening of little millet landraces for chemical composition. International Journal of Farm Sciences, 4(2), 33-38.

Kuta, D. D., Kwon-Ndung, E., Dachi, S., Ukwungwu, M., and Imolehin, E. D. (2003). Potential role of biotechnology tools for genetic improvement of "lost crops of Africa": the case of fonio (*Digitaria exilis* and *Digitaria iburua*). African Journal of Biotechnology, 2(12), 580-585.

Laparra, J. M., andHaros, M. (2016). Inclusion of ancient Latin-American crops in bread formulation improves intestinal iron absorption and modulates inflammatory markers. Food and Function, 7(2), 1096-1102.

Lee, S. H., Jouihan, H. A., Cooksey, R. C., Jones, D., Kim, H. J., Winge, D. R., andMcClain, D. A. (2013). Manganese supplementation protects against diet-induced diabetes in wild type mice by enhancing insulin secretion. Endocrinology, 154(3), 1029-1038.

Longvah, T., Anantan, I., Bhaskarachary, K., Venkaiah, K., andLongvah, T. (2017). Indian food composition tables (pp. 2-58). Hyderabad: National Institute of Nutrition, Indian Council of Medical Research.

Lozano R, Naghavi M, Foreman K. 2012. Global and regional mortality from 235 causes of death for 20 age groups in 1990 and 2010: a systematic analysis for the Global Burden of Disease Study 2010. Lancet;380:2095–128.

Lozano, R., Naghavi, M., Foreman, K., Lim, S., Shibuya, K., Aboyans, V., ... andRemuzzi, G. (2012). Global and regional mortality from 235 causes of death for 20 age groups in 1990 and 2010: a systematic analysis for the Global Burden of Disease Study 2010. The lancet, 380(9859), 2095-2128.

Magnesium in diet: MedlinePlus Medical Encyclopedia, https://medlineplus.gov/ency/article/002423.html accessed 20 April 2020.

Manuel Soriano-García, Imilla Ilnamiqui, Arias Olguín, Juan Pablo Carrillo Montes, Daniel Genaro Rosas Ramírez, José Silvestre Mendoza Figueroa, Erasmo Flores Valverde, María Rita Valladares Rodr guez. 2018. Nutritional functional value and therapeutic utilization of Amaranth, J Anal Pharm Res : 2018;7(5):596-600. DOI: 10.15406/japlr.2018.07.00288.

Marshall, F., andHildebrand, E. (2002). Cattle before crops: the beginnings of food production in Africa. Journal of World Prehistory, 16(2), 99-143.

Matiacevich, S. B., Castellión, M. L., Maldonado, S. B., andBuera, M. P. (2006). Water-dependent thermal transitions in quinoa embryos. Thermochimica Acta, 448(2), 117-122.

McKeown, N. M., Meigs, J. B., Liu, S., Wilson, P. W., and Jacques, P. F. (2002). Whole-grain intake is favorably associated with metabolic risk factors for type 2 diabetes and cardiovascular disease in the Framingham Offspring Study. The American Journal of Clinical Nutrition, 76(2), 390-398.

McKeown, N. M., Meigs, J. B., Liu, S., Wilson, P. W., andJacques, P. F. (2002). Whole-grain intake is favorably associated with metabolic risk factors for type 2 diabetes and cardiovascular disease in the Framingham Offspring Study. The American Journal of Clinical Nutrition, 76(2), 390-398.

Michala Jancurová, Lucia Minarovičová and Alexander Dandár, 2009, Quinoa – a Review, Czech J. Food Sci. 27(2) : 71–79.

Montoya-Rodríguez, A., de Mejía, E. G., Dia, V. P., Reyes-Moreno, C., andMilán-Carrillo, J. (2014). Extrusion improved the anti-inflammatory effect of amaranth (A maranthus hypochondriacus) hydrolysates in LPS-induced human THP-1 macrophage-like and mouse RAW 264.7 macrophages by preventing activation of NF-κ B signaling. Molecular Nutrition and Food Research, 58(5), 1028-1041.

Narasinga Rao BS, Vijayasarathy C, Prabhavathi T. 198. Iron absorption from habitual diets of Indians studied by the extrinsic tag technique. Indian J Med Res; 77:648-57.

Nestler JE, Jakubowicz DJ, Reamer P, Gunn RD, Allan G. 1999. Ovulatory and metabolic effects of D-chiro-inositol in the polycystic ovary syndrome. New Eng. J Med 340: 1314- 1320

Nestler, J. E., Jakubowicz, D. J., Reamer, P., Gunn, R. D., andAllan, G. (1999). Ovulatory and metabolic effects of D-chiro-inositol in the polycystic ovary syndrome. New England Journal of Medicine, 340(17), 1314-1320.

Nirmala, M., Rao, M. S., andMuralikrishna, G. (2000). Carbohydrates and their degrading enzymes from native and malted finger millet (Ragi, Eleusine coracana, Indaf-15). Food Chemistry, 69(2), 175-180.

Nishizawa, N., andFudamoto, Y. (1995). The elevation of plasma concentration of high-density lipoprotein cholesterol in mice fed with protein from proso millet. Bioscience, Biotechnology, and Biochemistry, 59(2), 333-335.

Oliveira, P. M., Zannini, E., and Arendt, E. K. (2014). Cereal fungal infection, mycotoxins, and lactic acid bacteria mediated bioprotection: From crop farming to cereal products. Food Microbiology, 37, 78-95.

Oomah, B. D. and Mazza, G. (1996). Flavonoids and antioxidative activities in buckwheat. Journal of Agricultural and Food Chemistry, 44(7), 1746-1750.

Ortmeyer, H. K., Huang, L. C., Zhang, L., Hansen, B. C., and and Larner Joseph. (1993). Chiroinositol deficiency and insulin resistance. II. Acute effects of D-chiroinositol administration in streptozotocin-diabetic rats, normal rats given a glucose load, and spontaneously insulin-resistant rhesus monkeys. Endocrinology, 132(2), 646-651.

Palomino, G., Hernández, L. T., andde la Cruz Torres, E. (2008). Nuclear genome size and chromosome analysis in Chenopodium quinoa and C. berlandieri subsp. nuttalliae. Euphytica, 164(1), 221-230.

Panghal, A., Khatkar, B. S., andSingh, U. (2006). Cereal proteins and their role in food industry. Indian Food Industry, 25(5), 58.

Panwar, P., Dubey, A., andVerma, A. K. (2016). Evaluation of nutraceutical and antinutritional properties in barnyard and finger millet varieties grown in Himalayan region. Journal of Food Science and Technology, 53(6), 2779-2787.

Platel, K., Eipeson, S. W., andSrinivasan, K. (2010). Bioaccessible mineral content of malted finger millet (*Eleusine coracana*), wheat (*Triticum aestivum*), and barley (*Hordeum vulgare*). Journal of Agricultural and Food Chemistry, 58(13), 8100-8103.

Quattrocchi, U. (2006). CRC world dictionary of grasses: Common names, Scientific Names, Eponyms, Synonyms, and Etymology-3 volume set. CRC Press.

Rajasekaran, N. S., Nithya, M., Rose, C., andChandra, T. S. (2004). The effect of finger millet feeding on the early responses during the process of wound healing in diabetic rats. Biochimica et Biophysica Acta (BBA)-Molecular Basis of Disease, 1689(3), 190-201.

Rao, M. V. (1986, October). The millets: their importance, present status and outlook. In Small millets in global agriculture. Proceedings of the first international small millets workshop, Bangalore, India.

Rao, N., Vijayasarathy, C., andPrabhavathi, T. (1983). Iron absorption from habitual diets of Indians studied by the extrinsic tag technique. Indian Journal of Medical Research.

Rao, P. U., andDeosthale, Y. G. (1988). In vitro availability of iron and zinc in white and coloured ragi (*Eleusine coracana*): role of tannin and phytate. Plant Foods for Human Nutrition, 38(1), 35-41.

Riley, K. W., Seetharam, A., andHarinarayana, G. (1989). Small millets in global agriculture: proceedings of the First International Small Millets Workshop, Bangalore, India, 29 Oct.-2 Nov. 1986.

Robinson, E., Blake, M. R., andSacks, G. (2020). Benchmarking Food and Beverage Companies on Obesity Prevention and Nutrition Policies: Evaluation of the BIA-Obesity Australia Initiative, 2017-2019. International Journal of Health Policy and Management.

Sadiq, I. Z., Maiwada, S. A., Dauda, D., Jamilu, Y. M., andMadungurum, M. A. (2015). Comparative nutritional analysis of black fonio (*Digitaria iburua*) and white fonio (*Digitaria exili*). International Research Journal of Biological Sciences, 4(6), 4-9.

Saleh, A. S., Zhang, Q., Chen, J., and Shen, Q. (2013). Millet grains: nutritional quality, processing, and potential health benefits. Comprehensive Reviews in Food Science and Food Safety, 12(3), 281-295.

Sharma CP. 2016 Overdraft in India's water banks: studying the effect of production of water intensive crops on ground water depletion. Master Thesis—Georgetown University, Washington DC.

Sharma, C. P. (2016). Overdraft in the Indian Water Banks Studying the Relation Between the Production of Water Intensive Crops and Groundwater Depletion in India (Doctoral dissertation, Georgetown University).

Sharma, K., and Chauhan, E. S. (2018). Nutritional composition, physical characteristics and health benefits of teff grain for human consumption: A review. The Pharma Innovation Journal, 7(10), 3-7.

Singh, P., and Raghuvanshi, R. S. (2012). Finger millet for food and nutritional security. African Journal of Food Science, 6(4), 77-84.

Skrabanja, V., Kreft, I., Golob, T., Modic, M., Ikeda, S., Ikeda, K., and Kosmelj, K. (2004). Nutrient content in buckwheat milling fractions. Cereal Chemistry, 81(2), 172-176.

Steadman, K. J., Burgoon, M. S., Lewis, B. A., Edwardson, S. E., andObendorf, R. L. (2001). Buckwheat seed milling fractions: description, macronutrient composition and dietary fibre. Journal of Cereal Science, 33(3), 271-278.

Tadera, K., Minami, Y., Takamatsu, K., and Matsuoka, T. (2006). Inhibition of α-glucosidase and α-amylase by flavonoids. Journal of nutritional science and vitaminology, 52(2), 149-153.

Tadesse, E. (1975). Tef (Eragrostis tef) cultivars: morphology and classification, Part II. Debre Zeit Agricultural Research Station. Bulletin, (66).

Takimoto, A., Nair, P. R., and Alavalapati, J. R. (2008). Socioeconomic potential of carbon sequestration through agroforestry in the West African Sahel. Mitigation and Adaptation Strategies for Global Change, 13(7), 745-761.

Tareke, B. (1975). A breakthrough in teff breeding technique. FAO Information bulletin on cereal improvement and production, 12, 11-13.

Tonini, A., and Cabrera, E. (2011). Opportunities for global rice research in a changing world (No. 2215-2019-1630).

Ugare, R., Chimmad, B., Naik, R., Bharati, P., andItagi, S. (2014). Glycemic index and significance of barnyard millet (*Echinochloa frumentacae*) in type II diabetics. Journal of Food Science and Technology, 51(2), 392-395.

United Nations Children Fund. From promise to impact: ending malnutrition by 2030; 2016 https://data.unicef.org/wp-content/ uploads/2016/06/130565-1.pdf. Accessed 20 feb 2020.

Verma, M. (2003). Nutritional quality evaluation product formulation glycemic and lipidemic response of buckwheat an underutilized plant food of Uttaranchal hills (Doctoral dissertation, GB Pant University of Agriculture and Technology, Pantnagar (Uttarakhand)).

Vietmeyer N.D., Borlaugh N.E., Axtell J., Burton G.W., Harlan J.R. andRachie K.O., (1996). (Ed) Fonio (Acha). Lost crop in Africa Chap. 3, BOSTID Publication, 58-75

Vietmeyer, N. D., Borlaugh, N. E., Axtell, J., Burton, G. W., Harlan, J. R., and Rachie, K. O. (1996). Fonio (Acha). Lost Crops of Africa, 1.

Vikaspedia (2019) https://vikaspedia.in/health/nutrition/nutritive-value-of-foods/nutritive-value-of-cereals-and-millets/small-millets-not-2018small2019-in-nutrition Accessed 19 feb 2020

Zevallos, V. F., Herencia, L. I., and Ciclitira, P. J. (2013). Quinoa, coeliac disease and gluten-free diet. State of the Art Report of Quinoa in the World in, 300-313.

2

Role of Millets in the Eradication of Malnutrition

Jiwan S. Sidhu and Tasleem A. Zafar

Department of Food Science and Nutrition, College of Life Sciences Kuwait University, P.O. Box 5969, Safat-13060, Kuwait

Introduction

Since times immemorial, cereals have always been the main source of the food for human population. Although the three major cereal crops, *i.e.,* wheat, maize and rice have dominated the scene of food production for human sustenance, but due to climatic changes, water shortages and weakening global food security, world's attention has been turning towards achieving a sustainable food supply through long-forgotten grains such as, various millets, which are also resistant to drought and heat. These grain crops have been utilized by the human population not only for food but also as fodder for their animals. Currently, our most important strategy is to exploit the genetic potential of underutilized cereal crops such as millets to achieve sustainable food supply for ever-increasing world population (Cheng, 2017). Now we need to develop our focus to identify such coarse grains with higher nutritional, agronomic and economic advantages over the traditional cereals like wheat, maize and rice, which can be genetically improved to serve as nutritious food for the human and a good fodder for the animals (Johns and Eyzaguirre, 2007). We do need to utilize the land more efficiently to produce a maximum amount of food per hectare. Interestingly, the cereals come second only to the growing of roots, tubers and vegetable crops (Ruiter *et al.*, 2018).

Millet grains are not only rich in calories, proteins, vitamins, minerals and dietary fiber, but are also rich in many health-promoting phenolic bioactive compounds, known for their antioxidant properties known for disease reduction and improvement in overall health and wellness of human beings (Shahidi and Chandrasekara, 2013) (Table 1 and 2). Reactive oxygen species are known to play a significant role in the development of many non-communicable diseases, such as, type-2 diabetes, cardiovascular diseases and even cancer. The bioactive antioxidant phenolics present in millets have been reported to reduce the oxidative stress as well as maintain the glucose levels in alloxan-induced type-2 diabetes in rats (Hegde *et al.*, 2005)

Table 1: Nutrient composition of major millets (g/100g, 12% moisture basis).

Millet	Energy (Kcals)	CHO (g)	Protein (g)	Fat (g)	CF (g)	Dietary Fiber (g)	Phytic P (mg)	Ash (g)
Sorghum	329	70.7	10.4	3.1	2.0	9.6	-	1.7
Pearl millet	363	67.0	11.8	4.8	2.3	7.8	-	1.8
Finger millet	336	72.6	7.7	1.5	3.6	19.1	209	2.7
Foxtail millet	351	63.2	11.2	4.0	6.7	-	-	-
Proso millet	364	63.8	12.5	3.5	5.2	-	-	-
Kodo millet	353	66.6	9.8	3.6	5.2	37.8	135	2.6

Adapted from : Gopalan *et al.*, 1989; Hegde *et al.*, 2005; Shahidi and Chandrasekar, 2013; Dias-Martins *et al.*, 2018

Table 2: Total phenolic acids content of a few selected millet grains (μg/g) of defatted meal.

Phenolic acid	Finger millet	Foxtail millet	Proso millet	Pearl millet
Gallic	5.0	4.5	6.2	5.4
Protocatechuic acid	119.8	11.8	72.0	1.6
p-Hydroxybenzoic	6.3	21.8	126.0	47.9
Vanillic	NA	118.7	168.6	16.0
Syringic	25.1	17.4	6.2	6.3
Chlorogenic	NA	NA	19.0	NA
Gentisic	NA	16.8	8.3	NA
Caffeic	15.9	38.3	339.2	30.4
Cinnamic	NA	NA	NA	NA
p-Coumaric	41.4	942.7	1235.2	91.9
Sinapic	0.8	16.2	18.9	12.8
trans-Ferulic	358.4	856.5	444.6	812.3

NA=Not Available (Adapted from:Shahidi and Chandrasekar, 2013)

The antioxidant phenolics in millets have been shown to increase on germination and show higher inhibitory effects on α-amylase and α-glucosidase thus preventing hyperglycemia in type-2 diabetics (Pradeep and Sreerama, 2015; Pradeep and Sreerama, 2017). Sharma *et al.* (2017a) while optimizing the germination conditions for Kodo millet (*Paspalum scrobiculatum*), not only reported a significant increase in bioactives but also the synthesis of six new valuable bioactive antioxidants. Using spectroscopic study, Kodo millet has been shown to possess higher free radical scavenging potential than other five different millets. Interestingly, these millets when fractionated into various mill fractions showed lowered antioxidant potential indicating the synergistic effect of all the phenolics in whole grain products (Hegde and Chandra, 2005). Sharma *et al.* (2017b) have analyzed the Kodo millet grown in Himachal Pradesh to possess not only higher amounts of protein, carbohydrates, minerals, dietary fiber, but also rich in antioxidant polyphenolics (such as, ferulic acid and cinnamic acid) and suggested its use as a good source of antioxidant-rich functional food for improved health benefits (Fig.1).

Recently, the focus has shifted to obtain nutritionally improved functional foods using a blend of millets and other foods of plant origin. Millet enriched diets have been reported to significantly reduce the blood pressure and myocardial injury induced by high salt in male rats and this research may have important significance for human subjects too (Wei *et al.*, 2018). Pearl millet flour when supplemented with Moringa (*Moringa oleifera*) seed flour, Showed a higher amino acid score, thus offering additional nutrition benefits (Nour *et al.*, 2018). A strong correlation (57.43%) between the metabolites of various millets cultivars grown both organically and conventionally with their cooking quality has been reported by Liang *et al.* (2018). They suggested the millet breeders to focus their efforts to improve nutritional characteristics and cooking quality also. Germination of millets is another strategy to improve the antioxidant capacity as well as general nutritional value. Li *et al.* (2007) have optimized the conditions for the germination of millet grains to obtain a water-soluble protein fraction with the highest capacity for hydroxyl radical inhibition.Germinated grains including sorghum, have been shown to be an excellent source of many bioactive phenolic compounds known to reduce the risk of type-2 diabetes and colon cancer (Donkor *et al.*, 2012). As most cereals, including millets, are known to be rich in mineral binders such as, phytic acid, phenolics and condensed tannins, the mineral bioavailability is a matter of concern. Various enzymatic degradation of these binders using phytase, laccase and tannase have been suggested to improve the bioavailability of these minerals, especially iron and zinc (Gabaza *et al.*, 2018). Considering the importance of millets in human nutrition, the role of a few important millets in the eradication of malnutrition as well approaches to produce a few of the health-promoting functional foods will be discussed.

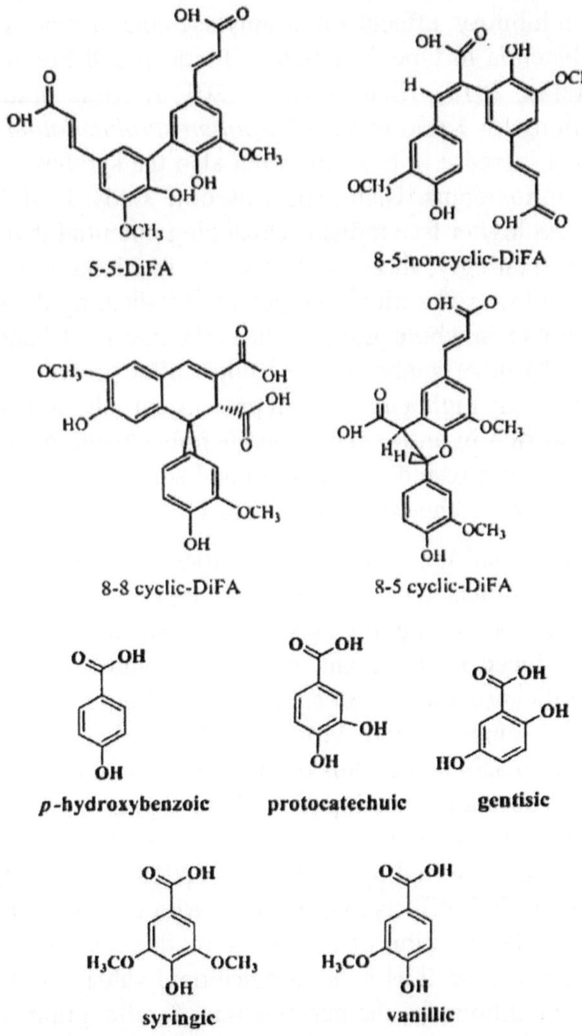

Fig. 1: Ferulic acid cross-linked compounds present in millet grains (Adapted from: Shahidi and Chandrasekar, 2015)

Pearl Millet (*Pennisetum glaucum* L.)

Pearl millet is a hardy crop that can grow and survive in poor sandy soils, with little water even under drought like conditions. This crop suits well under changing climatic conditions wherever water availability is coming under stress. Interestingly, the crop is quite responsive to improved agricultural practices, such as, breeding for higher yield potential, agronomic practices like planting time and plant density, use of nitrogenous fertilizers (Ullah *et al*., 2017), thus leading to higher grain yields (Table 3). Because of its heat and drought resistance, pearl millet crop being highly nutritious grain, gluten-free, rich in dietary fiber, having immense amounts of health-promoting phytochemicals, is also suitable to improve food security situation in many developing countries suffering from reduced irrigation water availability (Dias-Martins *et al*., 2018). Additionally, it has a low cost of cultivation and the grain can be germinated, extruded or milled into flour, to produce biscuits, pasta, snacks or non-dairy fermented beverages.

Table 3: Vitamin and mineral contents of a few selected millet grains (per 100g).

Constituent	Pearl Millet	Finger Millet	Foxtail Millet	Proso Millet	Kodo Millet	Sorghum
Calcium (mg)	42.0	30.0	31	8	35	25.0
Phosphorus (mg)	240	283	290	206	188	289
Iron (mg)	11.0	3.5	2.8	2.9	1.7	5.4
Magnesium (mg)	137	137	143	114	110	165
Sodium (mg)	10	11	1.3	5	4.8	2
Potassium (mg)	390	408	364	195	141	363
Copper (mg)	1.06	0.47	0.59	0.8	1.60	1.7
Zinc (mg)	3.1	2.3	3.51	1.7	0.70	1.7
Thiamin (mg)	0.38	0.41	0.59	0.41	0.15	0.38
Riboflavin (mg)	0.21	0.10	11	0.28	0.09	0.15
Niacin (mg)	2.80	5.10	3.2	4.5	1.2	4.30

Adapted from: Chandra *et al*., 2016; Ullah *et al*., 2017

Physicochemical and Functional Properties

The pearl millet flour blended with kidney beans flour, tigernut and xanthan gum has been evaluated for its rheological, functional and pasting characteristics by Awolu (2017). According to the studies, blend consisting of 85 percent of pearl millet, having kidney bean and tigernut flour can be used in place of 100 percent wheat flour for acceptable quality bread production. Marmouzi *et al.* (2018) have analyzed seeds of Moroccan variety of pearl millet and reported it to be rich in proteins (10.84±0.22%), oil (6.45±0.12%), minerals, phenolics, dietary fiber and the ethanolic fraction exhibited antibacterial activities against both the gram-positive and gram-negative bacteria. Siroha *et al.* (2019) have studied the physicochemical, rheological, morphological and *in vitro* digestibility of pearl millet starch after varying levels of acetylation and reported an increase in swelling power and solubility upto a level of 3.75 percent acetylation, beyond which it started decreasing. Peak setback and final viscosity of modified starch were higher than the native starch, with no difference in X-ray diffraction properties. Phenolic compounds from pearl millet grains have been shown to induce mitonase activated protein kinase phosphorylation and pro-apoptotic effects on the human osteosarcoma cells (Nani *et al*., 2019). Osteosarcoma, a bone tumor, has the most common occurrence among children and adults and is also reported to have resistance or low response to chemotherapy.

Processed Product Development

Among the coarse grains, pearl millet is the most commonly grown for edible purposes, as it is consumed by the poor people as a staple in India and many countries on the African continent. Because of the low levels of lysine and tryptophan, the protein quality of pearl millet grain is low, hence a number of approaches have been suggested to improve the protein quality and quantity of pearl millet. Interestingly, fermentation is one such process that is known not only to reduce the levels of anti-nutritional factors, but also improves the *in vitro* digestibility and its nutritive value (Ali *et al*., 2003). In a similar study earlier, Elyas *et al.* (2002) have reported a significant reduction in the polyphenols and phytic acid contents after fermentation for 36 Hours at room temperature.

The presence of polyphenolic compounds is reported to reduce the bioavailability of certain trace minerals (Adetota *et al*., 2019). According to the study involving the fortification of pearl millet porridge with moringa leaves, baobab fruit, citric acid and ascorbic acid, the bioavailability of iron was improved except in case of moringa leaves where zinc bioavailability was reduced, probably due to the presence of high levels of the polyphenolic compound in moringa leaves. The bioavailability of iron, calcium and magnesium improved significantly in the pearl millet porridge fortified with baobab fruits, this fruit probably being rich in citric acid and ascorbic acid. Obviously, the fortified pearl millet porridge can serve as a beneficial alternative to other breakfast foods being consumed by the poor section of the rural population in Asia and Africa.

Pasta products are becoming more popular in India now, but these products are preferably prepared from durum wheat or in some cases from bread wheat. Because pearl millet being gluten-free, offers an additional advantage to the celiac disease patients, but the pasta is less acceptable due to its color and astringent taste. Rathi *et al.* (2004) have investigated the influence of various treatments for depigmentation of pearl millet flour and its effect on the sensory quality, nutrient composition, protein and starch digestibility of pasta. Although, their depigmentation treatments lowered the protein and total dietary fiber contents, but improved the protein and starch digestibility, and consumer acceptability of the pasta. Recently, gluten-free pasta from depigmented pearl millet flour has been prepared by incorporating transglutaminase, sodium caseinate and whey protein powder (Kumar *et al.*, 2019). According to their results, the addition of transglutaminase resulted in good quality gluten-free pasta product with enhanced tenderness and chewiness.

Among the Read-to-Eat baked products, biscuits hold a place of prominence, as these are widely accepted, affordable and possess a long shelf-life. Adebiyi *et al.* (2016) have investigated the effect of malting and fermentation on the physicochemical properties including microstructure of pearl millet flour and biscuits. The biscuits prepared from malted and fermented pearl millet flour exhibited a slightly harder but "honey-comb" like texture compared with the control flour biscuits. In a follow-up study, Adebiyi *et al.* (2017) compared the sensory and nutritional quality of biscuits prepared from native, fermented and malted pearl millet flour. Compared with the native flour, the malting and fermentation were found to improve the crude fiber, protein and energy value of the pearl millet flour. The biscuits from malted and fermented pearl millet had improved nutritional, health-promoting effects as well as better consumers acceptability. Alyami *et al.* (2019) have investigated the glycemic, gastrointestinal and appetite responses to breakfast porridges made from pearl millet, finger millet, oats and rye. According to their findings, pearl millet porridge gave beneficial effects in terms of postprandial blood glucose.

The fermentation process during the preparation of a traditional Lohoh bread from pearl millet flour has been shown to significantly decrease the carbohydrate content with a concomitant increase in soluble sugars but no significant changes in protein and lipid contents (Osman, 2011). However, the fermentation process significantly reduced the phytic acid content, as well as the trypsin inhibitor and amylase inhibitor activities. In another study, the effect of dextran synthesized *in situ* by *Weissella confusaon* the nutritional quality and physical structure of whole grain pearl millet bread has been investigated by Wang *et al.* (2019). According to their results, 3.5 percent of dextran *in situ* producted a significant improvement in the dough extensibility, improvement in the specific loaf volume of bread and resulted in lower crumb firmness and also improved the nutritional quality of the bread, In terms of lowered glycemic index, improved the protein digestibility, increased phenolic content and antioxidant activity.

Foxtail Millet (*Setaria italica* L.)

Foxtail millet is also coarse grain from the grass family and has excellent drought resistance, needs much less water and can grow on poor soils. It is being cultivated on a wide area of the arid and semi-arid regions of the world. Foxtail grain comes in many colors such as, white, yellow and grey. Foxtail grain is known to rich in proteins, minerals and many antioxidant phytochemicals

(Yang *et al.*, 2017). According to their results, lutein and zeaxanthin pigments were the major pigments constituting about 55 to 65 percent of the pigments, with lutein being the predominant one.

Physicochemical and Functional Properties

In modern diets, a very important nutrient, *i.e.*, dietary fiber is getting lost during milling into finished products. Foxtail millet bran has been shown to be a good source of this important nutrient. Zhu *et al.* (2018) have investigated the physicochemical and functional properties of foxtail millet bran as a dietary fiber source. The water holding capacity and swelling power of foxtail millet bran was found to be 3.24 g/g and 2.06 ml/g respectively. The bran was also reported to possess good peanut oil binding (2.32 g/g), cholesterol binding (5.19 mg/g), bile salts (sodium cholate) absorption capacity (143.03 μmol/g), which indicated that foxtail millet bran has a good potential for use as a functional ingredient in food products. Tsuzuki *et al.* (2018) have utilized atmospheric pressure chemical ionization method combined with mass spectrometry (APCI-MS) to identify the steryl ferulates in whole foxtail millet grain as stigmastanyl ferulate (73%), stigmastanyl coumarate (10%) and the other constituted the remaining 17 percent of these ferulates. The yellow colored brains from foxtail millet were found to be much richer in lutein (13.58 mg/kg) and zeaxanthin (3.77 mg/kg) pigments than the white grain cultivar, thus offering higher nutritional benefits (Zhang *et al.*, 2019).

Processing Characteristics

The foxtail millet flour being rich in many nutrients and antioxidant phytochemicals, can be turned into a variety of processed functional food products. Bi *et al.* (2019) investigated the effect of roasting, boiling and freeze drying after boiling on the volatile flavor compounds in three varieties of Chinese foxtail millet. The roasting process increased the content of pyrazines giving "popcorn-like" or "smoky" odors, whereas boiling produced many unsaturated aldehydes, alcohols and benzene derivatives, giving "boiled rice" or "boiled potato" like odors. During germination of millets, a number of compounds are known to be generated. Bai *et al.* (2009) investigated the glutamate decarboxylase activity and the accumulation of γ-aminobutyric acid (GABA) in germinated foxtail millet grains under hypoxia condition but with the addition of glutamic acid, pyridoxal-5-phosphate and calcium chloride. GABA is an inhibitory neurotransmitter which plays a number of functions in mammals such as alleviation of pain and anxiety, regulation of blood pressure and heart rate, stimulates insulin secretion from the pancreas, thus effectively preventing type-2 diabetes. They were able to optimize conditions for the production of GABA (42.9 mg/100g) during germination of foxtail millet grains and suggested that the use of this approach to produce GABA-enriched functional foods for providing health benefits to consumers.

Food Applications

Foxtail millet is known to be rich in many phenolic compounds, such as ferulic acid, caffeic acid, sinapic acid, p-coumaric acid, while luteolin and kaempferol are the major flavonoids (Pradeep and Sreerama, 2018). They showed that these phenolics exhibited strong inhibitory activities towards α-amylase and α-glucosidase enzymes, thus offering a potential benefits as a functional food ingredient for regulating the post prandial hyperglycemia in diabetic persons. Using GC-MS technique, Zhang *et al.* (2018) have identified twelve aroma compounds in cooked foxtail millet grains. Longer cooking times resulted in higher concentration of these aroma compounds. Change in pH during cooking of porridge produced no difference in the production of aroma compounds, but more aroma compounds were produced during cooking of millet flour than the whole grains. The effect of different methods of drying (freeze drying; hot air drying; microwave drying) on the crystallinity and eating quality of gelatinized starch has been investigated by Wang *et al.* (2017). According to their results, the hot air drying and microwave drying increased starch crystallization, so a lower rate of rehydration, leading to the deterioration of eating quality of starch.

Foxtail millet grains upon germination are reported to improve its nutraceutical characteristics and antioxidant properties, thus serving as a better functional ingredient in healthy food product development (Sharma *et al.*, 2018a). The foaming capacity of β-glucans of the foxtail millet flour was also enhanced upon germination, thus the germinated foxtail grains would serve as a valuable ingredient in functional food development. Flour extracts obtained from germinated foxtail millet grains have been reported to exhibit higher amounts of phenolic compounds and flavonoids, GABA content and enhanced *in vitro* antioxidant capacity, thus showing a great potential for use in various functional food products (Sharma *et al.*, 2018b). In another similar study on germination of foxtail millet grains under high pressure soaking, Sharma *et al.* (2018c) have shown a significant improvement in the total phenolic content and antioxidant capacity of the flour produced from germinated grains. Moreover, the antinutritional components (such as phytates and tannins) decreased in the germinated grains during high pressure soaking process. Extrusion is an interesting technology to produce a range of processed food products and can use a variety of raw materials ranging from cereal, legumes, oilseed meals to vegetable powders. Kharat *et al.* (2019) have produced extruded products from foxtail, finger and pearl millets whole grain flours. Among all the grains tested, the foxtail millet flour gave the highest expansion ratio with the least energy input.

Health Food Applications

As per the latest observations, high blood pressure has been listed as the major public health problem in most of the countries of the world, and it is a leading cause of mortality and morbidity. Consumption of foxtail millet diet has been suggested to reduce the blood pressure and many other health risk markers in consumers with mild hypertension (Hou *et al.*, 2018). They suggested that the consumption of foxtail flour products can serve as an important source of nutraceuticals for the production of functional foods to reduce hypertension and other cardiovascular diseases. In an interesting study, Jaiswal *et al.* (2019) have successfully applied the genetic engineering approach to improve 10 important minerals in the foxtail millet grains. These nutritionally improved grains would be useful in supplying these essential minerals in the diet of poor income people in low income countries.

The digestion profile of carbohydrates present in food is known as an important factor associated with the higher risk of developing obesity, type-2 diabetes, and many other metabolic diseases. Ren *et al.* (2016) have investigated the *in vitro* digestibility of starch, degree of gelatinization and its effect of glycemic index of foxtail millets frozen food products. According to their findings, the frozen foxtail millet food products at -40°C increased the glass transition temperature, thus lowering the rate of starch digestion, which led to a lower glycemic index. The pregelatinized frozen stored foxtail millet products were suggested for long term starch-based intervention programs. Colorectal cancer (CRC) has become one of the leading causes of death in humans now and is responsible for 1 in 10 cancer cases deaths. As the cancer cells utilize glucose as a preferred source of energy through glycolysis for growth and proliferation, any attempt to inhibit glycolysis might be a better approach to improve the mortality and morbidity arising out of CRC. Shi *et al.* (2019) have isolated a bound polyphenol from foxtail millet bran that exhibits antiproliferative activity in HT-29 CRC through miR-149 mediated aerobic glycolysis. This bran derived bound polyphenol has been shown to upregulate the miR-149 gene expression which directly affected the 3'-UTR of c-myc, inhibiting the pyruvate kinase M2-mediated aerobic glycolysis, thus leading to improved intestinal health benefits.

Finger Millet (*Eleusine coracana*)

Finger millet known as ragi, is being grown in India with the State of Karnataka being the leading producer (58% of world production) of this grain. Despite India being a leading producer, very few persons in other parts of the country know about the benefits and nutritional significance of this highly useful grain. Finger millet is being grown as a staple in many countries of Asia and Africa. Although finger millet grain is gluten-free with low glycemic index values, immense nutritional and nutraceutical properties, this grain has not received the required importance it should have and remains underutilized in many countries.

Physicochemical Properties and Nutritional Contributions

Finger millets varieties come in many colors such as white, yellow, red, tan, violet or brown color, but only red colored cultivars are being grown extensively in the world. Some of the physical and functional properties of the grains and its flour obtained from finger millet cultivars being grown in sub-Saharan Africa have been reported by Ramashia *et al.* (2018). The physical properties such as color, kernel weight, true and bulk density, surface area, porosity, grain shape, water absorption capacity, digestibility of both grains and flour, reported by them, would be useful in assessing the quality of grains and the flour obtained therefrom. Chandra *et al.* (2016) have brought out an extensive review on the nutritional benefits of this important food grain. According to them, this grain is not only rich in proteins, dietary fiber, many vitamins and minerals, but is also an excellent source of many health-promoting phytochemicals and antioxidants. Because of the abundance of these phytochemicals, this grain offers a tremendous potential for use in the development of health-promoting functional foods, which can provide many beneficial properties such as antiulcer, antidiarrheal, anti-diabetic, anti-inflammatory, antimicrobial, anticarcinogenic, antioxidant properties. Moreover, the diet based on finger millet grains, which are rich in dietary fiber, will also offer benefits in the management of various metabolic disorders, such as, hypertension, fragility of vascular system, hypercholesterolemia, type-2-diabetes, oxidation of low-density lipoprotein cholesterol and is also reported to improve the overall gastrointestinal health of the subjects (Scalbert *et al.*, 2005).

Finger millet is known to be rich in many polyphenolics such as catechins, caffeic acid, quercetin, which provide immense health benefits to the consumers. Now a days, the consumption of plant-based foods is recommended for optimal health benefits, because these foods are rich in many bioactive phytochemicals. As all the phenolics are not equally absorbed, Hithamani and Srinivasan (2017) have co-administered piperine, an alkaloid from black pepper along with finger millet phenolics to assess their bioavailability. According to their findings, the levels of the finger millet-derived phenolics in plasma, kidney and small intestine increased with the co-administration of piperine. Finger millet phenolics were present in the brain tissues at all time intervals, with their concentrations being much higher when co-administered with piperine. However, the excretion of these phenolics in urine was also found to be higher, probably because of the higher absorption, leading to better retention, due to the presence of piperine along with the phenolics in the diet. Xiang *et al.* (2019a) have reported that the darker colored finger millet varieties contained higher amounts of phenolics and antioxidant activities than the varieties with lighter colored grains. In a similar another study from Xiang *et al.* (2019b) have reported that the main phenolic compounds showed a significant correlation with the antioxidant capacity from these finger millet samples.

The level of minerals in finger millet grain varies depending upon the genetic make-up of the cultivars. However, the bioavailability of these minerals is reduced due to the presence of a few antinutritional factors, such as, tannins, phytates and polyphenolics in the whole grains. A significant genetic correlation between calcium and other minerals has been reported in finger millet cultivars (Badigannavar and Ganapathi, 2018). According to Akbar *et al.* (2018), finger millet can be used as an excellent genetic model system to investigate the specific role of nutrients antinutrients interactions. Their results proposed that due to the diversion of more of ascorbic acid to tartaric acid, less of calcium oxalate is synthesized, indicating that finger millet may have a predominance of ascorbate-tartarate pathway instead of glyoxalate-oxalate conversion pathways.

Functional Foods from Finger Millet

The finger millet proteins, when hydrolyzed with trypsin or pepsin proteolytic enzymes, produced two unique peptides having potential antioxidant activity, probably due to the interaction of threonine and serine residues with the free radicals (Agrawal *et al.*, 2019). They suggested that these natural peptides from finger millet proteins can be used as a potential source of functional food ingredient for developing healthy food products. Consumption of finger millet containing arabinoxylan, a component of soluble dietary fiber, is known to reduce the high fat diet-induced metabolic disorders and gut bacterial dysbiosis (Sharma *et al.*, 2018). According to their findings, feeding of arabinoxylan, a non-starch polysaccharide from finger millet prevented the weight gain, improved glucose tolerance, serum lipids profile, inflammation and hepatic lipid accumulation. They proposed the use of arabinoxylan, a bioactive compound from finger millet, as a potential nutraceutical for the amelioration of high-fat diet-induced obesity.The elevated levels of oxidative stress markers (like glutathione, ascorbic acid, α-tocopherol, superoxide dismutase and catalase) and the decreased levels of antioxidants in the blood are reported to play a vital role in delayed wound healing in diabetic rats (Rajasekaran *et al.*, 2004). Their results supported the feeding of finger millet diet to the diabetic rats for about six weeks, improved the glucose levels and enhanced the antioxidant status leading to faster dermal wound healing.

Finger Millet Processing into Products

Threshing of grains from the finger millet crop stalks is a tedious process, so is the pearling of its grains for husk removal. Powar *et al.* (2019a) have designed a finger millet threshing drum by optimizing various parameters, such as feed rate, drum speed, concave clearance between the drum and outer casing, to improve threshing efficiency, pearling efficiency and reduced grain damage. They have reported a maximum threshing efficiency of 99 percent against a predicted value of 98 percent . The pearling efficiency of this machine was 86 percent against the predicted value of 85 percent, with a grain damage of 0.1 percent against a predicted value of 0.086 percent. In another study, Powar *et al.* (2019b) have also developed a finger millet pearling machine by optimizing various parameters, such as drum speed, concave clearance between the drum and the outer sieve casing, pearling sieve size to achieve the maximum pearling efficiency with a minimum grain damage. The pearling machine gave a maximum pearling efficiency of 99 percent against a predicted value of 98.95 percent. The minimum grain damage obtained was 0.5 percent against a predicted value of minimum grain damage of 0.73 percent, thus showing better performance by this pearling machine than even the precited values.

In many countries of Asia and Africa, finger millet can be consumed in various forms, such as flat bread, porridge or dumplings. As the whole grains from finger millet contains many anti-nutritional factors (such as, tannins, phenolics and phytates) which lower the bioavailability of many trace minerals (like iron and zinc), fermentation and cooking processes have been reported to improve the bioavailability of these essential minerals (Gabaza *et al.*, 2018). In a study, Bhol and Bosco (2014) have investigated the effect of malting of finger millet and its blending with red kidney bean flour to produce a pan bread formulation. The bread obtained from the malted finger millet flour (but having 20% wheat flour in the formulation) had better sensory quality and textural properties than the bread made with the incorporation of red kidney flour. However, the bread made with the addition of red kidney flour exhibited superior nutritional quality and mineral content than the bread made with the incorporation of malted finger millet flour.

Proso Millet (*Panicum miliaceum*)

Proso millet is suitable for cultivation in poor soils with high saline-alkaline content, has drought and high temperature resistance with highly efficacious water utilization and is also a short duration crop, suitable for food, feed and animal fodder purposes. So far not many attempts have been made in the area of its genetic improvement for increasing its grain yield and processing quality. Vetriventhan and Upadhyaya (2018) have collected about 200 highly diverse accessions of proso millet from 30 different countries and grew these grains in India with the objective of improving their agronomic performance and nutritional quality. They finally selected eighteen high-yielding, ten large-seeded and 26 with two or more nutrient-rich accessions, which can be tested for regional adaptation and the crop breeders can use these accessions for developing high-grain yielding with superior nutritional quality, for release as cultivars to the farmers for commercial cultivation.

Physicochemical and Nutritional Properties

Soaking is a preliminary step for the purpose of cooking proso millet grains before consumption, with the purpose of making nutrients present in the grains more available to the consumers. Balkrishna and Visvanathan (2019) have investigated the hydration kinetics of proso millet grains at different soaking temperatures. According to their results, among the various models tested to examine the moisture content of grains during the soaking temperature from 30° to 70°C, both the Page and Peleg models were more suitable to represent the hydration kinetics for the proso millet grains.

Proso millet grains are categorized according to amylose content of the starch, as a waxy (low amylose content) or non-waxy (high amylose content) cultivars. The amount of amylose in starch is reported to affect its swelling during cooking process giving more elastic and less viscous texture. Yang *et al.* (2018) have investigated the appearance, pasting characteristics, cooking and eating quality of waxy and non-waxy proso millet varieties, and found that amylose content of proso millet grain is positively correlated with peak time and temperature, peak viscosity, breakdown viscosity, setback viscosity. After cooking, the waxy varieties gave sticky, thinner porridge compared with thicker, harder porridge from non-waxy varieties. The presence of higher amounts of resistant starch led to slower digestibility of starch from non-waxy varieties than the waxy varieties of proso millet.

Dry heat treatment of proso millet flour and starch, heated to a temperature of 130°C for 2 to 4 hr, has been shown to affect the morphological, structural and pasting properties (Sun *et al.*, 2014). According to their results, dry heat treatment produced a significant effect on the pasting viscosity of flour than the starch. With the dry heat treatment, the onset of gelatinization, and the peak temperatures, increased significantly, but the endothermic enthalpy showed a decrease in this value. The gel structure of the flour and starch became more compact, and the particles became plumper than the native starch, when viewed under the scanning electron microscope. In another study, Singh and Adedeji (2017) have characterized the hydrothermally (HTM) and acid modified (AM) proso millet starch for its water binding capacity, swelling power and solubility. According to their findings, AM starch showed a reduction in water binding capacity, whereas in HTM starch the solubility swelling power was decreased, but the gelatinization temperature was increased to 87°C. 17°C compared to native starch (78.61°C). The AM treatment significantly decreased the peak viscosity, holding strength, final viscosity, setback and breakdown, but in the case of HTM starch, only reduction in breakdown was observed.

Use of proso millet proteins extracted by wet milling and ethanol extraction procedures has been employed to prepare nanoparticles of this protein to encapsulate curcumin for better delivery, has recently been reported (Wang *et al.,* 2018). The proso millet protein-curcumin nanoparticles were found to be spherical with a diameter of 250-350 nm, with entrapment efficiency of 11.2 to 78.9 percent . The proso millet proteins extracted by ethanol gave better encapsulation performance than the proteins obtained by wet milling process. The encapsulated curcumin showed better stability with no negative effect on its antioxidant activity, when assessed by DPPH and ABTS assays. According to their results, proso millet proteins can be a promising encapsulating material for the improved delivery of lipophilic compounds. In another study, Esfanjani *et al.* (2018) have suggested the Nano-encapsulation of phenolic compounds, as nano-emulsion, nano-scale phospholipids, or nano-structured lipid carriers, which provided more surface area, producing enhanced solubility, improved bioavailability, leading to controlled release of these nano-encapsulated phenolics.

Proso Millet Processed Products

The proso millet is usually used as bird feed, but recent necessity to produce gluten-free products has attracted more attention for the production of functional foods for humans. The unique property of proso millet proteins is a matter of concern because upon cooking the proso millet above 55°C, the digestibility of proteins is reduced by nearly 50 percent (Gulati *et al.*, 2018a). As per the studies on the microstructural changes in the protein bodies in pros millet grain, cooking did not bring about any noticeable change in the shape or size of protein bodies. When protein bodies cooked and digested with pepsin enzyme, were observed under the microscope, the uncooked protein bodies showed large cavities, but the cooked protein bodies had only tiny holes. So, the inability of the pepsin to digest the cooked protein bodies from the proso millet flour was clearly evident.

In a later study, Gulati *et al.* (2018b) investigated the effect of processing methods and solute interactions on the pepsin digestibility of cooked proso millet proteins. They used high pressure processing, germination of grains, roasting, autoclaving and treatment with transglutaminase enzyme, including heating with sucrose solution, sodium chloride and calcium chloride. None of these treatments succeeded in preventing the loss in pepsin digestibility, except gemination and transglutaminase. The use of concentrated solution of sucrose and sodium chloride were helpful in preventing the loss if pepsin digestibility. They suggested the use of a combination of enzymic modifications and cooking of proso millet flour with sucrose and salt the best way to use proso millet without any loss of protein digestibility during the development of functional foods for human use.

As proso millet grain does not contain gluten, it is finding acceptability among consumers suffering from celiac disease. Sanderson *et al.* (2017) have developed a couscous like products from decorticated proso millet for the north American consumers and evaluated it for its texture, flavor and overall acceptability. The consumers found the proso millet couscous to create a bitter, dry, tooth-packing and chewy product feeling in the mouth. Cordelino *et al.* (2019) have prepared a gluten-free pasta product from proso millet and analyzed the cooked pasta, along with a commercially available gluten-free pasta as well as a wheat pasta, for sensory quality and the *in vitro*starch and protein digestibility. The proso millet pasta had a lower protein digestibility than the wheat pasta. The consumers detected more graininess and starchiness in proso millet pasta than the commercial samples. Higher amylose content was found to correlate better with higher firmness and chewiness among proso millet pasta samples. The proso millet pasta with the lowest amylose and prolamin contents produced a poorer quality pasta product.

Sorghum (*Sorghum bicolor* L. Moench.)

Sorghum is a major food grain crop being grown in many parts of Africa and Asia for food and fodder purposes. It is traditionally processed into flour for use in flat bread or porridge production and supplies energy, proteins, vitamins, minerals and many health-promoting phytochemicals. A large population of the world derives most part of the proteins and energy from the food of plant origin, and sorghum grain is one the important one in that category. The recent advances in proteomics of drought resistance, grain development including proteins and amino acid composition of sorghum have been reviewed by Labuschagne (2018). This information is useful in developing newer sorghum varieties having higher protein digestibility and improved amino acid composition to provide necessary nutrients to the poorer sections of the population residing in African and Asian countries.

Chemical Composition and Phytochemicals

In Africa and Asia, sorghum grain is being used in a variety of foods, such as flour for flat bread, extruded snacks, porridge, cookies, pan bread and many other ethnic foods. Sorghum being gluten-free and rich in many health-promoting phytochemicals, these foods are now gaining popularity in many of the developing countries too. Many years ago, Awika and Rooney (2004) had published an extensive review about the high content of phytochemicals in sorghum grain and their effect on human health, especially in improving the cardiovascular health and in reducing the risk of certain types of cancer, as compared to other cereals. The presence of higher amounts of tannins in sorghum grain are reported to reduce the availability of calories, thus may help in reducing the epidemic of obesity. In a recent review, Girard and Awika (2018) have brought out the significance of diversity of bioactive phytochemicals present in sorghum grain, and their beneficial effects on human health, especially in terms of improved glucose metabolism and insulin sensitivity, lipid metabolism, reduced fat accumulation, reduced levels of oxidative stress and reduced inflammation in several human trials. They also highlighted some of the challenges to devise improved processing technologies to exploit the potential of sorghum phytochemicals for the development of health-promoting functional foods.

With the changing climate, severe drought-like conditions may become a reality in many parts of the Sub-Sahara and Asia, which may affect the yield as well as chemical composition of sorghum crop under acute moisture stress. Wu *et al.* (2017) have investigated the phenolic profile of sorghum grains when grown under different irrigation conditions such as full irrigation (100%), deficit irrigation (50%) or severe deficit irrigation (25%) conditions. According to their results, the phenolic content in sorghum grains significantly increased under deficit irrigation than the full irrigation conditions. This observation is very valuable to the sorghum breeders to evolve cultivars which can grow successfully under severe water stress conditions, giving not only higher grain yield but nutritionally rich in these health-promoting phytochemicals. Using HPLC-ESI-MS/MS, Luo *et al.* (2018) have identified polyphenolic compounds extracted from red sorghum bran. They had achieved better recovery by ultrasonic assisted extraction using aqueous ethanol and obtained a total polyphenolic content of 49.743 mg GAE/g. These phenolic compounds showed a very higher degree of DPPH radical scavenging activity, reducing power and antiproliferative activity in a dose dependent manner, thus showing that sorghum polyphenolics can be used as a potential source of natural antioxidants in the development of functional foods.

Malting and Fermentation of Sorghum

The malting of cereal grains is an age old process which give a product richer in amylases, protease, sweet aroma and tasty food products, apart from its use in the production of fermented beverages (whiskey and beer). Eburuche *et al.* (2019) have investigated the effect of malting on the physicochemical characteristics of white and red color sorghum varieties. According to their results, the increases in diastatic activity, moisture, crude protein and fat contents, significantly reduced the grain hardness (from 83.20 to 42.50 N in red, from 72.70 to 39.30 N in white grain) of the malted sorghum. Gabaza *et al.* (2108) reported a reduction of 22 to 88 percent of phytic acid in fermented sorghum and other millet grains, decrease in bound phenolic compounds but an increase in soluble phenolic compounds and bioaccessibility of both zinc (1.25 - 4.39 mg/100g dm) and iron (3.22 - 49.7 mg/100g dm). After the extraction of soluble components from malt, the leftover is called spent grains, which are reported to be very rich in many minerals, phenolics and dietary fiber. Miafo *et al.* (2019) have estimated free sugars, non-starchy polysaccharides-phenolic acid complexes from the bran, spent grains and sorghum seeds. The major sugars in non-starchy polysaccharides (NSP) in spent grains were mainly pentoses and were fourteen-folds in spent grains than the bran. Ferulic acid (56.31 – 100.73 mg/100g sample) was the major phenolic bound to the NSP in spent grains. They suggested the use of NSP from sorghum grains as an important nutraceutical for the production of functional foods from sorghum.

In a recent study, a few other processing conditions such as boiling (B), lactic acid bacteria (LABs) fermentation (F), fermentation and steaming (FS), fermentation, steaming and flaking (FSF), of whole sorghum grains on the chemical composition, antioxidant capacity, anti-nutrients and amino acid profile have been studied (Mohapatra *et al.*, 2019). The F sample gave a slight increase in protein and a decrease in fat content. In the case of fermented samples, the total phenolics were reduced by about 28 percent, DPPH scavenging activity and CUPRAC activity increased by 1.4 and 6 times, respectively. The FS sample had the highest reduction (58%) in trypsin inhibitor activity, whereas reduction in tannin content was between 30-39 percent in F, FS and FSF samples. The FSF sample showed the highest amount of amino acids (2.9 folds increase) with a three times the biological values compared with control sorghum sample. In another recent study, Gabaza *et al.* (2019) have reported the dominance of *Lactococci* in the bacterial communities of fermented sorghum, maize and millet slurries.

Product Development from Sorghum

Germination of grains has been practiced since many centuries with the major objective of achieving improved nutritional and functional properties. Elkhalifa and Bernhardt (2010) have investigated the effect of grain germination on the functional properties of sorghum flour. According to their results, the flour obtained from germinated sorghum grains had higher protein solubility, lower bulk density, increased oil and water binding capacity, higher emulsifying activity and stability, improved foaming capacity and stability, compared to the ungerminated one. Because of improved functional properties of flour obtained from germinated grains, they suggested its use for the development of newer food products from sorghum.

The effect of microwave roasting of sorghum grains has been reported not only to significantly enhance the total phenolic content as well as antioxidant properties, but also affected the CIE L*a*b* color values and decreased values for pasting properties such as the onset peak temperature, peak viscosity, breakdown viscosity, and setback values (Sharannagat *et al.*, 2019). Because of enhanced antioxidant properties, they suggested the use of roasted sorghum grains for the development of various functional foods and health-promoting bakery goods, especially useful to the consumers suffering from celiac disease. Chavez *et al.* (2017) have developed an extruded product from roasted sorghum grains and coffee powder blends which gave enhanced values for total phenolic compounds, phenolic acid and antioxidant capacity. With the increased amounts of coffee powder in the blend, product expansion, paste viscosity, water and solubility index were decreased, but the blends with coffee powder gave significantly higher antioxidant capacities than the sorghum grain alone. Sorghum flour is prone to develop rancidity and off flavors very quickly during storage. Meera *et al.* (2011) have suggested moist heat treatment for 15 minutes for the sorghum grain with the objective of preventing the rancidity problem during long term storage of flour.

Development of Baked Goods and Breakfast Cereals

As sorghum lacks good quality gluten proteins (like pan bread wheats), it is very difficult to produce pan bread with sufficient good loaf volume. But the production of other baked products, such as, cookies and pancakes is not that difficult. Rao *et al.* (2018) have investigated 12 sorghum cultivars for their cookie making quality, and reported that three cultivars (CSH23, CSH13R,

CV18R) were rich in dietary fiber and minerals and also gave cookies with the highest overall acceptability score among the cultivars tested. Abdulrahman *et al.* (2019) have prepared a sorghum-based Sudanese bread (*kissra*) fortified with Bambara groundnut (*Voandzeia subterranean*) seed flour (RGF) and analyzed it for nutritive value, protein quality and antioxidant activity, with sorghum bread as control. According to their results, the fortified *kissra* was found to be richer in proteins, fat, most minerals (except Mg and P), essential amino acids and energy value, but the tannins, phytic acid and trypsin inhibitor activity were significantly decreased. Protein quality, nutritional index, amino acid index, biological value, predicted protein efficiency ratio, *in vitro* protein digestibility, total phenolics, total flavonoids and DPPH free radical scavenging activity, were improved significantly for the RGF fortified bread when compared with the control. They recommended the fortification of *kissra* with RGF for improved nutritive value.

Breakfast is considered as one of the most important meals of the day, as it provides us with a boost of energy many essential nutrients to perform our daily duties in an efficient manner. Taking a good healthy breakfast has been associated with lower risk of many non-communicable diseases. Mineral availability has been reported to be low from sorghum grain-based breakfast cereals mainly because of the presence of phytic acid and tannins (Wu *et al.*, 2018). They prepared breakfast cereal flakes from three contrasting sorghum varieties and compared these with flakes made from whole wheat grain and analyzed the whole grain flours and cooked cereal flakes for tannins, phytates, and *in vitro* minerals availability (such as Fe, Zn and Ca). According to their results, the mineral availability for cooked breakfast cereals from all grains was significantly higher than their raw counterparts. Obviously, the mineral availability from cooked wheat cereal flakes was significantly higher than all the sorghum breakfast cereal flakes.

Role of Millets in Non-communicable Diseases Prevention

Effect on Starch and Protein Digestibility

The millets which require low fertility soil, less water and hardy crops to tolerate heat stress are being successfully grown in many countries of Asia and Africa, where these crops serve as an important source of calories and much needed proteins. Compared to other cereals such as wheat, maize or rice, millets are known to have lower starch and protein digestibility. A number of researchers have looked at the factors that produce lower starch and protein digestibility as well as the mechanisms or processing techniques which may be responsible for these effects. Annor *et al.* (2017) have reviewed the reasons for the lower starch and protein digestibility in millets. According to them structural characteristics of the starch, starch-protein-lipid interactions, fiber constituents and polyphenolics present in the millet grains are responsible for their hypoglycemic properties and lower protein digestibility. The starch hydrolysis rate is affected by the amount and type of fatty acids present in millet grains. The saturated fatty acids are less effective in reducing the starch hydrolysis than the unsaturated fatty acids. The processing treatments such as decortication, fermentation, germination, parboiling and the amounts of polyphenols, mostly affect the *in vitro* protein digestibility.

The parboiling process usually common in paddy processing, has recently been employed to enhance the decortication yield of millet grains and its possible effect on the phenolic acids as well as the *in vitro* starch and protein digestibility of a few selected (pearl and proso) millet products (Bora *et al.*, 2019).As reported by them, the parboiling treatment increased the decortication yield of pearl and proso millets by 37 percent and 28 percent , respectively, with a significant increase in their free and bound phenolic contents and DPPH radical scavenging activity. The parboiling process significantly ($p<0.05$) lowered the starch digestion, thus reducing the expected glycemic index. The *in vitro*protein digestibility of these millet products was also reduced by 14-17 percent by this treatment.

Effect on Non-communicable Diseases

Type-2 diabetes has been becoming a serious problem not only among the developed countries, but the number of diabetics is also surging in the developing countries. Laddha and Kulkarni (2019) have recently published an update on the role of tannins and vascular complications in diabetes. According to them, about 400 natural products have been investigated for their role in tackling diabetes and its complications. Some of the polyphenolics such as gallic acid, ellagic acid, catechin, epicatechin and procyanidins from many medicinal plants, nuts, spices, fruits and grains have been shown to control the progression of diabetes as well as its complications.

Another interesting area is the investigation of the role of xanthine oxidoreductase enzyme (XOR) in hyperuricemia, metabolic syndrome and risk of certain cancers.The unchecked activity of this enzyme is responsible for the overproduction of uric acid which can breakdown DNA and protein molecules leading the onset of many diseases (Pathak *et al.*, 2018). Many pharmaceutical companies are attempting to develop drugs with minimum side effects, to inhibit xanthine oxidoreductase enzyme with long-term treatment of hyperuricemia-related diseases. Now millets have attracted the attention of scientists to identify millet-derived compounds to target this enzyme. Among the thirty-two compounds chosen, the luteolin and quercetin from millets, have exhibited more binding affinity to this enzyme, with the luteolin-protein complex being energetically more stable than the quercetin. It is now well known that obesity and its related pathologies, such as type-2 diabetes and metabolic syndrome are associated with chronic inflammation and certain types of cancers. In a recent review, Battelli *et al.* (2019) have evaluated the role of xanthine oxidoreductase enzyme in metabolic syndrome and the risk of certain types of cancers, who reported a correlation of this enzyme with the obesity-associated metabolic syndrome. The production of reactive oxygen species (ROS) and reactive nitrogen species (RNS) and the uric acid produced by XOR leads to the development of hypertension, dyslipidemia and insulin resistance. This may then lead to cell proliferation and metastasis of cancer. Any drug developed from natural food components such as phenolics from millets could play an important relief for the persons suffering from debilitating diseases.

Development of Gluten-free Products

Millets are cereal species with small or tiny grains and are also known as coarse grains. These crops are very hardy, can grow on poor fertility soils, with minimum water, have shorter duration and have immense tolerance to high temperature and drought conditions. These millets are now being grown not only in Africa and Asia, but also in Europe, Australia, North and South America. Most important technological feature from the food science and nutrition point of view is that millets do not contain gluten protein which create consumption problems for the persons suffering from celiac disease. Interestingly, a number of products such as flat bread, pancakes, pan bread, cookies, biscuits, cakes, porridge, puffed and extruded products, can easily be produced from millets which are not only suitable for consumption by the persons suffering from celiac disease but are known to have low glycemic index and are extremely rich in many health promoting antioxidant phytochemicals. Regular consumption of millets and their processed products are known to prevent type-2 diabetes, cardiovascular diseases and certain types of cancer (Taylor, 2017; Awika, 2017). The biggest drawback of baked products produced from millets is that the quality is not as good that of wheat products. Scherf *et al.* (2018) have suggested the use of transglutaminase, plant, fungal, bacterial, animal or genetically engineered peptidases, to hydrolyze gluten proteins from wheat, barley, rye and oats to harmless peptide fragments for consumption of celiac people. The use of fungal peptidases, germinated wheat grains, lactic acid bacteria in producing high quality sour dough bread, pasta etc. All these products also meet the legal requirements of Codex Alimentarius limits of 20mg/kg of gluten proteins in gluten-free products.

Dietary fiber is another essential foodnutrient that millets are known to be rich sources. Kurek *et al.* (2018) have evaluated the physicochemical and nutritional properties of gluten-free pseudo cereal grains such as quinoa (*Chenopodium quinoa*), amaranth (*Amaranthus caudatus*) and a millet (*Panicum miliaceum*) and have suggested these grains to be a very good source of dietary fiber. Wiedemair and Huck (2018) have compared the performance of three hand-held infrared spectrometer for the measurement of total antioxidant activity in gluten-free grains and developed a model for use in non-milled samples of millets. Like the traditional wet chemical analysis methods, this method does not need sample preparation, as the whole grains can be analyzed for total antioxidant activity. To produce a stronger gluten network from gluten-free millets, Nemeth *et al.* (2019) have added pentosans extracted from rye flour and pyranose-2-oxidase enzyme to investigate their effect on the rheological properties of millet flour-based dough systems. From the SEM studies on dough and gel microstructures, they observed a significant association between the starch granules, making a macromolecular network, when rye pentosans and enzyme were incorporated into the dough system. As the traditionally prepared gluten-free products from rice flour and starch are being consumed by the people with celiac disease, they are reported to suffer from the deficiency of dietary fiber and many other essential nutrients (Arslan *et al.*, 2019). To overcome this concern, they have suggested the inclusion of dietary fiber from millets, pseudo cereals, fruits, legumes and pulses in developing such gluten-free products.

Gluten-free Processed Food Products

The market for gluten-free products is growing rapidly all around the world, as more and more persons are being diagnosed with celiac disease. The food technologists are facing a challenge to develop newer gluten-free products using unique nutritionally rich food ingredients. To prepare gluten-free muffins, Shaabani *et al.* (2018) have investigated the effect of using chickpea protein isolate (CPI), transglutaminase (MTG) enzyme and xanthan on the physical and rheological properties of batter prepared from millet flour. The use of xanthan, at all levels used, increased the specific loaf volume and porosity, and resulted in softer texture of muffins. Interestingly, xanthan increased specific volume at all levels used, but the CPI and MTG at first decreased specific volume, then increased it. They concluded that a good protein network can be formed in gluten-free muffin batter by using MTG and CPI depending upon the protein content of batter system.

The use of germinated millet grainsflour blends has been shown to produce gluten-free cookies with good consumer acceptability. Sharma *et al.* (2016)

have investigated the nutritional, sensory and *in vitro* antioxidant activity of gluten-free cookies prepared from flour blends of foxtail, barnyard and kodo millets. The cookies made from germinated millet grains gave higher values for proteins, total phenolics, antioxidant activity, less fat and carbohydrates than the raw flour samples. The cookie spread ratio from raw millet flour was higher compared to the cookies made from germinated millet flour blends, or wheat flour. Cookies made from germinated grain flour showed better consumer acceptability, higher phenolic content (45.43 mg/100g), DPPH activity (42.34%), dietary fiber (12.36g/100g), followed by cookies made from raw millet flours.

The incidence of celiac disease is about 1 percent at present but is increasing rapidly. As the celiac person have to follow gluten-free diet (products made from rice and starch) very strictly, it has been shown that they suffer from deficiency of dietary fiber and many other essential nutrients. To overcome this situation, biscuits are suggested as an alternative food, as these are much easier to prepare than pan bread (Cairano *et al.*, 2018). Additionally, gluten does not play much role on biscuit quality, millet and legume flour blends can easily be turned into biscuits with low glycemic index, thus providing all the necessary proteins, antioxidants and other nutrients. Malting of millets, fermentation and germination of these grains can further improve the quality of such biscuits.

Pasta being a very convenient food to prepare, has become extremely popular all over the world. Traditionally pasta products are made from durum wheat semolina, but the sue of millet flour has recently been suggested to lower the cost of this product. Gull *et al.* (2015) have investigated the addition of millet flours and carrot pomace on the cooking quality, color and texture of pasta prepared in a lab. As the substitution level of millet flour increased from 0 to 50 percent and carrot pomace power from 0 to 10 percent , the cooking solid loss increased, firmness decreased, with control pasta sample showing the lowest solids loss (7.66%), highest firmness (5.94 N), whereas the supplemented pasta has a cooking solid loss (10-24.40%)and firmness values (4.24 – 2.34 N). The millet flour and carrot pomace used in this study significantly affected the cooking quality characteristics, color and texture of the pasta products.

Prebiotics and Probiotics Millet Foods

Due to deficiency of micronutrients (vitamins and minerals) and macronutrients (energy and proteins), India has the largest number of undernourished children. Moreover, the children consume largely a vegetarian diet, which are poorer in some of these nutrients. It is suggested that most of these dietary deficiencies can be taken care of by diversifying their diets towards consumption of more of coarse grains (millets), legumes and leafy vegetables (Rao *et al.*, 2018).

Non-dairy Probiotic Drinks

As the millet grains are rich in phytochemicals, phytoestrogens, phenolics, antioxidants, phytic acid and sterols, a number of functional foods can easily be developed to offer beneficial effect to the consumers. Mridula and Sharma (2015) have developed a non-dairy probiotic drink (PD) using sprouted grains (wheat, barley, pearl millet and green gram), soymilk, sugar, stabilizers and *Lactobacillus acidophilus* NCDC14 strain. The probiotic count ranged from 9.10 to 11.06, 10.36 to 11.17, 10.36 to 11.51, and 10.36 to 11.32 log cfu/ml in wheat, barley, pearl millet, green gram based probiotic drinks. Considering the total solids, *in vitro* protein digestibility, acceptable quality non-dairy probiotic drink can be prepared for providing health benefits to the consumers.

In a similar way, Chavan *et al.* (2018) have developed a non-dairy fermented probiotic drink using germinated (*L. acidophilus*) and ungerminated cereals (barley, finger millet) and legume (*moth* bean) mixtures. To add more taste and flavor, sugar, cardamom, soymilk, almond and coconut milk were also added in varying proportions. Apart from sensory analysis of this drink, pH, acidity, bacterial counts and antioxidant activity were also estimated. The coconut milk containing non-dairy probiotic drink was ranked higher than the other drinks for overall acceptability. The potential of *L. plantarum* A6 strain to improve the nutritional quality of a cereal based fermented food has been investigated using genomic and transcriptomic basis by Turpin *et al.* (2018). They have identified genes in this bacterium for the synthesis of certain vitamins (folate, riboflavin, B_{12}) and for the bioaccessibility of iron. They suggested that genomic and transcriptomics can be used for examining the nutritional characteristics of *L. plantarum* A6 and its behavior in foods.

The effect of pearl millet barley based probiotic beverage on Shigella-induced pathogenicity in mice has recently been investigated by Ganguly *et al.* (2019). They prepared a probiotic beverage using *L. acidophilus* NCDC13, from whey-skim milk, germinated pearl millet flour, and liquid barley malt extract. The albino mice were orally challenged with *Shigella dysenteriae* and on 2nd, 5th and 8th day, colonization of pathogens and IgA were determined. The immune response was enhanced with higher secretion of IgA on 7th day onwards and the inhibition of pathogens was observed not only in the intestinal tract but also in all organs and feces.

The sour dough fermentation process has been shown to affect not only the rheology, sensory and shelf life attributes of baked products, but also their functional and nutritional properties (Gobbetti *et al.*, 2019). Certain aspects, such as potential to lower glycemic index, increased mineral bioavailability, and decrease in gluten content have now been well established. The recent review by them, reported the use of sourdough to reduce salt in bakery goods, management of irritable bowel syndrome (IBS) synthesis and release of certain bioactive compounds, metabolism of phenolics, use of non-conventional flours (legumes, pseudo-cereals, millets) and the use of mill fractions such as bran and germ.

A novel Kodo millet-based new food product, Nilamadana, fermented by using *Penicillium roqueforti*, has been developed by Dwivedi *et al.* (2015). Kodo millet is known to be rich in many phenolic compounds having strong antioxidant activities. This Kodo millet fermented product showed inhibition (17%) against pancreatic lipase and 55 percent towards the formation of advanced glycation end products (AGEs). Nilamadana showed reduced levels of sodium but enhanced levels of potassium and calcium, 13 percent reduction in cholesterol levels, with no toxic or mortality effects on the experimental rats. The results exhibited that fungal fermented Kodo millet food product has a potential for the prevention of certain lifestyle diseases.

Future Research Needs

The hardy nature of millet crops makes them grow successfully in poor fertility soils, with high water efficacy under drought conditions, and tolerance to high temperature. The millet grains are known to be rich in dietary fiber, many health-promoting phytochemicals, minerals and vitamins. The role of various millet constituents such as phenolics (mainly, luteolin and quercetin) in producing hypoglycemic effects, lowering of cholesterol, prevention of certain types of cancers, lowering of incidences of cardiovascular diseases, needs further investigation. Use of nano-encapsulation techniques needs further studies to make phenolic compounds more soluble, stable and bioavailable in the processed foods. More research is needed to produce acceptable functional foods from whole grain millets not only for the general public but also for the persons suffering from celiac disease. The mechanism involved in lowering the digestibility of starch and proteins present in millets needs further investigation. More genomic based research efforts are needed to elucidate the possibility of eliminating various antinutritional factors such as phytic acid and trypsin inhibitor, present in millet grains. Improved knowledge about the effect of different millet grains would help provide beneficial dietary advice and would, therefore, improve the health outcomes in general population of the world.

References

Abdualrahman, M. A. Y., Ma, H., Yagoub, A. E. A., Zhou, C., Ali, A. O., & Yang, W. (2019). Nutritional value, protein quality and antioxidant activity of Sudanese sorghum-based kissra bread fortified with bambara groundnut (*Voandzeia subterranea*) seed flour. Journal of the Saudi Society of Agricultural Sciences, 18(1), 32-40.

Adebiyi, J. A., Obadina, A. O., Adebo, O. A., & Kayitesi, E. (2017). Comparison of nutritional quality and sensory acceptability of biscuits obtained from native, fermented, and malted pearl millet (*Pennisetum glaucum*) flour. Food Chemistry, 232, 210-217.

Adebiyi, J. A., Kayitesi, E., Adebo, O. A., Changwa, R., & Njobeh, P. B. (2019). Food fermentation and mycotoxin detoxification: An African perspective. Food Control, 106, 106731.

Akbar, N., Gupta, S., Tiwari, A., Singh, K. P., & Kumar, A. (2018). Characterization of metabolic network of oxalic acid biosynthesis through RNA seq data analysis of developing spikes of finger millet (*Eleusine coracana*): Deciphering the role of key genes involved in oxalate formation in relation to grain calcium accumulation. Gene, 649, 40-49

Ali, M. A., El Tinay, A. H., & Abdalla, A. H. (2003). Effect of fermentation on the in vitro protein digestibility of pearl millet. Food Chemistry, 80(1), 51-54.

Alyami, J., Whitehouse, E., Yakubov, G. E., Pritchard, S. E., Hoad, C. L., Blackshaw, E., & Marciani, L. (2019). Glycaemic, gastrointestinal, hormonal and appetitive responses to pearl millet or oats porridge breakfasts: a randomised, crossover trial in healthy humans. British Journal of Nutrition, 122(10), 1142-1154.

Annor, G. A., Tyl, C., Marcone, M., Ragaee, S., & Marti, A. (2017). Why do millets have slower starch and protein digestibility than other cereals?. Trends in Food Science & Technology, 66, 73-83

Arslan, M., Rakha, A., Xiaobo, Z., & Mahmood, M. A. (2019). Complimenting gluten free bakery products with dietary fiber: Opportunities and constraints. Trends in Food Science & Technology, 83, 194-202.

Awika, J. M., & Rooney, L. W. (2004). Sorghum phytochemicals and their potential impact on human health. Phytochemistry, 65(9), 1199-1221.

Badigannavar, A., & Ganapathi, T. R. (2018). Genetic variability for mineral nutrients in indigenous germplasm lines of finger millet (*Eleusine coracana Gaertn.*). Journal of Cereal Science, 84, 1-6.

Balkrishna, S. P., & Visvanathan, R. (2019). Hydration kinetics of little millet and proso millet grains: effect of soaking temperature. Journal of Food Science and Technology, 56(7), 3534-3539.

Battelli, M. G., Bortolotti, M., Polito, L., & Bolognesi, A. (2019). Metabolic syndrome and cancer risk: the role of xanthine oxidoreductase. Redox Biology, 21, 101070.

Bhol, S., & Bosco, S. J. D. (2014). Influence of malted finger millet and red kidney bean flour on quality characteristics of developed bread. LWT-Food Science and Technology, 55(1), 294-300.

Bora, P., Ragaee, S., & Marcone, M. (2019). Effect of parboiling on decortication yield of millet grains and phenolic acids and *in vitro* digestibility of selected millet products. Food Chemistry, 274, 718-725.

Chandra, D., Chandra, S., & Sharma, A. K. (2016). Review of Finger millet (*Eleusine coracana* (L.) *Gaertn*): A power house of health benefiting nutrients. Food Science and Human Wellness, 5(3), 149-155.

Chavan, M., Gat, Y., Harmalkar, M., & Waghmare, R. (2018). Development of non-dairy fermented probiotic drink based on germinated and ungerminated cereals and legume. LWT, 91, 339-344.

Chávez, D. W., Ascheri, J. L., Carvalho, C. W., Godoy, R. L., & Pacheco, S. (2017). Sorghum and roasted coffee blends as a novel extruded product: Bioactive compounds and antioxidant capacity. Journal of Functional Foods, 29, 93-103

Cordelino, I. G., Tyl, C., Inamdar, L., Vickers, Z., Marti, A., & Ismail, B. P. (2019). Cooking quality, digestibility, and sensory properties of proso millet pasta as impacted by amylose content and prolamin profile. LWT, 99, 1-7.

Di Cairano, M., Galgano, F., Tolve, R., Caruso, M. C., & Condelli, N. (2018). Focus on gluten free biscuits: Ingredients and issues. Trends in Food Science & Technology, 81, 203-212.

Dias-Martins, A. M., Pessanha, K. L. F., Pacheco, S., Rodrigues, J. A. S., & Carvalho, C. W. P. (2018). Potential use of pearl millet (*Pennisetum glaucum* (L.) R. Br.) in Brazil: Food security, processing, health benefits and nutritional products. Food Research International, 109, 175-186.

Dwivedi, M., Vasantha, K. Y., Sreerama, Y. N., Haware, D. J., Singh, R. P., & Sattur, A. P. (2015). Nilamadana, new fungal fermented cereal based food. Journal of Functional Foods, 15, 217-224.

Eburuche, O. B., Attaugwu, R. N., Ufondu, H. E., & Uvere, P. O. (2019). Composition and hardness of malting red and white kaffir sorghum [*Sorghum bicolor* (L.) *Moench*] dried under the sun. Journal of Food Science and Technology, 56(7), 3513-3523.

Elkhalifa, A. E. O., & Bernhardt, R. (2010). Influence of grain germination on functional properties of sorghum flour. Food Chemistry, 121(2), 387-392.

Elyas, S. H., El Tinay, A. H., Yousif, N. E., & Elsheikh, E. A. (2002). Effect of natural fermentation on nutritive value and *in vitro* protein digestibility of pearl millet. Food Chemistry, 78(1), 75-79.

Esfanjani, A. F., Assadpour, E., & Jafari, S. M. (2018). Improving the bioavailability of phenolic compounds by loading them within lipid-based nanocarriers. Trends in Food Science & Technology, 76, 56-66.

Gabaza, M., Shumoy, H., Louwagie, L., Muchuweti, M., Vandamme, P., Du Laing, G., & Raes, K. (2018). Traditional fermentation and cooking of finger millet: Implications on mineral binders and subsequent bioaccessibility. Journal of Food Composition and Analysis, 68, 87-94.

Gabaza, M., Joossens, M., Cnockaert, M., Muchuweti, M., Raes, K., & Vandamme, P. (2019). Lactococci dominate the bacterial communities of fermented maize, sorghum and millet slurries in Zimbabwe. International Journal of Food Microbiology, 289, 77-87.

Ganguly, S., Sabikhi, L., & Singh, A. K. (2019). Effect of whey-pearl millet-barley based probiotic beverage on Shigella-induced pathogenicity in murine model. Journal of Functional Foods, 54, 498-505.

Girard, A. L., & Awika, J. M. (2018). Sorghum polyphenols and other bioactive components as functional and health promoting food ingredients. Journal of Cereal Science, 84, 112-124.

Gobbetti, M., De Angelis, M., Di Cagno, R., Calasso, M., Archetti, G., & Rizzello, C. G. (2019). Novel insights on the functional/nutritional features of the sourdough fermentation. International Journal of Food Microbiology, 302, 103-113

Gulati, P., Sabillón, L., & Rose, D. J. (2018). Effects of processing method and solute interactions on pepsin digestibility of cooked proso millet flour. Food Research International, 109, 583-588

Gull, A., Prasad, K., & Kumar, P. (2015). Effect of millet flours and carrot pomace on cooking qualities, color and texture of developed pasta. LWT-Food Science and Technology, 63(1), 470-474.

Hithamani, G., & Srinivasan, K. (2017). Bioavailability of finger millet (Eleusine coracana) phenolic compounds in rat as influenced by co-administered piperine. Food bioscience, 19, 101-109.

Hou, D., Chen, J., Ren, X., Wang, C., Diao, X., Hu, X., & Shen, Q. (2018). A whole foxtail millet diet reduces blood pressure in subjects with mild hypertension. Journal of Cereal Science, 84, 13-19.

Jaiswal, V., Bandyopadhyay, T., Gahlaut, V., Gupta, S., Dhaka, A., Ramchiary, N., & Prasad, M. (2019). Genome-wide association study (GWAS) delineates genomic loci for ten nutritional elements in foxtail millet (*Setaria italica* L.). Journal of Cereal Science, 85, 48-55

Kharat, S., Medina-Meza, I. G., Kowalski, R. J., Hosamani, A., Ramachandra, C. T., Hiregoudar, S., & Ganjyal, G. M. (2019). Extrusion processing characteristics of whole grain flours of select major millets (foxtail, finger, and pearl). Food and Bioproducts Processing, 114, 60-71.

Kumar, C. M., Sabikhi, L., Singh, A. K., Raju, P. N., Kumar, R., & Sharma, R. (2019). Effect of incorporation of sodium caseinate, whey protein concentrate and transglutaminase on the properties of depigmented pearl millet based gluten free pasta. LWT, 103, 19-26.

Kurek, M. A., Karp, S., Wyrwisz, J., & Niu, Y. (2018). Physicochemical properties of dietary fibers extracted from gluten-free sources: quinoa (*Chenopodium quinoa*), amaranth (*Amaranthus caudatus*) and millet (*Panicum miliaceum*). Food Hydrocolloids, 85, 321-330.

Labuschagne, M. T. (2018). A review of cereal grain proteomics and its potential for sorghum improvement. Journal of Cereal Science, 84, 151-158.

Laddha, A. P., & Kulkarni, Y. A. (2019). Tannins and vascular complications of Diabetes: An update. Phytomedicine, 56, 229-245.

Liang, K., Liang, S., Lu, L., Zhu, D., Zhu, H., Liu, P., & Zhang, M. (2018). Metabolic variation and cooking qualities of millet cultivars grown both organically and conventionally. Food Research International, 106, 825-833.

Luo, X., Cui, J., Zhang, H., Duan, Y., Zhang, D., Cai, M., & Chen, G. (2018). Ultrasound assisted extraction of polyphenolic compounds from red sorghum (*Sorghum bicolor* L.) bran and their biological activities and polyphenolic compositions. Industrial Crops and Products, 112, 296-304.

Marmouzi, I., Ali, K., Harhar, H., Gharby, S., Sayah, K., El Madani, N., & Faouzi, M. E. A. (2018). Functional composition, antibacterial and antioxidative properties of oil and phenolics from Moroccan *Pennisetum glaucum* seeds. Journal of the Saudi Society of Agricultural Sciences, 17(3), 229-234.

Meera, M. S., Bhashyam, M. K., & Ali, S. Z. (2011). Effect of heat treatment of sorghum grains on storage stability of flour. LWT-Food Science and Technology, 44(10), 2199-2204.

Miafo, A. P. T., Koubala, B. B., Kansci, G., & Muralikrishna, G. (2019). Free sugars and non-starch polysaccharides–phenolic acid complexes from bran, spent grain and sorghum seeds. Journal of Cereal Science, 87, 124-131.

Mohapatra, D., Patel, A. S., Kar, A., Deshpande, S. S., & Tripathi, M. K. (2019). Effect of different processing conditions on proximate composition, anti-oxidants, anti-nutrients and amino acid profile of grain sorghum. Food Chemistry, 271, 129-135.

Mridula, D., & Sharma, M. (2015). Development of non-dairy probiotic drink utilizing sprouted cereals, legume and soymilk. LWT-Food Science and Technology, 62(1), 482-487.

Nani, A., Belarbi, M., Murtaza, B., Benammar, C., Merghoub, T., Rialland, M., & Hichami, A. (2019). Polyphenols from Pennisetum glaucum grains induce MAP kinase phosphorylation and cell cycle arrest in human osteosarcoma cells. Journal of Functional Foods, 54, 422-432.

Németh, R., Bender, D., Jaksics, E., Calicchio, M., Langó, B., D'Amico, S., & Tömösközi, S. (2019). Investigation of the effect of pentosan addition and enzyme treatment on the rheological properties of millet flour based model dough systems. Food Hydrocolloids, 94, 381-390.

Osman, M. A. (2011). Effect of traditional fermentation process on the nutrient and antinutrient contents of pearl millet during preparation of Lohoh. Journal of the Saudi Society of Agricultural Sciences, 10(1), 1-6.

Pathak, R. K., Gupta, A., Shukla, R., & Baunthiyal, M. (2018). Identification of new drug-like compounds from millets as Xanthine oxidoreductase inhibitors for treatment of Hyperuricemia: A molecular docking and simulation study. Computational Biology and Chemistry, 76, 32-41.

Powar, R. V., Aware, V. V., & Shahare, P. U. (2019a). Optimizing operational parameters of finger millet threshing drum using RSM. Journal of Food Science and Technology, 56(7), 3481-3491.

Powar, R. V., Aware, V. V., Patil, S. B., & Shahare, P. U. (2019b). Development and evaluation of finger millet thresher-cum-pearler. Journal of Biosystems Engineering, 44(4), 194-206.

Pradeep, P. M., & Sreerama, Y. N. (2018). Phenolic antioxidants of foxtail and little millet cultivars and their inhibitory effects on α-amylase and α-glucosidase activities. Food Chemistry, 247, 46-55.

Rajasekaran, N. S., Nithya, M., Rose, C., & Chandra, T. S. (2004). The effect of finger millet feeding on the early responses during the process of wound healing in diabetic rats. Biochimica et Biophysica Acta (BBA)-Molecular Basis of Disease, 1689(3), 190-201.

Ramashia, S. E., Gwata, E. T., Meddows-Taylor, S., Anyasi, T. A., & Jideani, A. I. O. (2018). Some physical and functional properties of finger millet (*Eleusine coracana*) obtained in sub-Saharan Africa. Food Research International, 104, 110-118.

Rao, B. D., Kulkarni, D. B., & Kavitha, C. (2018). Study on evaluation of starch, dietary fiber and mineral composition of cookies developed from 12 sorghum cultivars. Food chemistry, 238, 82-86.

Rao, S., Santhakumar, A. B., Chinkwo, K. A., Wu, G., Johnson, S. K., & Blanchard, C. L. (2018). Characterization of phenolic compounds and antioxidant activity in sorghum grains. Journal of Cereal Science, 84, 103-111.

Rathi, A., Kawatra, A., & Sehgal, S. (2004). Influence of depigmentation of pearl millet (*Pennisetum glaucum* L.) on sensory attributes, nutrient composition, *in vitro* protein and starch digestibility of pasta. Food Chemistry, 85(2), 275-280.

Ren, X., Chen, J., Molla, M. M., Wang, C., Diao, X., & Shen, Q. (2016). In vitro starch digestibility and in vivo glycemic response of foxtail millet and its products. Food & Function, 7(1), 372-379.

Sanderson, E., Duizer, L. M., & McSweeney, M. B. (2017). Descriptive analysis of a new proso millet product. International Journal of Gastronomy and Food Science, 8, 14-18.

Sarma, S. M., Singh, D. P., Singh, P., Khare, P., Mangal, P., Singh, S., & Kondepudi, K. K. (2018). Finger millet arabinoxylan protects mice from high-fat diet induced lipid derangements, inflammation, endotoxemia and gut bacterial dysbiosis. International Journal of Biological Macromolecules, 106, 994-1003.

Shaabani, S., Yarmand, M. S., Kiani, H., & Emam-Djomeh, Z. (2018). The effect of chickpea protein isolate in combination with transglutaminase and xanthan on the physical and rheological characteristics of gluten free muffins and batter based on millet flour. LWT, 90, 362-372Sharma, N., & Niranjan, K. (2018a). Foxtail millet: Properties, processing, health benefits, and uses. Food Reviews International, 34(4), 329-363.

Sharma, S., Saxena, D. C., & Riar, C. S. (2016). Nutritional, sensory and in-vitro antioxidant characteristics of gluten free cookies prepared from flour blends of minor millets. Journal of Cereal Science, 72, 153-161.

Sharma, S., Saxena, D. C., & Riar, C. S. (2018b). Changes in the GABA and polyphenols contents of foxtail millet on germination and their relationship with in vitro antioxidant activity. Food Chemistry, 245, 863-870.

Sharma, N., Goyal, S. K., Alam, T., Fatma, S., Chaoruangrit, A., & Niranjan, K. (2018c). Effect of high pressure soaking on water absorption, gelatinization, and biochemical properties of germinated and non-germinated foxtail millet grains. Journal of Cereal Science, 83, 162-170.

Shi, J., Shan, S., Zhou, G., Li, H., Song, G., Li, Z., & Yang, D. (2019). Bound polyphenol from foxtail millet bran exhibits an antiproliferative activity in HT-29 cells by reprogramming miR-149-mediated aerobic glycolysis. Journal of Functional Foods, 56, 246-254.

Singh, M., & Adedeji, A. A. (2017). Characterization of hydrothermal and acid modified proso millet starch. LWT-Food Science and Technology, 79, 21-26.

Siroha, A. K., Sandhu, K. S., Kaur, M., & Kaur, V. (2019). Physicochemical, rheological, morphological and in vitro digestibility properties of pearl millet starch modified at varying levels of acetylation. International journal of Biological Macromolecules, 131, 1077-1083.

Sun, Q., Gong, M., Li, Y., & Xiong, L. (2014). Effect of dry heat treatment on the physicochemical properties and structure of proso millet flour and starch. Carbohydrate Polymers, 110, 128-134.

Taylor, J. R. (2017). Millets: Their unique nutritional and health-promoting attributes. In Gluten-free ancient grains (pp. 55-103). Woodhead Publishing.

Taylor, J., & Awika, J. (Eds.). (2017). Gluten-free ancient grains: cereals, pseudocereals, and legumes: sustainable, Nutritious, and health-promoting foods for the 21st century. Woodhead publishing.

Tsuzuki, W., Komba, S., Kotake-Nara, E., Aoyagi, M., Mogushi, H., Kawahara, S., & Horigane, A. (2018). The unique compositions of steryl ferulates in foxtail millet, barnyard millet and naked barley. Journal of Cereal Science, 81, 153-160.

Turpin, W., Weiman, M., Guyot, J. P., Lajus, A., Cruveiller, S., & Humblot, C. (2018). The genomic and transcriptomic basis of the potential of *Lactobacillus plantarum* A6 to improve the nutritional quality of a cereal based fermented food. International Journal of Food Microbiology, 266, 346-354.

Ullah, A., Ahmad, A., Khaliq, T., & Akhtar, J. (2017). Recognizing production options for pearl millet in Pakistan under changing climate scenarios. Journal of Integrative Agriculture, 16(4), 762-773.

Vetriventhan, M., & Upadhyaya, H. D. (2018). Diversity and trait-specific sources for productivity and nutritional traits in the global proso millet (*Panicum miliaceum* L.) germplasm collection. The Crop Journal, 6(5), 451-463

Wang, R., Chen, C., & Guo, S. (2017). Effects of drying methods on starch crystallinity of gelatinized foxtail millet (α-millet) and its eating quality. Journal of Food Engineering, 207, 81-89.

Wang, L., Gulati, P., Santra, D., Rose, D., & Zhang, Y. (2018). Nanoparticles prepared by proso millet protein as novel curcumin delivery system. Food Chemistry, 240, 1039-1046.

Wang, Y., Compaoré-Sérémé, D., Sawadogo-Lingani, H., Coda, R., Katina, K., & Maina, N. H. (2019). Influence of dextran synthesized in situ on the rheological, technological and nutritional properties of whole grain pearl millet bread. Food Chemistry, 285, 221-230.

Wiedemair, V., & Huck, C. W. (2018). Evaluation of the performance of three hand-held near-infrared spectrometer through investigation of total antioxidant capacity in gluten-free grains. Talanta, 189, 233-240.

Wu, G., Ashton, J., Simic, A., Fang, Z., & Johnson, S. K. (2018). Mineral availability is modified by tannin and phytate content in sorghum flaked breakfast cereals. Food Research International, 103, 509-514.

Xiang, J., Apea-Bah, F. B., Ndolo, V. U., Katundu, M. C., & Beta, T. (2019a). Profile of phenolic compounds and antioxidant activity of finger millet varieties. Food chemistry, 275, 361-368.

Xiang, Jinle, et al. (2019b): "Profile of phenolic compounds and antioxidant activity of finger millet varieties." Food Chemistry 275, 361-368.

Yang, Q., Zhang, P., Qu, Y., Gao, X., Liang, J., Yang, P., & Feng, B. (2018). Comparison of physicochemical properties and cooking edibility of waxy and non-waxy proso millet (*Panicum miliaceum* L.). Food Chemistry, 257, 271-278.

Zhang, B., Liu, J., Cheng, L., Zhang, Y., Hou, S., Sun, Z., & Han, Y. (2019). Carotenoid composition and expression of biosynthetic genes in yellow and white foxtail millet [*Setaria italica* (L.) Beauv]. Journal of Cereal Science, 85, 84-90.

3

Small Millets and Pseudo Cereals Distribution, Production, Productivity and Cultivation Methods

Pal Murugan Muthaiah, Janifer Raj Xavier, Gopal Kumar Sharma and Anil Dutt Semwal

Defence Food Research Laboratory, Siddarthanagar, Mysore-570011, India

Introduction

Small millets such as foxtail millet, proso millet, finger millet, barnyard millet, kodo millet, fonio millet, teff and little millet and pseudo cereals such as buckwheat,quinoa, amaranth and cockscomb and chia are the oldest crops and staple food for millions of people around the world. Small millets and pseudo cereals are the main source of fodder and feed material for livestock. These crops are widely grown in variety of agro ecological environments including tropical, subtropical and temperate and up to 2100 m above mean sea level (MSL) in relatively poor to marginal soils. They are relatively drought tolerant crops and even some extent to water logging condition also. They are less prone to pest and disease attack and respond well to organic manures and fertilizer application. They are rich in dietary energy, vitamin, minerals, antioxidants, amino acids fatty acids and dietary fibre. The bioactive compounds in small millets and pseudo cereals help in overcoming of several chronic diseases like isthemic strokes, cardiovascular diseases, cancers, obesity and Type-II diabetes (Jones *et al.*, 2000; Jones, 2006).

They also contribute to several securities (food, fodder, health, nutrition and ecological) and contributed livelihood security of small and marginal farmers. However, they have received less research and development attention (Irmgard and Bordoni, 2014) in the past due to development of high yielding wheat and rice varieties andhybrids to meet food requirements of ever increasing world population.

Distribution of Small Millets

Small millets are the major crops in developing countries especially in Asia and Africa and parts of Europe. Of the 195 countries in the world, millets are grown in 92 countries. Millets are grown in almost 33.5 million hectares (Table 1) of land all over the world and more than 95 percent area falls under Africa andAsian continents. Africa and Asia accounts for more than 97 percent of world millets production comprises pearl millet, finger millet, barnyard millet, foxtail millet, kodo millet, proso millet, fonio millet, teff and little millet (Table 2). According to FAO 2018, about 31 million tonnes of millets were produced globally. Pearl millets account nearly half of the global millet production. Small millets contribute for nearly 15 million tonnes of world millet production.

Among the small millets, foxtail millet accounts for 5 million tonnes, proso millet for 4 million tonnes and finger millet for over 3 million tonnes of world millet production. Very meager quantities of millets produced among the developed nations, especially for a high-value specialty market as birdseed. Even though, area and production of millets found to be higher in Africa, the productivity remained to be low compared to Asia. The per hectare productivity found to be higher in America due to mechanized and precision farming technologies followed on large land holding compared to subsistence farming in Asia and Africa.

Table 1: Area, Production and Productivity of Millets

S.No	Country	Area (ha)	Production (tonnes)	Productivity (hg/ha)
1.	Afghanistan	3515	11640	33115
2.	Angola	241059	51915	2154
3.	Argentina	1800	2254	12522
4.	Australia	35161	35922	10216
5.	Austria	2438	10864	44554
6.	Azerbaijan	86	2792	324593
7.	Bangladesh	32651	12713	3894
8.	Belarus	12180	19925	16359
9.	Benin	28856	26143	9060

S.No	Country	Area (ha)	Production (tonnes)	Productivity (hg/ha)
10.	Bhutan	1194	1755	14697
11.	Botswana	3273	1145	3499
12.	Bulgaria	2274	4898	21539
13.	Burkina Faso	1393878	1189079	8531
14.	Burundi	8837	9969	11281
15.	Cameroon	78975	104230	13198
16.	Central African Republic	9132	10137	11101
17.	Chad	1221948	756616	6192
18.	China, mainland	617149	1565965	25374
19.	China, Taiwan Province of	258	398	15461
20.	Congo	16109	14227	8832
21.	CÃ´te d'Ivoire	77673	64169	8261
22.	Croatia	95	207	21789
23.	Czechia	869	1016	11698
24.	Democratic People's Republic of Korea	73584	78486	10666
25.	Democratic Republic of the Congo	60897	40807	6701
26.	Eritrea	69544	23921	3440
27.	Ethiopia	515613	982958	19064
28.	France	10984	40914	37250
29	Gambia	114568	98894	8632
30.	Ghana	179332	181564	10124
31.	Greece	58	131	22499
32.	Guinea	167404	214747	12828
33.	Guinea-Bissau	12577	14788	11758
34.	Hungary	2500	3400	13600
35.	India	9107000	11640000	12781
36.	Iran (Islamic Republic of)	8333	17573	21088
37.	Iraq	2654	4130	15562
38.	Japan	256	203	7905
39.	Jordan	7	10	14315
40.	Kazakhstan	43420	40237	9267
41.	Kenya	115806	83554	7215
42.	Kyrgyzstan	40	86	21500
43.	Libya	4791	6682	13948
44.	Malawi	54230	31315	5774
45.	Maldives	4	4	10241
46.	Mali	2158250	1840321	8527
47.	Mauritania	11906	3293	2766
48.	Mexico	14	221	157857

S.No	Country	Area (ha)	Production (tonnes)	Productivity (hg/ha)
49.	Morocco	2153	4653	21609
50.	Mozambique	37665	16995	4512
51.	Myanmar	244113	247895	10155
52.	Namibia	256187	83515	3260
53.	Nepal	263497	313987	11916
54.	Niger	7033751	3856344	5483
55.	Nigeria	2795829	2240744	8015
56.	North Macedonia	33	44	13327
57.	Pakistan	455818	350010	7679
58.	Poland	13721	21938	15989
59.	Republic of Korea	1328	1608	12112
60.	Republic of Moldova	16	12	7500
61.	Russian Federation	187144	217200	11606
62.	Rwanda	11370	5116	4500
63.	Saudi Arabia	4907	8320	16956
64.	Senegal	817901	574000	7018
65.	Serbia	86	114	13297
66.	Sierra Leone	41769	44639	10687
67.	Slovakia	729	1129	15487
68.	Slovenia	454	585	12885
69.	South Africa	10225	5027	4916
70.	South Sudan	7262	5000	6885
71.	Spain	716	1759	24567
72.	Sri Lanka	6770	8060	11905
73.	Sudan	3753000	2647000	7053
74.	Switzerland	233	934	40086
75.	Syrian Arab Republic	2012	3477	17281
76.	Tajikistan	3	78	260000
77.	Togo	44106	26082	5913
78.	Turkey	2075	4397	21190
79.	Uganda	147625	209979	14224
80.	Ukraine	54800	80460	14682
81.	United Republic of Tanzania	332637	327467	9845
82.	United States of America	163090	326340	20010
83.	Uzbekistan	930	6289	67638
84.	Viet Nam	1862	3370	18101
85.	Yemen	97596	57836	5926
86.	Zambia	41463	32278	7785
87.	Zimbabwe	222032	38473	1733
88.	World	33560087	31019370	9243

Source: FAOSTAT-2018

Table 2: Continent wise millet area, production and productivity

Region	Area harvested (ha)	Production (tonnes)	Productivity (hg/ha)
Africa	22099632	15867785	7180
Asia	10971060	14381318	13108
Americas	164904	328815	19940
Europe	289329	405530	14016
Oceania	35161	35922	10216
World	33560087	31019370	9243

Source: FAOSTAT-2018

Global trade in millet is assessed to a range between 300,000 and 380,000 tonnes (Table 3), representing roughly 0.1 percent of world trade in cereals or 1.2 percent of world millet production. The major exporters are Asia and the United States, which together supply about two-thirds of recorded exports. A sizeable proportion (about 100,000 tons) of the recorded international trade is in proso millet, exported by the United States, Argentina and Australia to other developed countries. The European nations, accounts for more than 30 percent of global imports. Most of small millets produced in African continent consumed there itself. More than US$ 27982000 worth of millet traded worldwide.

Table 3: World Trade of Millet

Regions	Import Quantity(tonnes)	Import Value (1000 US$)	Export Quantity (tonnes)	Export Value (1000 US$)
Africa	59994	17847	29899	8541
Asia	173368	64640	91159	32083
Europe	125946	55553	134865	47730
Americas	22345	10700	111793	40766
Oceania	1740	1226	666	735
World	383394	149965	368382	129855

Source: FAOSTAT-2018

Foxtail millet

Foxtail millet (*Setaria italica* (L.) P. Beauv.) is one of the important small millet in terms of worldwide production and it is the sixth highest yielding cereals (Saleh *et al.*, 2013). It is one of the easily cultivated cereal grains belonging to the Setaria genus of Poaceae family and subfamily Panicoideae. The archaeological remains of foxtail millet found in northern China andEurope dating back some 7400–7935 years and 4000 years respectively (Lu *et al.*, 2014) clearly showed that foxtail millet is one of the world's oldest cultivated crops(Austin, 2006). Currently, foxtail millet is cultivated in 26 countries and ranks second in world production of millets. In terms of its yielding capacity, foxtail millet positions fourth among all millets (Balaravi, 2004). It is also known as organic products as cultivation of foxtail millet does not require use of pesticide and fertilizer (Sheahan, 2014).

Foxtail millet is also known by different names such as Italian millet, Italian foxtail, German millet, Siberian millet, and foxtail bristle grass. It grows 2–5 feet tall and can be cultivated in drier and cooler regions when compared with other millets. It perform well in sandy and loamy soils to produce 1 ton of forage on 2.5 inches of moisture, requiring approximately a third less moisture than maize. It can be cultivated at higher elevations of around 1500m and in lower lying lands close to sea bed. The crop matures in 75–90 days after transplanting/sowing and is considered to be economically important in the semi-arid tropics. It can also be cultivated under saline conditions. It is widely cultivated in China, Africa, India, Russia, USA and some parts of Europe. (Doust *et al.*, 2009; Pawar and Machewad, 2013; Pradeep and Sreerama, 2015).Foxtail millet is an excellent drought-resistant crop and thrive well in regions with erratic rainfalls.

Proso millet

Proso millet (*Panicum miliaceum*) is also known with many common names including proso millet, broomcorn millet, common millet, hog millet, Kashfi millet, red millet and white millet belongs to the Panicum genus of Poaceae family.Archeological studies revealed that the crop was first domesticated before 10,000 BCE in Northern China. (Lu *et al.*, 2009) The crop is extensively cultivated in China, India, Nepal, Russia, Ukraine, Belarus, the Middle East, Turkey, Romania and the United States, where about half a million acres are grown each year. The crop is notable both for its extremely short life span, with some varieties producing grain only 60 days after planting and its low water requirements, producing grain more efficiently per unit of moisture than any other grain species tested (Graybosch and Baltensperger, 2009).

Finger Millet

Finger millet (*Eleusine coracana L. Gaertn.*) belongs to the genus Eleusine and family Poacea is the fourth millet in terms of worldwide production after sorghum *(Sorghum bicolor)*, pearl millet *(Pennisetum glaucum)* and foxtail millet (*Setaria italica).* Finger millet is a principal food in many African and South Asian countries. In Africa, finger millet is second and represents 19 percent of millet production, after pearl millet (76%). Asian finger millet production keeps growing (by 50% in India during the last fifty years and by 8% per year in Nepal) while African production remains unchanged.Finger millet is thought to have been domesticated at the beginning of the Iron age in Africa and was introduced into India 3000 years ago before spreading to South-East Asia. It is widespread in warm temperate regions from Africa to Japan and Australia, but can also grow in colder regions as far north as Northern Ireland, during summer.

It is also known as a helpful famine crop, as it is easily stored for lean years. The grain is easily digestible, highly nutritious and multipurpose and can be cooked like rice, ground to make porridge or flour or used to make cakes. Sprouted grains are recommended for infants and elderly people. Finger millet too used to make liquor and beer, which yields by-products used for livestock feeding. Finger millet grain is not widely used to feed livestock because it is mainly a food grain, but it is of lesser quality for livestock than maize, sorghum and pearl millet. In India, it is occasionally used for feeding infant calves, growing animals, as well as sick and convalescing animals.

Out of the total minor millets produced, finger millet (*Eleusine coracana L. Gaertn.)* (ragi) accounts for about 85 percent of production in India (Divya, 2011). Finger millet is grown in India, Srilanka, Nepal and parts of Africa, Madgaskar, Malaysia, Japan and Uganda. In India, finger millet is cultivated over an area of 1.19 million hectares with production of 1.98 million tonnes giving an average productivity of 1661 kg per ha. Karnataka accounts for 56.21 and 59.52 percent of area and production of finger millet followed by Tamil Nadu (9.94% and 18.27%), Uttarakhand (9.40% and 7.76%) and Maharashtra (10.56% and 7.16%), respectively (http://www.indiastat.com).

Barnyard Millet

Barnyard Millet *(Echinochloa spp.)* is one of the oldest domesticated millets in the semiarid tropics of Asia and Africa. Two main species, *(Echinochloa esculenta (A. Braun) H. Scholz; syn. Echinochloa utilis)* Ohwi *et* Yabuno *(Japanese barnyard millet)* and *Echinochloa frumentacea)* Link; *syn. (Echinochloa colona var. frumentacea* (Link) Ridl) (Indian barnyard millet) are cultivated and grown as cereals *(*Saleh *et al., 2015).* The genus Echinochloa belongs to the tribe Paniceae, subfamily Panicoideae of the family Poaceae (Clayton and Renvoize, 1986).

Barnyard millet is widely grown in India, China, Japan, Malaysia and East Indies. It is a good substitute crop when the rice crop fails due to erratic climatic. It is also cultivated in some extent at Africa and United States of America. In India it is grown in Madhya Pradesh, Uttar Pradesh, Tamil Nadu, Andhra Pradesh, Karnataka, Maharashtra and Bihar (http://agritech.tnau.ac.in). It has a wide adaptation capacity and can be cultivated up to an altitude of 2000 m above mean sea level during summer season (Hilu, 1994).

Barnyard millet is primarily cultivated in two different agro-ecologies In India i.e one at mid hills of Himalayan region of Uttarakhand in the North and another in Deccan plateau region of Tamil Nadu in the south. Wild barnyard millet (*Echinochloa colona*) is generally found in paddy fields as weed and consumed as food during drought years in many states of India. It is the second significant small millet after finger millet with production and productivity 87 thousand tonnes and 857 kg/ha, respectively (Padulosi *et al*., 2009)

Kodo Millet

Kodo millet (*Paspalum scrobiculatum)* is also known as cow grass, rice grass and ditch millet. Native Paspalum or Indian Crown Grass is highly drought tolerant crop. The grain of kodo millet is coarsest of all food grains. It originated in tropical Africa and domesticated in India 3000 years ago. It is mainly cultivated in India, Nepal, Philippines, Indonesia, Vietnam, Thailand, and in West Africa. In India, it is cultivated in Rajasthan, Uttar Pradesh, Tamil Nadu, West Bengal, Madhya Pradesh and Andhra Pradesh over an area of 1.96 lakh ha with total production of about 0.84 lakh tonnes and with productivity of 429 kg/ha during the year 2015-16 (Prabhakar *et al*., 2017). Tamil Nadu and Madhya Pradesh have the maximum share in the production and promotion of kodo millet.

Little Millet

Little millet (*Panicum miliare*) is extensively grown in temperate and tropic zone of Asia i.e. china, East Asia, India and Malaysia. Little millet is one of the oldest crops domesticated in India and is important minor millet grown in dry lands. Little millet is cultivated about 0.5 million ha in India. The crop is well known in Tamil Nadu, Karnataka, Andhra Pradesh, Maharashtra, Jharkhand, Madhya Pradesh, Orissa and Gujarat and grown quite extensively in many parts of the country. Its colloquial names are kutki, samai, samalu and same. The crop is strongly associated with tribal agriculture and grown as an important catch crop (Chadha, 2009). It cultivation is restricted to only certain cultural occasions in certain parts of the country (Mallikarjun *et al*., 2013). The crop can withstand both drought and water logged conditions. It can be cultivated up to 2000 m above sea level. It is rich in dietary fibre and has been reported to have 37 percent to 38 percent. Little millet is one of the oldest crops domesticated in India and is important minor millet grown in dry lands.

Fonio Millet

Fonio millet is a traditional cereal grain native to West Africa. It has been grown in dry savannas of West Africa for years and in some West African communities including Nigeria, Mali, Guinea and Burkina Faso. Farmers in the

West Africa earmark about 300,000 hectares each year for Fonio production, and the crop feeds about 3-4 million people (Vietmeyer *et al.*, 1996). Among the millet, Fonio is the highly nutritious and a good source of amino acids, cyteine and methionine, which are not found in some common cereals (Kuta *et al.*, 2003). White and black Fonio are widely used two species. White fonio found in Senegal, Chad and Nigeria and particularly raised in plateau located in the central Nigeria popularly called "*acha*" and also in the surrounding areas. The black fonio cultivation is restricted to the Bauchi and Jos of Nigeria along with other regions of the northern Benin and Togo (Vietmeyer *et al.*,1996) . Even though fonio was considered to be one of the "Lost Crops of Africa" and was wholly disregarded by scientists the grain is now re-explored and chosen for improvement as a species for cultivation. The area, production and productivity of fonio millet are presented in Table 4. The perusal of table clearly showed that fonio has been cultivated in Guinea, Nigeria and Burkina Faso as important stable food crop. Highest-level of productivity is noticed in CÃ´te d'Ivoire.

Table 4: Area, Production and Productivity of Fonio millet

County	Area (ha)	Production (tonnes)	Yield (hg/ha)
Benin	1869	1167	6244
Burkina Faso	14133	11270	7974
CÃ´te d'Ivoire	14947	20556	13752
Guinea	614538	477432	7769
Guinea-Bissau	939	738	7862
Mali	46476	27868	5996
Niger	10341	6400	6189
Nigeria	170378	83403	4895
Senegal	3692	2742	7427
World	877313	631576	7199

Source: FAOSTAT-2018

Teff

Teff is a warm-season annual cereal and the only cultivated species in the genus Eragrostis. Ethiopia is considered the center of origin of teff (National Researach Council, 1996). Similar to wheat and rice, 100 g of teff grains provide 357 kcal (Cheng *et al.*, 2017). It is rich in iron, calcium and fiber (FAO, 2015). Teff provides 11-13 proteins and an excellent source of essential amino acids, especially lysine: the amino acid that is most often deficient in grains (Ayalew *et al.,* 2011). Teff grains are low on the glycemic index, which makes them suitable for people with Type 2 diabetes. The grains are also gluten-free and useful for people who suffer from gluten intolerance or celiac disease (Davison and Laca, 2009)

In Ethiopia, teff is a major staple and main crop in terms of cultivation area and production value (O'connor, 2016). Teff accounted for about 24 percent of the nationwide grain cultivated area and nearly half of the small holding farmers grew it (Bachewe, 2015). Ethiopia produces more than 90 percent of world teff production. Due to its increasing popularity, teff cultivation also started in Australia, China, India, South Africa and the US.

Distribution of Pseudo Cereals

Amaranth

Amaranth belongs to family Amaranthaceae of cosmopolitan genus Amaranthus. The word Amaranthus derived from the Greek word "Anthos" (Flower) means everlasting or un-wilting. The genus amaranth is mainly comprised of about 400 species and the division of species is based on their utilization method into grain (*A.caudatus, A.cruentus* and *A.hypochondriacus*) and vegetable amaranthus (*A.dubius, A.tricolor* and *A.cruentus*). Grain Amaranth species are of new world origin, *A. caudatus* from Andean Peru and Ecuador, *A.cruentus* and *A. hypochondriacus* from Mexico and Central America. Currently, the grain amaranths are grown from the temperate to tropical zone.

It has a great amount of genetic diversity, phenotypic plasticity and is extremely adaptable to adverse growing conditions, resists heat and drought, has no major disease problem and is among the easiest of plants to grow in agriculturally marginal lands. The crop is known for its excellent nutritional value and therapeutic nature (Balasubramanian and Karthikeyan, 2016) compared with other crops, this pseudo cereal is rich in protein (17–19% of dry weight) with double the amount of essential amino acids than wheat grain protein (Becker *et al.*, 1981; Beressani *et al.*, 1987). It is a rich source of important amino acids namely alanine, valine, leucine, arginine, phenylalanine, tryptophan, isoleucine serine etc. The protein of amaranth does not contain gluten therefore it can be used in the diet of patients suffering from celiac diseases.

Buckwheat

Buckwheat (*Fagopyrum* sp.) is known as a nutritional powerhouse as nearly 75 percent its of protein is available to the human body and making it one of the best sources of plant-based protein. It belongs to the polygonaceae with accepted botanical name of species namely *Fagopyrum esculentum Moench, Fagopyrums agittatum Gilib* and *Fagopyrum tataricum (L.) Gaertn.* buckwheat thought to be originated in china and cultivated in china as early as the 2nd- 1st centuries BC (Li and Yang, 1992). The cultivation of Buckwheat occurred in the 5th and 6th centuries AD (Krotov, 1963). It spreaded to Europe via Russia.From Europe, buckwheat spreads to North America along with immigrants from several countries as it was often used on newly cleared land. Ohnishi (1988) found that the cultivation of common buckwheat in the Himalayas was limited to approximately 500-2500 m in altitude. Above 2500 m, Tartary buckwheat replaces it in cropping patterns. The area, production and productivity of buckwheat is presented in Table 5 The perusal of table clearly showed that buckwheat has been cultivated in all the continents.

China and Russia are major producer of buckwheat in the world with production of 1134988 tonnes and 979914 tonnes, respectively. Globally, buckwheat has been cultivated 3001495 ha of land with production of 2905294 tonnes. The per hectare productivity found to be higher in France and it is relatively low input crop performs well even in marginal soil. It also add nutrient to soil as cover crop or rotational crop. It grows quickly, within a month maturity rate that allows it to kill off most competing weeds. It is a moderate yielding plant. Buckwheat also used as vegetable crop as well as source of nectar for honey production (Campbell, 1997).

Table 5: Area, Production and Productivity of Buckwheat

County	Area (ha)	Production (tonnes)	Yield (hg/ha)
Belarus	18324	18495	10093
Bhutan	2337	3924	16792
Bosnia and Herzegovina	1920	1350	7031
Brazil	48491	66173	13646
Canada	12700	15000	11811
China, mainland	1331500	1134988	8524
Croatia	601	564	9384
Czechia	878	2259	25726
Estonia	2909	1443	4960
France	57367	188071	32784
Georgia	104	112	10753
Hungary	900	850	9444
Japan	63900	29000	4538
Kazakhstan	95753	82704	8637
Kyrgyzstan	17	38	22353
Latvia	25600	25300	9883
Lithuania	52670	53426	10144
Nepal	10296	11472	11142
Poland	78201	93599	11969
Republic of Korea	2042	1624	7951
Republic of Moldova	55	37	6727
Russian Federation	979914	931713	9508
Slovakia	531	652	12279
Slovenia	3475	3225	9281
South Africa	573	230	4014

County	Area (ha)	Production (tonnes)	Yield (hg/ha)
Ukraine	113300	137010	12093
United Republic of Tanzania	16804	17595	10471
United States of America	80332	84440	10511
World	3001495	2905294	9679

Buckwheat is used in many different food preparations in different countries. It is mainly consumed as a noodle soba in japan. Buckwheat flour is generally mixed with wheat flour for the preparation of pancakes, biscuits, noodles, cereals, porridge and meat extender in Europe, America and Russia. In Southeast Asia, buckwheat is a staple food in many hilly areas. The flour is used to make unleavened bread chapattis. Buckwheat is used to make alcoholic drinks the liquor prepared from Tartary buckwheat being ascribed medicinal qualities. In China, it has been reported that buckwheat is used for the production of vinegar.

Cockscomb

Coleus, (*Celosia sps*) is a small group among the edible and ornamental plants belonging to Amaranthacea. The generic name "Celosia" derived from Greek word "kelos" meaning "burned" and refers to the flame-like flower heads. *Celosia cristata* is commonly known as cockscomb, since the flower resembles rooster (cock) head. It belongs to the class Magnoliopsida, order Caryophyllales and family Amaranthaceae. The most common regions known and recognized for cockscomb are Africa, South America, India and some parts of Asia. The most common names that prevails worldwide for *Celosia cristata* are cockscomb, crested celosia, yellow toreador, red cockscomb, foxtail amaranth, fire-flame bush, shinaji tea and woodfordia.

The family Amaranthacea constitutes nearly 165 genera, 2040 species and is considered to be abundantly rich species lineage among the flowering plant order of caryophyllales. Most of the plants amongst the family Amaranthacae consist of annual or perennial plants, herbs or shrubs, small trees and vines. This family can be identified from the tropical region to cool temperate regions. Nativity of Amaranthacea is tropical and sub-tropical areas of Central America, Africa and Australia. The agricultural habitats these families are habituated to temperate and sub-tropical regions. Native people in tropical and subtropical countries as well as temperate climate used Amaranthacea species as *Triterpene saponins* are found to be present generously in Amaranthacea species.

Chia

Chia (*Salvia hispanica* L.), also known as super food, super seed, miracle seed magical seed and the food of endurance runner is an herbaceous plant belongs to family Lamiaceae and genus *Salvia* (Arctos Specimen Database, 2018). The minute, mighty chia seed was known as "the food of the ancient Mayans, Aztecs and Tarahumara Indians." In fact, Mayan and Aztec warriors used chia seeds as their sole food source when traveling long distances and during war. Chia seed was extensively used as an offering to the gods in religious ceremonies or used as a form of payment of taxes. The oil extracted from the seeds has been used to produce cosmetics. (Mariana *et al.,* 2019) It is also reported that the branches, leaves and roots of chia were used less commonly to fight respiratory infections (Sosa, 2016).

The *Salvia* genus consists more than 900 species and extensively distributed throughout several regions of the world, including Southern Africa, Central America, North America, South America and South-East Asia (Takano, 2017).

The chia plant is originated in northern Guatemala and Southern Mexico and today it is cultivated in Australia, Bolivia, Colombia, Guatemala, Peru, Argentina and Mexico. Mexico is the world's largest producer of chia (Busilacchi *et al.*, 2013).

Mountainous regions of temperate and subtropical regions account for the largest cultivation of chia seeds (Capitani *et al.*, 2012). Climatic conditions, geographic location, nutrients and soil conditions play a major role in chemical composition and nutritional value of chia seeds (Ayerza, 2009; Ayerza and Coates, 2009). Ayerza and Coates (2011) reported that cultivation of chia crop in colder and higher region resulted higher content of omega-3 unsaturated fatty acids

Cultivation of chia in India is started at small scale in Karnataka and Rajasthan states. It is expected that cultivation may increase in India owing to health benefits of chia consumption to overcome to chronic diseases such as obesity, cardiovascular diseases, diabetes and cancer and higher product price.

Quinoa

Quinoa (*Chenopodium quinoa Willd.*) is a ancient crop of American continent. Scientific evidences showed that quinoa is originated from the Andean regions of South America (Matiacevich *et al.*, 2006). Archeological discoveries have indicated that quinoa was utilized before 3000 BC in Northern Chile and before 5000 BC in Ayacucho region of Peru. It was widely cultivated in Andean region till the conquest of South America by Spanish. The traditional cultivation area of quinoa was replaced by Wheat and barley by European rulers, which resulted in cultivation of quinoa in rural area for domestic consumption alone.

According to FAO statistics 2018, Peru and Bolivia are the leading producers and exporters of quinoa to the other nations in the world followed by Ecuador. Peru has the higher level of productivity followed by Ecuador. Quinoa grows well from the sea level to high altitudes in wide range of acidic soils. Based on this characteristic Valencia-Chamorro (2003) classified Quinoa into five eco types namely sea-level, valley, subtropical, salar, and antiplanic. The crop tolerates extreme environments from -1 °C to +35 °C with frost resistant characteristic.

The nutritional superiority of quinoa has been acknowledged since ancient times in the Inca Empire. At present, quinoa has been appreciated for its nutritional benefits all over the world for its protein, mineral and vitamin contents. The protein content of quinoa seeds varies from 8 percent to 22 percent, which is higher on average than that in common cereals such as rice, wheat and barley. Quinoa also high quantity of vitamins and mineral and seed lipids. It can be effectively used as rice replacement as breakfast cereals. According to a field crops article by the University of Wisconsin and the University of Minnesota, quinoa means "mother grain" in the Incan language.

Table 6: Area, Production and Productivity of Quinoa

County	Area (ha)	Production (tonnes)	Yield (hg/ha)
Bolivia	111605	70763	6340
Ecuador	2048	2146	10479
Peru	64660	86011	13302
World	178313	158920	8912

Agronomy of Small Millets

The ultimate aim of crop production is to get maximum production per unit area. To obtain higher yield, it is important to follow good agronomic practices or cultural method which includes land preparation, seed treatment, seed rate, method and time of sowing weed management, water management, insect and pest management and time of harvest in coordination with existing climatic condition. The details of cultivation methods for small millets are given in Table 7.

Climatic requirement: Small millets are grown mostly in warm and dry climate. They are the crop of tropical and subtropical climate and can be cultivated from mean sea level to an altitude of 2100 MSL. They are highly drought tolerant and therefore, can be grown in areas where rainfall is scanty and erratic. They well thrive in areas receiving only 40 to 50 cm annual rainfall. Proso millet evade even drought by early maturity.

Soil: Small millets can be cultivated from gravelly and stony upland poor soils to loam soils. Deep, loamy, fertile soils rich in organic matter are preferred for satisfactory growth. Well-drained soils with adequate moisture supply are required for uninterrupted growth of this crop. Most of the small millets cannot tolerate waterlogged condition except teff. They also perform well in acidic and alkaline soils.

Seed rate: Seeds of small millets are tiny nature and handful of small millets contain around 1000 number of seeds. 4-10 kg of seed is required for line sowing and 12-15 kg required for broadcasting per hectare of land.

Seed treatment: Small millets seeds were treated with ceresan at the rate of 3 g/kg of seed to avoid pest and diseases as prophylactic measure.

Sowing Time: Mostly small millets were sown during June to July in Kharif season and September to October in Rabi season. The crops can be also sown during late season also depending upon the climatic condition. Small millets are the sustainable crop substitute in case of failure of wheat, rice and major millets due to erratic rainfall and climatic condition.

Method of Sowing: Small millets were sown either through boadcasting Broadcasting method or line sowing. Being tiny, small millets seeds were mixed with sand and broadcasted for the uniform coverage in the field. It will also helps in avoiding weed growth.

Spacing: Mostly a row-to-row spacing of 15 cm and plant-to-plant spacing of 5-10 cm were maintained to better yield. In case of broadcasted field thinning will be carried out to avoid plant-to-plant competition for nutrients and other resources

Table 7: Agronomic practices of small millets

Details	Barnyard Millet	Finger millet	Fonio Millet	Foxtail millet	Kodo Millet	Little Millet	Proso millet	Teff
Climatic requirement	Warm anddry climate Scanty and erratic rainfall 40 to 50 centimeter annual rainfall	Crop of tropical and subtropical climate and can be cultivated up to an altitude of 2100 m Ideal temperature of growth 26º-29º C	Crop of tropical and subtropical climate Perform well in drought and low fertility soil Thrives well from 20ºC at higher altitudes to 25–30ºC near sea level	Grown in tropics as well as temperate regions both under low and moderate rainfall. Can be grown even at an altitude of 2000 meters and 50-75 cm annual rainfall.	Warm anddry climate Scanty and erratic with 40 to 50 cm annual rainfall.	Crop of tropical subtropical and temperate climate Can be cultivated up to an altitude of 2000 m MSL Withstand drought andwaterlogging	Crop of warm climate Evade drought by early maturity Drought tolerant	Warm anddry climate
Soil	Gravelly and stony upland poor soils to loam soils. Deep, loamy, fertile soils, rich in organic matter,	Alluvial, loamy and sandy soil with good drainage Tolerate certain level alkalinity Can be cultivated up to an altitude of 2100 m	Grows well in sandy to stony, poor, degraded soils	Grows well on well-drained loamy soils. Will not tolerate water-logged soils or extreme drought	Gravelly and stony upland poor soils to loam soils. Deep, loamy, fertile soils, rich in organic matter,	Alluvial, loamy and sandy soil with good drainage	Perform well in well drained loam and sandy soil	Low desert sand to waterlogged clays Perform well acid and alkaline soil
Seed rate	8-10 kg/ha (line sowing) 12-15 kg/ha (broadcasting)	8-10 kg/ha (line sowing) 4 kg/ha (transplanting)	6-8 kg/ha (line sowing) 25-30 kg/ha (broadcasting)	8-10 kg/ha (line sowing) 15 kg/ha (broadcasting)	10 kg/ha (line sowing) 15 kg/ha (broadcasting)	8 kg/ha (line sowing) 15 kg/ha (broadcasting)	8 kg/ha (line sowing) 10 kg/ha (broadcasting)	4-10 kg of seed
Seed treatment	Ceresan at the rate of 3 g/kg of seed	thiram at the rate of 2.5 g/kg of seed		Ceresan at the rate of 3 g/kg of seed	Ceresan at the rate of 3 g/kg of seed	Ceresan at the rate of 3 g/kg of seed	Ceresan or at the rate of 2.5 g/kg of seed	

Details	Barnyard Millet	Finger millet	Fonio Millet	Foxtail millet	Kodo Millet	Little Millet	Proso millet	Teff
Sowing time	Kharif- June to July Rabi- September to October	Kharif- June to July Rabi- September to October Summer-January to February	Mid May to July	Kharif- June to July Rabi- September to October	Kharif- June to July Rabi- Septem-ber to October	Kharif- June to July Rabi- September to October	Kharif- June to July Rabi- October Summer-April	July to august
Method of Sowing	Broadcasting and line sowing	Broadcasting and line sowing	Broadcast-ing and line sowing	Broadcasting and line sowing	Broadcasting and line sowing	Broadcasting and line sowing	Broadcasting and line sowing	
Spacing	25x10cm	25x10 cm	15-20 X 10 cm	25-30X8-10 cm	22.5x10cm	22.5x18-0cm		
Manures and fertilisers	5-10 FYM/ha 40:20; 20 NPK	5-10 FYM/ha 6030:30 NPK	-	5-10 FYM/ha 40:20; 20 NPK	5-10 FYM/ha 40:20; 20 NPK	-10 FYM/ha 40:20; 20 NPK	4-6 FYM/ha 60:30:30NPK	
Varieties:	VL-Madira-181, VL-Madira-172, Pratap sanwa-1, kancahn, VLM-172 and VL-Madi-ra-2007.	BirsaGourav, A-404, VL-149, VL-124, Goda-vari, Ratnagiri, GujratNagli, PR-202 and Indaf-8		Pant Setaria-4, TNAU-43, HMT-100-1, SIA-326, PS-4, K-2, K-3 and Krishna devaraya	JK-13, JK-48, GK-2, Vamban, IPS 147-1, JK-62, JK-76, GPUK-3 and Kherapa.	Tarini, Ko-lab ,Sabara TNAU-63Jawahar Kutki-36.CO-3	GPUP-21, CO (Pv)-5, TNAU-145, TNAU-151 ,Pratap che-na-1,PRC-1 and TNAU-164.	Magna, Sergeg-na,Bunign, kuncho, Qoreta, Bost, Package, Oromia, Shenkore,Cross37, Kora. Gemechis, Dagim.
Water man-agement	Does not require frequent irriga-tion If dry spells once in 25 days (I) Panicle Initia-tion (II)	Does not require frequent irrigation Critical stages of Irrigation: Tilling, flowering	Mostly culti-vated as rain fed crop	Does not require frequent irriga-tion First irrigation at 25-30 DAS Second irriga-tion at 40-45 DAS	Once 4-7 days	First irrigation at 25-30 DAS Second irrigation at 45-50 (DAS).	Does not require frequent irriga-tion Critical stages of Irrigation: tillering Heavy irrigation should not be done	

Details	Barnyard Millet	Finger millet	Fonio Millet	Foxtail millet	Kodo Millet	Little Millet	Proso millet	Teff
Important weeds	Grassy weeds:*Echinochloacolonum, Enhinochloacrusgulli* (sawan), *Dactylocteniumaegypticum* (makra), (kodo), *Setariaglauca*(banra), *Cynodondactylon* (doob), *Phragmiteskarka* (narkul), *Cyperusrotundus* (motha), *Sorghum halepanse*(banchari Broad-leaved weeds: *Celosia argentia* (chilimil), *Commelinabenghalensis* (kankoua), *Phylanthusniruri* (hulhul), *Solanumnigrum* (makoi) and *Amaranthusviridis* (chaulai).	Grassy weeds: *Echinochloacolonum, Enhinochloacrusgulli* (sawan), *Dactylocteniumaegypticum* (makra), (kodo), *Setaria glauca* (banra), *Cynodondactylon* (doob), *Phragmites karka* (narkul), *Cyperus rotundus* (motha), *Sorghum halepanse*(banchari Broad-leaved weeds: *Celosia argentia* (chilimil), *Commelinabenghalensis* (kankoua), *Phylanthusniruri* (hulhul), *Solanum nigrum* (makoi) and *Amaranthus viridis* (chaulai).	striga *(Striga-hermonthica)*	-NA-	Grassy weeds: *Echinochloacolonum, Enhinochloacrusgulli* (sawan), *Dactylocteniumaegypticum* (makra), (kodo), *Setaria glauca* (banra), *Cynodondactylon*(doob), *Phragmites karka* (narkul), *Cyperus rotundus* (motha), *Sorghum halepanse*(banchari Broad-leaved weeds: *Celosia argentia* (chilimil), *Commelinabenghalensis* (kankoua), *Phylanthusniruri* (hulhul), *Solanum nigrum* (makoi) and *Amaranthus viridis* (chaulai).	Grassy weeds: Echinochloacolonum, Enhinochloacrusgulli (sawan), Dactylocteniumaegypticum (makra), (kodo), Setaria glauca (banra), Cynodondactylon (doob), Phragmites karka (narkul), Cyperus rotundus(motha), Sorghum halepanse (banchari Broad-leaved weeds: Celosia argentia (chilimil), Commelinabenghalensis (kankoua), Phylanthusniruri (hulhul), Solanum nigrum (makoi) and Amaranthus viridis (chaulai).	Grassy weeds: *Echinochloacolonum, Enhinochloacrusgulli* (sawan), *Dactylocteniumaegypticum* (makra), (kodo), *Setaria glauca*(banra), *Cynodondactylon* (doob), *Phragmites karka* (narkul), *Cyperus rotundus* (motha), *Sorghum halepanse*(banchari Broad-leaved weeds: *Celosia argentia* (chilimil), *Commelinabenghalensis* (kankoua), *Phylanthusniruri* (hulhul), *Solanum nigrum* (makoi) and *Amaranthus viridis* (chaulai).	Asendabo(Wobalo), Yewofgomen, Sheket, Aluma, wofankure, Engicha, Wofzerir, Sirana, Angago. Mech, Quni, jere, Guri, Migreserie, Enboye, Kumundo, Mormi, Menj, Abatebo, (vernacular Name used in Ethiopia)

Details	Barnyard Millet	Finger millet	Fonio Millet	Foxtail millet	Kodo Millet	Little Millet	Proso millet	Teff
Weed Control	2 hand weeding Post-emergence application: 2, 4-D sodium salt (80%) at the rate of 1.0 kg a.i./ha at 20-25 DAS Pre-emergence spray: Isoproturon at the rate of 1.0 kg a.i. /ha	2 hand weeding Post-emergence application: 2, 4-D sodium salt (80%) at the rate of 1.0 kg a.i./ha at 20-25 DAS Pre-emergence spray: Isoproturon at the rate of 1.0 kg a.i. /ha	2 hand weeding	2 hand weeding Application : 2, 4-D sodium salt (80%) at the rate of 1.0 kg a.i./ha at 20-25 DAS Pre-emergence spray: Isoproturon at the rate of 1.0 kg a.i. /ha	Two hand weeding at an interval of 15 days Post-emergence application of 2, 4-D sodium salt (80%) at the rate of 1.0 kg a.i./ha at 20-25 DAS. Isoproturon at the rate of 1.0 kg a.i. /ha as pre-emergence spray is also effective in weeds control.	Two hand weeding at an interval of 15 days Post-emergence application of 2, 4-D sodium salt (80%) at the rate of 1.0 kg a.i./ha at 20-25 DAS. Isoproturon at the rate of 1.0 kg a.i. /ha as pre-emergence spray is also effective in weeds control.	Hand weeding at an interval of 15 days Post-emergence application of 2, 4-D sodium salt (80%) at the rate of 1.0 kg a.i./ha at 20-25 DAS. Isoproturon at the rate of 1.0 kg a.i. /ha as pre-emergence	2 hand weeding 2, 4-D at the rate of 1litre/ha for broad leave weed
Diseases and its control measure	Downy mildew- Spray of 0.2 percent solution of Mancozeb 75 WP Smut - Seed treated with thiram or ceresan at the rate of 2.5 g/kg of seed and soaking seeds in hot water at 55 °C for 7-12 minutes.	Blast *(Piricularia fungi-)*Seed treatment with thiram or ceresan at the rate of 2.5 g/kg of seed Downy mildew and seedling blight *–spray* 0.2 percent Zineb/ Mancozeb 75 WP	Fungal	Downy mildew- Spray of 0.2 percent solution of Mancozeb 75 WP Smut - Seed treated with thiram or ceresan at the rate of 2.5 g/kg of seed and soaking seeds in hot water at 55 °C for 7-12 minutes.	Rust: spraying of 0.2 percent solution of Mancozeb 75 WP Head smut; Seed treated with thiram or ceresan at the rate of 2.5 g/ kg of seed and soaking seeds in hot water at 55 °C for 7-12 minutes	Downy mildew- Spray of 0.2 percent solution of Mancozeb 75 WP Smut - Seed treated with thiram or ceresan at the rate of 2.5 g/kg of seed and soaking seeds in hot water at 55 °C for 7-12 minutes.	Head smut (*Sphacelorhe-cadestruens)* *:sead treatment* or ceresan at the rate of 2.5 g/kg of seed	Tef rust (Uromy-ceseragrostidis Tracy) Head smudge(Hel-minthosporium sps)

Details	Barnyard Millet	Finger millet	Fonio Millet	Foxtail millet	Kodo Millet	Little Millet	Proso millet	Teff
	Rust- Spray of Mancozeb (75 WP) at the rate of 2 kg/ha.			Rust- Spray of Mancozeb (75 WP) at the rate of 2 kg/ha.		Rust- Spray of Mancozeb (75 WP) at the rate of 2 kg/ha.		
Insect-pests management	Stem borer- Apply Phorate at the rate of15 kg/ha (10% granules) in the soil at the time of field preparation. Termites- Using Methyl parathion (2%) dust at the rate of 20-25 kg/ha before sowing	Army worms and cut worms: Dusting of Malathion 5 percent at the rate of 24 kg/ha or Endosulfan 4 percent at the rate of 12kg/ha Leaf aphid Spraying of Dimethoate (0.05%) or Quinalphos (0.05%) or Endosulphan (0.07%)	Whitebacked-planthopper (*Sogatellafurcifera)* Insecticidal soaps	Shoot fly: Phorate at the rate of15 kg/ha (10% granules) in the soil at the time of field preparation	Shoot fly: Phorate at the rate of15 kg/ha (10% granules) in the soil at the time of field preparation or Carbofuran (Furadan) 3 percent granules at the rate of 30 kg/ ha in furrows or as broadcast before sowing. Ants stem borer; 20-25 kg per hectare 5 percent Malathion or 2 percent methyl parathion dust in the soil before sowing	Shoot fly: Phorate at the rate of15 kg/ ha (10% granules) in the soil at the time of field preparation	Shoot fly: Phorate at the rate of15 kg/ha (10% granules) in the soil at the time of field preparation or Carbofuran (Furadan) 3 percent granules at the rate of 30 kg/ha in furrows or as broadcast before sowing.	

Details	Barnyard Millet	Finger millet	Fonio Millet	Foxtail millet	Kodo Millet	Little Millet	Proso millet	Teff
Harvesting Time	90-120 days of sowing	120-135 days of sowing	70-130 days of sowing	80-100 days of sowing	Kharif season: September or October Rabi season: January to February.	120-135 days of sowing	65-75 days of sowing	120-180 days of sowing
Yield	Yield: Grain-12-15 quintals/ ha Straw-20-25 quintals/ha.	Yield: Grain-20-25 quintals/ ha Straw-60-80 quintals/ha	Yield: Grain 8-11 quintals/ ha Straw -18-20 quin-tals/h	Grain: 15-18 quintal/ha Straw: 2040 quintals/ha	Grain: 15-18 quintal/ha Straw: 30-40 quintals/ha	Yield: Grain-12-15 quintals/ha Straw-20-25 quin-tals/ha.	Grain: 15-18 quintal Straw:50-60 quintals/ha	Yield: Grain-15-22 quintals/ha Straw-20-25 quintals/ha.

Source. www.krishisewa.com,www.agrifarming.in, kvk.icar.gov.in andwww.agrifarming .in

Field Preparation for Small Millets

The main objectives of field preparation are to eliminate and control of unwanted plants and weeds to reduce competition with the established main crop and to provide favorable environments for sowing, allowing germination, emergence and good plant development. Proper field preparation also helps in maintenance of fertility and productivity over the long term by preserving the soil organic matter and avoiding erosion by enabling mixing of fertilizers, lime, or agro-chemical products into the soil. It also breaks hard pans or compressed layers to increase water infiltration through the soil whilst avoiding erosion

Timely field preparation enables timely sowing which guarantees higher yield. Land preparation should ensure completely burial of all crop residue, organic wastes and weeds. Summer ploughing is beneficial to eradicate the weed seeds and hibernating insects and disease organisms by exposing them to the heat of summer. Preliminary ploughing should be carried out at optimum moisture content to get good tilth and should avoid when moisture is in excess. For rainfed crop, with onset of rains in May-June, the field is ploughed once or twice to obtain a good tilth. Harrowing of soil should invariably follow after each ploughing to reduce the clod size. After the initial ploughing, the subsequent ploughings and harrowings are carried out when the moisture content of the clods are reduced. The numbers of ploughings are to be minimized to reduce the cost of cultivation. Tillage operations should be repeated when the weed seeds are just germinated. When the soils are heavily infested with perennial weeds like Cynodon (dhub grass) or Cyperus (nut grass), deep ploughing is needed. Moisture is a critical element in good seedbed preparation and is essential for the successful establishment of the crop.

Manures and fertilizers: Addition of organic manures is always beneficial since it helps to improve the water retention capacity of soil in addition to providing essential nutrients to the crop plants. The crop should be manured with 5- 20 tonnes/ha farmyard manure depending upon the soil fertility about a month before sowing. Application of required quantity of nitrogen, phosphorous and potassium fertilizers after proper soil testing enable proper growth of millet plants and to get higher yield .The fertilizers may be applied at the time of sowing in furrows.

Water management

Small Millets sown during Kharif season do not require any irrigation. However, if dry spell prevails for longer period, then first irrigation at 25-30 days after sowing and second irrigation at 40-45 days after sowing must be given to boost the yields. Excessive rainwater should be drained out get proper growth.

Weeds and Weeds management: *Echinochloa colonum, Enhinochloa crusgulli* (sawan), *Dactyloctenium aegypticum* (makra), *Elusine indica* (kodo), *Setaria glauca* (banra), *Cynodondactylon* (doob), *Phragmites karka* (narkul), *Cyperus rotundus* (motha), *Sorghum halepanse* (banchari) are common grass weeds found in small millets cultivation areas. The broad-leaved weeds are *Celosia argentia* (chilimil), *Commelina benghalensis* (kankoua), *Phylanthus niruri* (hulhul), *Solanum nigrum* (makoi) and *Amaranthus viridis* (chaulai). It is crucial to control weeds in the initial stages of plant growth. Generally two weeding at an interval of 15 days are sufficient. Weeding may be done with hand hoe or wheel hoe in line-sown crop. Pre-emergence spray of isoproturon at the rate of 1.0 kg a.i. /ha and post-emergence application of 2, 4-D sodium salt (80%) at the rate of 1.0 kg a.i./ha at 20-25 DAS, will help in effective weeds control.

Diseases and its control measure

Downy mildew, rust smuts and blights are the common diseases found in small millets. They can be effectively controlled by spraying proper fungicides and following certain intercultural operations.

Rust: Rust of small millets caused by *Puccinia substriata* var. *indica* and *Uromyces eragrostidis* Tracy (Teff) Rust like brown pustules are visibly seen on leaves. This disease hampers photosynthesis and cause considerable loss in yield. Spraying 0.2 percent solution of Mancozeb 75 WP can effectively control it.

Head smut (*Sphacelothe cadestruens*) is a common disease in small millets. The affected ears heads were full of black masses covered with a thin yellow membrane. This is seed born disease can be controlled by treating the seed with thiram or ceresan at the rate of 2.5 g/kg of seed and soaking seeds in hot water at 55 °C for 7-12 minutes

Downy mildew (*Sclerospora* sp.) is a major destructive disease found in small millets . It is a seed borne disease can be effectively by treating seeds with thiram or ceresan@ 2.5 g/kg of seed and soaking seeds in hot water at 55 °C for 7-12 minutes and spray of 0.2 percent solution of Mancozeb 75 WP.

Blast caused by *Pyricularia* sps fungi is other important disease of small millets. It produces rounded or lenticular spots with a central grey to pale-olive area underneath. When these spots extended across the leaf base, the distal part withered. It leads loss in crop yield. Seed treatment with thiram or ceresan at the rate of 2.5 g/kg of seedwill effectively control the occurrence of disease.

Insect-pests management

Stem borer, shoot fly, armyworms, aphids and termites are the common pest of small millets.

Shoot fly can be controlled by application of phorate at the rate of15 kg/ ha (10% granules) in the soil at the time of field preparation or carbofuran (Furadan) 3 percent granules at the rate of 30 kg/ha in furrows or as broadcast before sowing.

White ants and stem borer can be controlled by applying 20-25 kg per hectare 5 percent Malathion or 2 percent methyl parathion dust in the soil before sowing.

Army worms and cutworms can be controlled by dusting of Malathion 5 percent at the rate of 24 kg/ha or Endosulfan 4 percent at the rate of 12kg/ha when symptoms are noticed.

Leaf aphid can effectively controlled by spraying of dimethoate (0.05%) or quinalphos (0.05%) or endosulphan (0.07%) and stem borer by spraying the crop with dimethoate (0.05%) or monocrotophos (0.04%) helps in pest control.

Harvesting Time: Small millets crops mature from 40 days to 120 days after sowing or transplanting .In Kharif season, the crop becomes ready for harvest in the month of September or October in northern India and in Rabi season, it is harvested from January to February. Proso millet is very short duration millet and mature with 40-70 days depending upon the climatic condition

Yield: With improve package and practices, one can obtain 8-25 quintals of grain and 20-50 quintal straw per hectare.

Agronomy of Pseudo Cereals

The agronomic practices followed for the cultivation of Pseudo cereals such as amaranth, buckwheat, chia, cockscomb and quinoa to obtain optimum yield are given Table 7.

Climatic Requirement: Pseudo cereals also known as nutritional powerhouses are grown well in both rainy and summer season mostly in warm and dry climate. They are the crop of tropical and subtropical climate and can be cultivated from mean sea level to an altitude of 2100 MSL. Quinoa, the super grain and buckwheat can be cultivated even at altitude 4000 MSL. However, compared to quinoa, buckwheat is much sensitive freezing temperature.

Soil: Pseudo cereals performs well in well drained loamy to sandy loam and medium clay soil rich in organic matter, are preferred for satisfactory growth. Well-drained soils with suitable moisture supply are required for uninterrupted growth of this crop. Pseudo cereals cannot tolerate waterlogged condition. They also perform well in acidic and alkaline soils. Buck wheat showed higher level tolerance to soil acidity and perform well in pH below 4.5 also.

Seed rate: Seeds of amaranth, cockscomb are much tiny in nature and required at the rate of 2 kg per hectare of land .15-20 kg of seed is required for cultivation of buckwheat and quinoa per hectare. The Chia also known as sabja in hindi language requires 6 kg of seed for cultivation in one hectare

Seed treatment: Seeds of amaranthus and cockscomb were treated with Bavistin at the rate of 2 g/kg of seed to avoid pest and diseases as prophylactic measure.

Sowing Time: Mostly pseudo cereals were sown during June to July in kharif season and- September to October in rabi season. The crops can be also sown during late season also depending upon the climatic condition. Amaranth and Cockscomb were sowed mid March in summer, mid July during Kharif and October for rabi season. The crops can be cultivated throughout depending upon water availability.

Method of Sowing: Pseudo cereals were sown either through Broadcasting method or line sowing. Being tiny, amaranth and cockscomb seeds were mixed with sand and broadcasted for the uniform coverage in the field. It will also helps in avoiding weed growth.

Spacing: Mostly a row-to-row spacing of 30-60 cm and plant-to-plant 10 cm were maintained to better yield. In case of broadcasted field thinning will be carried out to avoid plant-to-plant competition for nutrients and other resources

Manures and fertilizers: Addition of organic manures such as farmyard manure helps in improve the soil fertility as well as water retention capacity. The field should be manured with 10- 20 t/ha FYM depending upon the soil fertility about a month before sowing. Application of required quantity of nitrogen, phosphorous and potassium fertilizers after proper soil testing enable proper growth of millet plants and to get higher yield. The fertilizers can be applied during sowing in furrows.

Water management

Most of the pseudo cereals require frequent irrigation to avoid wilting of the crop. Irrigation at an interval of 7-10 days is necessary to maintain crop growth.

Crops	Amarnth	Buck Wheat	Chia	Cockscomb	Quinoa
Climatic requirement	Grows well in both summer and rainy season Widely distributed in both tropical and sub tropical regions	Grows well in cool and moist climate Perform well in high altitude also Sensitive to freezing temperature High temperature can cause flower blasting	Crop of tropical and subtropical region Grows well from 400-2000 above MSL	Grows well in both summer and rainy season 6-8 hours sunlight required	Grows well in both tropical and sub tropical regions Grows well from sea level 4000 above MSL Ideal temperature 18-20 c
Soil	Comes up well in well-drained loamy soil rich in organic matter. Ideal is pH is 5.5-7.5	Well drained soil with moderate fertility Grows well in pH below 4.5 Higher tolerance to soil acidity	Perform well in light to medium clay as well as sandy soil	Aerated porous soil with good drainage	Grows well in Sandy loam Ideal is pH is 5.5-9.0
Seed rate	2 kg/ha for direct sowing.	20 kg/ha	6 kg / ha	2 kg/ha for direct sowing	15-20 kg/ha
Seed treatment	Bavistin@2g/kg of seed	Nil	Nil	Bavistin@2g/kg of seed	-
Sowing time	Mid March-Summer Mid July -Kharif Throughout the year-irrigated cultivation	May to Mid august-Kharif	Jun-July-Kharif Oct-Nov -Rabi	Jun- Kharif Oct- Rabi	Mid May-Kharif
Method of Sowing	Broadcasting	Seed sowing at adepth of ½ to inch	Seed broadcasting Seedling transplanting		Seed sowing at a depth of 1to 3 cm
Spacing	30 X 10	30X 10 cm	60cm X 70-80 cm	40-60 X	25-50 X10 cm
Manures and fertilizers	20 -25 tonnes FYM 50:25:20 kg NPK	10-12 tonnes FYM 20:10:10 kg of NPK	15-20 tonnes FYM 90:60: 75 kg NPK/ha		20-30 tonnes FYM 100: 50 : 50 kg NPK/ha

Crops	**Amarnth**	**Buck Wheat**	**Chia**	**Cockscomb**	**Quinoa**
Varieties:	CO-1(*A. dubius*) Co. 2 (*A. tricolor*) Co. 3 (*A. tristis*) ChhotiChaulai (*A. blitum*) BadiChaulai (A. *tricolor*) PusaKirti (A. *tricolor*) PusaKiran PusaLalChaulai (*A. tricolor)* 2Arka Arunima •	Common buckwheat Tartary Buckwhaet	Black Chia White Chia	Giant red crested cockscomb, Lemon lime celosia Cockscomb yellow Cockscomb snake cobra Dwarf Cockscombs	White Quinoa Dark red Quinoa Purple Quinoa black. Quinoa
Water management	Drought tolerant crop Irrigation 4-6 days interval in summer	- Irrigation 6-10 days interval	Irrigation7-10 days interval	Frequent irrigation	Frequent irrigation to avoid wilting of crop
Important weeds		It completely cover the field and out compete weeds	Not available	Not available	Not available
Weed Control	2-3 hand hoeing	Not required	Hand hoeing andherbicidal application	2-3 hand hoeing	102 hand hoeing

Crops	Amarnth	Buck Wheat	Chia	Cockscomb	Quinoa
Diseases and its control measure	Leaf spot-*Cercosporaspp and Alternaria Sps : control at the rate of2 percent foliar spary Pseudomonas fluorescence and soil application of 2 percent Trichiderma viridie White rust -Albugo candida* : control at the rate ofsparay of Dithane M-45 or any other copper fungicide	Minor level infection by leaf spot and *Rhizotoania root rot*	Viral infection on leaves	Leaf spot-*Cercospora sps and Alternaria sps : control at the rate of2 percent foliar spary Pseudomonas fluorescence and soil application of 2 percent Trichiderma viridie*	Downey Mildew -*Peronosporafarinosa* : Application of fungicide metalxyl
Insect-pests management	Leaf webber, caterpillars and ants-control by spraying of malathion at the rate of 1.5 to 2 ml per litre of water	Minor level infection by aphids	Minor level infection by aphids	Leaf webber, caterpillars and ants-control by spraying of malathion at the rate of 1.5 to 2 ml per litre of water	Leaf eating caterpillars and leaf minors – control spray of quinolphos 2 ml/lit of water. Aphids –control sparyof Dimethoate at the rate of 2ml/ lit of water.
Harvesting Time	80-90 days after transplanting	80-90 days after sowing	100-150 days after showing/ transplanting	90 days after transplanting	90-120 days after showing/ transplanting
Yield	1-2.5 tonnes depending upon the variety	0.8-1.5 tonnes /ha	2-2.8 tonnes /ha	1-2.0 tonnes depending upon the variety	1.5-5 tonnes/ha

Source: www.krishisewa.com,www.agrifarming.in, kvk.icar.gov.in andwww.agrifarming .in

Weed management

It is essential to control weeds in the initial stages of plant growth. Generally two weeding at an interval of 15 days are sufficient. Weeding may be done with hand hoe or wheel hoe in line-sown crop. Pre-emergence spray of isoproturon at the rate of 1.0 kg a.i. /ha and post-emergence application of 2, 4-D sodium salt (80%) at the rate of 1.0 kg a.i./ha at 20-25 DAS, will help in effective weeds control.

Diseases and its control measure

Leaf spot caused by *Cercospora* spp and *Alternaria* sps one of major diseases of amaranth and cockscomb. The leaf spot diseases affect the photosynthesis and in turn resulted in poor yield. They can be effectively controlled by foliar spraying 2 percent Pseudomonas fluorescence and soil application of 2 percent Trichiderma viridie and following certain intercultural operations.

White Rust caused by Albugo candida is another major disease of amaranth and cockscomb. Symptoms of infection include chlorosis on leaf surfaces, white blister-like growths on the underside of leaves and on the stems of the plant, and swelling of the roots. Spraying of dithane M-45 or any other copper fungicide can effectively control it.

Downy mildew (*Peronospora farinosa*) is a major disease found in quinoa. Application of fungicide metalxyl effectively control the downy mildew infection in quinoa.

Buck wheat and chia show resistance to most of the plant diseases and very minor level infection by leaf spot , rhizoctonia root rot and some viral infection on leaves.

Insect-Pests Management

Leaf webber, caterpillars and white ants are major pest of amaranth and cockscomb and they can be controlled by topical application of of malathion at the rate of 1.5 to 2 ml per litre of water.

Leaf eating caterpillars and leaf minors and Aphids are major pest of quinoa crop and they can be controlled by sparying quinolphos 2 ml/l and dimethoate at the rate of 2ml/l of water respectively.

Buck wheat and chia showed very minor level infection by leaf aphids. Economic damage found to be very low by leaf aphids.

Harvesting Time: Amaranth, cockscomb and buckwheat crops mature from 80-90 days after sowing or transplanting. They are relatively short duration crops compared chia and quinoa as they 100-150 days for maturity.

Yield: With improved package of practices farmer can obtain 0.8 to 5 tonnes per hectare.

Conclusion

Small millets and pseudo cereals are widely distributed crops and cultivated in more than 90 countries in the world. They are highly adapted to extreme climatic condition and edaphic factors and mature earlier than the common cereals. They are also known as nutrient powerhouses as they rich in vitamin, minerals, and phytochemicals such as polyphenols, tocopherols, phytosterols, and dietary fiber. They are low glycemic index foods with low or nil quantity of gluten Small millets and pseudo cereals are staple food for millions population They are future crop to meet the food need of ever increasing world population.

References

Afifa, S., Kamili ,Hakeem, Tahir, Saleem., Tariq, Sultan., andM.S.Trumboo (2020) https://kvk.icar.gov.in/API/Content/PPupload/k0034_2.pdf . Accessed on 15 sep 2020

Arctos Specimen Database.,(2018). Collaborative collection management solution. Retrieved from http://arctos.database.museum/name/Salvia%20hispanica# ArctosPlants Accessed: September, 10, 2018.

Austin, D. F. (2006). Fox-tail millets (*Setaria*: Poaceae)—abandoned food in two hemispheres. Economic Botany, 60(2), 143-158.

Ayalew, A., Kena, K., and Dejene, T. (2011). Application of NP Fertilizers for better production of Teff (Eragrostis tef (Zucc.) Trotter) on different types of soils in southern Ethiopia. Journal of Natural Sciences Research, 1(1), 1-11.

Ayerza, R. (2009). The seed's protein and oil content, fatty acid composition, and growing cycle length of a single genotype of chia (*Salvia hispanica* L.) as affected by environmental factors. Journal of Oleo science, 58(7), 347-354.

Bachewe, F. N., Koru, B., andTaffesse, A. S. (2015). Productivity and efficiency of smallholder teff farmers in Ethiopia (No. 79). International Food Policy Research Institute (IFPRI).

Bala Ravi S (2004). Neglected millets that save the poor from starvation. LEISA India 6(1): 34- 36.

Balasubramanian, T., and Karthikeyan, M. (2016). Therapeutic effect of amaranthus hybridus on diabetic nephropathy. J Develop Drugs, 5, 147.

Becker, R., Wheeler, E. L., Lorenz, K., Stafford, A. E., Grosjean, O. K., Betschart, A. A., andSaunders, R. M. (1981). A compositional study of amaranth grain. Journal of Food Science, 46(4), 1175-1180.

Bressani, R., Gonzalez, J. M., Elias, L. G., andMelgar, M. (1987). Effect of fertilizer application on the yield, protein and fat content, and protein quality of raw and cooked grain of three amaranth species. Plant Foods for Human Nutrition, 37(1), 59-67.

Busilacchi, H., Bueno, M., Severin, C., Di Sapio, O., Quiroga, M., andFlores, V. (2013). Evaluación de Salvia hispanica L. cultivadaen el sur de Santa Fe (República Argentina). Cultivostropicales, 34(4), 55-59. Retrieved from http://scielo.sld.cu/pdf/ctr/v34n4/ctr09413.pdf

Campbell, C. G. (1997). Buckwheat: Fagopyrum esculentum Moench (Vol. 19). Bioversity International.

Capitani, M. I., Spotorno, V., Nolasco, S. M., andTomás, M. C. (2012). Physicochemical and functional characterization of by-products from chia (*Salvia hispanica* L.) seeds of Argentina. LWT-Food Science and Technology, 45(1), 94-102.

Chadha, K.L. (2009) Handbook of Agriculture. Indian Council of Agricultural Research (ICAR), New Delhi, 76-82.

Cheng, A., Mayes, S., Dalle, G., Demissew, S., and Massawe, F. (2017). Diversifying crops for food and nutrition security–a case of teff. Biological Reviews, 92(1), 188-198.

Clayton, W. D., andRenvoize, S. A. (1986). *Genera graminum.* Grasses of the world. Genera graminum. Grasses of the World., 13.

Coates, W. (2011). Protein content, oil content and fatty acid profiles as potential criteria to determine the origin of commercially grown chia (*Salvia hispanica* L.). Industrial Crops and Products, 34(2), 1366-1371.

Davison, J., and Laca, M. (2010). Grain production of 15 teff varieties grown in Churchill County, Nevada during 2009.

Divya, G. M., Krishnamurthy, K. N., and Gowda, D. M. (2013). Growth and instability analysis of finger millet crop in Karnataka. Mysore Journal of Agricultural Science, 47(1), 35-39.

Doust, A. N., Kellogg, E. A., Devos, K. M., and Bennetzen, J. L. (2009). Foxtail millet: a sequence-driven grass model system. Plant physiology, 149(1), 137-141.

FAO STAT 2018- http://www.fao.org/faostat/en/#data/QC accessed on 24 sep 2020

FAO. Analysis of price incentives for Teff in Ethiopia Technical notes series, MAFAP, by Assefa B. Demeke M., Lanos B, 2015 Rome.

Grancieri, M., Martino, H. S. D., andGonzalez de Mejia, E. (2019). Chia seed (*Salvia hispanica* L.) as a source of proteins and bioactive peptides with health benefits: A review. Comprehensive Reviews in Food Science and Food Safety, 18(2), 480-499.

Graybosch, R. A., and Baltensperger, D. D. (2009). Evaluation of the waxy endosperm trait in proso millet (*Panicum miliaceum*). Plant Breeding, 128(1), 70-73.

Hilu, K. W. (1994). Evidence from RAPD markers in the evolution of Echinochloa millets (*Poaceae*). Plant Systematics and Evolution, 189(3), 247-257.

Hoeschle Zeledon, I., and Bordoni, P. (2002). Global Facilitation Unit for Underutilized Species.

http://agritech.tnau.ac.in/agriculture/millets_kudiraivali.html, Accessed on 16 sep 2020

https://www.linkedin.com/pulse/finger-millet-ragi-eleusine-coracanal-lets-promote-building-osk-reddy, Accessed on 1 Nov 2020

Jagdish reddy (2020). Buckwheat cultivation kuttu farming practises. https://www.agrifarming.in/buckwheat-cultivation-kuttu-farming-practices. Accessed on 16 sep 2020

Jagdish reddy (2020). Quinoa farming information guides for beginners https://www.agrifarming.in/quinoa-farming-information-guide, accessed on 05 Sep 2020.

Jones, J. M. (2006). Grain-based foods and health. Cereal Foods World, 51(3), 108.

Krishisewa Production technology of amarnthus (*Amaranthus* sps) -2020-https://www.krishisewa.com/articles/production-technology/147-amaranthus.html, accessed on 10 Sep 2020.

Krotov, A.S. (1963). Buckwheat. Moscow-Leningrad, Izdateljstvosel Skanozjaahstvennij Literatury.

Kuta, D. D., Kwon-Ndung, E., Dachi, S., Ukwungwu, M., andImolehin, E. D. (2003). Potential role of biotechnology tools for genetic improvement of "lost crops of Africa": the case of fonio (*Digitariaexilis* and *Digitariaiburua*). African Journal of Biotechnology, 2(12), 580-585.

Kuta, D. D., Kwon-Ngung, E., Dachi, S., Bakare, O., andOgunkanmi, L. A. (2005). Optimization of protocols for DNA extraction and RAPD analysis in West African fonio (*Digitariaexilis* and *Digitariaiburua*) germplasm characterization. African Journal of Biotechnology, 4(12).

Li, Q. (1992). Preliminary investigation on buckwheat origin in Yunnan, China. In Proc. 5th Intl. Symp. Buckwheat, Taiyuan (pp. 44-48).

Lu, H., Zhang, J., Liu, K. B., Wu, N., Li, Y., Zhou, K., and Li, Q. (2009). Earliest domestication of common millet (*Panicum miliaceum*) in East Asia extended to 10,000 years ago. Proceedings of the National Academy of Sciences, 106(18), 7367-7372.

Mallikarjun,Y., Kamatar, Hemalatha S,, Meghana, D, R,, Sharanappa, Talawar,. andRama, K, Naik,. (2013), Evaluation of Little Millet (*Panicum sumatrense*) Land Races for Cooking andNutritional Composition, Current Research in Biological and Pharmaceutical Sciences, 2 (1), Jan-Feb 2013, 07-11

Matiacevich, S. B., Castellión, M. L., Maldonado, S. B., & Buera, M. P. (2006). Water-dependent thermal transitions in quinoa embryos. Thermochimica Acta, 448(2), 117-122.

National Research Council. (1996). Lost crops of Africa: volume I: grains. National Academies Press.

Ntui, V. O., Thirukkumaran, G., Azadi, P., Khan, R. S., Nakamura, I., andMii, M. (2010). Stable integration and expression of wasabi defensin gene in "Egusi" melon (*Colocynthiscitrullus* L.) confers resistance to Fusarium wilt and Alternaria leaf spot. Plant cell reports, 29(9), 943-954.

O'Connor, A. (2016). Is teff the new super grain. The New York Times.

Ohnishi, O. (1988). Population genetics of cultivated common buckwheat, Fagopyrum esculentum Moech. VII. Allozyme variability in Japan, Korea and China. Jpn. J. Genet. 63:507-522

Padulosi, S., Mal, B., Bala Ravi, S., Gowda, J., Gowda, K. T. K., Shanthakumar, G., and Dutta, M. (2009). Food security and climate change: role of plant genetic resources of minor millets. Indian Journal of Plant Genetic Resources, 22(1), 1.

Panicum miliaceum". Germplasm Resources Information Network (GRIN). Agricultural Research Service (ARS), United States Department of Agriculture (USDA). Accessed 29 Sep 2020.

Pawar, V. D., & Machewad, G. M. (2006). Processing of foxtail millet for improved nutrient availability. Journal of food Processing and Preservation, 30(3), 269-279.

Prabhakar, Prabhu, C,. Ganiger, Boraiah B., Sujata Bhat., Nandini, C., Kiran, Tippeswamy V., and Manjunath H. A. (2017) Improved Production Technology for Kodo millet, Technical Bulletin - 2/2017-18

Pradeep, P. M., and Sreerama, Y. N. (2015). Impact of processing on the phenolic profiles of small millets: Evaluation of their antioxidant and enzyme inhibitory properties associated with hyperglycemia. Food Chemistry, 169, 455-463.

Ravi, S. B. (2004). Neglected millets that save the poor from starvation. Leisa India, 6(1), 1-8.

Sakamma, S., Umesh, K. B., Girish, M. R., Ravi, S. C., Satishkumar, M., and Bellundagi, V. (2018). Finger millet (*Eleusine coracana* L. Gaertn.) production system: Status, potential, constraints and implications for improving small farmer's welfare. J Agric Sci, 10(1), 162-179.

Saleh, A. S., Zhang, Q., Chen, J., andShen, Q. (2013). Millet grains: nutritional quality, processing, and potential health benefits. Comprehensive Reviews in Food Science and food Safety, 12(3), 281-295.

Saleh, A., Zhang, Q., Chen, J. and Shen, Q. 2013. Millet grains: Nutritional quality, processing, and potential health benefits. Comprehensive Reviews in Food Science and Food Safety. 12:281-295. https://doi.org/10.1111/1541-4337.12012

Sheahan, C. M. (2014). Plant guide for Foxtail millet (Setariaitalica). USDA-Natural Resources Conservation Service, Cape May Plant Materials Center.

Sood, S., Khulbe, R. K., Kumar, A., Agrawal, P. K., andUpadhyaya, H. D. (2015). Barnyard millet global core collection evaluation in the submontane Himalayan region of India using multivariate analysis. The Crop Journal, 3(6), 517-525.

Sosa, A., Ruiz, G., Rana, J., Gordillo, G., West, H., Sharma, M., and Torre, R. R. R. (2016). Chia crop (*Salvia hispanica* L.): its history and importance as a source of polyunsaturated fatty acids omega-3 around the world: a review. J Crop Res Fert, 1, 1-9.

Takano, A. (2017). Taxonomic study on Japanese Salvia (*Lamiaceae*): Phylogenetic position of S. akiensis, and polyphyletic nature of S. lutescens var. intermedia. PhytoKeys, (80), 87.

Valencia-Chamorro S.A (2003): Quinoa. In: Cabal-lero B.: Encyclopedia of Food Science and Nutrition. Vol. 8. Academic Press, Amsterdam: 4895-4902.

Vietmeyer, N. D., Borlaugh, N. E., Axtell, J., Burton, G. W., Harlan, J. R., andRachie, K. O. (1996). Fonio (Acha). Lost Crops of Africa, 1.

Yerza (h). R., andCoates, W. (2009). Influence of environment on growing period and yield, protein, oil and α-linolenic content of three chia (*Salvia hispanica* L.) selections. Industrial Crops and Products, 30(2), 321–324. https://doi.org/10.1016/j.indcrop.2009.03.009

4

Little Millets (*Panicum sumatrance*) Chemical Composition, Nutritional Quality and Processing

Parveen Kumari and B.S. Khatkar

Department of Food Technology, Guru Jambheshwar University of Science and Technology, Hisar, Haryana, India

Introduction

Millets, represents groups of small seeded cereals and grasses cultivated in Asian countries from ancient time and used for domestic purpose as feed and food. Being capable to grow under adverse environmental conditions (high temperature, dry land with limited rain) millets are the main food commodity of low income groups in various developing countries around the world. Due to the unique characteristics, millets occupy a special place among the cereals. In some developing countries millets are the main contributor to food security and potential health benefits. Minor millets impart nutrients to diet being good source of essential amino acids, micro-nutrients and vitamins (Ahmed *et al.*, 2013). Studies by various researchers concluded that consumption of minor millets reduces the incidence of diabetes and disorders related to gastro-intestinal track (Saleh *et al.*, 2013). Being a good source of calcium, phosphorous and iron millets consumption can be helpful for menopausal women as it balance the mineral loss (Prabha1 and Selvi, 2016). Millets have wide application in multigrain cereal products, in gluten free diets and in preparation of traditional foods such as idli, chakli, dosa, porridges and papads (Singh and Sarita, 2016). Besides these conclusive qualities, millets have an important place in culture and social celebration of local population. Now-a-days with the advancement in research, millets are growing as functional and nutraceutical foods. Millets are basically divided into two groups: major millets (sorghum and pearl millet) and minor millets (little millet, foxtail, finger, kodo, barnyard and proso millet). India is the largest producer of minor millets in world (Ahmed *et al.*, 2013).

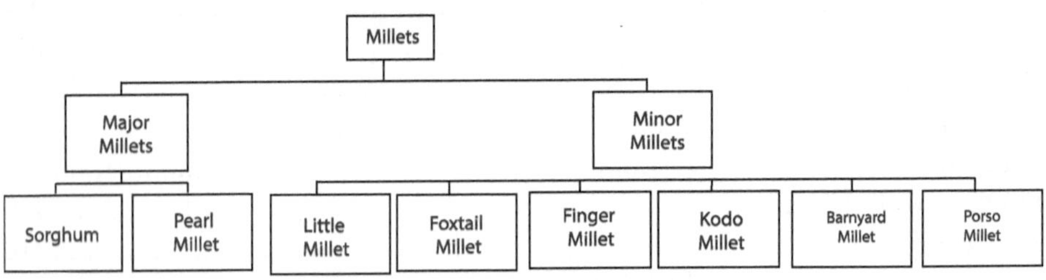

Fig. 1: Classification of millets

Little millet (*Panicum sumatrance*), member of family poaceae is commonly named as Kutki, Moraiyo, Shavan (Hindi), Samalu (Telugu), Sava, Vari, Halvi (Marathi), Gajro, Kuri (Gujarati), Samai (Tamil), Sama (Bengali), Suan (Oriya) and Saame (Kannada) in different regions of India. Little millet is indigenous to India, Nepal, Africa and other dry lands and can be grown on plains and hilly areas without much care. Madhya Pradesh, Tamil Nadu and Karnataka are the main little millet growing states in India. Little millets is used as staple food by the population of low income in some countries and as animal feed around the world. Plant of little millet attains a height of 30-90 cm and it's oblong panicle is approx 14-40 cm in length. Similar to other millets it can survive under water logged and at dry land conditions. *Panicum sumantrense* is classified into two groups a) psilopodium, b) sumatrense on the basis of their inflorescence morphology. Sumatrense is further divided into two races named Nana (subraces Compacta and Laxa) and Robusta (subraces Erecta and

Laxa). It is a nutrient rich but one of the underutilized millet. While nowadays it is getting popularity as a nutrient for malnourished population as well as raw material for malt foods.

Recently, concept of functional and nutraceutical food is popular among researcher and health conscious population. Little millet is comparable to other cereals such as wheat and rice being rich source of protein, carbohydrate, curde fibre and fat. It is also a good source of phytochemicals like phenolic compounds, flavonoids, phytate and tannins and most of these phytochemicals are credited for antioxidant profile of millets. This underutilized millet is a rich source of vitamin B and minerals such as iron, zinc, calcium and potassium. As it provides essential nutrients to diet can be helpful in reducing the obesity and it's high fiber content makes it a better replacement of rice in diet. Suma and Urooj (2012) reported importance of bound phytochemicals in prevention of prostrate, colon, breast and other cancers related to digestive systems. Slow digestion process of complex carbohydrate of little millet is helpful for the diets of diabetic patients. Being a rich source of good cholesterol little millet is recommended in diet of growing kids and to strengthen the body (Gayatri, 2015). Studies by various scientists emphasized on potential of indigenous millets being used as medicine for the treatment of diabetes and post-menopausal syndrome, in diets as health food and in fortified foods.

Chemical Composition

Nutritional profile of little millet is comparable or superior to regularly consumed cereal like wheat and rice being good source of dietary fibre, protein, fat, carbohydrate and minerals. Little millets occupy special place among millets because of their guten-free, non-acid forming and non-sticky nature. Experts and nutritionist recommended millets as a regular part of diet for health conscious people. Non–starchy polysaccharides (cellulose, hemicelluloses and pectinacious material), free sugar (glucose, fructose and sucrose) and starch (amylase and amylopectin) are the main fractions of carbohydrates in millets (Himanshu *et al.*, 2018). A major portion of dietary fiber is contributed by cell wall material and aleurone layer of grain (Ushakumari and Malleshi, 2007). Dietary fiber and non–starchy polysaccharides are the main constituents of little millet carbohydrates, are helpful in preventing constipation and lowering down blood cholesterol. Fibre content of it is about 38 times and 6.3 times higher than rice and wheat, respectively (Hadimani and Malleshi, 1993). Presence of complex carbohydrates aids in slow digestion, resulting in gradual release of sugar which might serve as boon for diabetic patients (Itagi *et al.*, 2013). Carbohydrate quality and quantity are main deciding factors for end use application and to depict it's suitability and functionality in formulation of diets for specific purpose (Kamatar *et al.*, 2013).

Proteins are main building blocks of human body and are required for mental and physical growth and well being, especially of growing kids. Main staple cereals are lacking in some of essential amino acid, like lysine, methionine and tryptophan (Nirgude *et al.*, 2014). Functionality and quality of protein is dependent on amino acid profile and their alignment. Albumins, globulins, glutelin and glutelin-like cross linked prolamin are the major protein fractions in millets. Major portion of millets protein is constituted by prolamin and glutelin followed by globulin and albumin parts (Himanshu *et al.*, 2018). Geervani and Eggum (1989) reported variations in protein digestibility of millets that varied from 95 to 99.3 percent. However, net protein utilization and biological value of millets protein is lower than the pearl millet while its digestible energy is higher than the pearl millet. Millets are rich source of sulphur containing essential amino acids (Obilana and Manyasa, 2002). Methionine, lycine and cystine are the main amino acids present in little millet. Amino acid profile of little millet protein is well balanced and might serve as an important part of vegetarian diet. It works as boon for the person suffering from celiac disease being gluten free in nature (Usha *et al.*, 2011). Alpha amylase is having key role in starch metabolism of cereal grains in growth phase and germination. It also has wide application in breweries, starch saccharification and in supplemented foods (Adewale *et al.*, 2006). Usha *et al.* (2011) reported that germinated little millet is a rich source of alpha amylase which has great potential in industrial application with respect to temperature (50^0C) and pH (5), stability against different activators and inhibitors.

Millets fats are generally known as invisible fats and dominantly present in endosperm and bran portion of seeds. Fat in millets exits in the form of neutral lipids (triacylglycerols), glycolipids and phospholipids (Lysophosphatidylcholine). Other than that stigmasterol and campesterol are also reported in same ratio in both free and esterified form. Cerebrosides (ceramide monohexosides), sterol glycoside, esterified sterol glycoside, monogalactosyldiacylglycerol and digalactosyldiacylglycerol are the major glycolipids in millets (Osagie and Kates, 1984). Oleic, palmitic, linoleic and linolenic are the main characterizing fatty acids reported in millets (Sridhar and laxminarayan, 1992). Little millets are good source of fat, rich in essential and polyunsaturated fatty acids specially linolenic acid. Fatty acid profile of little millet is similar to kodo and finger millet. Little millet and kodo millet showed same ratio of linoleic and oleic acid (Bora and Marcone, 2013). Monogalactosyl monoglycerides and digalactosyl monoglycerides are also present in limited amount in little millet (Sridhar and Lakshminarayana, 1992). It is comparable to kodo millets and oat groats with respect to fatty acid composition. Little millet is an excellent source of good cholesterol and can be recommended in the diet of growing kids and also helpful to strengthen body (Gayatri, 2015).

Reported mineral content of millets are comparable and even quantitatively superior than other cereals crops like maize, rice, sorghum and wheat. Millets are considered unique due to their richness in minerals and dietary fibers (Devi *et al.*, 2011). Coloured varieties of little millets are observed to have higher mineral contents (Hadimani and Malleshi, 1993). Millets are rich in phosphorus, iron, calcium and magnesium. Phosphorus is an important part of ATP (adenosine triphosphate), while magnesium diminishes the chances of heart attack and migraine by helping in blood vessels relaxation and maintaining normal blood pressure (Devi *et al.*, 2011). Finger millets reported to have higher calcium content and might serve as better substitute of rice based food in diet of growing girls as it helps in maintaining the nitrogen balance along with improved retention of calcium (Babu *et al.*, 1987). Among the millets, little millets is reported to have higher iron content while other millets like kodo showed better copper content. Overall, mineral content was observed higher in minor millets compared to other cereal grains including sorghum. Dehulling is an important part of millet processing that some time results in loss of mineral matter specially magnesium, calcium and sodium while iron and potassium contents are not affected significantly. Extent of loss is dependent on type of cultivar and millets. Various cultivars of little millets contained fair amount of iron, zinc, manganese and copper (Kundgol *et al.*, 2013). Iron content of little millet is 13 and 1.8 times of rice and wheat, respectively. Millets rich in calcium, phosphorus and iron might serve as boon for women under menopausal stage (Gayatri, 2015).

Some of important vitamins like vitamin E and vitamin B – complex are present abundantly in millets except vitamin B_{12}. Little millet reported to have higher amount of thiamin compared to wheat and rice. Level of vitamin C is very limited in mature seeds of minor millets. The α-tocopherol exists in the form of γ-isomer in millets. Millets contained lesser quantity of it compared to corn and soybean oil but activity of α-tocopherol as vitamin is much higher than other tocopherols. In millets activity of γ-isomer is less than 10 percent α-tocopherol activity. About 13 percent of total niacin reported in millets exits in water extractable form (Bieri, 1974). Quantity of niacin in little millet is much higher than other cereals crops (Himanshu *et al.*, 2018). Feasibility of traditional breeding is less for high provitamin A producing millet species because of lesser diversity in beta-carotene content among the millets cultivars (Upadhyaya *et al.*, 2011). Refining process affected significantly level of vitamin A and E in crude fat of common millets. Study reported almost complete loss in activity of vitamin A after refining process.

Phytochemicals such as phenolic compounds and flavonoids are main contributors of antioxidant activity that are present in abundant in little millet and credited for enhancement in its nutritional and therapeutic values. (Pradeep and Guha, 2011). Environment conditions, cultural practices and genetic build of cultivars greatly affected the composition and amount of phenolic compounds. Pradeep and Sreerama (2017) reported that cultivars of little millets are rich source of phenolic compounds. Amount and functional properties of different bioactive compounds such as saponins, flavanoids, alkaloids and carotenoids are reported to be different in diverse studied cultivars (Prabha and selvi, 2016). Little millets were reported to have higher phenolic compounds compared to foxtail millets. Phenolic are mainly benzoic acid derivatives and further divided into two major groups: hydroxyl benzoic acid and derivatives of hydroxycinnamic acid. Group one generally presents in bound form like gallic acid, vanilic, syringic acid and are main component of hydrolysable tannins and lignins. The 2nd class includes ferulic, caffeic, p-coumaric and sinapic acid that are mainly attached to cell wall structural units (Chetan *et al.*, 2008). Notable content of extractable and bound phenolic compounds with flavonoids were reported in little millets cultivars. Hydroxycinnamic acid, hydroxyl benzoic acid, ferulic, *p*-coumaric acids and their derivatives are the prominent phenolic acids in little millets (Chandrasekara and Shahidi, 2011). These bound and soluble phenolic compounds exhibited strong antioxidant activities and showed potent inhibition capacity against enzymes (amylase and glucosidase). Phenolic might play an important role in prevention of diseases that are mainly contributed by activity of free radicals (Revankar *et al.*, 2018). Techniques like enzymatic hydrolysis, fermentation and alkali treatment aids to better release of bounded phenolic acid from whole millet grains during processing. Colored cultivars of little millets having anthocyanin pigment may be preferred for the development of functional and nutraceutical food with natural colorants. Overall, phenolic compounds whether bound or in soluble form helps in prevention of diabetes and in regulation of blood glucose level.

Bioavailability of Nutrients

Bioavailability of nutrients in millets is limited due to presence of antinutrients like tannins and phytate. Millets richness in dietary fiber is also one of the important limiting factor affecting bioavailability of minerals. Rao *et al.* (1983) reported that absorption of millet iron was lesser than commonly consumed wheat and rice in humans. Rao and Deosthale (1988) reported that pigmented cultivars of finger millets showed reduced bioavailability of iron compared to white varieties and that might be due the presence of higher tannins content in coloured varieties. Ionizable form of iron is more readily absorbed in body. Germination of seeds or reduction of tannins by different methods might serve in enhancement of iron bioavailability. Absorption of minerals and vitamins depend on body requirements. However, higher concentration of nutrients in millets nullified their lower bioavailability (Himanshu *et al.*, 2018). Starch digestibility in whole millets grain is greatly affected by presence of dietary fiber, enzyme inhibitors, lipids, protein content and antinutrients (Singh *et al* ., 2010). Generally, fat and protein is concentrated in germ and pericarp of grain so removal of outer layer in milling might be helpful in increasing the starch digestibility (Annor *et al.*, 2013). There is a great need of studies on processing innovations in cooking methods of millets to increase bioavailability of vitamin B and minerals. Effect of different physiological status on nutrients bioavailability is greatly needed.

Nutritional Attributes

Little millet (P*anicum sumatrance*) is a rich source of nutritionally valuable nutrients like protein, dietary fiber, fat, carbohydrate, vitamins and minerals compared to rice and wheat (Devi *et al.*, 2011). Analysis of physical parameters is required to judge grain quality for processing. Rao *et al.*(2019) reported significant differences among the physical parameters of minor grains. Little millet has bulk density, thousand kernel weight, length, width and thickness of grain, as 821 kg/m^{3,} 1.83 g, 2.21 mm, 1.87 mm and 1.44 mm, respectively. Quality and quantity of protein is quite comparable to other cereals but carbohydrate content is low. Some minor millets like foxtail, common and barnyard millets have high fat content that is held responsible for low storage stability (Himanshu *et al.*, 2018). Nutritional composition of little millet is presented in the Table 1. Estimation of moisture content is necessary to evaluate processing conditions and storage quality of any food commodity. As depicted in Table 1 moisture content varied from 3.46 to13.35 percent among the cultivars of little millets. Kundgol *et al.*(2013) reported significant variation in moisture, ash and fat content among little millet cultivers from various locality while fiber and protein content showed no significant differences. Physical parameters like shape and size significantly not affect nutrients viz. protein, fiber, ash and fat. Little millet is considered a good source of fat rich in good cholesterol and level of fat varied from 1.49 to 7.8 percent. Phenolic compounds are credited for the antioxidant profile of food commodities. Little millets are good source of antioxidants. Study by Pradeep and Sreerama (2017) reported variation in phenolic compounds of varieties of little millet grains. A significant variation was observed in total flavonoid content while total phenolic content showed no significant difference among the cultivars of little millets.

Table 1: Nutritional Composition of little millets

Parameters	Quantity (/100gm)
Moisture (g)	3.46-13.35
Protein (g)	7.45-15.0
Fat (g)	1.49-7.8
Ash(g)	2.5-5.4
Total carbohydrates (g)	60- 76.07
Reducing sugar (g)	5.2
Non-reducing sugar (g)	70.87
Total dietary fiber (g)	4.0-38.00
Soluble dietary fiber (g)	5.30
Insoluble dietary fiber (g)	10.50
Phenolic compounds(mg)	9.64-42.6
Energy (Kcal)	303-383.2
Minerals	
Calcium (mg)	12-153
Iron (mg)	7.6-20.0
Copper (mg)	0.19-4.0
Magnesium (mg)	120-266
Zinc (mg)	1.47-11.0
Phosphorus (mg %)	220-455
Sodium(mg)	6-8.1
Vitamins	
Thiamine(mg)	0.08-0.30
Riboflavin(mg)	0.09-0.21
Niacin(mg)	1.29-3.2
Vitamin B6 (mg)	0.22
Folate (µg)	49
Ascorbic acid (g)	3.7
Amino acids	
Tryptophan(mg)	60-181
Lysine(mg)	110-747
Methionine(mg)	180-226
Phenyl alanine (mg)	330-542
Threonine (mg)	190-558
Valine (mg)	350-679
Leucine(mg)	760-879
Iso-leucine (mg)	370-582
Lysine (g/16 g N)	1.86-2.48
Fatty acids	
Linoleic acid (mg)	2834

Parameters	Quantity (/100gm)
Oleic acid (mg)	1433
Palmitic acid (mg)	1284
Stearic acid (mg)	220

Source: Himanshu *et al.*, 2018; Pradeep and Sreerama 2017; Prabha and Selvi, 2016; Chandel *et al.*, 2014; Kundgol *et al.*, 2013

Pradeep and Guha (2011) reported total phenolic content 429.9 mg GAE/100gm in little millet extracted using 1 percent HCL methanol solution at 60 ^{0}C.The chemical composition of little millets with respect to carbohydrate (60- 76.07 %), protein (7.45-15.0 %), dietary fiber (4.0-38.0 %), minerals (0.39- 1.7 %), phenolic compounds (9.64-42.6 mg/100g) and energy (303-383.2 Kcal) has been reported by various researchers (Pradeep and Sreerama, 2017; Prabha and Selvi, 2016; Kundgol *et al.*, 2013).

As given in Table 1, little millet contained appreciable amount of essential amino acid and fatty acid. Millets are rich source of iron (7.6-20.0 mg/100g), calcium (12-153 mg/100g), phosphorus (220-455 mg/100g), magnesium (120-266 mg/100g) and zinc (1.47-11.0 mg/100g). Ash content in little millet (1.82 %) is comparable to rye (1.96 %) and sorghum (1.87 %) (Sanaa *et al.*, 2005). Variation in nutrient is might be influenced by quality of soil. High level of iron and phosphorus might be helpful for people having low body mass. Little millets are reported to have fair amount of vitamins viz. thiamine, riboflavin, niacin and folate while it is poor source of vitamin C (Table 1). Health goodness of little millet is due to the presence of phenolic compounds, high fiber, mineral, fat and protein content. Regular consumption of millets might be helpful in prevention of chronic diseases like cardiovascular, cancer and diabetes mellitus (Prabha and Selvi, 2016).

Processing Innovations

Processing refers to conversion of grain into edible form to improve it's eating quality. Food processing is the way to increase nutritional attributes, enhance digestibility, bioavailability of nutrients and to reduce anti-nutrients. Utilization of millets and cereals as food significantly depends on their processing (Jaybhaye *et al.*, 2014). Millets are having great potential to be processed into different products at traditional and industrial levels including small, medium and extensive industries. Millet can be processed into various food products using processing techniques such as milling, decortication, germination, roasting, popping, fermentation and extrusion.

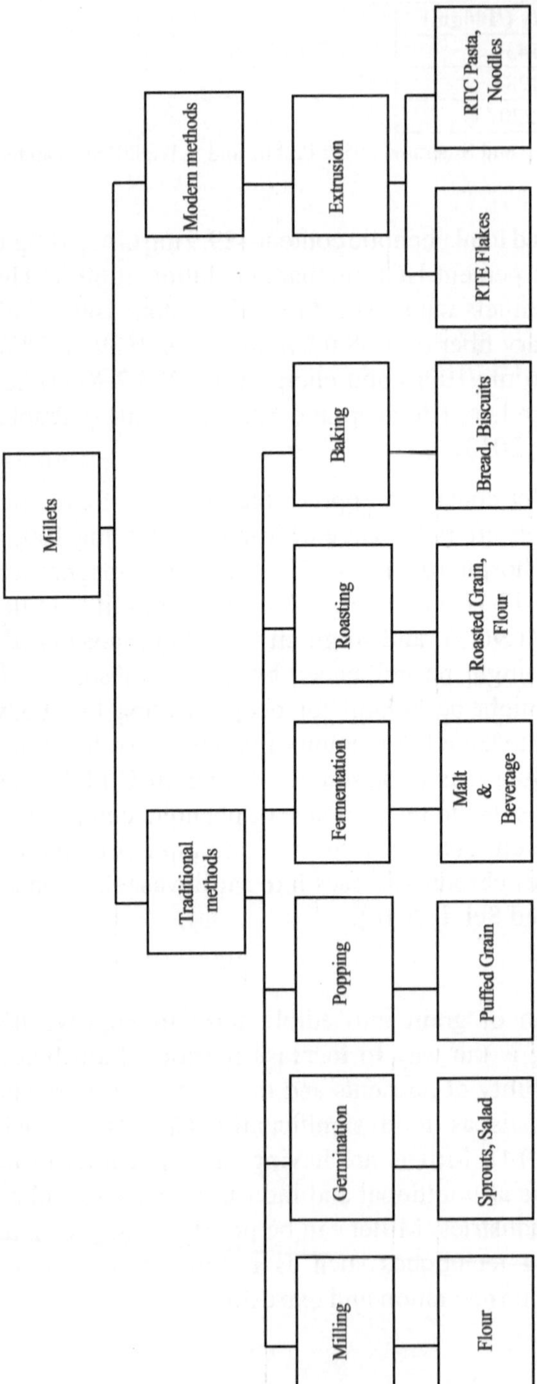

Fig. 2: Processing techniques of millets

Decortication

In millets decortication or dehulling is a common practice from ancient time. Earlier it was carried out at house-hold level by hand pounding but with advancement in technology modified rice milling machines are used for this purpose. Centrifugal sheller also applied for dehulling of small millets. Husk fraction varied from 1.5 to 29.3 percent in millets (Hadimani and Malleshi, 1993). Unlike cereals, decortication of minor millets is not easier due to their small seeded structure. Hydrothermal treatment aids to easy removal of hull by hardening the endosperm. It further helps in texture softening during cooking within 5 min. Study on pasting and rheological properties of finger millet dough supported it's application in diversified foods (Shobana and Malleshi, 2007). However, dehulling significantly influenced the nutrient content in finger millet. Hama *et al.*(2011), studied the effect of decortication on nutritional profile of pearl millet and sorghum. Protein and fat contents were not affected significantly while reduction was observed in total mineral, fiber, total phenolic content and antioxidant activity. Mineral content and dietary fiber were greatly affected due to their location in pericarp and aleurone layer of grain. Application of pearl millet as functional food was limited due to nutrient loss. Chandrasekara *et al.*(2012) also reported reduction in polyphenol, phytic acid and tannin content while dehulling aids to protein digestiability. Although, decortication significantly affected the nutrient but the process is necessary to improve edible and sensory quality of millet based products.

Milling

Availability of convenient and ready to use millet based food is limited due to absence of proper milling techniques. Colour pigmentation, coarse seed covering and astringent taste with low storage stability are the main constrains for millets extensive utilization. Processes like debranning, pearling and chemical treatments are helpful to overcome these problems associated with millets and these techniques also increase nutritional value of products with consumer acceptance (Akingbala, 1991). Yield of milling is greatly influenced by shelling index and milling efficiency. Mandhyan *et al.* (1987) reported that milling efficiency of little millet (83.38%) was greater than kodo (76.43 %) and *Panicum miliaceus* L. (74.10%). Bran content and head yield among minor millets and pearl millet varied from 5.0 to 11.00 percent, 63.20 to 90.00percent, respectively. Millets retained fair amount of dietary fiber 9-16 percent after milling (Hadimani and Malleshi, 1993). Chapati made from pearl millet flour showed lower content of phytic acid and polyphenols while significant improvement was noted in protein and starch digestibility (Chowdhury and Punia, 1997). As reported by various researchers, most of the milling operations are done manually therefore, convenient and mechanized milling technology is required to increase yield with better retention of nutrients.

Popping or Puffing

Popping or puffing of the cereals grain is one of the traditional processing method used in preparation of expanded and puffed grains, legumes and other ready to eat snack. Starch is the main ingredient in processing of millet grains for ready to eat puffed products. Puffing or popping process induces structural changes in starch that affects expansion of millet grains to produce crisp products with additional textural properties. Ushakumari *et al.* (2004) reported that puffing or popping is traditional as well as modern method of breakfast cereals preparation and can be successfully applied in the preparation of ready to eat products from foxtail millet to increase it's table value. High temperature short time (HTST) treatment was applied on dehulled finger millet for the preparation of ready to eat products. Results concluded that shape factor (0.52 to 0.58) and moisture content (about 40%) were the critical factors for preparation of popped millet product with desired expansion ratio (Ushakumari *et al.*, 2007). Fat and fiber content of popped products were lower as compared to than raw millet while carbohydrate content was significantly higher. It might be due to higher concentration of fat and fiber in outer seed layer that is removed during decortication (Choudhury *et al.*, 2011). Popping of millet grains enhanced their protein and carbohydrate digestibility by inhibiting some enzymes and enzyme inhibitors (Subramanian *et al.*, 1986). It might be helpful in improvement of sensory characteristics like color, aroma, appearance and taste of the manufactured food products. Novel techniques for optimization of puffing conditions is need of hour to develop millet based ready to eat expanded products with health promoting nutrients commercially. Nirmala *et al.* (2000) reported that HTST process increased bioavailability of minerals, improved carbohydrate and protein digestibility and reduced anti-nutrients such as tannin and phytates in puffed products. Iron enriched ready to eat little millets flake were prepared using continuous batch processing methods by Kotagi *et al.* (2013). Garden cress seed were added at 5 percent level to millet flour to prepare flakes. Kavita *et al.* (2015) reported that ready to cook little millets flakes showed higher dietary fiber content compared to decorticated seeds. Glycemic load of RTC flakes was 9.24. Presently, air puffing machines are available for the mass production of millet based popped or puffed products (Verma and Patel, 2013).

Extrusion

Extrusion, one of the advanced processing method involves application of steam for cooking (HTST) particularly for starch and protein based formulations. Its advantages are high productivity without effluents, increase protein digestibility, high quality and uniformity in products also add versatility to diet (Dahlin and Lorenz, 1992). Balasubramanian *et al.* (2012) used finger, pearl millet and soybean flour for the preparation of blended snack food. The above prepared product showed desired quality for texture, appearance and bulk density. The characteristics of extruded products like hardness, crispiness, expansion ratio and bulk density were significantly affected by barrel temperature of extruder. A formulation of millet (finger and pearl) and milk based composite was developed for the preparation of ready to cook extruded product from maida and composite (50:50) having adequate nutrition composition and acceptable sensory properties (Devi and Narayanasamy, 2013). Extrusion process aids to mineral, protein and starch digestibility. Therefore, it can serve as rich source of nutrient for majority of population suffering from deficiency of protein and energy intake.

Soaking, Germination and Malting

Soaking is a popular and traditional technique preferred for the reduction of anti-nutrients like phytic acid, phytase activity and to enhance mineral bio-availability. Lestienne *et al.* (2007) studied the leaching and degradation of phytase activity, phytates, iron and zinc in soaked whole seed, decorticated seeds and millet flour. The results indicated that leaching of phytates and phytase was faster in dehulled and milled millet. Eyzaguirre *et al.* (2006) concluded that soaking of pearl millet reduced iron content by 25 percent along with degradation of phytates in combination with milling and cooking process. Irradiation process followed by cooking in soaked millet significantly reduced the protein digestibility (ElShazali *et al.*, 2011). Soaking and cooking can be an alternative way to reduce anti-nutrients of millets to improve their nutritional quality as well as to increase their bioavailability. Germination of millets reduced tannin content. Germination improved *in vitro* starch, mineral and protein digestibility of pearl millet (Shahidi and Chandrasekara, 2013). Pradeep and Guha (2011) stated that process of germination, roasting and steaming significantly affected the nutraceutical and antioxidant potential of little millets. Total phenolic compounds showed increase from 453.3 to 521.0 mg GAE/100g in roasted millet grains followed by steamed and native millets. Therefore roasted little millet has great potential as functional food in millet based food preparations.

Millet malt is an important ingredient for infant food formulations, health food, amylase rich and supplementary foods. Malting significantly improves mineral availability and sensory attributes (Sangita and Srivastav, 2000). Germination is an important step for malting. Pelembe *et al.* (2002) reported optimum germination condition 25-30 °C for 3-5 days for finger millet germination conditions. Reduction was observed in phytic acid and tannin content while increase was noted in iron and zinc content due to germination. In little millet malting affected significantly antioxidant and nutraceutical properties (Pradeep and Guha, 2011). Malting in pearl and finger millet resulted decrease in protein content and increase in protein efficiency ratio and mineral bioavailabiliy (Krishnan *et al.*, 2012).

Fermentation

Fermentation process not only aids better taste but also increasesd the level of nutrients like protein, minerals, fiber, vitamins and protein digestibility. It also reduces anti-nutrients in food grains (Maha *et al.*, 2003; Verma and Patel, 2013). Various fermented products such as dosa, idli and dhokla are important part of breakfast and evening meal in different parts of India. Although millets are rich in protein but lysine and tryptophan content is limited, thus fermentation might serve to improve protein quality. Khetarpaul (2003) recorded improvement in protein digestibility to the content of 90 percent by fermentation pearl millet seeds. Pearl millet was fermented using micro-organism viz. *L. brevis, L. fermentation, S. diastaticus and S. cerevisiae.* Fermentation was carried out in single, mixed and sequential culture by incubating sample at 30°C for 72 hrs. No significant change was observed in ash and protein content of fermented millet while starch digestibility was improved significantly. (Arora *et al.*, 2011) reported increase in niacin, thiamine, protein fractions, total lysine, soluble dietary fiber, sugars and availability of calcium, zinc and iron content in blends after germination and fermentation pearl millets (probiotic). Fermentation could be better source for increasing starch digestibility, improve mineral and protein bioavailability in various formulations like weaning formulations, cutlets, vermicelli and biscuits from fermented pearl millet flour (Maha *et al.*, 2003). Malting significantly improved mineral (Fe and Zn) while reduction was noted in phytic acid and tannin content in finger millet (Pradeep and Guha, 2011).

Baking

Baked products are popular around the world due to their versatile nature, being economical and have longer shelf life suitable for easy marketing (Patel and Rao, 1996). Baking of millets not only improves the nutritional status in terms of micronutrients, dietary fiber but also provides market for millets based value added bakery products (Verma and Patel, 2013). Millet based product might serve as an alternative for wheat based baked products and also provides gluten free biscuit and bread. Different millets flour (finger, little and foxtail millets) can be added to other cereal flours to prepare various products viz. biscuits, cakes, nan-khatai, muffins and cheese. Mannuramath *et al.* (2015) reported the preparation of functional bread by adding little millet flour (10%, 30% and 50%) in wheat flour with good nutritional value. Hemalatha *et al.* (2006) prepared biscuits by incorporating little millet flour (10-50%) in wheat flour. Results showed increase in dietary fiber and mineral content. Krishnan *et al.* (2011) developed biscuits formulation from seed coat powder of finger millet having higher strength, crisp texture with better calcium, fiber and protein content.

Processing of millets not only enhanced their taste and digestibility but also improved their nutritional as well as economic value (Ushakumari *et al.*, 2004). Processing of millets into healthy ready to eat products would increase their market value by increasing popularity among public.

Traditional and Modern Foods and Beverages

Little millet, a small seeded crop is preferred over other cereals being good source of nutrients. Quality parameters of little millet make it suitable for processing and also aid in preparation of various traditional as well as modern millet based food products. Millets can be consumed in various forms such as powder, cooked, germinated and as salads. Traditionally, millets are consumed in the form of porridges, roti and idli. With the advancement in processing and awareness of health conscious population, market of millet based food is increasing day by day. Food consumption pattern is evolving with time. Earlier, main emphasis was on hunger satisfaction, survival, maintenance of good health and avoidance of adverse health effect but in current scenario main emphasis in on consumption of functional and nutraceutical food having wide health applications.

Patil *et al.* (2015) reported that ready to cook flakes of little millet have good nutritional value with respect to dietary fiber content. One serving of 30 g flakes provided sufficient nutrients i.e. minerals (0.13 g), iron (9.67 mg), fat (0.13 g), protein (2.25 g) and no trans fat. Flakes showed medium glycemic index range varied from 41.57 to 61.80 among normal population. Flakes showed six months storage life with moisture and fatty acid content of 11.82 percent and 18.02 percent, respectively. RTC flakes of little millets not only add convenience to traditional cooking but also provide health variant to diet with good storage life.

Iron enriched little millet flakes prepared by blending (5%) with garden cress seed showed significantly higher iron content 65.83 mg/100g of flakes compared to control (61.42 mg/100g). Fat content was 114.28 percent and protein content of Millet Garden Cress seed (MGC) flakes was13.15 percent, higher than controlled flakes. Total dietary fiber and carbohydrate content were recorded lesser in MFC flakes. Storage life of MGC flakes was 5 months while the control flakes showed shelf life of more than six months (Kotagi *et al.*, 2013).

Mannuramath *et al.* (2015) prepared functional bread by incorporating little millet flour in wheat flour. Developed functional bread showed significant decrease in weight, loaf volume, specific volume and height with increase in proportion of little millet flour. Quality evaluation of bread with 30 percent little millet flour (LMF) replacement showed significant increase in micronutrients like Iron (94%), Copper (70%), Zinc (29%), phosphorus (28%) and dietary fiber (19%). Study concluded that bread with 30 percent addition of LMF can be recognized as functional food having potential to be used as health food for management of metabolic disorder related to diet consumption. Study concluded that there is a great scope for the commercial production of LMF fortified functional bread with acceptable nutritional and sensory quality. Hemalatha *et al.* (2006) studied the formulation of biscuits by replacement with little millet flour and concluded that addition of 30 percent LMF produced biscuits with significantly higher content of ash (1.52%), fiber (1.40%), calcium (25.80%), iron (4.21%) and thiamine (0.39%). Storage study showed increase in moisture and acidity of LMF fortified biscuit with enhancement of storage period.

Porridge prepared using millets is a traditional trademark in German, Russian and Chinese cuisines. Mudde, thick porridge of finger millet is popular among urban and rural population of Karnatka. Millet is an important food for the population under Himalayan foot hills. Fortification of different millet flour (finger, pearl and little millet) in roti, idli, dosa, pakora, vedai, sweet halwa, porridge, paddu, samai dosa and paysam is a common practice in millet growing areas. 'Kodo ko jaanr', 'Koozh' and 'Chhang' are the famous alcoholic beverages in eastern Himalaya regions, Tamil Nadu and ladakh region, respectively. Drinks are prepared from finger millets seeds and flour (Ilango and Antony, 2014). Ambali, refers to naturally fermented finger millet beverage. Malt prepared from millets has been main ingredient in the preparation of beverage since long. Mahewu, a traditionally prepared non-alcoholic drink popular in Zimbabwe is manufactured using finger millet and sorghum (Gadaga *et al.*, 1999). Various types of liquids products made from millets are consumed all over India. Modern foods like baby vita or ragi health drink prepared using finger millet are commercially available. Extrusion technology can be applied for the preparation of instant beverage using pearl millet malt that helps in significant reduction of starch peak viscosity (Obilana *et al.*, 2014). Similarly, little millet has huge potential to be used in various formulations and health based modern food products. Weaning food blend prepared from sorghum (42%), legumes (20%) and additives such as oil, sugar, skim milk powder and vanillin (28%) contained fair protein (16.6% to 19.3%), good fiber (0.9% to 1.3%), high iron content (5.3 to 9.1 mg/100 g), energy (405.8 to 413.2 kcal /100 g) and significantly improved lysine content (Asma *et al.*, 2006).

Anti-Nutrients in Little Millets

Anti-nutrients are the compounds that offer hindrance to the bioavailability of some nutrients in body. Being rich source of iron and zinc millet are recommended in diet of infants and growing kids. Bioavailability of these micro-nutrients is the main concern in some of developing countries. Anti-nutrients like phytates, dietary fiber and some phenolic compounds are main factors behind reduced bioavailability of these nutrients (Camaro and Amro, 2003). Some processing techniques like germination, decortications, fermentation and malting might be helpful in reduction of anti-nutrients and increase in bioavailability of nutrients in millets (Sharma and Kapoor, 1996). Millets are good source of fiber, phytates and phenolic compounds. Phytic acid chemically named as 1,2,3,4,5,6 hexakis-dihydrogen phosphate and is mainly concentrated in bran part of cereals, legumes, oil seeds and nuts. Phytic acid is generally not affected by landraces, change in locality, size and colour. Sridevi and Yenagi (2007) studied antioxidant properties of different major, minor cereals and millets grains and their processed forms. Study concluded that little millets showed lowest value of phytic acid while brown ragi showed maximum content. Lorenz (1983) reported 27-53 percent reduction in phytate content in millet on dehulling process. Dehulling reduced phytic phosphorus content was 39 percent in little millet, 23 percent in barnyard millet, 12 percent in common millets and 25 percent in kodo millets. Minerals cations are entrapped by phytic acids (phosphorus) in protein storage vacuoles (Raboy, 2009). Various studies on bioefficacy on biofortified crops by different researchers confirmed the antagonistic role of IP6 in mineral adsorption. Hama *et al.* (2012) reported that presence of high phytate content in various studied cultivars of pearl millets affected bioavailability of iron significantly while absorption of zinc was not affected. Similarly, greater content of polyphenols in iron fortified pearl millets also diminished the bioavailability of iron (Tako *et al.*, 2015). Research on reduction of anti-nutrients during growth phase might be promising plan to improve bioavailability of nutrients in millets (Vinoth and Ravindhran, 2017). Rao (2003) reported absence of phlabotannins in four different cultivars of little millets.

Health Benefits

Millets are enjoying good status over other cereal grains being rich source of phenolic compounds, flavanoids, alkaloids, saponins, essential fatty acids, dietary fiber, tannins, reducing sugar, protein, carbohydrate and essential amino acids. Phytochemicals, bioactive components of plant origin have immense potential as health promoter and disease prevention. Minor millets are important grains rich in health promoting nutrients. The term functional food is assigned to food having potential diseases preventive phenomenon against diseases such as cancer, diabetes and cataract. Nutraceutical refers to bioactive compounds such as vitamins, fatty acids, minerals and essential amino acids having protective action against degenerative diseases in extracted or isolated forms. Various epidemiological studies on millet based diet concluded that regular consumption of millets reduced chances of diseases like diabetes, heart disease and hypertension. Millets diet is considered safe for the people having gluten allergy and suffering from celiac diseases. Millets are easily digestible due to their non-acid forming and non-allergenic nature (Saleh *et al.*, 2013).

Little millet is rich in nutraceutical like phenolics, GABA (gama-amino butyric acid), phytates, sterols, resistant starch and lignans. These bioactive nutraceutical in millet based diets might serve as health promoters by their synergistic or additive effect. GABA, is an essential inhibitory neurotransmitter having numerous health appliances in central nervous system of mammalian (Shahidi and Chandrasekara, 2013). Prathapan *et al.* (2011) reported metal chelating behaviour, enzyme inhibitory and free radical scavenging activities of nutraceutical present in little millets. Free-radical are main inducers of rancidity in oil based foods. Furthermore, these reactive oxygen species participate in biological activities and attacks lipids, DNA, proteins and other bio-molecules to induce degenerative chronic diseases. Antioxidants act to retard the activities of free radicals by inactivating or scavenging them. Natural antioxidants are gaining popularity over synthetic ones due to health hazards related with consumption of synthetic antioxidants. Millets are attracting the attention of researcher and food manufacturer being good source of natural antioxidant. Various studies on millet phenolic compounds claimed their diverse role as anti-inflammatory, anti-carcinogenic agent and in vasodilatory actions, mostly attributed to their free radical scavenging, metal chelating and antioxidant activities (Arpornsuwan *et al.*, 2010). Chethan *et al.* (2008) studied the effect of different phenolic compounds viz. gallic, ferulic, vanillic, protocatechuic p-coumaric acid on cataract eye lens inhibition and results concluded effective inhibition of the same by phenoilcs of finger millet. Millet consumption in regular diet reduced the chances of diabetes, cardiovascular diseases and duodenal ulcer (Saleh *et al.*, 2013).

In little millets dietary fibers with some phenolic compounds are main contributor in preventing diabetes in millet consuming population. Recent studies have concluded low glycemic index is exhibited due to higher fiber content in little millets (Itagi *et al.*, 2013). Consumption of functional bread of little millets might be helpful in management of metabolic disorders (Mannuramath *et al.*, 2015). Millets are rich in phosphorus, magnesium and iron. Magnesium is helpful in reducing the migraine effect, heart attack and by relaxing blood vessels it maintained blood pressure that further aids to better blood flow and reduces the chances of cardiovascular diseases. Magnesium works as cofactor for approx 300 enzymes. Moreover, it lowers down the triglycerides and increased insulin sensitivity. Phosphorus is key element of ATP energy bullets and also required for proper functioning of cell membranes and nervous system. Millets might serve as good source of iron in women at different physiological stage like pregnancy, lactating and menopausal stage. Little millet including other millets might be included in choice of heart – healthy foods due to their high magnesium content. Essential fatty acid composition and high level of polyunsaturated fat along with sufficient level of essential amino acid make it good diet for growing kids. Furthermore it increased level of good cholesterol in body.

Ferulic acid, one of the important phenolic compound in millet exhibited various health claims by reducing total cholesterol, increasing viability of sperms and bioavailability of vitamin E. It also has protective action skin damages induced by UV-radiation and work as anti-inflammatory actions (Castelluccio *et al.*, 1995). Shobana *et al.* (2007) reported that formulations for diabetic food based on millers showed less GI compared to rice based formulations. There is a great need of commercial products from millets having slow digestibility similar to Isomaltulose or Palatinose (slow digestible carbohydrates). Roasting process significantly influences the nutraceutical properties of little millet by increasing their phenolic content that aids to increase in antioxidant activity. Roasted little millet offers a good source of healthy food rich in nutraceutical (Pradeep and Guha , 2011).

Conclusions and Future Perspectives

Millets are considered food for poor population in most of developing countries. Survival of millets under adverse growing conditions aids to national food security during natural disasters. Little millet is rich source of protein, dietary fiber, minerals, and essential amino acids along with phenolic compounds having nutraceutical properties. Millets have great potential to be used in wide varieties of food products. Scientific studies on utilization of highly nutritious little millets to combat malnutrition are very limited. Novel innovative techniques are required for anti-nutrient reduction. Fortification of cereals with millets might serve better source of some nutrients like lysine, dietary fiber to produce nutritionally balanced food products. Millet processing and utilization for food product preparation has potential prospects for nutritional value, quality parameters, health applications along with cereals replacers but scope and utilization is still very meager. Further research is needed to study bioavailability of micronutrients like mineral and vitamins. It seems that literature on scientific reasoning, characterizing health benefits and development of health foods require further studies on commercial applications of millets. Studies may also focus on application and consumption of millets in developed countries to explore their commercial utilizations. There is a great need to meet diversified requirement of little millet based food among health conscious population. Therefore, it is highly desirable that global efforts need to step up to promote utilization of highly nutritious millet in food basket of consumers.

References

Adewale, I.O., Agumanu, E.N., Otith- Okoronkwo, F.I. (2006). Comparative studies on α-amylases from malted maize (Zea mays), finger millet (Eleusine coracana) and sorghum (Sorghum bicolor). Carbohydrate Polymers . (66): 71-74.

Ahmed, S.M., Saleh, Q.Z., Jing, C., Qun, S. (2013). Millet Grains: Nutritional Quality, Processing-and Potential Health Benefits. Comprehensive Reviews in Food Science and Food Safety. 12(3): 281- 295.

Akingbala, J.O. (1991). Effect of processing on flavonoids in millet (Pennisetum americanum) flour. Cereal Chemistry. 68: 180-183.

Annor, G.A., Marcone, M., Bertoft, E., Seetharaman, K. (2013). In Vitro Starch Digestibility and Expected Glycemic Index of Kodo Millet (*Paspalum scrobiculatum*) as Affected by Starch-Protein-Lipid Interactions. Cereal Chemistry, 90(3): 211-217.

Arora, S., Jood, S., Khetarpaul, N. (2011). Effect of germination and probiotic fermentation on nutrient profile of pearl millet based food blends. British Food Journal. 113(4): 470-481.

Arpornsuwan, T., Changsri, K., Roytraku, S., Punjanon, T. (2010). The effects of the extracts from Carthamus tinctorius L. on gene expression related to cholesterol metabolism in rats. Songklanakarin Journal of Science and Technology. 32:129–136.

Asma, M.A., El Fadil, E.B., El Tinay, A.H. (2006). Development of weaning food from sorghum supplemented with legumes and oil seeds. Food Nutrition Bulletin. 27(1): 26-34.

Babu, B.V., Ramana, T., Radhakrishnan, T.M. (1987) Chemical composition and protein content in hybrid varieties of finger millet. Indian Journal of Agricultural Sciences. 57:520-2.

Balasubramanian, S., Singh, K.K., Patil, R.T., Kolhe, K.O. (2012). Quality evaluation of millet-soy blended extrudates formulated through linear programming. Journal of Food Science and Technology. 49(4): 450-458.

Bieri, J.G., Evarts, R.P. (1974). Vitamin E activity of γ-tocopherol in the rat, chick and hamster. The Journal of Nutrition. 104:850-7.

Bora, P. Marcone, M., Seetharaman, K. (2013). Nutritional properties of differrent millet types and their selected products. University of Guelph.

Camaro, E., Amaro, M.A. (2003). Nutritional aspect of zinc availability. International Journal of Food Science and Nutrition. 54:143-151.

Castelluccio, C., Paganga, G., Melikian, N., Bolwell, G.P., Pridham, J., Sampson, J. (1995). Antioxidant potential of intermediates in phenylpropanoid metabolism in higher plants. FEBS Letters. 368:188-92.

Chandrasekara, A., Naczk, M., Shahidi, F. (2012). Effect of processing on the antioxidant activity of millet grains. Food Chemistry. 133: 1-9.

Chandrasekara, A., Shahidi, F. (2011). Antioxidant phenolics of millet control lipid peroxidation in human ldl cholesterol and food systems. Journal of the American Oil Chemist's Society. DOI: 10.1007/s11746-011-1918-5.

Chethan, S., Dharmesh, S.M., Malleshi, N.G.(2008). Inhibition of aldose reductase from cataracted eye lenses by finger millet (*Eleusine coracana*) polyphenols. Bioorganic and Medical Chemistry. 16:10085-90.

Chethan, S., Sreerama, Y. N., Malleshi, N. G. (2008). Mode of inhibition of finger millet malt amylases by the millet phenolics. Food Chemistry 111.1: 187-191.

Choudhury, M., Das, P., Baroova, B. (2011). Nutritional evaluation of popped and malted indigenous millet of Assam. Journal of Food Science and Technology. 48(6):706–11.

Chowdhury, S., Punia, D. (1997). Nutrient and antinutrient composition of pearl millet grains as affected by milling and baking. Nahrung 41(2):105–7.

Dahlin, K., Lorenz, K. (1992). Protein digestibility of extruded cereal grains. Food Chemistry. 48: 13-18.

Devi, M.P., Narayanasamy, S. (2013). Extraction and dehydration of millet milk powder for formulation of extruded product. IOSR Journal of Environmental Science, Toxicology and Food Technology. 7(1): 63-70.

Devi, P. B., Vijayabharathi, R., Sathyabama, S., Malleshi, N.G., Priyadarisini, V.B. (2011). Health benefits of finger millet (*Eleusine coracana* L.) polyphenols and dietary fiber: a review. Journal of Food Science and Technology. DOI: 10.1007/s13197-011-0584-9

ElShazali, A.M., Nahid, A.A., Salma, H.A., Elfadil, E.B. (2011). Effect of radiation process on antinutrients, protein digestibility and sensory quality of pearl millet flour during processing and storage. International Food Research Journal. 18(4): 1401- 1407.

Eyzaguirre, R. Z., Nienaltowska, K., De Jong, L. E., Hasenack, B. B., & Nout, M. R. (2006). Effect of food processing of pearl millet (*Pennisetum glaucum*) IKMP-5 on the level of phenolics, phytate, iron and zinc. Journal of the Science of Food and Agriculture, 86(9), 1391-1398.

Gadaga, T.H., Mutukumiraa, A.N., Narvhusb, J.A., Feresu SB (1999). A review of traditional fermented foods and beverages of Zimbabwe. International Journal of Food Microbiology. 53: 1-11.

Gayathri, A. (2015). Little millet nutrition, health benefits and facts you must know

Geervani, P., Eggum, B.O (1989). Nutrient composition and protein quality of minor millets. Plant Foods for Human Nutrition. 39:201-8.

Hadimani, N.A., Malleshi, N.G. (1993). Studies on milling, physico-chemical properties, nutrient composition and dietary fiber content of millets. Journal of Food Science and Technology. 30: 17-20.

Hama, F., Icard-Verniere, C., Guyot, J.P., Picq, C., Diawara, B., Mouquet-Rivier, C. (2011). Changes in micro and macronutrient composition of pearl millet and white sorghum during in-field versus laboratory decortication. Journal of Cereal Sciences. 54:425–33.

Hama, F.,Vernière, C.I., Guyot, J.P., Rochette, I., Diawara, B., Mouquet-Rivier, C. (2012). Potential of non GMO biofortified pearl millet (*Pennisetumglaucum*) for increasing iron and zinc content and their estimated bioavailability during abrasive decortication. International Journal of Food Science and Technology. 47: 1–9. doi: 10.1111/j.1365-2621.2012.03017.x

Hemalatha, G., Amutha, S., Malathi, D., Vivekanadan, P., Rajannan, G. (2006). Development of little millet (*Panicum sumatrense*) substituted biscuits and characterization of packaging requirement. Tropical Agricultural Research. Vol 18.

Himanshu, K. Chauhan, M. Sonawane, S.K., Arya, S.S. (2018). Nutritional and Nutraceutical Properties of Millets: A Review. Clinical Journal of Nutrition and Dietetics.1:1-10.

Ilango, S., Antony, U. (2014). Assessment of the microbiological quality of koozh, a fermented millet beverage. African Journal of Microbiological Research. 8(3): 308-312. DOI: 10.5897/AJMR2013.6482.

Itagi, S., Naik, R., Yenagi, N. (2013). Versatile little millet therapeutic mix for diabetic and non diabetics. Asian Journal of Food Science and Technology. 4(10):33-35.

Izadi, Z., Nasirpour, A., Izadi, M., Izadi, T. (2012). Reducing blood cholesterol by a healthy diet. International Food Research Journal. 19(1):29–37.

Jaybhaye, R.V., Pardeshi, I.L., Vengaiah, P.C., Srivastav, P.P. (2014). Processing and Technology for Millet Based Food Products: A Review. Journal of Ready to Eat Food. 1(2):32-48.

Kamatar, M.Y., Hemalatha, S., Meghana, D.R., Talawar, S., Naik, R.K. (2013). Evaluation of Little Millet (*Panicum sumatrense*) Land Races for Cooking and Nutritional Composition Current Research in Biological and Pharmaceutical Sciences. 2 (1):07-11

Khetarpaul, N. (2003). Improvement of nutritional value of pearl millet by fermentation and utilization of the fermented products. In: Recent Trends in Millet Processing and Utilization, CCS Haryana Agriculture University, Hisar, India, pp: 67-73.

Kotagi, K., Chimmad, B., Naik, R., Kamatar, M.Y. (2013). Nutrient enrichment of little millet (*Panicum miliare*) flakes with garden cress seeds. International Journal of Food and Nutritional Sciences. 2(3):36-39.

Krishnakumari, S., Thayumanavan, B. (1995) Content of starch and sugars and in vitro digestion of starch by alpha-amylase in five minor millets. Plant Foods for Human Nutrition. 48: 327–333.

Krishnan, R., Dharmaraj, U., Malleshi, N.G. (2011). Quality characteristics of biscuits prepared from finger millet seed coat based composite flour. Food Chemistry. 129: 499-506.

Krishnan, R., Dharmaraj, U., Malleshi, N.G. (2012). Influence of decortication, popping and malting on bioaccessibility of calcium, iron and zinc in finger millet. LWT-Food Science and Technology. 48(2): 169-174.

Kundgol, N.G., Kasturiba B., Math, K.K., Kamatar, M. Y., Usha M. (2013). Impact of Decortication on Chemical Composition, Antioxidant Content and Antioxidant Activity of Little Millet Landraces. International Journal of Engineering Research & Technology. 2(10).

Lestienne, I., Buisson, M., Lullien-Pellerin, V., Picq, C., Trèche, S. (2007). Losses of nutrients and anti-nutritional factors during abrasive decortication of two pearl millet cultivars (*Pennisetum glaucum*). Food Chemistry. 100:1316–1323.

Lorenz, K. (1983). Tannins and phytate content in proso millets (*Panicum miliaceum*). Cereal Chemistry.

Maha, A.M. Ali, Tinay, A.H., Abdalla, A.H. (2003). Effect of fermentation on the in vitro protein digestibility of pearl millet. Food Chemistry. 80: 51-54.

Mandhyan, B.L., Sharma, S.K., Tyagi, H.R. (1987). Study of physical properties, milling characteristics and cooking time of minor millets. Journal of Food Science and Technology. 24: 280-282.

Mannuramath, M., Yenagi, N., Orsat, V. (2015). Quality evaluation of little millet (Panicum miliare) incorporated functional bread. Journal of Food Science and Technology. 52(12):8357–8363.

Narinsinga Rao, B.S. (1988). Dietary fibre in Indian diets and its nutritional significance. Bulletin of Nutrition Foundation of India. 9:1-5.

Nirgude, M., Babu, B.K., Shambhavi, Y., Singh, U. M., Upadhyaya, H. D., Kumar, A. (2014). Development and molecular characterization of genic molecular markers for grain protein and calcium content in finger millet (*Eleusinecoracana* (L.) Gaertn.). Molecular Biology Reports. 41:1189–1200. doi: 10.1007/ s11033-013-2825-7

Nirmala, M., Subba Rao, M.V.S.S.T., Muralikrishna, G. (2000). Carbohydrates and their degrading enzymes from native and malted finger millet (Ragi, Eleusine coracana, Indaf-15). Food Chemistry. 69: 175-180.

Obilana, A. B., & Manyasa, E. (2002). Millets. In Pseudocereals and less common cereals (pp. 177-217). Springer, Berlin, Heidelberg.

Obilana, A. B., & Manyasa, E. (2002). Millets. In Pseudocereals and less common cereals (pp. 177-217). Springer, Berlin, Heidelberg.

Obilana, A.O., Odhav, B., Jideani, V.A. (2014). Functional and physical properties of instant beverage powder made from two different varieties of pearl millet. Journal of Food Nutrition Research. 2(5): 250-257.

Osagie, A.U., Kates, M. (1984). Lipid composition of millet (*Pennisetum americanum*) seeds. Lipids. 19:958-65.

Ostlund, R. E. (2002). Phytosterols in human nutrition. Annual Review of Nutrition. 22: 533-549.

Patel, M.M. and Rao, V. (1996). Influence of untreated, heat and germinated black flours on biscuit making quality of wheat flour. Journal of Food Science and Technology. 33(1): 53-56.

Patil, K. B., Chimmad, B.V., Itagi, S. (2015). Glycemic index and quality evaluation of little millet (*Panicum miliare*) flakes with enhanced shelf life. Journal of Food Science and Technology. 52(9):6078–6082.

Pelembe, L.A., Erasmus, C., Taylor, J.R. (2002). Development of protein–rich composite sorghum– cowpea instant porridge by extrusion cooking process. LWT - Food Science and Technology (Lebensmittel – Winssenschaff und Technologie). 35(2): 120-127.

Prabha, K.C., Selvi, S. (2016). Nutrient and antioxidant evaluation of four underutilized minor millets. International Journal of Current Microbiology and Applied Sciences. 5(7):224-233.

Pradeep, P.M., Sreerama, Y.N. (2017). Phenolic antioxidants of foxtail and little millet cultivars and their inhibitory effects on α-amylase and α-glucosidase activities. Food Chemistry. doi:https://doi.org/10.1016/j.foodchem.2017.11.103

Pradeep, S.R., Guha, M. (2011). Effect of processing methods on the nutraceutical and antioxidant properties of little millet (*Panicum sumatrense*) extracts. Food Chemistry. 126: 1643-1647.

Prathapan, A., Singh, M.K., Anusree, S.S., Soban Kumar, D.R., Sundaresan, A., Raghu, K.G. (2011). Antiperoxidative, free radical scavenging and metal chelating activities of boerhaavia diffusa l. Journal of Food Bio-Chemistry. 35:1548–1554.

Raboy, V. (2009). Approaches and challenges to engineering seed phytate and total phosphorus. Plant Science. 177:281–296. doi: 10.1016/j.plantsci.2009.06.012

Rao, B. D., Sharma, S., Kiranmai, E., Tonapi, V.A. (2019). Effect of processing on the physico-chemical parameters of minor millet grains. International Journal of Chemical Studies. 7(1): 276-281

Rao, N.B.S. (2003). Bioactive phytochemicals in Indian foods and their potential in health promotion and disease prevention. Asia Pacific Journal of Clinical Nutrition. 12(1):9–22.

Rao, N.B.S., Vijayasarathy C, Prabhavathi T. (1983). Iron absorption from habitual diets of Indians studied by the extrinsic tag technique. Indian Journal of Medical Research. 77:648-57.

Rao, P.U., Deosthale, Y.G. (1988). In vitro availability of iron and zinc in white and coloured ragi (Eleusine coracana): Role of tannin and phytate. Plant Foods for Human Nutrition. 38:35-41.

Revankar, J., Divya, K., Shamnani, A., Podili, K. (2018). Antioxidant activities of Pearl millet (Pennisetum glaucum) and Little millet (*Panicum sumatrense*) in different *in vitro* models. International Journal of Bioassays. 7.2:5595-5601.

Saleh, A.S., Zhang, Q., Chen, J., Shen, Q. (2013). Millet Grains: Nutritional Quality, Processing, and Potential Health Benefits. Comprehensive Reviews in Food Science and Food Safety. 12(1): 281-295.

Sanaa, R., El-sayed, M., Abdel-Aal, Maher. (2005). Antioxidant activity and nutrient composition of selected cereals for food use. Food Chemistry. 98: 32-38.

Sangita, K., & Sarita, S. (2000). Nutritive value of malted flours of finger millet genotypes and their use in the preparation of burfi. Journal of food science and technology (Mysore), 37(4), 419-422.

Sarita, Singh, E. (2016). Potential of Millets: Nutrients Composition and Health Benefits. Journal of Scientific and Innovative Research. 5(2): 46-50

Shahidi, F., Chandrasekara, A. (2013). Millet grain phenolics and their role in disease risk reduction and health promotion: A review. Journal of Functional Foods. 5:570–581.

Sharma, A., Kapoor, A.C. (1996). Levels of antinutritional factors in pearl millet as affected by processing treatments and various types of fermentation. Plant Foods for Human Nutrition. 49: 241-252.

Shobana, S., Malleshi, N.G. (2007). Preparation and functional properties of decorticated finger millet (Eleusine coracana). Journal of Food Engineering. 79(2):529–38.

Shobana, S., Singh, R.U., Malleshi, N.G., Ali, S.Z., (2007). Glycemic response of rice, wheat and finger millet based diabetic food formulations in normoglycemic subjects. International Journal Food Science Nutrition. 58(5): 363-372.

Singh, J., Dartois, A., & Kaur, L. (2010). Starch digestibility in food matrix: a review. Trends in Food Science & Technology. 21(4):168-180.

Sridevi, Yenagi, N. (2007). Evaluation of antioxidant properties of grains and processed regional cereals of Karnataka. Karnataka Journal for Agricultural Science. 21(4):602-603.

Sridhar, R., Lakshminaryana, G. (1992). Lipid class contents and fatty acid composition of small millets: little (*Panicum sumentrase*) kodo (Paspalum scrobiculatum), and barnyard (*Echinocola colona*). Journal of Agriculture and Food Chemistry. 40:2131-2134.

Subramanian, V., Jambunathan, R., & Ramaiah, C. D. (1986). Physical and chemical characteristics of pearl millet grains and their relationship to roti quality. Journal of Food Science, 51(4), 1005-1008.

Suma, P., & Urooj, A. (2012). Antioxidant activity of extracts from foxtail millet (Setaria italica). Journal of Food Science and Technology, 49(4), 500-504.

Tako, E., Beebe, S.E., Reed, S., Hart, J.J., Glahn, R.P. (2015). Higher iron pearl millet (*Pennisetumglaucum* L.) provides more absorbable iron that is limited by increased polyphenolic content. Nutrition Journal. 14:11. doi: 10.1186/1475-2891-14-11

Upadhyaya H. D., Ravishankar C. R., Narasimhudu Y., Sarma N. D. R. K., Singh S. K., Varshney S. K., (2011). Identification of trait-specific germplasm and developing a minicore collection for efficient use of foxtail millet genetic resources in crop improvement. Field Crops Research. 124 459–467.

Usha, B., Veni, K.G., Kumar, M. D., Hemalatha, K.P.J. (2011). Partial characterization of α-amylase from germinating little millets (*Panicum sumatrense*). Journal of Phytology 2011, 3(1): 01–08

Ushakumari, S.R., Rastogi, N.K., Malleshi, N.G. (2007). Optimization of process variables for the preparation of expanded finger millet using response surface methodology. Journal Food Engineering. 82: 35-42.

Ushakumari, S.R., Shrikantan, L., Malleshi, N.G. (2004). The functional properties of popped, flaked, extruded and roller dried foxtail millet (*Setaria italica*). International Journal of Food Science and Technology. 39: 907-915.

Verma, V. and Patel, S. (2013). Value added products from nutri-cereals: Finger millet (*Eleusine coracana*). Emirates Journal Food Agriculture. 25(3): 169-176; doi: 10.9755/ejfa.v25i3.10764.

Vinoth, A. Ravindhran, R. (2017). Biofortification in Millets: A Sustainable Approach for Nutritional Security. Frontier in Plant Science. 8:29. doi: 10.3389/fpls.2017.00029.

5

Foxtail Millet (*Setaria italica*) Chemistry, Nutrition and Processing Technologies

Dadasaheb D. Wadikar

DRDO-Defence Food Research Laboratory
Siddartha Nagar, Mysore-570011

Chemistry and Nutritional Attributes

Foxtail millet (*Setaria italica)* is one of the oldest cultivated millet with several varieties throughout the world with grains varying in colour from pale yellow, to gray, white and red. Archaeologists say that foxtail millet is so old that no wild plant of the species is known to exist today. Foxtail millet is known with several names in different parts of world and India (Table 1). Foxtail millet is thought to be native to China and is commonly cultivated in several regions of Africa, India, Russia, USA and Europe. It is an annual grass with erect and robust culms. Plant height is around 60-150 cm while the leaves are 20-40 cm long and 1.5-3 cm wide with lanceolate shape. Leaf sheaths are small, glabrous or pubescent and 1-3 mm. Leaf lamina is dense, erect or pendent, lobed and 6-40 × 0.5-5 cm. It has elliptic, ovate or sub-globose spikelets of 2-3 mm. The Fig. 1 shows the foxtail millet crop standing in the field and the harvested grain of the millet (Health Benefits Times, 2018; Sheahan, 2014).The seeds of foxtail millet are small and measure around 2 mm in diameter. It is encased in a thin, papery hull which is easily removed upon threshing. A thousand of the seeds weighs approx 2 g. The kernel consists of distinct husk, bran and endosperm tissues. Embryo is a distinct tissue, but its proportion in the kernel is around 2 percent. Normally, husk accounts to 15 to 20 percent of the kernel whereas the bran amount to about 5 percent and the endosperm to about 75 percent of the kernel. The seed coat and husk of foxtail is generally of single entity with glossy yellow colour. The husk is non-edible and unusually hard to digest similar to the husk in paddy, whereas the bran is a edible component but separated to prepare milled millets for food uses. To prepare edible items out of millets, the husk is separated by milling and along with that generally, the bran is also separated similar to milled rice. Hulling does not affect the nutrient value as the germ stays intact through this process (www.smallmillets.in). The nutritive value of foxtail millet is quite rich when compared to other millets. It is twice richer in protein, four times richer in mineral and fat and thrice richer in calcium as compared to rice. (Gopalan *et al.,* 1989) It has been reported that this millet is a rich source of nutrients and contains 60–70 percent dietary carbohydrates, 6–15 percent protein, 1.5–5 percent fat, 6–12 percent dietary fibre, 2–4 percent minerals and several other phytochemicals. The engineering properties of foxtail millet (*Setaria italic* L) variety HMT 1001 (Sunil *et al.*, 2016) determined at moisture content of 11.67 ± 0.078 percent (wb), revealed that mean values for length, width and thickness were 2.169, 1.590 and 1.454 mm, respectively. The average value for geometric mean diameter, sphericity, weight of hundred grains, bulk density, true density and porosity were 1.703 mm, 0.785, 0.245 g, 737.127 kg.m-3, 1260.132 kg.m-3 and 41.471 percent, respectively. The angle of repose mean value was 26.782°. The co-efficient of static friction on four types of structural material was found to be ranging from 0.30578 (acrylic sheet), 0.35606 (glass), 0.36396 (galvanized iron sheet) and 0.44522 (card board). The mean value of hardness recorded was 32.051 N. Chemistry and Nutritional Attributes

Foxtail millet is a cereal grain with a moderate protein content (about 11-12% DM). Protein variability is quite large; low (about 9% DM) as well as very high values (over 20% DM) have been recorded (Mosse *et al.*, 1989). Foxtail millet protein is superior to that of wheat or corn in terms of content of essential amino acids, It contains indispensable amino acid lysine in the range of 0.20 - 0.30 percent. Millet lacks gluten, the wheat protein that makes dough prepared from wheat flour elastic; hence millet flour is not suitable for leavened. Foxtail millet protein may improve insulin sensitivity and cholesterol metabolism through an increase of adiponectin concentration.

Table 1: Name of Foxtail Millet

Foxtail millet (Setaria italic)			
Other Scientific Names	**English Names**	**Names in India**	
Panicum italicum L,	Italian millet	Hindi	*Kakum, Kangani*
Setaria italica (L.) P. Beauv. var. stramineofructa (F.T. Hubbard) L.H. Bailey,	Hay Millet	Bengali	*Kaon*
	Foxtail Bristle Grass	Gujarati	*Kang*
		Marathi	*Kang or Rala*
Setaria italica (L.) P. Beauv. var. metzgeri (Koern.) Jáv,	Hungarian millet	Oriya	*Kanghu, Kangam*
		Punjabi	*Kangni*
	Chinese millet	Tamil	*Tenai*
Setaria italica (L.) P. Beauv. subvar. metzgeri (Koern.) F.T. Hubbard,	German millet	Telugu	*Korralu*
Chaetochloa italica (L.) Scribn	Green millet		

Source: Millet Network India, MNI (2018); Health Benefits Times, HBT (2018)

Fig. 1a. Foxtail Millet Plants

Fig. 1b. Foxtail Millet Seeds

Adiponectin concentration is a protein of human milk, which probably protects against obesity and diabetes (type II) (Choi *et al.,* 2005). Therefore foxtail millet protein would serve as a food component in obesity related diseases. Millet flour is used in making flat cakes and breads (Robert Ronzio, 2004). In general, foxtail millet is better source of protective nutrients comparted to major cereals.

Carbohydrate/Starch

The free sugars found in this millet are glucose, fructose, sucrose and raffinose and their contents ranges from 1-1.4 percent with sucrose (0.3-1.2%) being the predominant sugar (Malleshi *et al.,* 1986). Foxtail millets have total starch ranging from 64-79 percent (Geervani and Eggum, 1989; Krishnakumari and Thayumanavan, 1995) Amylose and amylopectin content of foxtail millet carbohydrate vary from 16-28 percent and 72-84 percent respectively Foxtail millet is a great source of starch, making it a high-energy food. The starch contained in foxtail millet is more digestible than maize starch. Some foxtail millet varieties contain 100 percent amylopectin. The millet grain contains about 65 percent carbohydrate, a high proportion of which is in the form of non-starchy polysaccharides and dietary fibre which help in prevention of constipation, lowering of blood cholesterol and slow release of glucose to the blood stream during digestion. Foxtail millet have a low Glycemic Index compared to most other cereals. Since it exhibits a low Glycemic Index, it increases satiety by decreasing hunger because it digests slowly. Type of starch (amylose/amylopectin) also influences the Glycemic index. Millets have a high amylose content which is why they are slow absorbing and sustainably release sugar to our blood and enable to work for long durations. (Choi *et al.,* 2005). Nagaprabha and Bhattacharya (2015) reported on the textural characteristics of gels, formed with foxtail millet flour at different concentrations (9-13%, dry solid basis) along with selected cations and hydrocolloids. Different concentrations of monovalent (NaCl) and divalent ($CaCl_2$ and $FeSO_4$) cations, and hydrocolloids such as gelatin, gellan, starch, xanthan and agar were incorporated in gels. The gel forming ability markedly improved with an increase in the concentrations of $CaCl_2$ and $FeSO_4$, while NaCl had a marginal effect. Gels prepared with 11 percent of foxtail millet alone were found to sensorially acceptable while addition of 0.5 percent $CaCl_2$ or 0.2 percent $FeSO_4$ respective with 10 percent foxtail millet also resulted in acceptable gels. Among the hydrocolloids, starch and gellan had marked effects on gel formation, and textural and sensory attributes. Foxtail millet has a good potential for development of new ranges of gelled health benefiting convenience products with nutraceutical property, and the addition of cations helps to achieve acceptable structural integrity. Nagaprabha *et al.* (2018) further reported on combination of starch-starch gels using green gram and foxtail millet. Both green gram and Foxtail millet formed well-set gels at 9 percent concentration of starch.

The total mineral content of foxtail millet is 3.3 percent (Table 2) and is much higher than the more commonly consumed cereal grains including sorghum, however de-hulling of the grain, like in other millets, causes considerable nutrient losses. Foxtail millet is also an excellent source of protein and fiber. It is a relatively coarse cereal grain with a crude fiber content is approximately 7.7 percent (Table 2). In general, millets are rich sources of insoluble (IDF) and soluble (SDF) dietary fibre and has comparable or even higher total dietary fibre (TDF) than other cereals. Decorticating significantly decreases millet TDF. Studies on foxtail millet have reported the IDF as 18-30 percent and SDF as 0.6-2 percent in whole form and decortication decreased the amount of IDF to 1.5-3 percent and SDF to 0.3-0.9 percent. Millet type, variety and extent of decortication have an important effect on the IDF and SDF content (Geervan and Eggum, 1989).

Fat and Fatty Acids

The fat content in foxtail millet ranges from 1 to 5.0 percent in widely cultivated varities (Table 2). Most of the lipids in millets are present as free lipids (60-70%) followed by bound and structural lipids. Bound and structural lipids of foxtail millets were reported to be 1.3-5 percent and 0.4-0.9 percent respectively. Linoleic (38-40%), oleic (27-37%), palmitic (16-22%) and linolenic (1-4%) are the major fatty acids found in millets. Unsaturated fatty acids account for more than 85 percent of the total fatty acid content in millets. (Sridhar and Laxminarayan, 1992)

The major fatty acids in foxtail millet were linoleic acid (66.5%), followed by oleic acid (16.1%), palmitic acid (7.42%), stearic acid (6.3%), and linolenic acid (2.48). Fatty acids including linoleic acid, palmitic acid, stearic acid and linolenic acid varied significantly in different foxtail millet varieties, except oleic acid. Several researches showed that millets with good fat composition can be obtained through breeding techniques or cultivation management. (Aixia Zhang *et al.,* 2015)

Millet pericarp and germ have considerable amount of lipids hence the total lipid content and fatty acid profile can be affected by the extent decortication of millets (Liang et al., 2010). Thus, millets can be considered as rich source of unsaturated fatty acid in our diet. Lipid content and types of fatty acids can affect the storage quality, sensory properties and *in vitro* starch digestibility of millets(Nantanga *et al.,* 2008; Annor *et al*., 2013).

Vitamins and Phytonutrients

Foxtail millet grains are rich in important vitamins *viz*., thiamine, riboflavin, folic acid and niacin (Table 2). Foxtail millet contains high amount of Vitamin A (32 mg) and vitamin E (31 mg) along with folic acid (15 mg). It contains approx. 81mg Magnesium per 100g of sample. Magnesium helps to relax blood vessels, enhances nutrient delivery by improving the blood flow and maintains the blood pressure and thus further protects the cardiovascular system. Magnesium increases insulin sensitivity and lowers triglycerides. It also acts as a co-factor for more than 300 enzymes. This millet is a rich source of calcium (300-350 mg/100 g) and probably a good source of iron too. Foxtail millet is an excellent source of phosphorus which is an important

Table 2: Chemical and nutritional parameters of Foxtail millet

Proximate Composition	**Moisture %**	**Protein (N x 6.25) (g)**	**Fat (g)**	**Ash (g)**	**Crude fibre (g)**	**Carbohydrate (g)**	**Energy (kcal)**					
	11.2	12.3	4..3	3.3	8.0	63.2	351					
Lipid Profile	Linoleic acid	Oleic acid	Palmitic acid	Stearic Acid	Linolenic Acid							
	66.5	16.1	7.4	6.3	2.9							
Vitamin Profile	Carotene (Vit. A)	Thiamine (Vit B1)	Riboflavin (Vit B2)	Niacin (Vit B3)	Pantothenic acid (Vit B5)	Folic acid (Vit B9)	Tocopherol (Vit E)					
	32 µg	0.59 mg	0.11mg	3.2 mg	0.82 mg	15 mg	31 mg					
Mineral Profile (mg/100g edible portion)	Calcium (Ca)	Iron (Fe)	Phosphorus (P)	Magnesium (Mg)	Sodium (Na)	Potassium (K)	Copper (Cu)	Chromium (Cr)	Manganese (Mn)	Molybdenum (Mb)	Zinc (Zn)	
	31mg	2.8mg	290mg	81 mg	4.6 mg	250 mg	1.4 mg	0.03 mg	0.6mg	0.07 mg	2.4 mg	
Amino Acid	Arginine	Cystine	Histidine	Isoleucine	Leucine	Lysine	Methionine	Phenylalanine	Tyrosine	Threonine	Tryptophan	Valine
mg/g	220	100	130	480	1040	140	180	420	--	190	60	430

(Hulse et al., 1980; United States National Research Council/National Academy of Sciences, 1982; USDA/HNIS, 1984; http://www.fao.org/3/T0818E/T0818E0b.htm; [a]Gopalan et al., 1989; HBT, 2018)

mineral for energy production and is an essential component of ATP – the energy store of the body. It also forms an essential part of nervous system and cell membranes. Lignans present in foxtail millet are converted to mammalian lignans and entero-lactone by the healthy gut micro flora in our body which is thought to protect against breast cancer as well as heart diseases (Adlercreutz and Mazur, 1997).

Protein and Amino Acids

Protein in millets has three main fractions: Fraction I- albumin+globulin, Fraction II-true prolamin+prolamin like, Fraction III- true glutelin+glutelin like. The albumin and globulin fraction forms 8.5-16.26 percent, prolamin fraction forms 30 percent to 60 percent, while glutelin forms 15.2 percent of the total protein (Table 3). The millet protein has well balanced amino acid profile and is a good source of methionine, cystine, leucin and lysine.

Table 3: Protein Fractions of Foxtail Millet

Fraction	**Foxtail millet**	
	Range	Mean
Albumin+globulin	11.6-29.6	17.1
Prolamin	47.6-63.4	56.1
Cross-linked prolamin	6.4-17.6	8.9
Glutelin-like	5.2-11.9	9.2
Glutelin	-	6.7
Residue	-	2.0
Total	-	98.0

Source: Monteiro.*et al.*, 1982

Foxtail millet which is rich in essential amino acid lysine can be used as a supplementary amino acid source to most cereals and the millet protein characterization showed that its protein concentrate is a potential functional food ingredient. These essential amino acids are of special benefit to those who depend on plant food for their protein nourishment. Foxtail millet which is rich in essential amino acid lysine can be used as a supplementary protein source to most cereals and the millet protein characterization showed that its protein concentrate is a potential functional food ingredient (Mohammad *et al.,* 2009)

The chemical score for amino acids of foxtail millet has been reported as 41. The amino acid content of more than 400 mg/g of protein has been reported (Table 2) in case of isoleucine, leucine, valine and phenylalanine. Protein quality and digestible energy in dehulled foxtail millet (%) was reported by Geervani and Eggum,1989 as true digestibility 95.0, biological value 48.4, net protein utilization 46.3 and digestible energy 96.1.

With increase in grain protein, the lysine content of the protein decreased. The protein was found to be high in leucine. Naren and Virupaksha (1990) observed that the prolamin was relatively rich in the sulphur amino acid methionine and that the sulphur status of the soil affected the synthesis of prolamin in the grain. In studies on in vitro protein digestibility, 90.5 to 96.9 percent of the protein in foxtail millet was digestible by pepsin, and 89.7 to 95.6 percent by papain (Monteiro *et al.,* 1988). Poor digestibility with trypsin (21.6 to 36.9 percent) was improved by prior treatment with acid. Heat treatment or lysine supplementation improved the protein quality (Geervani and Eggum, 1989).

Despite of its excellent nutritional quality, Foxtail millet has received less attention compared to some major cereals like rice, wheat and corn. Recently, millets are gaining importance in the production of gluten-free and low GI food products. Many areas of the world mainly grow millets and form major component of their daily food consumption. In addition to be nutritionally rich, the advantage of growing minor millets is that it is a rain-fed crop which forms part of a multi-cropping system, in the sense that it is mostly grown along with legumes and oilseeds. On the darker side, these are underutilized and neglected crops owing to their lower preference driven by affluence, longer time and efforts involved in processing of the millets and the lower cooking quality. If these problems could be solved, their high nutritional value can make them doubly valuable as food for farming families and a potential source of income.

Processing- Traditions and Innovations

Foxtail millet kernel is covered with hard-cellulosic husk unlike the naked millets such as sorghum and pearl millet. The foxtail millet grain has to be dehusked as the husk is not digestible, hence removal of the husk layer becomes the primary task of processing so as to make it fit for human consumption. This can be done by traditional/ manual pounding as well as using mechanical dehusking assemblies/ equipments.To convert the foxtail millet grain ready for consumption several basic operations such dehulling, soaking, and cooking are must. These operations have different effects on the nutritional profile as well as kernel/grain characteristics.The simplest and most common approach is dehusking and making it into foxtail millet rice. It is called so, because similar to paddy and rice; post-dehusking the foxtail millet can be cooked/ steamed as rice and consumed. To remove the husk from the grain, one can use two forces – impact and shear. A stone grinding mill, manual or motor powered, employs the sheer force while manual pounding or centrifugal hulling machines use the impact force. Hand pounding is a time-consuming process and the uniformity in the final product cannot be assured, however this ancient practice although good, is rarely used now, unless specifically required at household level. Another dehusking approach used is Buchi Method Processing: This is the second-best method to dehusk the seed. It is normally done through a blunted blade driven mixer called buchimixie. Large scale or commercial production involves mechanical dehusking. Large scale processing of small millets compromises on the nutritional value of the millet rice output by obliterating the bran layer. But the pest infestation problem continues to be severe and most processors resort to chemical methods of cleaning their products pest free. At the other end of the supply chain, the lack of small-scale processing has adversely affected the availability of the processed millets for use by the farming communities themselves a rural household cannot afford to buy millet rice from the market. Small scale processing machines and process flows have been developed. But the inherent variations in the harvested grains characteristics are significant. This problem magnifies when the small millets are aggregated for processing. So, in that sense, there is an inherent advantage in small scale processing. But a big stumbling block in achieving good quality millet output, i.e., clean minute millet rice with minimum bran loss, is the lack of skilled operators who understand the grains and are trained in using the right tools and machines (Sridharya Millet Planet, SMP, 2019).

The hulling process protocol involves pre and post dehusking steps (Dhan Foundation, 2016).

Pre-hulling is nessary for removing dust, light materials other than grains, large and small foreign materials like pebbles, mud balls, straw, etc. from the millet grains and separating large grains from smaller grains in a batch.

Post-hulling operations have to be carried out for separating millet husk, husk rich flour and un hulled grains from hulled fractions and separating shattered kernels and broken kernels of different sizes.

Devisetti *et al.* (2014) reported on effect of dehulling on dietary fiber content of foxtail millet and the results revealed that that loss of dietary fibers and anti-nutritional factors as they were concentrated in the husk and bran of foxtail millet. Pawar and Machewad (2006) studied effect of dehulling, soaking, and cooking on foxtail millet and reported that maximum decrease in protein, polyphenols, phytate phosphorus, phytic acid, iron and zinc increase in bioavailability of ionizable iron and soluble zinc. Rajeswari *et al.* (2015) in an alkaline cooking experiments conducted on foxtail millet observed a higher retention of protein, ash, carbohydrate and polyphenolic content lower fat, DPPH radical scavenging and total antioxidant activity. The dehusked foxtail millet grain is very commonly consumed as rice. In addition, milling of foxtail millet is other option often practiced to make it into flour. Milling involves dehulling and polishing of cereal grains. With the modern types and advanced in food processing techniques such as polishing of grain improve their acceptability while thermal processing techniques such as extrusion or microwave processing enhances commercial utility of the grains making them marketable.

Food extrusion is a process in which food or feed ingredients are forced to flow under one or several conditions of mixing, heating and shear, through a die that forms and/or puff-dries the ingredients. A continuous process by which moistened, expansile, starchy and/or proteinaceous materials are plasticized and cooked by a combination of moisture, pressure, temperature, and mechanical shear (Riaz, 2000). In the extruder, thermo-mechanical cooking of feed material occurs at high temperature under pressure and shear stress is generated in the screw barrel assembly. The cooked melt is then texturized and shaped in the die. The thermo-mechanical action during extrusion brings about starch gelatinization, protein denaturation and inactivation of enzymes, microbes and many anti nutritional factors. All these occurs in a shear environment, resulting in a plasticized continuous mass (Battacharji *et al.*, 1997).The expansion occurs based on the pressure gradients and the diameter.

Kharat *et al.* (2015) studied on development of foxtail millet based extruded product and processing conditions to get the desired product were optimized. Barrel temperature of 118.23°C and 400 rpm of screw speed with 15.88 percent of feed moisture produced high-quality foxtail millet based extrudates. Ushakumari *et al.* (2004) reported on extrusion of foxtail millet using twin-screw pilot-scale extrusion cooker at temperatures 150 °C–160 °C. Extruded millet illustrated a porous crunchy texture with an accountable solubility index and an expansion ratio of 1.8 Zhao and Wei (2014) conducted an extrusion test on foxtail millet based fiber-free flour, fat free flour, protein-free flour, whole meal, and starch. They observed and concluded that, presence of fiber, fat, and protein in extrudates resulted in the loss of thin wall, roundness, and integrity, with lower specific volume. Wadikar *et al.* (2014) reported extrusion of foxtail millet along with other millets such as finger millet and great millet to get nutritionally improved product using the multi-millet combinations. Kharat *et al.* (2019) studied the extrusion processing characteristics of whole grain flours of three major millets (finger, foxtail and pearl). Impacts of the extrusion processing conditions, including moisture (15, 17.5 and 20%), screw speed (300, 350 and 400 rpm) and barrel temperature (110, 120 and 130°C), on the physicochemical properties of the millet extrudates was evaluated. Quality parameters of extrudates including, expansion ratio (ER), water solubility index (WSI) and water absorption index (WAI) were measured. Variations in the structural properties were confirmed by macrostructural observation using scanning electron microscopy (SEM). Foxtail extrudates showed the highest expansion ratio (4.41) followed by pearl millet (3.78) and finger millet extrudates (3.18). WAI values for the foxtail millet, pearl millet and finger millet extrudates were 4.18 g/g, 2.69 g/g and 3.54 g/g respectively. Overall, the foxtail millet gave the best response, with the highest ER value achieved at 130°C, with least SME input. The results highlight the potential use of whole millets flours in the development of direct expanded extruded foods.

Popping or puffing is another HTST process wherein the conditioned grains are heated at a very high temperature for short times causing the grain to explode.Popping is a HTST treatment given to millet grains to prepare an expanded product. The popped grains were found to have better nutritional properties, with an increase in crude protein content, decrease in the total fat content, improved starch and protein digestibility due to the separation of starch granules from protein matrix on popping, thus making it more receptive for enzymatic digestion. Maximum popping yield was obtained at 230 °C, along with better quality and degree of doneness. (Chaudhary *et al.,* 2011). Among various foxtail millet processing, Ushakumari *et al.* (2004) reported on decortication of foxtail millet while Pawar and Machewad (2006) revealed that even the simple steps of dehulling, soaking, and cooking decreased the anti-nutrient content by approximately 50 percent and increased the *in vitro* protein digestibility by 32.71 percent.

Fermentation and hydrothermal treatment of foxtail millet flour resulted in changes in physicochemical properties. The crude protein content increased and total carbohydrate content decreased, nutritional quality of starch was improved and favorable modifications in the pasting characteristics were observed (Amadou *et al.,* 2014). Antony *et al.* (1996) reported on fermentation of foxtail millet. The results suggested that the millet fermented could form a useful food supplement, particularly in the diet of the urban population consuming milled polished rice. Fermentation by endogenous microflora increased the total soluble sugars and reducing sugars with a simultaneous decrease in the starch content. The protein extractability and albumin/globulin fractions were improved. The beneficial long-chain fatty acid profile of raw flour was retained. Acetic and butyric acid were the major short chain fatty acid. The fermented millet could be consumed after boiling in the form of porridge with salt/sugar and/or buttermilk. Alternatively, it could be mixed with other cereals and steamed products such as 'idli', 'dhokla' or' appam'. Yeung *et al.* (2005) reported on Non thermal process and quality changes of foxtail millet yakju-a Korean fermented food or beverage by microfiltration. Sensory quality of foxtail millet Yakju produced by non thermal treatment was better than that of thermal treatment.Chung *et al.* (2005) used ultrafiltration where in when foxtail millet Yakju was filtered through 0.45 μm nitrocellulose membrane followed by various ultrafiltration membranes with different molecular weight cut-off values, the permeate flux of Biomax 100K membrane with initial flux of 201.5 L/h was the highest among the membranes used. These membrane filtration treatments resulted in complete removal of microorganisms as well as decrease in turbidity and solid content without changes in other chemical properties.

Germination or malting of cereal grains may result in some biochemical modifications and produce malt with improved nutritional quality that can be used in various traditional recipes. It has been found that germination of proso-millet grains increased the free amino acids and total sugars and decreased the dry weight and starch content. Increase in lysine, tryptophan and nonprotein nitrogen were also noticed(Parameswaran and Sadasivam, 1994).

Germination is a biochemical process in seeds that involves interconversion and production of new compounds, which adds to the functional properties of the cereals. There is growing public interest in the process of germination because it yields natural, minimally processed and highly nutritious products that canal so help in lowering the risk of conditions like hyperglycemia and hyperlipidemia. One of the most common effects of germination on millets is the increased yield of hydroxyl radical inhibiting water soluble protein (Bai *et al.,* 2009). This enhanced nutritional property on malting were observed after 12 hr of steeping and 48 hr of germination at 25 °C.

There have been several conditions mentioned for optimized germinating of millets based on the requirements of downstream processing to the target product. Choudhury *et al.* (2011) optimized the germination process for yellow and purple varieties of foxtail millet, suggesting that a soaking period of 20 h and germination period of 72 h at temperatures of 25°C– 30°C produced foxtail millet with negligibly reduced protein content, significantly lowered fat and carbohydrate content and increased starch and protein digestibility.

Sharma *et al.* (2016) studied the nutritional quality of native and germinated foxtail millet and found that the germinated grains were nutritionally superior. Germination of foxtail millet seeds enhanced and modified its composition by increasing the availability of higher quantity of bioactive compounds such as total phenolics, antioxidants, total flavonoids, dietary fiber, proteins, minerals and decreasing the anti-nutritional factors. The initial tannin and phytate contents of foxtail millet was found to be reduced by more than 50 percent after the germination process. There was substantial enhancement of total phenolics and flavonoids during the germination process. Laxmi *et al.*(2015) reported that steeping of foxtail millet for 12 h followed by germination for 48 h increased its nutritional properties significantly. Non-significant change in protein content reduction in fat content and FFA and total carbohydrates; increase in TSS, TA, DPPH scavenging activity, mineral, and vitamin content. Pawar and Pawar (1997) reported that steeping of grains in water for 16 h and germinating for 48 hours resulted in maximum malt yield. The effects of wet heat treatment (boiling and pressure cooking), dry heat treatment (roasting) and germination on various properties of foxtail millet flour have also been reported (Nazni *et al.,* 2016). There was no significant difference in the swelling power and solid loss between raw and processed flours, while the solubility was found to be the highest for germinated foxtail millet flours. Moreover, the anti-nutritional factor was found to be minimum in the germinated samples. Sharma *et al.* (2018) reported on germination along with ultrasonic assisted extraction induced a significant beneficial effect on the characteristics of polyphenolic components profile, GABA contents and in vitro antioxidant capacity of the foxtail millet flour extracts. The total antioxidant activity (29.0-45.23 mg AAE/g) and reducing power (0.53-0.76 μg/ml) increase during germination were due to quantitative increase in phthalic acid; hex-3yl-ester, hexadecenoic acid methylester etc., whereas increase in DPPH (48.32-59.62%) and hydrogen peroxide scavenging activities (35.44-63.07 mM Trolox/g) were attributed to increase in hexadecanoic acid methylester; 9,12-Octadecadienoicacid ethylester and synthesis of new compounds like pentadecanoicacid, 1,4-methyl-methyl ester etc. The metal chelating abilities (34.92-57.38 mg EDTA/g) and *in vitro* antioxidant activity increase due to overall increase in phenolics, flavonoids along with GABA contents. Synthesis of additional polyphenolic components viz. astaxanthin, propanoicacid, 1-monolinoleoylglycerol trimethylsilylether, 9,12,15-octadecatrienoicacid etc. as a result of germinated explored the possible potential utilization of germinated foxtail millet grains in various functional and convenience food formulations.

Foxtail Millet Based Products

Most of the cereal or millet-based products are either convenience mixes or ready to eat products which hold market potential. There are several products which are consumed at household level and may be available in commercial eateries/restaurants. The products such as foxtail millet flour and rice are available on market shelves and are the products of milling industry. The products such as dosa, idli, bread, roti and dumpling can be prepared using foxtail millet semolina or flour. The institutes such as Indian Institute of Millet Research (IIMR), Hyderabad, Central Food Technological Research Institute (CFTRI), Mysuru and Defence Food Research Laboratory (DFRL), Mysuru offer very low cost and/or free technologies of millet-based products such as instant dosa mix, upma mix and several others including bakery products.

Dharmaraj *et al.* (2015) reported that foxtail millet needed pre-treatment to obtain uniform sized semolina fractions. The pre-treated millet was evaluated for dehusking efficiency, hardness, color and milling yield of the semolina fractions. It was observed that steaming was beneficial compared to roasting and dry heat treatment. The millet steamed for 15 min was polished to 2 different grades and milled to prepare coarse and fine semolina. The yield of the coarse semolina was 75 percent and that of the fine semolina was 18 percent. It was observed that fat, protein, dietary fiber and ash contents decreased with increase in the degree of polishing. The fine semolina fractions contained slightly higher fat and protein contents compared to their coarse counterpart. The solubility index of coarse fractions was higher than that of the fine fractions at ambient temperature and the trend was reversed at 97°C. The gelatinization temperature of fine fraction was comparatively higher than that of the coarse semolina while viscosity parameters were higher for the coarse semolina compared to the fine fraction. The studies indicated that the millet with 3 percent polishing was suitable for the preparation of semolina with retained nutrient content.

The use of foxtail millet as weaning food ingredient has been reported by Thathola and Srivastava (2011). The combination of foxtail millet flour , malted barnyard millet flour, roasted soybean flour and skim milk powder provided high nutritional quality and acceptable rheological and sensory properties.

The use of foxtail millet along with soy grits for the preparation of instant upma mix(Ketki *et al.,* 2017,) gave encouraging results of sensory evaluation with mix prepared from 75 percent foxtail millet, 20 percent wheat semolina and 5 percent soy grits were well accepted. The proximate composition of cooked upma exhibited moisture content from 37.50 to 40.60 percent, protein 11.20 to 12.90 percent, fat 6.38 to 8.14 percent, fibre 1.15 to 7.47 percent, ash 2.00 to 3.80 percent, carbohydrate content 59.87 to 70.16 percent and energy value 245.62 to 256.90 Kcal/100 g. Composite flours for down stream processing into different products using foxtail millet as one of the major ingredients have been reported. A combination of wheat flour and foxtail millet for bread reported by Chhavi and Sarita (2012), while Deshpane and Poshadri (1995) reported ready to eat extrudates from combination of Foxtail millet flour, rice flour, Bengal gram meal, amaranth seed flour, cow pea meal and another combination of foxtail millet flour, broken rice, chickpea, rice bran. Both combinations resulted in extrudates with good sensory properties and rich in nutrients. One can attempt different version of composite flours using foxtail millet. Shahaji *et al* (2018) reported that composite mix of rice, maize and foxtail millet for extruded product and that was rich in protein content (7.02%) and fiber content (2.35%), hence it has a good nutritive value. As the fat content (1.42%) was low, thus it can be a good diet food also. Ferriola and Stone (1998) reported on millet flakes wherein foxtail millet whole grain was used. Kumar *et al.* (2017) reported that development and quality evaluation of retort processed ready to eat foxtail millet halwa. Although, they emphasized on being a gluten free product the advantage of the product is it's nutritional quality and the long shelf life making it commercially viable option for millet entrepreneurs.

Oh and Jung (2015) reported *Omegisool* a traditional low alcoholic fermented beverage based on foxtail millet. The beverage has sweet and sour taste and reported often usage for quenching thirst and fatigue relief in Korea. It is prepared by mixing foxtail millet powder cakes (omega cake), Nuruk and water and allowing to ferment in a pot for 3-5 days. The Nuruk is prepared by spontaneous solid fermentation of barley grains at room temperature resulting in growth of various wild types of microorganisms, filtrate of the fermented broth same is used to make *Omegisool.* Jeong *et al.* (2014) reported on comparison of the physicochemical characteristics and sensory qualities of Sikhye (Korean traditional rice beverage) prepared with foxtail millet, proso millet and sorghum. Products had significant different sensory characteristics and but in general high antioxidant activity. Syama *et al.* (2015) reported on probiotic dahi incorporating raw and roasted foxtail millet flour.

Thathola *et al.* (2011) reported two preparations made from foxtail millet (Setariaitalica) viz. biscuit and burfi, a sweet. A 45 percent of foxtail millet flour was added to 55 percent of refined wheat flour for biscuit making and 57 percent of foxtail millet flour was added to 43 percent of bengal gram flour for burfi. The GI of biscuits and burfi was 50.8 and 37.5, respectively. It was further evaluated for its nutritional benefits in diabetic patients. The study concluded that the supplementation of low GI foxtail millet-based biscuits and burfi to lead to a significant decrease in serum glucose, serum lipids (serum cholesterol, TG and VLDL) and GHb in type 2 diabetics. Thus, millets do have a potential for a protective role in the management of diabetes And could be a recommendation for the beneficial effect.

Future Applications

Foxtail millet has lot of potential to use in place of wheat as well as rice due to its starch properties. There are several bakery and extruded products with industrial level production as they hold commercial potential. The fermented products from foxtail millet including beverages could see future industrial application.

References

Adlercreutz H and Mazur W. (1997) Phyto-ostrogens and western diseases. Ann Med 29-95.

AixiaZhang, Xiaodong Liu, Guirong Wang, Huijun Wang, Jingke Liu, Wei Zhao, Yuzong Zhang. (2015). Crude Fat Content and Fatty Acid Profile and Their Correlations in Foxtail Millet. Cereal Chem, 92 (5): 455–459.

Amadou, I.; Gounga, M.E.; Shi, Y.H.; Le, G.W. (2014) Fermentation and heat-moisture treatment induced changes on the physicochemical properties of foxtail millet (*Setariaitalica*) flour. Food Bioprod. Process. 92, 38–45.

Annor, G. A., Marcone, M., Bertoft, E., & Seetharaman, K. (2013). In Vitro Starch Digestibility and Expected Glycemic Index of Kodo Millet (*Paspalumscrobiculatum*) as Affected by Starch-Protein-Lipid Interactions. Cereal Chemistry, 90(3), 211-217.

Antony, U.; Sripriya, G.; Chandra, T.S. (1996) The effect of fermentation on the primary nutrients in foxtail millet (*Setariaitalica*). Food Chem. 56, 381–384.

Bai, Q.; Chai, M.; Gu, Z.; Cao, X.; Li, Y.; Liu, K. (2009). Effects of components in culture medium on glutamate decarboxylase activity and gamma-aminobutyric acid accumulation in foxtail millet (*Setariaitalica* L.) during germination. Food Chem. , 116, 152–157

Bhattacharjee S, Chakaraborty P and Mukherjee S (1997). Physic-chemical characteristics of extruded snack prepared from rice and chickpea by single screw extrusion. Journal of Food Science and Technology, 34: 320-323.

Chhavi, A.; Sarita, S. (2012) Evaluation of composite millet breads for sensory and nutritional qualities and glycemic response. Malays. J. Nutr. 18, 89–101.

Choi YY, Osada K, Ito Y, Nagasawa T, Choi MR, Nishizawa N. (2005). Effect of dietary protein of Korean foxtail millet on plasma adiponectin, HDL-cholesterol and insulin levels in genetically type 2 diabetic mice. Bioscience, Biotechnology and Biochemistry 69:31-37

Choudhury, M.; Das, P.; Baroova, B. (2011) Nutritional evaluation of popped and malted indigenous millet of Assam. J. Food Sci. Technol., 48, 706–711.

Chung, J.H. Mok, C.K. ; Lim, S.B.; Park, Y.S. (2005) Quality Improvement of Foxtail Millet Yakju by Ultrafiltration Process . Food Engineering Progress, 9(2):133-138.

Deshpande, H.W.; Poshadri, A. (2011) Physical and sensory characteristics of extruded snacks prepared from foxtail millet based composite flours. Int. Food Res. J. 18, 730–735

Devisetti, R.; Yadahally, S.N.; Bhattacharya, S. (2014) Nutrients and antinutrients in foxtail and proso-millet milled fractions: Evaluation of their flour functionality. LWT - Food Sci. Technol. 59, 889–895.

Dhan Foundation (2016) Assessment of Existing Small Millet Processing Equipment in India. Scaling up Small Millet Post-harvest and Nutritious Food Products Project. PP-1-17

Dharmaraj, U., Sathyendra Rao, B.V., Sakhare, S.D. Inamdar, A.A. (2015) Preparation of semolina from foxtail millet (*Setaria italica*) and evaluation of its quality characteristics. J. Cereal Sci. 68, 1–7.

FAO, http://faostat.fao.org/site/339/default.aspx

Ferriola, D.; Stone, M. (1998) Sweetener effects on flaked millet breakfast cereals. J. Food Sci. 1998, 63, 726–729.

Geervani, P., &Eggum, B. O. (1989). Nutrient composition and protein quality of minor millets. Plant Foods for Human Nutrition, 39(2), 201-208.

Gopalan C, Ramashastri BV, Balasubramanium SC-(1989), Revised & updated BS Narsinga Rao, YG Deosthale and KC Pant (2004). Nutritive value of India Foods, National Institute of Nutrition, ICMR Hyderabad.

HBT (2018) https://www.healthbenefitstimes.com/foxtail-millet/

Hulse JH, Laing EM, Pearson OE. (1980). Sorghum and the millets: their composition and nutritive value. New York: Academic Press. p 1–997.

ICRISAT; FAO, (1996). The world sorghum and millet economies: facts, trends and outlooks. ICRISAT/FAO

Jeong, M.S.; Ko, J.Y.; Song, S.B.; Lee, J.S.; Jung, T.W.; Yoon, Y.H.; Oh, I.S.; Woo, K.S. (2014) (2014) Physicochemical characteristics of Sikhye (Korean traditional rice beverage) using foxtail millet, proso millet, and sorghum. J. Korean Soc. Food Sci. Nutr. 43, 1785–1790

KetkiDhumketi, Dr.Alpana Singh and Dr. LPS Rajput (2017). Suitability of foxtail millet semolina and soy grits for the formulation of instant upma mix. International Journal of Chemical Studies. 5(5): 75-79

Kharat S. Ilce G. Medina-MezabRyanJ.Kowalskic, ArunkumarHosamania Ramachandra C.T. SharanagoudaHiregoudara, Girish M Ganjyalc (2019) Extrusion processing characteristics of whole grain flours of select major millets (foxtail, finger, and pearl).Food and Bioproducts Processing. Volume 114, March 2019, Pages 60-71.

Kharat, S.; Hiregoudar, S.; Beladhadi, R.V. Optimization of extrusion process parameters for the development of foxtail millet based extruded snacks. Karnataka J. Agric. Sci. 2015, 28, 301–303.

Krishnakumari, S., and Thayumanavan, B. (1995). Content of starch and sugars and in vitro digestion of starch by α-amylase in five minor millets. Plant Foods for Human Nutrition, 48(4), 327-333.

Kumar R., Harish S., Vijayalakshmi Subramanian, Sunny Kumar S., Nadanasabapathi S. (2017) Development and quality evaluation of retort processed RTE functional gluten free foxtail millet halwa. Croat. J. Food Sci. Technol. (2017) 9 (2)

Laxmi, G.; Chaturvedi, N.; Richa, S. (2015). The Impact of malting on nutritional composition of foxtail millet, wheat and chickpea. J. Nutr. Food Sci., 5, 2.

Liang, S., Yang, G., & Ma, Y. (2010). Chemical characteristics and fatty acid profile of foxtail millet bran oil. Journal of the American Oil Chemists' Society, 87(1), 63-67.

Malleshi, N. G., Desikachar, H. S. R., &Tharanathan, R. N. (1986). Free sugars and non-starchy polysaccharides of finger millet (*Eleusinecoracana*), pearl millet (*Pennisetumtyphoideum*), foxtail millet (Setariaitalica) and their malts. Food chemistry, 20(4), 253-261.

MNI-Millet Network of India. 2018. Deccan Development Society FIAN, Indiahttp://www.swaraj.org/shikshantar/millets.pdf

Mohamad T.K., Zhu K, Issoufu A, Fatmata T, Zhou H. (2009) Functionally, in vitro digestibility and physicochemical properties of two varieties of defatted foxtail millet protein concentrates. International Journal of Molecular Sciences. 10 (12): 5224-5228

Monteiro, P. V., Gopal, D. H., Virupaksha, T. K., and Ramachandra, G. (1988). Chemical composition and in vitro protein digestibility of Italian millett (*Setaria italica*). Food chemistry, 29(1), 19-26.

Monteiro, P.V.; Virupaksha, T.K.; Rao, D.R. (1982) Proteins of Italian millet: Amino acid composition, solubility fractionation and electrophoresis of protein fractions. J. Sci. Food Agric. 33, 1072–1079

Mosse, J; Baudet J; and Huet, J.C. (1989). Relationship between amino acids composition and nitrogen of foxtail millet (Setariaitalica) grain of different varieties. J Sci Food Agri, 46 (4): 383-392

Nagaprabha, P. and Bhattacharya, S. (2015) Textural characterization of foxtail millet gels: effect of cations and hydrocolloids. J. Food Sci. Technol., 53, 257–268.

Nagaprabha, P. Rajesh Devisetti, and Sila Bhattacharya (2018) Physicochemical and microstructural characterisation of green gram and foxtail millet starch gels. J Food Sci Technol. 55(2): 782–791.

Nantanga, K.K., Seetharaman, K, de Kock HL and Taylor J (2008). Thermal treatments to partially pre-cook and improve the shelf life of whole pearl millet flour. Journal of Science of Food and Agriculture, 88 (11), 1892-1899

Naren, A. P., & Virupaksha, T. K. (1990). Alpha-and beta-setarins: methionine-rich proteins of Italian millet (*Setaria italica* (L.) Beauv.). Cereal chemistry (USA).

Nazni, P. and Shobana, D.R. (2016) Effect of processing on the characteristics changes in barnyard and foxtail millet. J. Food Process. Technol. 7, 566.

Oh, Yeong Ji; Jung, Dong Sun (2015). Evaluation of probiotic properties of Lactobacillus and Pediococcus strains isolated from Omegisool, a traditionally fermented millet alcoholic beverage in Korea. LWT - Food Science and Technology, 63 (1).

Parameswaran and Sadasivam (1994) Changes in the carbohydrates and nitrogenous components during germination of prosomillet (*Panicum miliaceum*). Plant Food Hum Nutr, 45(2):97-102

Pawar, V.D. and Machewad, G.M. (2006) Processing of foxtail millet for improved nutrient availability. J. Food Process. Preserv., 30, 269–279

Pawar, V.S. and Pawar, V.D. (1997) Malting characteristics and biochemical changes of foxtail millet. J. Food Sci. Technol., 34, 416–418.

Rajeswari, J.R.; Guha, M.; Jayadeep, A. and Rao, B.V.S. (2015) Effect of alkaline cooking on proximate, phenolics and antioxidant activity of foxtail millet (*Setariaitalica*). World Appl. Sci. J., 33, 146–152.

Riaz M. N. (2000)Extruders in Food Applications, Technomic Publishing Co.

Robert Ronzio. (2004). The encyclopedia of nutrient and good health, 2nd edition, viva books Pvt. Ltd., New Delhi.

Shahaji kadam, Kailash Chandra Yadav and Raviraj Saini, (2018). Development and quality evaluation of foxtail millet (*Setariaitalica*) based extruded product using twin screw extruder. Journal of Pharmacognosy and Phytochemistry; 7(4): 658-663

Sharma S., Dharmesh C. Saxena, Charanjit S. Riar (2018). Changes in the GABA and polyphenols contents of foxtail millet on germination and their relationship with in vitro antioxidant activity. Food Chemistry, 245(15)863-870.

Sharma, S.; Saxena, D.C.; Riar, C.S.; Yildiz, F. (2015). Antioxidant activity, total phenolics, flavonoids and antinutritional characteristics of germinated foxtail millet (*Setariaitalica*). Cogent Food Agric., 1, 1081728.

Sheahan, C.M. (2014). Plant guide for foxtail millet (Setariaitalica). USDA-Natural Resources Conservation Service, Cape May Plant Materials Center, Cape May, NJ

SMP-Siridhanya Millet Planet. (2019) https://www.milletplanet.org/processing-millets

Sridhar R and LakshinarayanaG 1992. Lipid class contents and fatty acid composition of small millets: Little (*Panicum sumatrense*), kodo (*Paspalumscrobiculam*) and barnyard (*Echinocloacolona*). J Agric Food Chem. 40:2131-2134

Sunil, C.K., Venkatachalapathy, N., Shanmugasundaram, S., Akash Pare and Loganathan, M., (2016). Engineering Properties of Foxtail Millet (*Setaria Italic* L): Variety- HMT 1001. International Journal of Science, Environment and Technology, 5(2): 632 – 637.

Syama, A.; Hiremath, J.P.; Kempanna, C. (2015). Sensory and physico-chemical attributes of probiotic dahi incorporated with foxtail millet (*Setaria italica*) flour. Indian J. Nutr. Diet., 52, 177–183.

Thathola, A.; Srivastava, S. and Singh, G. (2011). Effect of foxtail millet (*Setaria italica*) supplementation on serum glucose, serum lipids and glycosylated hemoglobin in type 2 diabetics. Diabetol. Croat., 40, 23–28.

United States Department of Agriculture/Human Nutrition Information Service (USDA/HNIS). (1984). Composition of foods: cereal grains and pasta. Agriculture Handbook No. 8-20. Washington, DC.

United States National Research Council/National Academy of Sciences. (1982). United States-Canadian tables of feed composition. Washington, DC: National Academy Press.

Ushakumari, S.R.; Latha, S. and Malleshi, N.G. (2004) The functional properties of popped, flaked, extruded and roller-dried foxtail millet (*Setariaitalica*). Int. J. Food Sci. Technol., 39, 907–915.

Wadikar, D. D., Kangane, S. S., Parate, V., and Patki, P. E. (2014). Optimization of multi-millet ready-to-eat extruded snack with digestibility and nutritional perspective. Ind. J. Nutr., 1, 104.

Yeung Joo, K.; Young Ju, O. and Jeong Sam, K. (2005). Non-thermal process and quality changes of foxtail millet yakju by micro filtration. J. Korean Soc. Food Sci. Nutr., 34, 277–284.

Zhao, X. and Wei, Y. 2014. Effect of sequent extracting major components in foxtail millet on the properties of its expanded extrudates. Cereal Feed Ind., 8, 13–17.

6

Barnyard Millet (*Echinocholoa frumentacea*); Chemistry, Nutritional Attributes, Processing Innovations, Traditional and Modern Foods and Beverages, Anti-Nutrition and Health Benefits

***Sangeeta*[1], *Payal Garg*[1] *and Pankaj Kumar*[2]**

[1]*Food Processing Department, Guru Nanak College, Budhlada, Mansa Punjab*

[2]*Department of Microbiology, Dolphin (PG) Instt of Biomedical and Natural Sc., Dehradun, Uttarakhand*

Introduction

The non-glutinous, easily digestible and alkaline forming characteristics make millets superior in nutritional point of view. Among various cultivated millets in India, barn-yard millet, little millet, brown top millet, guinea millet and kodo millet are the most important minor millets (Nazni and Karuna, 2016). The Barnyard millet, Sawa millet or Japanese barnyard produces a crop within six weeks and considered as the fastest growing minor millets among the all minor millet (Veena *et al.*, 2005). Japanese millet or Barnyard has several existing varieties. The wild ancestor of barnyard millet is barnyard grass and is believed to have been directly domesticated from this grass. Two main cultivated species of barnyard millet are *Echinochloa frumentacea* (sawa, Indian barnyard, or billion dollar grass) and *E. utilis* (Japanese barnyard). These species are grown widely as a minor millets in India, Pakistan, China, Japan, Nepal, and Africa (Gomashe, 2016). Barnyard millet (Echinochloa frumentacaea) is known as the hardiest millet, because of its well thriving capability under adverse agro-climatic conditions and highest yielding quality in low agriculture input. Several other names of barnyard millet are Japanese barnyard millet, *sanwa, sawan, sanwank, ooda* and *oadalu* (Ugare *et al.*, 2014). It is mainly cultivated as a substitute crop in emergencies or on marginal lands, when the major crops like rice and other crops will not grow well or fails. Some countries like India, Japan and China grow it as rice substitute, whereas it is grown for grain in Australia and many other Asian countries. The length of barnyard millet grain is 2-3 mm and width is 1-2 mm approximately (Gomez *et al.*, 2003). Barnyard millet is considered as fastest growing multipurpose crop because it yields food and forage in short time even under the unfavorable agroclimatic conditions and low input of agriculture (Ugare *et al.*, 2014). Barnyard millet is very high in fibre content and similar to proso millet in term of fat and protein. Barnyard millet has good amounts of dietary fiber and macronutrients, which makes it superior grain in nutritional point of view. The utilization of barnyard millet is limited, in spite of the fact that it is nutritionally rich and superior to cereals like many other minor millet. The grains of barnyard millet are cooked in water or milk like rice and also used as functional food (Sharma *et al.*, 2016). Recently, barnyard millet is gaining attention as important grain because of its nutritionally superior quality as well as good cooking and sensory properties (Manimekalai *et al.*, 2018).

Chemistry and Nutritional Attributes

The nutritional food components and quality are the key elements, which play a vital role in maintaining overall physical health of human body. Nutritional balance is the viable force for human health and body development and helps in maximizing the genetic potential of human body (Saleh *et al.*, 2013). Therefore, the deep-rooted problems of malnutrition and food insecurity can only be solved by seriously considering the importance of dietary quality (Singh and Raghuvanshi, 2012). Millets are nutritionally rich, almost similar to that of major cereals (wheat and rice) and also have cultivating advantages (Parameswaran and Sadasivam, 1994). Millets are reported to be high in methionine and also have good amount of other essential amino acids except threonine and lysine. Other micronutrients and phytochemicals are also present in appreciable amount in millets (Mal *et al.*, 2010; Singh and Raguvanshi, 2012). Due to non-glutinous nature and other nutritional merits, barnyard millet is the most recommended food for patients enduring from various diseases like diabetes, celiac disease and allergic diseases including atopic dermatitis etc. (Watanabe, 1999; Ugrare *et al.*, 2014). Person suffering from celiac disease are allergic to wheat-based food products because of intolerant to gluten. Like other small millets, barnyard millet is proved to be beneficial for celiac disease patients due to the absence of gluten. Barnyard millet is nutritionally superior, rich in dietary fibre (soluble and insoluble fractions), minerals (Iron, Zinc and Magnesium), niacin, antioxidants, highly digestible protein and low in phytic acid content (Hadimani and Malleshi, 1993; Singh *et al.*, 2010; Anonymous, 2014). Due to the high fibre content, barnyard millet exhibits hypoglycemic effect (Gupta *et al.*, 2015). Additionally, it has low amount of carbohydrate content, which is slowly digestible in nature. All these nutritional quality attributes of barnyard millet makes it a natural designer food or the gift of nature for the modern mankind engaged in sedentary activities (Hadimani and Malleshi, 1993; Watanabe, 1999; Veena *et al* ., 2003; Ugare, 2008; Manimekalai *et al.*, 2018). Various researchers analyzed the nutritional composition of barnyard millet and reported it as an important nutritional crop. Barnyard millet contains carbohydrate (55-74%), protein (8.4-15%), fat (2.2-4.5%) and crude fibre (6.6-13.6%) as reported in literature. The approximate nutritional composition of barnyard millet as investigated by different researcher is represented in Table 1. Significant difference in physico-chemical characteristics and nutrient composition with in different varieties of barnyard millet was reported in literature (Veena *et al.*, 2005). The cooking time of observed varieties ranged between 8.5 to 12.0 min and good hydration and physical characteristics was observed in some varieties. Among nine observed varieties of barnyard millet variation in nutritional components was observed viz. 8.1-9.8 percent moisture content, 3.2-4.5 percent fat, 8.4-15.5 percent protein, 2.7-4.2 percent total minerals, 1.2-1.5 mg/100g iron, 17.1-32.7 mg/100g calcium, and 51.5-59.5 percent starch contents. Starch and protein digestibility (*In vitro)* ranged between 20.7-28.8 percent and 74.9-84.7 percent, respectively. The observed values for dietary fibre content were fairly high with 3.5-4.6 percent soluble and 6.1-10.5 percent insoluble fractions (Veena *et al.*, 2005).

Table 1: Nutritional composition of barnyard millet (g/100g dry basis, Except Where Stated Otherwise) as described by different sources

Nutrient Source	Hulse *et al* ., 1980	Veena *et al* ., 2005	Ugare *et* al ., 2014	Anonymous, 2014	Taylor, 2017
Moisture	-	8.1-9.8	8.7	-	-
Protein	11	8.4-15.5	10.5	6.2	15
Fat	3.9	3.2-4.5	3.6	2.2	-
Crude fiber	13.6	-	6.6	9.8	-
Total minerals	-	2.7-4.2	2.0	4.4	-
Total carbohydrates	55	-	68.8	65.5	74
Calorific value (100g)	300 kcal	-	398.0 kcal	-	1877 kJ
Total dietary fiber	-	-	12.6	-	-
Soluble TDF	-	3.5 - 4.6	4.2	-	-
Insoluble TDF	-	6.1-10.5	8.4	-	-

Saleh *et al.* (2013) also reported that barnyard is rich in protein (11%), crude fiber (13.6%) and Fe (18.6 mg/100g). The proximate macronutrient composition of the millets within varieties does not vary extremely and comparable to that of other cereals generally. The major dietary mineral and vitamin (mg/100g dry basis) as reported by different researcher is represented in Table 2. According to the reported data, slight variation in dietary minerals but almost similar vitamin content in barnyard millet was observed by different researchers. The amino acid (lysine), carbohydrate in term of starch, ash, total lipid and total fatty acid present in barnyard millet as reported by different researchers is given in Table 3 (Sridhar and Lakshminarayana, 1992; Taylor, 2017). Unlike most of the millets which have low lysine content, barnyarn millet has high lysine content i.e. 1.4 g/100g (Table 3) (Taylor, 2017). Lipids are the main components that differ considerably between different millet species. Among major millets, foxtail and pearl millet have appreciably higher lipid contents and little millet and Japanese barnyard millet are considerably high in lipids among the minor millets (Taylor, 2017).

Table 2: Major Dietary Minerals and Vitamins (mg/100g dry basis) present in Barnyard Millet as per available data

Nutrient Source	Fe	Mg	Zn	Ca	P	Vit. B_1	Vit. B_2
Taylor, 2017	7.6	190	0.9	210	240	0.30	0.09
Anonymous, 2014	15.0	-	-	11	280	-	-
Geervani and Eggum, 1989b	(2.3-12.9)	(167- 213)	(0.4-1.4)	(58-361)			
Hulse et al , 1980	18.6	-	-	22	-	0.33	0.10

Both de-hulled and de-hulled heat-treated barnyard millet grains have lower values of glycemic index (GI) than glucose (Fig.1) according to the data available in literature (Ugare *et al.*, 2014). The average GI of the de-hulled

Table 3: Lysine (Amino acid), carbohydrate (starch), total lipid, ash and total fatty acid (g/100g dry basis, except where stated otherwise) present in barnyard millet (Taylor, 2017; Sridhar and Lakshminarayana, 1992)

Species	**Lysine (g/100g Protein)**	**Ash**	**Carbohydrate (Essentially Starch)**	**Total Lipids**	**Palmitic (16:0)**	**Stearic (18:0)**	**Oleic (18:1)**	**Linoleic (18:2)**	**Linolenic (18:3)**
					Total Fatty Acids (%)				
Barnyard millet	1.4	1.3	74	5.2 (4.7–5.7)	11 (6-17)	1	27 (24–30)	58 (48–68)	1

barnyard millet grains is 50.0 ranging from 45.2-54.8, while, relatively lower values i.e. 41.7 GI (average) ranging 38.4-45.3 was observed for the dehulled heat treated grains (Ugare *et al.*, 2014). Barnyard millet based health mix have higher GI-59.80 (Surekha *et al.*, 2004) as compared to the plain grain meal which exhibited significantly lower GI-41.7 (Ugare *et al.*, 2014). Thus, barnyard millet is a promising grain for the formulation of designer foods especially for patients suffering from diabetes.

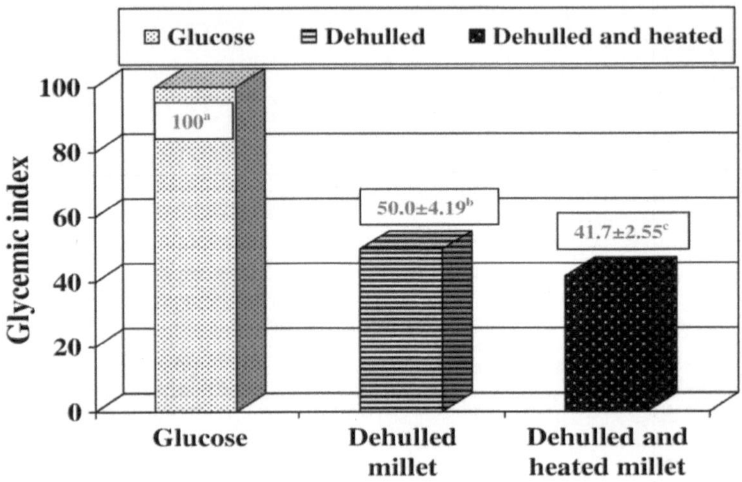

Fig 1: Comparative Glycemic index of de-hulled, de-hulled heated barnyard millet and glucose (Ugare *et al* ., 2014)

Processing Innovation

The absence of proper processing technologies, like primary processing, to make them ready to use and secondary as well as tertiary processing, to develop value added ready to eat millet based food items are the major issues which limits their utilization, diversification and betterment in economic status (Malleshi, 2014). The inherent anti-nutritional components of millets decreases the bioavailability of nutrients like minerals and their digestibility. Thus primary processing of millets is necessary to improve their nutritive value, consumer acceptance and diversification in use. Different methods of processing barnyard millet and their effects on product nutritional quality as reported by various researchers are represented in Table 4. An important determining factor in food choice is the increasing consumer concern about health and nutrition (Nehir and Simsek, 2012). Like other millets, processing of some barnyard millet varieties is a difficult and challenging task, but nutritional quality as well as health beneficial effect of barnyard millets, and increasing demand for healthy foods provides great scope and opportunities in its processing and encourages process mechanization and new product development technologies.

Like other grains, dehulling of barnyard millet is the necessary pre-processing step to make it suitable for human consumption. After dehulling, main steps involved in processing of barnyard millet are dehusking, debranning or decortication and milling (Singh *et al.*, 2010). Dehulling of barnyard millet reduces the phytin-phosphorus content up to 23 percent, as reported by Rao and Deosthale (1983), and improves protein and minerals biological availability (Erdman, 1981). In traditional methods, manual beating of grains in mortar with pestle repeatedly is required for dehusking/dehulling. The grains of barnyard millet are sturdily encased inside the bracts (palea and lemma), so the traditional mortar pestle method of milling consumes lots of time and labour. To reduce the time consumption, and requirement of human labour, mechanization of millet processing is required. This not only reduces the requirement of large manpower involved in manual dehusking but also secures the other inherent advantages of controlled mechanized process as well as cope with the large demands of market for a clean dehusked product (Lohani *et al.*, 2014). Nowadays, various machineries like centrifugal-rice-sheller cum polisher and cone polisher are being used for the dehusking and polishing of millets (FAO, 1995). Vivek Millet thresher 1 was developed in year 2003 and has the threshing and dehulling capacity of 30–35 kg barnyard millet grains/h and 5-6 kg barnyard millet/hour separately by using sieves of different sizes (Singh *et al.*, 2003). Since the year 2005, the threshing of barnyard as well as other millets planted at Vivekananda Parvatiya Krishi Anusandhan Sansthan farm is being done by using Vivek Millet Thresher-1 which proves the importance of this machine (Gupta *et al.*, 2015). The advanced dehuller machine with optimized parameters was developed for barnyard millet, to make the milling process mechanize, to increase the efficiency and to reduce consumption of specific energy and grains breakage (Singh, 2010). Before recommending the process technology for commercial utilization, the basic technology of polishing and shelling has to be worked out properly at laboratory scale (Lohani *et al* ., 2014). The moisture content of grain affects the optimum shelling efficiency of sheller (SATAKE rice sheller) and 8-10 percent moisture content (dry basis) is the recommended optimum moisture content for dehusking of barnyard millet grain in SATAKE rice sheller, as reported by Lohani *et al.* (2014). Higher the moisture content, poor is the shelling efficiency and 10 percent moisture content on dry basis is sufficient to obtain optimum shelling efficiency and to achieve highest yield, less nutritive loss (Protein, ash, fat and fibre content) with minimum broken of brown rice grain (Lohani *et al.*, 2014). Like other minor millet barnyard millet can be dehusked in rubber roll or centrifugal shellers and the debranning of husked material can be done in rice milling machinery. The separation of husk and bran in pure form, with about 55 percent head grains yield and the husk free bran containing about 15 percent oil can be possible to by adapting this technology. The deoiled barnyard millet bran can be used as animal feed and in specialty foods as dietary fibre source. The milled barnyard grains can be used in many forms similar to rice, eg., Ready-to-Cook products, grits or semolina after size reduction, cooked as discrete grains like rice, and in the form of flour after pulverization etc. Various value added bakery products can be prepared by using composite of barnyard millet and refined wheat flour (Seetharam and Krishne, 2008). Like other minor millets, secondary processing can also be given to barnyard millet like flaking, extrusion cooking, roller drying and popping etc. to prepare flakes, pre-gelatinized food material, expanded Ready-to-Eat food products, roller dried ready to cook products (soup powder, porridge) and popped grains, respectively (Malleshi, 1986; Srivastava *et al.*, 2003; Jaybhyae and Srivastav, 2010[a] Jaybhyae and Srivastav, 2010b). Parboiling and malting are also the pre-processing methods which can be used for making nutritionally superior products from barnyard millet (Chaturvedi and Rawat, 2018). These processing particularly malting resulted in biochemical modifications which improves the nutritional quality and reduce anti-nutrients content. Chaturvedi and Rawat (2018) evaluated the quality of flour prepared from malted barnyard millet (soaking 24h) followed by germination (72h) and developed value added rusk and reported that there is an improvement in term of nutritional, functional and sensory quality. Thus processing methods like malting can play a vital key role to expand the spectrum of malted barnyard millet flour in bakery products and significantly contribute to national food security and potential health benefits.

Traditional and Modern Foods and Beverages

Change in processing technology, customer behavior and preference towards food play a vital role in expanding the area under millets, ensuring food security, maintaining ecological balance, malnutrition prevention and increase their scope for utilization on industrial scale. A lot of studies on millets processing reported promising results in successful utilization of minor millets for the development of various traditional as well as conventional value added health food. Exploratory studies on the development of processed products from barnyard millet or other like flaked, popped, extruded, roller dried, and puffed products; weaning foods; malted, fermented, and composite flours, etc. have been carried out and have promising results (Malleshi, 1989; Saleh *et al.*, 2013).

Table 4: Process Technologies and their effect on quality of barnyard millet

Process Technology	Process Details	Effects of processing
Dehulling (Erdman, 1981; Rao and Deosthale, 1983)	-	Reduces phytin-phosphorus content by 23 percent Improve bioavailability of protein and minerals
Dehusking (Lohani *et al.*, 2014)	8-10 percent moisture content (dry basis) in SATAKE rice Sheller	Less nutritive loss (Protein, ash, fat and fibre content)
Malting (Sharma *et al* ., 2016; Taylor, 2017)	Steeping followed by germination (33°C/36h)	Increases TP (168%) Increases Free TP (180%) Increases Bound TP (45%) Increases F (148%) Increases Free F (380%) Increases Bound F (35%)
	Germination (25°C/48 h)	Increases TP (209%) Increases F (22%)
Germination (Sharma *et al.*, 2016)	-	Significant increase in dietary fibres and minerals Decrease in Phytates
Malting (Chaturvedi and Rawat, 2018)	Soaking (24h) followed by germination (72h)	Improves nutritional, functional and sensory quality
Thermal processing (Sharma *et al.*, 2016; Taylor, 2017)	Steeping whole grain in autoclave (20 min)	Reduces TP (5%) Reduces F (22%)
	Microwave processing of Dry whole grain (700 W for 3 min)	Reduces TP (8%) Increases F (4%)
Note: TP: Total Phenolics; F: Flavanoids		

Nutritional and health perspective importance of barnyard millet are clearly revealed by the rich amount of its nutritional components like protein, minerals, dietary fiber and antioxidants etc. To overcome the amino acid deficiency of minor millets (barnyard, Italian, French, kodo and little millet grains), fortification of minor millets with lysine (0.6 g/100 g dry matter) was done and reported the increase of net protein utilization and biological value (Geervani and Eggum, 1989a.). Thus to diversify the food security basket may be helped by the incorporation of barnyard millet and its products in daily food habits (Saleh *et al.*, 2013). In India, barnyard millet is regionally dominant food and fodder crop. Traditionally, barnyard millet after dehusking is cooked and consumed like rice or in the form of kheer (Gupta *et al.*, 2009). Barnyard millet based pudding is quite a famous dish in hilly area. In southern India, muruku, dosa and idli prepared from barnyard millet are very popular traditional foods. In various traditional foods and beverages such as popped grain, porridges, and other snacks etc., barnyard millet serves as a major food component, and used as an ingredient together with other millets for multigrain and gluten-free cereal products. Recently, different type of value added products such as multigrain atta, multigrain biscuits, vermicelli, sweets, ready mixes etc. are being developed by using barnyard millet (Sood *et al.*, 2015). It is highly suitable and recommended as health food for infants, pregnant women and diabetics. In addition, fodder of barnyard millet is important as cattle feed because of its high palatability and use for making hay or silage (Gomashe, 2016). Data available on the processed products from barnyard millets as described by various researchers is discussed in this section.

Arora and Srivastava (2002) developed barnyard, finger millet, fenugreek seeds and legumes based khichadi and laddu having high acceptability and 51.59 percent and 79.40 percent energy of total energy was reported to be provided by carbohydrate content of khichadi and laddu respectively. Barnyard and Kodo millet based chappathi, dosa, noodles and rusk were developed by Poongodi *et al.* (2003), and reported that 20 percent for noodles and 30 percent for rusk, chapathi and dosa were the acceptable level of millet flour to be incorporated. Veena *et al.* (2003) suggested the use of barnyard millet in common cereal based traditional foods for improved nutrition and other quality attributes. Incorporation of barnyard millet, up to 75 percent in idlies and rotis and 50 percent for chakli, was reported safe and acceptable. Weaning food developed by using 30 percent of barnyard and foxtail millet flour each (malted), 25 percent soybean flour (roasted) and 15 percent skim milk powder had 18.37 g/100g protein and 398 kcal/100g energy according to the data reported by Thathola and Srivastava (2002). The sensory quality and acceptability of this barnyard millet based weaning mix were reported significantly similar to that of marketed weaning mix. The use of barnyard millet flour as refined wheat flour replacement (upto 50%) in biscuits lowers the glycemic index (50.17) as compared to that of wheat flour biscuits (73.58) whereas the nutrient composition remains almost same (Srivastava and Singh, 2003). The barnyard millet flour can be substituted possibly with wheat flour up to 20 percent level in bread formulation and resulted in increase in ash content, protein and loaf volume of dough; decrease in gluten content and sedimentation value color of crust and bread shape remain unaffected but crumb color changed to dull brown from creamish white, as reported by Kamaraddi and Shanthakumar (2003). Srivastava *et al.* (2003) prepared different type of ladoos from the popped grain of barnyard, foxtail and little millet. Open iron pan popping method using common salt (1:20) at 240-260 °C for 15-25 s was used for popping of millets. Wide range of nine point hedonic sensory score of ladoos (5.0 - 8.1) was reported. This might be due the incorporation of jaggery, roasted groundnut and coconut powder in ladoos at different level. Addition of millet flour (barnyard, foxtail and finger millet) up to 10-30 percent increase the nutritional composition of their composite flour with wheat. Milled barnyard flour reported to have improved level of protein, total ash, and crude fat as compared to whole barnyard millet flour (Singh *et al.*, 2005). Acceptability of cookies developed by barnyard millet flour and garden cress seeds have been reported to be increased at higher level by increasing the proportion of barnyard millet flour (Balloli, 2011). Barnyard millet (*Echinochloa frumentacea*) based ready-to-eat extruded puffed snack food was developed by Jaybhyae and Srivastav (2010a; 2010b). For the production of puffed or popped millet grains on large scale, modern air puffing machine has been developed (Verma and Patel, 2013). Barnyard millet flour biscuits were reported to be better in term of acceptability than crude refined wheat flour biscuits (Anju and Sarita, 2010). Barnyard millet flour based cookies by incorporating pulse flour (10%) and vegetable powder (10%) with acceptable sensory attributes were developed by Surekha *et al.*(2013a). Different types of barnyard millet flour based noodles by incorporating pulse flour and vegetable powder in different proportion were developed and reported to have low glycemic index and increased nutrients (Surekha *et al.*, 2013b). Among all age groups, noodles and cookies are the most preferred food item and have longer keeping quality and commercial importance. Therefore, barnyard millet can be a promising ingredient for the formulation of various value added food products having low glycemic index, because of its low (58.56%) but slowly digestible carbohydrate content (25.88%) with related health benefits (Veena, 2003; Jaybhaye *et al.*, 2014). Barnyard millet based cold extruded puffed microwave processed ready to eat food products useful during fasting with comparable sensory attributes has been developed (Dhumal *et al.*, 2014). Anti-diabetic effects of barnyard and other minor millet based food products (Ready to Eat) are due to of their low glycemic index (40-50) have been demonstrated by various researchers (Jaybhaye *et al.*, 2014). Takhellambam *et al.*(2016) explored the minor millets (barnyard, little, ragi and proso), for the development of Ready-to-Cook millet flakes to meet the modern consumers need and reported that the resulted millet flakes have good amounts of nutrients and comparable to commercial flakes of oats and rice. Chandraprabha *et al.* (2017) developed highly acceptable vermicelli and uppuma, from the composite flour of barnyard millet (40%), whole wheat flour (58%) and Ekanayakam root powder (2%), with a mean sensory score of 7.75 and 7.73. Shunmugapriya *et al.* (2020) developed standardize minor millets based milk (barnyard, kodo, little and finger millet) by the method of enzymatic extraction, and reported that this millet milk met the standards of plant-based milk in terms of physical properties and have improved antioxidant activity, high *in vitro* protein and starch digestibility and reduced antinutritional factor. This type of processing approaches can be proved promising for the development of nutritionally sound minor millets based milk at home scale and commercial level.

Thus, the reported data revealed that minor millets especially barnyard millet offers a great range of opportunities for its application in the form of diversified products along with better quality attributes in term of nutrition and sensory.

Anti-Nutrition and Health Benefits

The potential role of millets as functional foods received attention because of their health promotive phytochemicals. Millets consumption reduces the risk of various disease like heart disease, improves digestion system, protects from diabetes, lowers cancer risks, increases immunity and respiratory health, detoxification of body, improves neural and muscular systems protects against several degenerative diseases (Parkinson's disease and metabolic (Scalbert *et al* ., 2005; Manach *et al.*, 2005; Chandrasekara and Shahidi, 2011). The important health beneficial nutrients are present in millets, like lipids oligosaccharides, resistant starch, antioxidants (phenolic acids, flavonoids, avenanthramides, phytosterols and lignans etc.) (Edge *et al.*, 2005). Barnyard millet has attracted little attention and literature on physical, functional and nutritional quality is limited. The published work available on minor millets especially barnyard millet is very less. The health benefits and anti-nutritional factor of barnyard millet are discussed in this section, as per the available literature.

In comparison to other minor millets and rice, barnyard millet could be considered as an effective hypolipidimic, hypoglycemic and faecal bulking agent. They are alkaline forming and non allergenic in nature and therefore easily digestible in nature (Saleh *et al.*, 2013). The low and slowly digestible carbohydrate content of barnyard millet makes it the precious gift of nature for the modern mankind engaged in sedentary activities. The fatty acid mainly present in barnyard millet is linoleic followed by palmitic and oleic acid. Due to high degree of amylase retrogradation, barnyard millet facilitates the resistant starches formation in higher amounts. Hence, barnyard millet is recommended potentially for the patients suffering from diabetes mellitus and cardiovascular disease. Bioactive components and anti-nutritional compounds present in different varieties of barnyard millet were analyzed and compared with different varieties of finger millet as represented in Table 5 (Panwar, *et al.*, 2016). Higher polyphenolic and tannin content in barnyard millet is reported and have approximately two time higher total polyphenolic and 1.5 times higher in tannin as compared to finger millet. Different varieties of barnyard millet have 20.30 to 27.80 mg GAE/g total polyphenols and 3.25 to 3.96 mg/g tannin content. Saponins content ranged from 8.38-10.26 mg diosgenin equivalent/g in varieties of barnyard millet (Banno *et al.*, 2004; Panwar *et al.*, 2016).

Table 5: Comparative data on the bio-active and anti-nutritional components of different barnyard and finger millet varieties (Panwar *et al.*, 2016)

Source	Bio-active components					Anti-nutritional components	
	Variety	Total phenolic content	Tannin content	Saponin content	Ortho-di-hydroxy phenol	Phytic acid	Trypsin-inhibition
			(mg/g)		(µg/g)	(mg/g)	(%)
Barnyard –millet	VL21	27.80^{a}	3.96^{a}	10.26^{a}	107.14^{c}	3.62^{e}	8.07^{b}
	VL29	20.35^{b}	3.42^{c}	8.53^{b}	122.85^{ab}	3.30^{g}	8.37^{a}
	VL172	20.30^{b}	3.25^{d}	8.38^{b}	117.86^{b}	3.43^{f}	7.85^{c}
	VL207	27.55^{a}	3.62^{b}	10.11^{a}	130.71^{a}	3.70^{d}	7.87^{c}
	PRJ1	27.70^{a}	3.34^{d}	8.42^{b}	127.14^{ab}	3.37^{g}	8.24^{ab}
Finger-millet	VL146	12.55^{c}	2.57^{e}	2.13^{c}	59.28^{def}	5.56^{b}	8.06^{abc}
	VL149	11.35^{cd}	2.62^{e}	2.63^{c}	64.28^{d}	5.58^{a}	8.02^{abc}
	VL204	12.10^{c}	2.05^{g}	2.32^{c}	53.57^{f}	5.54^{c}	8.36^{a}
	VL315	10.35^{d}	2.18^{fg}	2.23^{c}	62.85^{ef}	5.57^{ab}	8.24^{b}
	VL324	11.45^{cd}	2.33^{f}	2.52^{c}	59.28^{ef}	5.58^{a}	8.34^{a}

Note: All recorded values are mean of three determinations; Different letters in the same column indicate significant difference ($P \leq 0.05$) using LSD

Various biological health benefits like anti-inflammatory, anti-HIV, anti-diabetic and anti-atherosclerotic have been reported to be exerted by saponins. Hence, saponins can be the desirable attribute of functional food due to its hypolipidemic, hepatoprotective, gastro-protective and other protective functions as represented in Table 6. Different varieties of barnyard millet are also reported to be higher in Ortho-dihydroxy phenol content (107.14-130.71µg catechol/g) as compared to finger millet, which are the important polyphenolic components of plants having a wide range of bioactive quality attributes.

Health Benefits

Various health beneficial effects of barnyard millet are discussed below in detail as reported by various researchers.

Diabetes Mellitus

Diabetes mellitus is not a single disorder, but a diagnostic term for a group of chronic metabolic disorder characterized by abnormal glucose homeostasis and disturbances in carbohydrate, lipid, and protein metabolism. It is considered as the one among most common endocrine diseases. It results in deficiency in insulin production (type-1) or combined resistance to the action of insulin and insulin secretory response (type-2). Millets help in preventing the problem of diabetes because of their magnesium content in significant amount which increases the insulin and glucose receptors efficiency in the body (Saleh *et al.*, 2013). According to the data reported, rats that were fed barnyard millet treated starch in diet have lower blood glucose, triglycerides and serum cholesterol, in comparison to rats fed rice and other minor millets in diet (Kumari and Thayumanavan, 1997).

Table 6: Phytochemicals (Phenolic) and related health aspects of the barnyard millets (Taylor, 2017)

Millet	Flavonoids	Total phenolics	Some significant phenolic phytochemicals	Related health aspects
Barnyard millet	29-58 mg/100g	29–80 mg/100 g	Vanillic acid, Ferulic acid, Sinapic acid, Naringenin, Kaempferol, Apegenin	Hypolipidemic effects and Antihyperglycemic effects

Barnyard millet and foxtail millets have low glycemic index (40-50) among other minor millets. Barnyard millet and finger millet based products with legumes and fenugreek seeds reduce the glycemic index (Sharma and Raghuram, 1990; Arora *et al.*, 2003). The blood glucose and lipid levels are reported to be reduced very effectively by using barnyard millet (Ugare *et al.*, 2014). The important attribute of millets is their high dietary fibre content, which have hypoglycemic and hypocholestremic effects, apart from contributing to high satiety value.

Alternate autoclaving (5 cycle) and cooling (4°C) increase the resistant starch of minor millets. In comparison to rice and other minor millets resistant starch of barnyard millet lowers the blood glucose, serum cholesterol and triglycerides to much extent according to the data reported by a study on rats (Kumari and Thayumanavan, 1997). Healthy products from dehulling and roasting of barnyard millets are reported to be effective in type-2 diabetics and resulted in slight reductions in triglycerides and serum glucose and an increase in HDL. Roasting treatment substantially reduced the glycemic index of barnyard millet (Ugare *et al.*, 2014).

Cardiovascular Diseases

Like other millets, barnyard millet is a promising source of magnesium, thus help in the reduction of blood pressure and heart strokes risk especially in case of atherosclerosis. Potassium present in millets reduces the risk of cardiovascular disease by acting as a vasodilator and helps in keeping blood pressure low. Millets also have plant lignans present in them which protects the body against heart disease and certain type of cancers because of their ability to convert into animal lignans in the presence of digestive microflora. Millets are the rich source of dietary fiber and play a vital role in lowering body cholesterol by LDL elimination from the body and increasing HDL effects. Various studies reported that barnyard millet starch (treated) in comparison to rice and other minor millets, lowers blood glucose, triglycerides and serum cholesterol of rats to higher extent (Kumari and Thayumanavan, 1997)

Gastrointestinal Disorders

Increase in the retention of nutrient and reduction in the risk of gastrointestinal problems like gastric ulcers or colon cancer can be achieved by regulating digestion process. Millets are good source of fiber content, thus helps in the reduction or elimination of various digestive system problems like excess gas, cramping, bloating and constipation. Gluten free nature of barnyard millet can be helpful for person susceptible to gluten because of celiac disease. Celiac disease is an immune mediated enteropathic disease usually triggered by the gluten ingestion in susceptible individuals (Catassi and Fasano, 2008). Replacing cereals like wheat, rye and barley based food products which are high in gluten with gluten free grains based food products, may be helpful for Celiac patients by adhering to gluten free diet (Thompson, 2009). Millets have considerable potential for the developments gluten free foods and beverages because of their gluten free nature, therefore, can meet the growing demand for gluten free food products which are suitable for celiac disease patients (Taylor *et al.*, 2006; Adebowale *et al.*, 2011; Chandrasekara and Shahidi, 2012a and 2012b).

Detoxification (Antioxidant Properties)

Like other millets, barnyard millet is also a promising source of antioxidant compounds which are beneficial in the detoxification of body by neutralizing the free radicals, cleaning up of other body toxins especially those in kidney and liver. Polyphenols like quercetin, ellagic acid, cucurmin and various other beneficial catechins can be helpful in cleaning the body from toxins and any other foreign agents by promoting proper excretion and neutralizing enzymatic activity in different body organs. Therefore, polyphenols and barnyard millets are getting tremendous attention due to their health beneficial role (Tsao, 2010). Barnyard millet, kodo millet, little millet, finger millet, foxtail millet and sorghum were screened for their free radical quenching ability (1, 1, diphenyl-2-picrylhydrazyl) by electron spin resonance (Hegde and Chandra, 2005). Therefore, barnyard millet and other millets may serve as a natural antioxidants source different in food applications and proved to be as a nutraceuticals and functional food ingredient in reduction of disease risk and health promotion.

Antinutitional Compounds

Barnyard millet serves as good source of natural antioxidants because of various phytochemicals present. However various health beneficial phytochemicals like phenolic acids, flavonoids and tannins etc are present in barnyard millet, but also have some anti-nutrients such as trypsin inhibitors, α-amylase inhibitor, tannins and phytate in it (Kulkarni *et al.,* 1992). Nowadays antinutritional components present in plant foods are also in great focus to understand their potential for health beneficial effects (Yabuno, 2001). Various studies reported the protective effect of phytates against oxidative stress by chelating iron involved in Fenton's reaction, and antioxidants effect of some phenolic compounds and tannins (Yabuno, 2001). The phytic acid of different varieties of barnyard millet ranged from 3.30-3.70 mg/g (Table 5) (Panwar *et al.,* 2016). As compared to other millets like finger millet, barnyard millet contains lower amount phytic acid content. (Lorenz, 1983; Panwar, Dubey and Verma, 2016). Phytic acid of barnyard millet is comparatively lower than legumes, oil seeds, cereals, nuts and pollens (1–5%) (Petry *et al.*, 2015). Antinutritional activities of phytic acid has long been recognized as its ability to bind, precipitate and decrease the availability of di-and trivalent cationic minerals (Phillipy, 2003); inhibition of various digestive enzymes (α-amylase, pepsin, and trypsin) (Ravindran, 1992). Hence, phytic acid should be present in lower concentration and is a desirable trait as shown by barnyard millet. The trypsin inhibitor activity of different barnyard millet varieties as percent inhibition of trypsin enzyme varied from 7.85-8.37 percent and significantly similar to that of finger millet varieties (Panwar *et al.,* 2016).

Inspite of the fact that phenolic phytochemicals of millets have considerable antidiabetic-related effects (Panwar *et al.,* 2016) but some phenolics, including simple phenolics, tannins, and anthocyanins which are present in rich amount in seed coat of millets have powerful inhibitory action against the major human starch digestive enzymes (α-glucosidase and α-amylase) as reported by various researchers (Shobana *et al.,* 2009; Pradeep and Sreerama, 2015). As discussed in previous section that barnyard millet contains phytochemicals, such as phenolic acids, flavonoids and tannins etc up to the extent which serve as good source of natural antioxidants (Kulkarni *et al.*, 1992; Panwar, *et al.,* 2016). Chemical components having antinutrient effect like α-amylase inhibitor, trypsin inhibitors, phytate and tannins are also present in barnyard millet (Panwar *et al.*, 2016) and are in focus to understand their potential health benefits like protection against oxidative stress effect of phytates and antioxidants effect of tannins etc (Yabuno, 2001).

Conclusion

The research carried out to date on nutritional quality, innovations in processing, traditional and modern foods and beverages, anti-nutrition and health benefits of barnyard millet and related food products is presented in this chapter. Based on the data available related to barnyard millet research carried out, it can be clearly understood that many health-promoting components such as dietry fiber, vitamins, minerals, and phytochemicals are present in barnyard millet which also have great potential towards several health benefits. Like other cereals such as rice and wheat, barnyard millet is also a good source of protein, fat and crude fiber and apart from mineral and vitamins but contains comparatively low and slowly digestible carbohydrates as good quality attribute. It serves as good source of natural antioxidants like phytochemicals, such as flavonoids, phenolic acids, and tannins. Although barnyard millet is nutritionally superior than other major cereals and has great potentials to be included in normal as well as therapeutic diet formulations, yet its utilization is very less. Therefore, valuable efforts for making value added products from barnyard millet that provides convenience, texture, taste, high acceptability and shelf-stability at economical cost for fulfilling the poor people requirements is necessary to increase its utilization. Novel processing methods are also required to enhance the micronutrients bioavailability and to improve the quality and diversification of barnyard millet based diets.

References

Adebowale, A. R. A., Emmambux, M. N., Beukes, M., and Taylor, J. R. (2011). Fractionation and characterization of teff proteins. Journal of Cereal Science, 54(3), 380-386.

Anju, T., and Sarita, S. (2010). Suitability of Foxtail millet (*Setaria italica)* and Barnyard millet (*Echinochloa frumentacea)*for development of low glycemic index biscuits. *Mal. J. Nutr.,* 16(3), 361-368.

Anonymous (2014). Status Paper on Coarse Cereals, Directorate of Millets Development, Department of Agriculture and Cooperation, Ministry of Agriculture, Government of India.

Arora, P., Sehgal, S., and Kawatra, A. (2003). Content and HCl-extractability of minerals as affected by acid treatment of pearl millet. Food Chemistry, 80(1), 141-144.

Arora, S., and Srivastava, S. (2002). Suitability of millet based food products for diabetics. Journal of Food Science and Technology 39, 423— 428

Ballolli, U., Chimmad, B.V., Bharati, P., and Nagnur, S. (2011). Development of Nutty Barnyard millet (*Echinochloa frumantacea* Link) cookies. NAIP: *National Symposium Recapturing Nutritious Millets for Health and Management of Diseases,* UAS, Dharwad, December: 16-17, pp-42.

Banno, N., Akihisa, T., Tokuda, H., Yasukawa, K., Higashihara, H., Ukiya, M., and Nishino, H. (2004). Triterpene acids from the leaves of Perilla frutescens and their anti-inflammatory and antitumor-promoting effects. *Bioscience, Biotechnology, and Biochemistry, 68*(1), 85-90.

Catassi C and Fasano A. (2008). Celiac disease. In: Gallagher E, editor. Gluten-free cereal products and beverages. Burlington, MA: Elsevier. p 1–27.

Chandraprabha, Sharon, C.L., Panjikkaran, S.T., Aneena, E.R., and Beena, C. (2017). Development and nutritional qualities of vermicelli prepared from barnyard millet and Ekanayakam root bark. Journal of Pharmacognosy and Phytochemistry, 6(6), 2359-2362

Chandrasekara, A., and Shahidi, F. (2011). Antiproliferative potential and DNA scission inhibitory activity of phenolics from whole millet grains. Journal of Functional Foods, 3(3), 159-170.

Chandrasekara, A., and Shahidi, F. (2012)[a]. Antioxidant phenolics of millet control lipid peroxidation in human LDL cholesterol and food systems. Journal of the American Oil Chemists' Society, 89(2), 275-285.

Chandrasekara, A., and Shahidi, F. (2012)[b]. Bioaccessibility and antioxidant potential of millet grain phenolics as affected by simulated in vitro digestion and microbial fermentation. Journal of Functional Foods, 4(1), 226-237.

Chaturvedi,.N., and Rawat, D., (2018). Quality Evaluation of Malted Barnyard Millet Flour and Development of Value Added Rusk. International Journal for Research in Applied Science and Engineering Technology. 6(6), 631-636

Dhumal, C. V., Pardeshi, I. L., Sutar, P. P., and Jayabhaye, R. V. (2014). Development of potato and barnyard millet based ready to eat (RTE) fasting food. Journal of Ready to Eat Food, 1(1), 11-17.

Edge, M. S., Jones, J. M., and Marquart, L. (2005). A new life for whole grains. Journal of the American Dietetic Association, 105(12), 1856-1860.

Erdman Jr, J. W. (1981). Bioavailability of trace minerals from cereals and legumes. Cereal Chemistry.

F. A. O. (1995). Sorghum and millets in human nutrition. FAO Food and Nutrition Series, 27.

Geervani, P., and Eggum, B. O. (1989)[a]. Effect of heating and fortification with lysine on protein quality of minor millets. Plant Foods for Human Nutrition, 39(4), 349-357.

Geervani, P., and Eggum, B. O. (1989)[b]. Nutrient composition and protein quality of minor millets. Plant Foods for Human Nutrition, 39(2), 201-208.

Gomashe, S. S. (2016). Barnyard millet: Present Status and Future thrust Areas. Millets and Sorghum: Biology and Genetic Improvement; John Wiley and Sons: Hoboken, NJ, USA, 184-198.

Gomez, M., Ronda, F., Blanco, C. A., Caballero, P. A., and Apesteguía, A. (2003). Effect of dietary fibre on dough rheology and bread quality. European Food Research and Technology, 216(1), 51-56.

Gupta, A., Mahajan, V., Kumar, M., and Gupta, H.S. (2009) Biodiversity in the Barnyard millet (Echinochloa frumentacea, Link, Poaceae) germplasm in India. Genetic Resources and Crop Evolution, 56, 883–889.

Gupta, A., Sood, S., Agrawal, P. K., and Bhatt, J. C. (2015). B 29: An Easy Dehulling Barnyard Millet (*Echinochloa frumentacea* Link) Genotype. National Academy Science Letters, 38(1), 21-24.

Hadimani, N. A., and Malleshi, N. G. (1993). Studies on milling, physico-chemical properties, nutrient composition and dietary fibre content of millets. Journal of Food Science and Technology (India), 30(1), 17-20.

Hegde, P. S., and Chandra, T. S. (2005). ESR spectroscopic study reveals higher free radical quenching potential in kodo millet (*Paspalum scrobiculatum*) compared to other millets. Food Chemistry, 92(1), 177-182.

Hulse, Laing and Pearson, 1980; United States National Research Council/National Academy of Sciences, 1982; USDA/HNIS. 1984

Jaybhaye, R. V. and Srivastav, P.P. (2010[a]). Oven toasting of barnyard millet based ready-to-eat (RTE) snacks: Process parameter optimization and sensory evaluation. Proceedings of International Conference on 'Food Technology – Edition II' (30 – 31 October, 2010) IICPT, Tanajavur (T.N.).

Jaybhaye, R. V. and Srivastav, P.P. (2010[b]). Optimization of process parameters for development of millet based puffed snack food. International Conference on 'Food and Health' (10–12th November, 2010)

Jaybhaye, R. V., Pardeshi, I. L., Vengaiah, P. C., and Srivastav, P. P. (2014). Processing and technology for millet based food products: a review. Journal of Ready to Eat Food, 1(2), 32-48.

Kamaraddi, V., and Shanthakumar, G. (2003). Effect of incorporation of small millet flour to wheat flour on chemical, rheological and bread characteristics. Recent trends in millet processing and utilization. CCS Hisar Agricultural University, 74-81.

Kulkarni, L. R., Naik, R. K., and Katarki, P. A. (1992). Chemical composition of minor millets. Karnataka Journal of Agriculture Science, 5(3), 255-258.

Kumari, S. K., and Thayumanavan, B. (1997). Comparative study of resistant starch from minor millets on intestinal responses, blood glucose, serum cholesterol and triglycerides in rats. Journal of the Science of Food and Agriculture, 75(3), 296-302.

Lohani, U. C., Pandey, J. P., Anupama, S., Shahi, N. C., and Khan, C. (2014). Effect of moisture content on shelling characteristics of barnyard millet (*Echinochloa frumentacea*). International Journal of Basic and Applied Agricultural Research, 12(1), 99-103.

Lorenz, K. (1983). Tannins and phytate content in proso millets (*Panicum miliaceum*). Cereal Chemistry, 60, 424-426.

Mal, B., Padulosi, S., and Ravi, S. B. (2010). Minor millets in South Asia: learnings from IFAD-NUS Project in India and Nepal. Bioversity International, Maccarese, Rome, Italy and the MS Swaminathan Research Foundation, Chennai, India, 185.

Malleshi, N. G. (1989). Processing of small millets for food and industrial uses. Small millets in global agriculture, 325-339.

Malleshi, N. G. (2014). Post-harvest processing of millets for value addition. Retrived here: http://isites. harvard. edu/fs/docs/icb. topic868074. files/Millet% 20processing% 20and% 20b otanical% 20names. doc.

Manach, C., Mazur, A., and Scalbert, A. (2005). Polyphenols and prevention of cardiovascular diseases. Current opinion in lipidology, 16(1), 77-84.

Nazni, P., and Karuna, T. D. (2016). Development and Quality Evaluation of Barnyard Millet Bran Incorporated Rusk and Muffin. J Food Ind Microbiol, 2(116), 2.

Nehir El, S., and Simsek, S. (2012). Food technological applications for optimal nutrition: an overview of opportunities for the food industry. Comprehensive Reviews in Food Science and Food Safety, 11(1), 2-12.

Panwar, P., Dubey, A., and Verma, A. K. (2016). Evaluation of nutraceutical and antinutritional properties in barnyard and finger millet varieties grown in Himalayan region. Journal of Food Science and Technology, 53(6), 2779-2787.

Parameswaran, K. P., and Sadasivam, S. (1994). Changes in the carbohydrates and nitrogenous components during germination of proso millet, *Panicum miliaceum*. Plant Foods for Human Nutrition, 45(2), 97-102.

Petry, N., Boy, E., Wirth, J., and Hurrell, R. (2015). The potential of the common bean (Phaseolus vulgaris) as a vehicle for iron biofortification. Nutrients, 7(2), 1144-1173.

Phillipy, B. Q. (2003). Inositol phosphate in foods. Advance Food Science Nutrition Research, 45, 1-60.

Poongodi, T., Vijayakumar, J. B., Nazni, P., and Rajeshwari, M. (2003). Value addition for minor millets and its glycemic load among normal and type 2 diabetic subjects. In 39th National Nutritional Conference of NSI, Hyderabad.

Pradeep, P. M., and Sreerama, Y. N. (2015). Impact of processing on the phenolic profiles of small millets: Evaluation of their antioxidant and enzyme inhibitory properties associated with hyperglycemia. Food Chemistry, 169, 455-463.

Rao, D. S., and Deosthale, Y. G. (1983). Mineral composition, ionisable iron and soluble zine in malted grains of pearl millet and ragi. Food Chemistry, 11(3), 217-223.

Ravindran, G. (1992). Seed protein of millets: amino acid composition, proteinase inhibitors and in-vitro protein digestibility. Food Chemistry, 44(1), 13-17.

Saleh, A. S., Zhang, Q., Chen, J., and Shen, Q. (2013). Millet grains: nutritional quality, processing, and potential health benefits. Comprehensive Reviews in Food Science and Food Safety, 12(3), 281-295.

Scalbert, A., Manach, C., Morand, C., Rémésy, C., and Jiménez, L. (2005). Dietary polyphenols and the prevention of diseases. Critical Reviews in Food Science and Nutrition, 45(4), 287-306.

Seethatram, A., and Krishne, G. K.T. (2008). Production and utilization of small millets in India, (In) Food uses of small millets and avenues for further processing and value addition. Project coordination cell, All India Coordinated Small Millets Improvement Project, ICAR, G.K.V.K. Campus, Bangalore

Sharma, R. D., and Raghuram, T. C. (1990). Hypoglycemic effect of fenugreek seeds in non-insulin dependent diabetic subjects. Nutrition Research, 10(7), 731-739.

Sharma, S., Saxena, D. C., and Riar, C. S. (2016). Analysing the effect of germination on phenolics, dietary fibres, minerals and γ-amino butyric acid contents of barnyard millet (Echinochloa frumentaceae). Food Bioscience, 13, 60-68.

Shobana, S., Sreerama, Y. N., and Malleshi, N. G. (2009). Composition and enzyme inhibitory properties of finger millet (*Eleusine coracana* L.) seed coat phenolics: Mode of inhibition of α-glucosidase and pancreatic amylase. Food Chemistry, 115(4), 1268-1273.

Singh, K. P., Kundu, S., and Gupta, H. S. (2003). Development of higher capacity thresher for ragi/kodo. In Proc (pp. 8-9).

Singh, K. P., Mishra, H. N., and Saha, S. (2010). Moisture-dependent properties of barnyard millet grain and kernel. Journal of Food Engineering, 96(4), 598-606.

Singh, P., and Raghuvanshi, R. S. (2012). Finger millet for food and nutritional security. African Journal of Food Science, 6(4), 77-84.

Singh, P., Gurumukh, S., Srivastava, S., and Agarwal, P. (2005). Functional characteristics of blends containing wheat flour and millet flour. Beverage Food World Feb, 28

Sood, S., Khulbe, R. K., Gupta, A. K., Agrawal, P. K., Upadhyaya, H. D., and Bhatt, J. C. (2015). Barnyard millet–a potential food and feed crop of future. Plant Breeding, 134(2), 135-147.

Sridhar, R., and Lakshminarayana, G. (1992). Lipid class contents and fatty acid composition of small millets: little (Panicum sumatrense), kodo (*Paspalum scrobiculatum*), and barnyard (*Echinochloa colona*). Journal of Agricultural and Food Chemistry, 40(11), 2131-2134.

Srivastava, S., and Singh, G. (2003). Processing of millet for value addition and development of health food. In Proceeding of National Seminar on Recent Trend in Millet Processing and Utilization, CCS HAU, Hissar, India (pp. 109-116).

Srivastava, S., Dhyani, M., and Singh, G. (2003). Popping characteristics of Barnyard and foxtail millet and their use in preparation of sweets. Recent Trends in Millet Processing and Utilization, Hisar CCS HAU, Hissar, India (pp.38-40).

Sudharshana, L., Monteiro, P. V., and Ramachandra, G. (1988). Studies on the proteins of kodo millet (*Paspalum scrobiculatum*). Journal of the Science of Food and Agriculture, 42(4), 315-323.

Surekha, N., Chimmad, B. V., and Shantkumar, G. (2004). Barnyard millet based mix for better health. Proc 36th Annual Meet of Nutrition Society of India, Mysore, Nov, 5-6.

Surekha, N., Ravikumar, S., Naik, Mythri, S., and Rohini, D. (2013) Barnyard Millet (*Echinochloa Frumentacea* Link) Cookies: Development, Value Addition, Consumer Acceptability, Nutritional and Shelf Life Evaluation. IOSR Journal Of Environmental Science, Toxicology and Food Technology (IOSR-JESTFT), 7(3), 01-10.

Takhellambam, D.R., Chimmad, V.B., and Prkasam, N.J. (2016). Ready-to-cook millet flakes based on minor millets for modern consumer. Journal of Food Science and Technology, 53(2), 1312–1318.

Taylor, J. R. (2017). Millets: Their Unique Nutritional and Health-Promoting Attributes. In Gluten-Free Ancient Grains (pp. 55-103). Woodhead Publishing.

Taylor, J. R., Schober, T. J., and Bean, S. R. (2006). Novel food and non-food uses for sorghum and millets. Journal of Cereal Science, 44(3), 252-271.

Thathola A and Srivastava S (2002). Physico-chemical properties and nutritional traits of millet-based weaning foods suitable for infants of the Kumaon hills, northern India. Asia Pacific Journal Clinical Nutrition, 11(1): 28-32.

Thompson, T. (2009). The nutritional quality of gluten-free foods. Gluten-free Food Science and Technology, 42-51.

Tsao, R. (2010). Chemistry and biochemistry of dietary polyphenols. Nutrients, 2(12), 1231-1246.

Ugare, R. (2008). Health benefits, storage quality and value addition of barnyard millet (*Echinochloa frumentacaea* Link) (Thesis sciences). 18. Dharwad: University of Agricultural Science

Ugare, R., Chimmad, B., Naik, R., Bharati, P., and Itagi, S. (2014). Glycemic index and significance of barnyard millet (*Echinochloa frumentacae*) in type II diabetics. Journal of Food Science and Technology, 51(2), 392-395.

Veena, B. (2003). Nutritional, functional and utilization studies on barnyard millet. M.H.Sc. Thesis, University of Agricultural Sciences, Dharwad.

Veena, B., Chimmad, B. V., Naik, R. K., and Malagi, U. (2010). Development of barnyard millet based traditional foods. Karnataka Journal of Agricultural Sciences, 17(3).

Veena, B., Chimmad, B. V., Naik, R. K., and Shantakumar, G. (2005). Physico-chemical and nutritional studies in barnyard millet. Karnataka J Agril Sci, 18(1), 101-105.

Verma, V., and Patel, S. (2013). Value added products from nutri-cereals: Finger millet (Eleusine coracana). Emirates Journal of Food and Agriculture, 169-176.

Watanabe, M. (1999). Antioxidative phenolic compounds from Japanese barnyard millet (Echinochloa utilis) grains. Journal of Agricultural and Food Chemistry, 47, 4500–4505

Yabuno, T. (2001) Taxonomy and phylogeny of the genus Echinochloa. In: Natural history of genus Echinochloa. Zennoukyo Shuppan, Tokyo, Japan, pp 15–30.

7

Finger Millet (*Eleusine coracana*) Chemistry, Nutrition and Processing Technologies

Dadasaheb D. Wadikar

DRDO-Defence Food Research Laboratory
Siddartha Nagar, Mysore-570011

Introduction

The millets are a group of highly variable small seeded grasses, widely grown around the world as cereal crops or grains for both human food and fodder. These are important crops in semi-arid and tropical regions of the world (in particular in Afro-Asian countries) due to their resistance to pests and diseases, short growing season and productivity under heat and drought conditions. Finger millet (*Eleusine coracana* L.) is a small seeded annual cereal which belongs to the grass family, Poaceae. The height of a mature plant ranges from 30-150 cm in the cool, high-altitude regions of Africa and Asia, where it is grown for its seeds. The seeds, which may be white, light brown or dark brown, are consumed in a variety of forms including as unleavened bread made from milled flour. Various types of porridge and alcoholic beverages are also prepared from the seeds. Cultivated finger millet was domesticated about 5000 years ago from the wild subspecies in the highlands that range from Ethiopia to Uganda. Domesticated finger millet was then also cultivated in the low-lands of Africa. This was introduced into India around 3000 years ago, with the result that India is now another center of diversity for finger millet. (FAO, 2019). Finger millet is commonly known as "ragi" in India. The ragi word is originated from a Sanskrit word "raga" meaning red. This can be grown in almost all types of soils and climatic conditions. It is known as poor man's food because of its long sustenance as it can be stored safely for many years without infestation by insect and pests. This property makes it a very necessary famine reserve food. Presently it is cultivated in India, Nepal, Sri Lanka, East China Bangladesh, Kenya, Tanzania etc. The various commonly used names of finger millet in Indian states are summarized in Table 1.

Table 1: Various Names of Finger millet

Finger millet (*Eleusine coracana* L.)		
English Names	**Names in Indian Languages**	
Finger millet **African Millet** **Ragi,** **Koracan**	Bengali	*Marwa*
	Gujarati	*Nagli, Bavto*
	Hindi,	*Ragi, Mandika, Marwah*
	Kannada	*Ragi*
	Malayalam	*Ragi, Moothari*
	Marathi	*Nagli, Nachani*
	Oriya	*Mandia*
	Punjabi	*Mandhuka, Mandhal*
	Tamil	*Ragi, Keppai, kelvaragu*
	Telugu	*Ragi chodi*

Fig. 1. Finger millet Crop (Ragi)

Fig. 2. Finger millet Grain (Ragi)

The International Crops Research Institute for the Semi-Arid Tropics (ICRISAT), a member of the CGIAR consortium, partners with farmers, governments, researchers and NGOs helps farmers grow nutritious crops, including finger millet. In India, Indian Institute of Millets Research (IIMR) is a premier agricultural research institute engaged in basic and strategic research on sorghum and other millets under Indian Council of Agricultural Research (ICAR). IIMR coordinates and facilitates sorghum research at national level through All India Coordinated Research Projects on Sorghum, Pearl Millet and Small Millets and provides linkages with various national and international agencies. The headquarters of All India Coordinated Research Project on Small Millets (includes finger millet.) is located at the GKVK Campus of UAS Bengaluru. Established in 1986, it has the responsibility to plan, coordinate and execute the research programs (AICRP trials, FLD, TSP and NEH Programs) to augment the production and productivity of the six small millets. The mandated research in the project is carried out through network of 12 centers located in SAUs in the states, ICAR institutes and 19 cooperating centers. Project Coordinating Unit at Bangalore is also a Nodal center for DUS testing and CRP on Agro-Biodiversity Project (http://www.aicrpsm.res.in/). The centers (Table 2) where some work on finger millet in India is done are:

Table 2: AICRP Centers on Finger millet

1.	**Centre**	**Location**	**Mandate**
2.	**Bangalore**	UAS,GKVK, Bangalore, Karnataka	All Small Millets
3.	**Berhampur**	RARS, OUAT, Berhampur, Orissa	Finger millet, Little millet
4.	**Coimbatore**	TNAU, Coimbatore	Finger millet & other small millets
5.	**Dholi**	RAU, Dholi (Muzaffarpur), Bihar	Proso millet, Finger millet, Barnyard millet
6.	**Jagdalpur**	IGKVV, Jagadalpur, Chhattisgarh	All Small Millets
7.	**Kolhapur**	MPKVV, Kolhapur, Maharashtra	Finger millet and other small millets
8.	**Mandya**	UAS, Mandya, Karnataka	Finger millet
9.	**Mysore**	CFTRI, Mysore, Karnataka	Finger millet and other small millets
10.	**Ranchi**	BAU, Jharkhand	Finger millet, Little millet
11.	**Ranichauri**	GBPAU&T, Uttarakhand	Finger millet, Barnyard millet
12.	**Vizianagaram**	ANGRAU, Vizianagaram, AP.	Finger millet

Eleusine coracana is the most common species of finger millet cultivated for food use, whereas *Eleusine indica* and *Eleusine africana* are wild and semi-wild species. Its grain is having distinct morphological features. It is a small seeded grain; its kernel is not having a true caryopsis but a utricle. Its pericarp (glumes) are not fused with the seed coat or testa thus its pericarp can be easily removed by rubbing or by soaking in water, sometimes it gets detached from the seed during threshing. Large differences in color, appearance and size of this millet kernel have been observed among varieties. As grain color varies from white to orange, deep brown, purple to almost red. But brick red is the most common color of its seed. Their kernel shape may be spherical, globular and oval, and varies in size from 1 mm to 1.8 mm. Its endosperm is white in color. Its seed coat contains five layers and these are attached tightly to the endosperm McDonough *et al.* (1986). The special qualities of finger millet is that it's a gluten-free cereal with excellent malting qualities, high in calcium and iron and can be stored for up to two years without harmful pesticides, acting as a food reserve during the lean season (FAO, 2019).

Chemistry and Nutritional Attributes

Historically, finger millet is reported as rich source of carbohydrates (72%) and comprises (1.04%) free sugars, (65.5%) starch and non-starchy polysaccharides Malleshi *et al.* (1986) and (11.5%) dietary fiber Gopalan *et al.* (2009). Over the years several varieties have been studied for various aspects of agronomy, post-harvest processing and nutrition. In general the average chemical/nutritional profile of finger millet grains is summarized in Table-3. Wankhede *et al.* (1979) studied the carbohydrate profile of a few varieties of finger millet and reported 59.5 percent to 61.2 percent starch, 6.2 percent to 7.2 percent pentosans, 1.4 percent to 1.8 percent cellulose and 0.04 percent to 0.6 percent lignins. Premavalli and Wadikar (2012) summarized data on nutrients in 76 varieties of finger millet and reported that on an average the finger-millet grain comprise of the major nutrients as carbohydrates from 61 to 88.9 percent, protein 5.23 to 12.8 percent and lipids 1 to 4.5 percent reflecting the varietal difference probably due to the hybridization, cultivation environment and practices. The composition of carbohydrates though contributes mainly to starch content, also has considerable quantum of 15 to 20 percent unavailable carbohydrates as dietary fiber component and small quantities of 2 to 4.5 percent free sugars. The starch component with 50 to 60 percent has 14.2 to 17.9 percent amylase and 82.1 to 86.1 percent amylopectin fractions.

Finger millet is having excellent nutritional value. As it contains 6 percent to 8 percent protein, 1 percent to 1.7 percent fat, starch 65 percent to 75 percent, minerals 2 percent to 2.25 percent and dietary fiber 18 percent to 20 percent. The seed coat of this millet is rich source of phenolic compounds, minerals and dietary fiber Shobana *et al.*(2009). The endosperm is soft and fragile and is divided into three parts viz., peripheral, corneous and floury. Floury endosperm comprises about 83 percent of whole grain. Its endosperm is mostly filled with starch granules. The shape of starch granules may be spherical, polygonal and lenticular. Finger millet starch comprises amylose and amylopectin in the range of 25: 75. Among the oligosaccharides, raffinose is reported in the range of 0.02±0.006 percent Longwah *et al.* (2017).

Table 3. Chemical and nutritional parameters of Finger millet

Proximate Composition	**Moisture**	**Protein (N x 6.25) (g)**	**Fat (g)**	**Ash (g)**	**Crude fibre (g)**	**Carbohydrate (g)**	**Energy (kcal)**					
	11.2	7.7	1.5	2.6	3.6	72.6	336					
Lipid Profile	Linoleic acid	Oleic acid	Palmitic acid	Stearic Acid	Linolenic Acid							
	66.5	16.1	7.4	6.3	2.9							
Vitamin Profile[a]	Carotene (Vit. A)	Thiamine (Vit B1)	Riboflavin (Vit B2)	Niacin (Vit B3)	Pantothenic acid (Vit B5)	Pyridoxine (Vit B6)	Biotin (Vit B7)	Folic acid (Vit B9)	Vitamin C	Vitamin D2	Tocopherol (Vit E)	Vit K1
	32 µg	0.37 mg	0.17mg	1.34 mg	0.29 mg	0.5mg	0.88ug	15 mg	0.11mg	41.46ug	0.16 mg	3ug
Mineral Profile[a] (mg/100g edible portion)	Calcium (Ca)	Iron (Fe)	Phosphorus (P)	Magnesium (Mg)	Sodium (Na)	Potassium (K)	Copper (Cu)	Chromium (Cr)	Manganese (Mn)	Molybdenum (Mb)	Zinc (Zn)	
	350mg	3.9mg	290mg	81 mg	4.6 mg	250 mg	1.4 mg	0.03 mg	0.6mg	0.07 mg	2.4 mg	
Amino Acid[a]	Arginine	Cystine	Histidine	Isoleucine	Leucine	Lysine	Methionine	Phenylalanine	Tyrosine	Threonine	Tryptophan	Valine
mg/g	220	163	170	275	594	181	194	325	-	263	191	413

Source: Hulse. Laing and Pearson. 1980: United States National Research Council/National Academy of Sciences. 1982. USDA/HNIS. 1984. http://www.fao.org/3/T0818E/T0818E0b.htm, [a]Gopalan *et al.*, 1989, Premavalli and Wadikar, 2012, Longwah *et al.*, 2017, [$]HBT (2018) .

Protein and Amino Acids

The study on 136 varities of finger millet from 1954 showed the major important nutrient protein was in the range of 4.5 to 12.8 percent. Though new hybrid varieties are cultivated, the impact is not much on protein levels. Amongst the protein fractions, prolamine also known as elusinin is found to be one of the major fractions and others being albumin, globulin, glutelin which marginally varies with the varietal difference. But the prolamines are similar to those of prolamines of maize, sorghum and teff. The amino acid pattern of finger millet protein as reported by many researchers reflects that it contains majority of the essential amino acids except the low levels of lysine and threonine. It is a good source of glutamic acid and the lysine present satisfies 50 percent of WHO recommendations (Premavalli and Wadikar, 2012). Protein content of this millet ranges between 6 percent to 8 percent, although low protein content of 5 percent and high protein content of 12 percent have been reported in different varieties Hulse *et al.*(1980). Prolamins content is about 35-50 percent, albumins and globulins constitute 8 percent to 15 percent of total proteins. Its amino acid composition is good as it contains (2.5%) lysine, (13%) tryptophan, (2.9%) methionine, (3.1%) threonine and (4%) leucine and isoleucine.

Lipids and Fatty Acids

Lipids of this millet are mostly triglycerides and these are known to reduce the incidence of duodenal ulcer. Finger millet is having a lipid content of 1.5 percent. Oleic (49%), linoleic (25%) and palmitic acids (25%) are predominant fatty acids of this millet. About 72 percent of total lipids are present as neutral lipids, 13 percent as glycolipids and 6 percent as phospholipids Mahadevappa VG, Raina PL (1978). Amongst 121 varieties studied so far (Premavalli and Wadikar, 2012), the lipid levels range from 0.7 to 4.6 percent. The lipid classes were normal as neutral, glycol and phospholipids with slight lower content of phospholipids, but followed similar pattern in all the varieties. However, the quantity varied very marginally amongst varieties, but not significant. Fatty acid composition of finger millet as studied more widely compared to lipid classes have plamitic, stearic, oleic, linoleic and linolenic acids as common acids with others in small quantities reflecting 21 to 30 percent essential fatty acids. Fatty acid profile of the finger millet in general is provided in Table 3, however Longwah *et al.* (2017) reported that the average contents of Total Saturated Fatty Acids (TSFAs), Mono Unsaturated Fatty Acids (MUFAs) and Poly Unsaturated Fatty Acids (PUFAs) in finger millet were 317±17.0mg, 585±36.3mg and 431±20.7mg respectively. Among the phytosterols (mg/100g) the Campesterol (5.67±0.15); Stigmasterol (2.55±0.03) and β-Sitosterol (57.36±1.80) have been reported.

Table 4. Protein Fractions of Finger millet (Virupaksha *et al.*, 1975)

Fraction	Finger millet	
	Range	Mean
Albumin+globulin	17.3-27.6	22.4
Prolamin	24.6-36.2	32.3
Cross-linked prolamin	2.5-3.3	2.78
Glutelin-like	-	-
Glutelin	12.4-28.2	21.2
Residue	16.1-25.3	21.3
Total	74.7-83.9	78.7

Dietary fiber

Dietary fibre generally refers to the 'unavailable carbohydrates' and is mainly classified as soluble and insoluble fractions. It's one of the most important components after calcium for which finger millet has its significance in human diet. The health benefits associated with high fibre foods are delayed nutrient absorption, increased fecal bulk, lowering of blood lipids, prevention of colon cancer, barrier to digestion, mobility of intestinal contents, increased faecal transit time and fermentability characteristics (Tharanathan and Mahadevamma, 2003). The dietary fibers are classified (Table 5) into two categories with specific composition and varied physiological function.

Table 5. Conventional classification of dietary fibres as soluble and insoluble fibres with major physiological effects (Cho and Dreher, 2001; Chawla and Patil, 2010; Dai and Chau, 2016).

Chemical constituents from plant cell walls		Major physiological effects and mechanisms
Soluble fibre	Non-cellulosic polysaccharides, oligosaccharides, pectins, β-glucans, gums	Delay gastric emptying, regulate blood glucose levels, lower serum cholesterol levels, due to its effects on increasing viscosity of gut content and colonic fermentation
Insoluble fibre	Cellulose, hemicellulose, lignin	Shorten bowel transit time, improve laxation due to its bulking capacity; support the growth of intestinal microflora (esp. probiotic species) due to its fermentation in the large intestine

Total dietary fiber content of this millet is 22.0 percent which is relatively higher than other common cereal grains e.g. 12.6 percent wheat, 4.6 percent rice, 13.4 percent maize and 12.8 percent sorghum. Dietary fibers are categorized as water soluble and water insoluble. Chethan S and Malleshi NG (2007) reported 15.7 percent insoluble dietary fiber, 1.4 percent soluble dietary fiber, in finger millet grain. While as Shobana S and Malleshi NG (2007) reported 22.0 percent total dietary fiber, 19.7 percent insoluble dietary fiber and 2.5 percent soluble dietary fiber in finger millet. Dietary fiber of finger millet contributes to 15-20 percent as studied in 23 varieties with insoluble fiber 14 to 19.2 percent and soluble fiber 0.6 to 0.89 percent (Premavalli and Wadikar, 2012).

Vitamins

Finger millet contains both lipo-soluble and water-soluble vitamins viz., thiamin, riboflavin, niacin and apparently ascorbic acid and tocopherols Obilana AB and Manyasa E (2002). Water-soluble B-vitamins of finger millet are concentrated in the aleuronic layer and germ, while as lipo-soluble vitamins are mainly located in the germ. Finger millet contains small quantities of B-series vitamin and the studies confine to earlier years up to 1970's. Among 59 varieties studied, thiamine has the range from 0.19 to 0.67 mg, riboflavin 0.02 to 0.25 mg and niacin 0.1 to 2.2 mg. Considering the low level of vitamins and expected further losses during processing, serious investigations on vitamins profile has not been dealt by the researchers (Premavalli and Wadikar, 2012). However, the average vitamins in finger millet (Table 2) although lower in content need to be retained in downstream processing to get full benefits of the grain. Vitamin C (0.11mg) and Vitamin D have also been reported to be present in finger millet. Longwah *et al.* (2017) reported that alpha (α) tocopherol 0.09 and gamma (γ)-tocopherol 0.66 mg /100g, total carotenoids content 0.09 mg/100 g, 154±25.6 ug contributed by lutein (25.53±5.82 ug), zeaxanthin (1.45±0.23 ug) and β-carotene (1.53±0.25 ug)

Minerals

Finger millet is a good source of minerals as minor nutrients and particularly calcium and iron has been studied exhaustively. Amongst 223 varietal samples studied for minerals, the broader range of calcium was 200 to 1116 mg with an exception of lowest 10 mg calcium in Tanzanian variety. Iron was found to be in the range of 1.3 to 37 mg except highest 53 mg iron in the Ethiopian variety. The other elements reported covers phosphorous, potassium, magnesium at considerable levels and sodium, Zinc, manganese, copper at low levels. The earlier work of 1920-30's reported iodine content of 101 µg per kg which is highest among food grains. Calcium content is high in white varieties and iron is found higher in brown varieties. Besides this, the other observation is that calcium is high in base level grown brown varieties, while, iron is high in hilly grown varieties (Premavalli and Wadikar, 2012). The average mineral content of the finger millet given in Table 2 reveals that it is a treasure source nutritional elements.

Non-Nutritional Components

Finger millets contains phytates, tannins, oxalate, trypsin inhibitor, phenols and dietary fiber which are considered as antinutrients due to their metal chelating, protein binding and enzyme inhibiting properties. In 45 varieties reported phytates are present in Indian varieties in the normal range of 115.9mg to 248mg except the high levels in Sri Lankan varieties, 470 to 490 mg and Kenyan varieties 851 to 1419 mg. Such a board range reflects on the varietal difference and country of cultivation. The various processing methods generally adopted during product development reduced the phytates from 5 to 94.5 percent depending on the method, in turn bioavailability of minerals improved by 7.4 to 88.3 percent. Phenolic compounds comprise of tannins, phenolic acids and traces of flavonoids where tannins provide colour to the grain. In 147 varieties studied tannins ranged from 0.14 to 3.37 percent in colored varieties and traces in white variety. While, 34 varieties studied phenol contents reported is 0.053 to 2.44 percent. The processing methods have shown the reduction of phenolics by about 50 percent maximum with improved invitro protein digestibility. The processing of grains also reduces or eliminates the trypsin inhibitors and the oxalates found in finger millet are as found in fruits and vegetables which do not pose any nutritional problem (Premavalli and Vasudish, 2012). Among the organic acids, Fumaric acid (0.20±0.060), Malic acid (0.75±0.35), Succinic Acid (1.89±0.56) and Tartaric acid (1.60±0.26) are present in the finger millet (Longwah *et al.*, 2017). They further reported phytate content of 306±6.9 mg/100 gm among the 5 varieties. Wadikar *et al.* (2006) reported that average content of phytic acid in hilly varieties (VL 146, VL-149 and VL-204) and base varieties (MR-1, HR-911, GPU-28, INDAF-5, INDAF-8, INDAF-9, INDAF-11) was 224.6mg and 194.9mg, while the tannin content ranged between 0.34 percent and 0.53 percent, respectively (Table 6).

Table 6: Phytate phosphorus and tannin contents in ragi varieties (Wadikar *et al.*, 2006)

Varieties	Phytate phosphorous, mg			Tannin as % tannic acid		
	Raw	Puffed	% Reduction*	Raw	Puffed	% Reduction*
Hill region						
VL-146	241.4	139.7	42.1	0.32	0.31	3.7
VL-149	212.3	142.6	32.8	0.36	0.35	2.8
VL-204	220.1	168.6	23.4	0.32	0.31	3.7
Mean ±SD	224.6 ± 15.08	150.3± 15.90		0.34± 0.02	0.32± 0.021	-
Base region						
MR-11	115.9	62.6	46.0	0.62	0.49	22.0
HR-911	262.6	153.6	41.5	0.54	0.42	20.9
GPU-28	199.2	157.1	21.1	0.51	0.42	17.0

Varieties	Phytate phosphorous, mg			Tannin as % tannic acid		
	Raw	Puffed	% Reduction*	Raw	Puffed	% Reduction*
INDAF-5	194.0	97.8	49.6	0.56	0.46	17.9
INDAF-8	127.9	96.0	25.0	0.53	0.46	13.2
INDAF-9	215.8	144.9	32.9	0.62	0.52	16.2
INDAF-11	248.9	236.2	5.1	0.23	0.20	13.0
Mean ±SD	194.9 ± 55.80	135.5 ± 56.80	-	0.53 ± 0.099	0.43 ± 0.084	-
for hilly region and n=7 for base region; *: Reduction due to puffing						

Traditional and Modern Processing Technologies for Finger Millet

Processing is normally employed for enhancement of food grains quality through transforming them into edible form and reducing the antinutrients present in them. Consumption of millets can be improved via their processing into several forms such as rice flour, roasted, sprouting, popped, porridges, salted ready-to-eat grains and fermented foods. Harvesting of finger millet after the maturity is done commonly with stalks and in certain cases only pinnacles (ear heads) are removed. The harvested finger millet ear heads / stalks are dried and thrashed to remove the dried grains. The grains are further dried and stored till further use in dry places as the finger millet rains have very good shelf life. Prior to human consumption, finger millet is usually pulverized and whole meal is utilized for the preparation of traditional foods such as dumplings, mudde and ragi roti. In addition to traditional foods, it is also processed to prepare popped, malted, and fermented products. Noodles for diabetic patients were successfully developed from finger millet by Shukla and Srivastava (2011).

Several traditional and modern processing techniques used for finger millet are described below:

Milling

For milling purpose, finger millet grains are first cleaned to remove foreign materials such as stones, chaffs, stalks etc. then passed through abrasive or friction mills to separate out glumes (non-edible cellulosic tissue) and then pulverized or milled into coarse or fine flour. Normally it is pulverized in stone mill or iron disc or emery coated disc mills. Sometimes pearling or decortications are used to dehusk the finger millet grains; it results in pulverization of both the seed coat and endosperm. Hence finger millet is invariably pulverized along with seed coat to prepare whole meal. Centrifugal sheller can also be used to dehull/decorticate the small millets. The flour as such is sold as raw ragi flour in the market.

Decortication

Removal of outer coat of the seed is referred as decortication. This method is used for debranning of all cereals, but it is not effective for finger millet owing to its seed coat intactly attached to fragile endosperm. However hydrothermal processing is used to decorticate finger millet this involves (hydration, steaming and drying) which hardens the endosperm of grain and enables it to withstand mechanical impact. The decorticated finger millet could be cooked as such as rice is cooked. This is a very recent process developed for finger millet Change to (Malleshi NG 2006). Liu *et al.* (2012) stated finger millet can be cooked like rice with soft texture within 5 minutes and can also be consumed as main course meal due to its edible quality and sensory characters. Study was conducted on different methods of decortications such as traditional (hand-pounding methods) verses mechanical methods (Abrasive methods) in the laboratory using finger millet. The nutritional composition and characteristics of traditional and mechanical decorticated millets were compared and result indicate the improved nutritional composition with reduction in antinutritional factors in mechanical decorticated millets and they have also shown in results that no significant differences were found in traditional decortications method (Hama *et al.*, 2011).

Decortications have a significant effect on reduction of phytate phosphorus and polyphenols contents by 39.8 percent and 74.7 percent, respectively (Kakade and Hathan, 2015) which ultimately enables the nutritional bioavailability of protein and minerals. Shobana and Malleshi (2007) stated that scutellum and aleurone cells of finger millet losses phytate phosphorous during decortications. McDonough *et al.* (2000) reported reduction in amount of tannin and polyphenols content in testa layer of finger millet.

The hydrothermal treatment given to cereals and millets prior to decortications in order to achieve maximum recovery of grains without nutrients losses is referred as parboiling. It's more often and commonly used in case of rice. Parboiling treatment was initially developed in some Asian countries to cut down milling losses. It is done through soaking, steaming and drying process. Parboiling helps to penetrates the water-soluble nutrients from outer layer of grain seed coat to inner layer of grain and increases endosperm hardness than un-parboiled grains. Thus, the nutrients present in the outer layer of grains do not get removed during decortications process (Kik, 1957; Kakade and Hathan, 2015).

Usha and Malleshi (2011) reported that decortication increases the carbohydrate digestibility by 5 percent and protein by 8 percent. The presence of partially gelatinized starch backed up by the loss of granular rigidity could be the reason of increasing carbohydrate digestibility and increased protein digestibility may be due to the reduction in polyphenol content (34%). Removal of seed coat during decortication reduced dietary fiber and polyphenol content resulted in increased in protein and carbohydrate digestibility.

Puffing or Popping

Popping is one of the traditional methods to prepare popped finger millet which is usually converted to flour for further use. In this process, finger millet grains and conditioned by mixing with 3-5 percent additional water to raise the moisture content. The soaking / tempering is done for 2-4 hrs and grains are subjected to high temperature and short time (HTST) treatment by agitation in sand to about 230°C. This quick heat transfer causes grain explosion resulted into popped grains of finger millet. This process results in development of highly desirable aroma because of maillard reaction between sugars and amino acids. Popped finger millet is a precooked ready to eat product. It can also be pulverized and mixed with protein rich sources to prepare ready nutritious supplementary foods (Premavalli *et al.,* 2003). However, popping contaminates the product with particles of sand which is used as a heat transfer media and thus affects its eating quality. To overcome this drawback air popping in a suitable mechanical device has been successfully explored. But this method lacks the characteristics aroma compared to that using sand (Malleshi and Desikachar, 1981). Popped finger millet can be prepared at household community or at industrial level. Wadikar *et al.* (2006) reported that puffing of finger millet from conditioned grains and the reduction in anti-nutrient compounds (Table-6) such as phytic acid (reduction by 21-49%) and tannins (reduction by 3-22%). Sneha *et al* .(2018) reported about the reduction in anti-nutrients in white finger millet varieties (GE 6834-1, KMR-344 and DHWFM 11-3) as result of processing particularly popping as well as malting. Popping resulted in reduction by 11 percent to 22 percent with different soaking times in the three white varieties. Majumdar *et al.* (2006) reported the effect of puffing on calcium (Ca) and iron (Fe) contents of 3 hill grown (5500 ft msl) and 8 base grown (2300 ft msl) finger millet (ragi) varieties and their levels in ragi based food mixes. Hilly varieties had slightly lower Ca but higher Fe content. Puffing reduced the Ca and Fe contents by about 1.3 and 2.03 percent, respectively.

Fig. 3. Finger-millet seeds after popping

Total ash content for WRFMF (Whole Raw Finger Millet Flour), GFMF (Germinated Finger Millet Flour) and PFMF (Popped Finger Millet Flour) were 2.8±0.17, 2.7±0.10 and 2.88±0.08 (g/100 g). Finger millets are good source of protein and in the same line protein content was 6.3±0.20 in WRFMF, 8.8±0.30 in GFMF, 7.1±0.3 g/100g in PFMF. Mineral analysis of processed forms of finger millet revealed that calcium and iron content increased significantly during germination. While during popping of finger millet, calcium content decreased and iron content increased significantly (Chauhan and Sarita, 2018).

Malting

Germination, steeping, sprouting and drying are the most commonly used terms to pass on to the method of malting (Kakade and Hathan, 2015) and is known to enhance nutrient value and to lower down the anti-nutritional content of finger millet. Malting of finger millet is commonly practiced for specialty foods. During this process bioavailability of proteins, carbohydrates and minerals are enhanced. Some B-group vitamins are synthesized and concentration of anti-nutritional factors is also reduced. Malting involves soaking of viable seeds in water to hydrate and to facilitate sprouting. These sprouts are then kiln dried. Finally, the rootlets are separated from the grain manually by rubbing with hand. All these operations influence the quality of malt. Seed germination is most important step because during this process the hydrolytic enzymes are developed these cause endosperm modification and increases nutritional properties. Malting of finger millet has been successfully utilized for developing various health foods such as infant food, weaning food, milk-based beverages and confectionary products (Malleshi NG, 2007). Germination of finger millet is the most familiar method in India and nutrition profile of germinated finger millet is measured better then germinated maize and germinated sorghum (Singh and Raghuvanshi, 2012). Malleshi and Desikachar (1986) and Senappa (1988) stated that amylase activity of finger millet is higher than sorghum and other millets which take 4-5 days to reach to maximum amylase activity. Germination protects the grains from fungal infection and has dominant effect on number of biochemical changes which enhances the nutritional value of malted products. It leads to expansion of α and β-amylase throughout germination process, which develops desirable aroma during roasting/kilning and makes it a supreme grain for malt foods (Varma and Patel, 2013). Rao (1994) reported reduction of iron content in white and brown finger millet from 12.0 to 2.8 mg/100 g and 4.4 to 1.8 mg/100 g respectively. It was also reported that malted white finger millet and brown finger millet had significantly higher amount of soluble zinc (25 & 81%) and ionisable iron (55 & 27.1%). Further, reduction of 54 percent tannin content in brown finger millet and 65 percent and 58 percent reduction of phytin and phosphorus in white and brown finger millet were also found in their study. Deosthale (2002) and Mamrio *et al.* (2001) stated malting leads to reduction of 41 to 33 percent phytate in finger millet. Singh and Raghuvanshi (2012) stated that sprouted finger millet contains 5 mg iron, 230 mg phosphorus and 323.85 mg calcium. Germination of finger millet for 48 h has improved significantly in vitro extractability of calcium, zinc and iron (Mamiro *et al.*, 2001) Malted finger millet contains 88.3 percent ionisable iron which is much higher than that of raw finger millet contain (7.4% ionisable iron) at pH 7.5 (Deosthale, 2002). Malting and germination reported to have a significant improvement in protein content. It was found that malting of finger millet carried out at 48 h/30°C increase the protein content from 6.42 g/100 g to 7.32 g/100 g (Hejazi and Orsat, 2016). While 25 percent improvement in protein availability of germinated (24 h) finger millet (from 14 to 17.5%) was reported by Swami *et al.* (2013). Among the amino acids, it was reported (Malleshi and Desikachar, 1986) that an increase in lysine & methionine content from 3.5 to 4.0 mg/100 and 1.3 to 1.5 mg/100 g, respectively in germinated finger millet was observed. Malting also increases threonine, tryptophan and protein content of finger millet. Mbithi *et al.* (2000) have reported an increase in amount of the sulfur containing amino acids such as cysteine and methionine during sprouting of finger millet. They also observed that tannin content of finger millet was reduced from 914 ± 14.4 mg/100 g to 20 percent, 45 percent, 62 percent and 72 percent after germination of 0, 24, 48, 72 h respectively. Nirmala *et al* (2000) showed that Indaf-15 finger millet variety has a good malting potential because of high levels of reducing sugars produced and high amylase activity during malting. Kalpana Platel *et al.* (2010) have shown the influence of malting on the bio-accessibility of minerals in finger millet with reference to iron, zinc, calcium, copper and manganese and a comparative study on wheat and barley has been reported.

Fermentation

Fermentation is a metabolic process which converts complex material into simpler form with the help of micro-organism. Davidek *et al.* (1990) has reported changes during fermentation including destruction of inhibitors, breakdown of protein and increase in amino nitrogen. The traditional, naturally fermented finger millet product is called Ambali and it's very commonly used. However, the apart from ragi dumplings, ragi idli and dosai are other commonly prepared fermented products from ragi. Varma and Patel (2013) stated fermented foods like Idli and Dosa are well-liked foods in several parts of India as breakfast and even as evening meals in southern parts. Ragi is broadly utilized as one of the core ingredients in many of fermented food products which not only improves the taste but also enriches the food with fiber, calcium and protein content due to the reduction in antinutrients content. Fermentation has great impact on enhancement of Net Protein Utilization (NPU), niacin, thiamin, riboflavin and biological value (BV) of finger millet (Rajyalakshmi and Geervani, 1990; Aliya and Geervani, 1981). Use of endogenous grain microflora in finger millet flour during fermentation shows a significant reduction in the amount of antinutrient factors such as trypsin inhibitor activity (TIA) by 32 percent, tannins by 52 percent and phytates by 20 percent (Antony and Chandra, 1998). It has been reported that fermentation also increases mineral availability and accessibility (Singh and Raghuvanshi, 2012) such as calcium 20 percent, zinc 26 percent, phosphorous 26 percent, iron 27 percent. Basappa *et al.* (1997) reported significantly increased concentration of amino acid profile in fermented finger millet such as niacin-4.2 mg/100g, pantothenic acid-1.6 mg/100g, riboflavin 0.62 mg/100g, which is higher than raw finger millet. Onyango *et al.* (2004) prepared high energy density fermented uji from different combinations of maize, finger millet, sorghum and cassava using alpha- amylase and extrusion. It was observed that fermentation alone was not able to reduce viscosity of uji but addition of 0.1-2.1 ml/100 ml alpha-amylase to the fermented slurry or extrusion of the fermented and dried flour at 150-180°C and a screw speed of 200 rpm reduced viscosity of uji from 6000-7000 to 1000-2000 cp with acceptable energy densities (0.6-0.8 kcal/g) for child feeding.

Baking

Baking can be defined as the process which transforms dough, basically made of flour and water (other ingredients such as sugars, fat, egg, leavening agent and other additives will depend on each specific product), into a food with unique sensorial features (Purlis, 2010). Bakery products are important ready-to-eat processed foods. Baking is a method of cooking food that uses dry heat, normally in an oven. Baking involves simultaneous heat and mass transfer. The purpose of baking is to alter the sensory properties of food to improve palatability and to extend the range of tastes, aromas and textures in foods produced from similar raw materials. The different types of baked foods are prepared on the basis of different ingredients they are Bread, Cookies, Biscuits, Bars, Desserts, Muffins, Pizza, Snack cakes, Bun, Bread roll, Cracker, Pastry, Pudding etc. Several baked product using finger-millet are listed in Table 8.

Extrusion

Extrusion technology is one of the most widely used novel approaches of transforming ingredients into the value added products. Extrusion cooking is work with the principle of combined efforts of shear force along with high pressure and temperature which is responsible to modify starch properties. With the help of shear energy, exerted by the rotating screw and additional heating of the barrel, the food material is heated to its melting point or plasticizing point (Leszek and Van Zuilichem, 2011). It is one of the contemporary food processing technologies applied for preparation of variety of snacks, specialty and supplementary foods (Harper and Jansen, 1985) and offers advantages of preparation of Ready-to-Eat Foods of desired shape, size, texture and sensory characteristics at very low processing cost (Gay, 2001). Extrusion cooking also reduces the anti-nutritional factors, renders the product microbiologically safe and enhances the consumer acceptability (Nibedita and Sukumar, 2003). There are several reports about millet extrusion giving good results with flour with moisture ranging from 12-18 percent and a barrel temperature from 100-120°C providing excellent expansion index by means of porous, crunchy and smooth surface texture. The extrusion process being a high temperature short time (HTST) process, the temperature and equilibrium moisture of the flour plays an important role in gelatinization process which in turn influences the texture, physical characteristics such as expansion ratio, appearance and determines the sensory quality of the product. The extrusion of multiple millets is an interesting phenomenon as the starches of different nature are co-extruded and the product parameters are influenced by synergetic effects (Wadikar *et al.*, 2014). Several products from finger-millet using cold/ hot extrusion can be prepared including pasta and noodles. Based on the above technologies several products traditionally prepared at household level and some with commercial potential have been summarized in Table 8.

Table 7: Effect of processing on dietary fiber fractions of finger millet (Roopa & Premavalli 2012)

Processing	Total Dietary Fiber (%)	Insoluble Dietary Fiber (%)	Soluble Dietary Fiber (%)
Native flour	19.8±0.5	18.1±00.4	0.7±0.9
Cooking	14.9±0.2	13.0±0.2	1.9±0.2
Pressure cooking	15.7±0.5	14.0±0.2	1.6±0.1
Autoclaving	15.4±0.1	13.9±0.2	1.5±0.1
Re-autoclaving	14.7±0.2	12.8±0.1	1.9±0.1
Puffing	20.3±0.2	19.6±0.2	0.8±0.2
Roasting	14.7±0.1	13.1±0.4	1.6±0.2
Baking	9.7±0.4	8.6±0.2	1.2±0.3
Frying	12.2±1.0	11.1±0.9	1.1±0.3
Germination	10.7±0.4	8.9±0.3	1.8±0.1
Malting	12.0±0.4	8.8±0.1	3.3±0.1
Toasting (Roti)	13.6±0.3	12.6±0.2	1.1±0.3
Toasting (Dosa)	11.1±0.2	9.8±0.1	1.3±0.1

Table 8. Various products based on Finger millet prepared at household level and some with commercial potential

No	Product	Details
1	**Pasta/Noodlcs/ Vermicelli**	Varma and Patel (2013) reported different combinations of prepared noodles like wholly made up of finger millet, soy and wheat flour blended in the proportion of 1:4:5 and finger millet flour & wheat flour in the proportion of 1:1. A number of additional combinations of blends can also be explored in the preparation of noodles maintaining food values of ingredients and their accessibility in mind of consumers. Shukla and Srivastava (2011) reported fortification of 30-50 percent of finger millets with wheat flour for preparation of noodles has beneficial advantages for diabetic patients due to low glycemic response compared to control. Krishnan & Prabhasankar (2010) reported on pasta from finger millet & raw banana wherein influence of sprouted finger millet on pasting, microstructure, sensory, and nutritional was studied. Pasta prepared with addition of 20 percent finger millet resulted in improved dietary fiber, minerals and antioxidant profile (Gull *et al.*, 2015
2	**Biscuit, Bread,**	Krishnan *et al.* (2011) developed biscuit from finger millet seed coat. On the basis of sensory evaluation, they found that 10 percent of SCM (seed coat matter) from native and hydrothermally processed millet, 20 percent from malted millet could be used in composite biscuit flour. Saha *et al.* (2011) has formulated biscuits from composite flours having two proportion of 70:30 and 60:40 (w/w) finger millet: wheat flour, which were then examined for biscuit quality and dough features. Results shows that a composite flour of 60:40 (w/w) finger millet: wheat flour was very good mainly related to biscuit quality. Singh eat al. (2012) reported on bread prepared from millet-based composite flours.

No	Product	Details
3	**Muffins, Cake**	Shukla K and Srivastava S (2011) reported on muffins prepared by replacing wheat flour with 0 percent, 20 percent, 40 percent, 60 percent, 80 percent, and 100 percent FMF (finger millet flour), emulsifiers and hydrocolloids. Sudha *et al.* (1998) formulated muffins by replacing wheat flour with finger millet flour at a level of 0 to 100 percent (finger millet), hydrocolloids and emulsifiers. Impact of finger millet, hydrocolloids and emulsifiers on the microscopy of batter, quality characteristics and rheology properties of muffins were also evaluated. Results showed that blend of additives with 60 percent finger millet flour considerably improved the volume and quality attributes of muffins. Desai *et al.* (2010) reported on finger millet based cake and improvement in nutritional quality of cake using malted finger millet with respect to the fiber content and minerals content.
4	**Papads**	Flattened and dried ragi dough products which are toasted or deep fried and used as adjuncts with a meal Kamat SS (2008).
5	**Extruded Product**	Wadikar *et al.* (2012) reported extruded products such as multi-millet snacks, using finger millet, sorghum etc. Expanded finger millet has also been recently developed by Ushakumari SR (2009) from the decorticated finger millet. Its dietary fiber content is lower than that of popped finger millet as it is prepared from decorticated finger millet which is devoid of seed coat.
6	**Instant Products (Beverage / Soups)**	Mugocha *et al.* (2000) optimized the bacterial cultures formulation to produce the composite finger millet and skimmed milk based powder gruel. They inoculated composites powder having 0 to 100 percent finger millet by volume at various temperatures (30-45^{0}C). Results showed that using an incubation temperature of 45^{0}C and a storage temperature of 7^{0}C thick product of yoghurt like consistency can be obtained regardless of amount of finger millet gruel (0-50%). Roller dried finger millet-based soup mixes were develpoed by Guha M, Malleshi NG (2006).
7	**RTE nutritious supplementary foods**	Ready-to-eat nutritious supplementary foods from popped finger millet flour were also prepared by Malleshi NG (2007). This is a whole-grain product rich in macronutrients, micronutrients, dietary fiber, and usually mixed with vegetable or milk protein sources such as popped bengal gram, milk powder, and oil seeds sweetened with jaggery or sugar.
8	**Convenience Mixes (Idli, dosa, upma, halwa etc)**	Millet dosa mix, Millet beverage mix, Millet kheer mix, Millet chapathi mix, Several convenience mixes such as Millet roti mix, Millet dosa mix, Millet Pakoda mix, Millet halwa mix, Millet dhokla mix, Ragi cookie mix, Millet Jamun mix, Millet Nippatu Mix, Millet Ada mix, Millet halbai mix (Premavalli *et al.* 2006) Vasudish *et al.* (2017) reported that the technology of following products has been transferred to several entrepreneurs: Sweetened Millet Mix, Spiced Millet Mix, Millet idli Mix, Millet Roti mix, Millet chapthi mix, millet jamun mix etc. Multi grain based halwa mixes (a sweet dish prepared with flour, sugar, and clarified butter) and composite mixes (Itagi *et al.*, 2013).

No	Product	Details
9	**Beverages**	'Kodo ko jaanr' is the most common fermented alcoholic beverage prepared from dry seeds of finger millet in the Eastern Himalayan regions of the Darjeeling hills and Sikkim in India. Chhang is also a fermented finger millet beverage popular in Ladakh region in India. Koozh is another fermented beverage made with pearl or finger millet flour and rice, and consumed by ethnic communities in Tamil Nadu (Ilango and Antony, 2014).

Major Health Benefits

The grain finger millet is top sought grain due to its high calcium content. Finger millet is also rich in a mixture of soluble and insoluble dietary fibers or roughage that are resistant to breakdown during digestion and help to prevent gastro-intestinal disorders, colon cancer, coronary heart disease, and diabetes (Anderson *et al.*, 2009). The dietary fiber and polyphenols in finger millet are known to offer several health benefits such as anti-diabetic, antioxidant, hypo-cholesterolaemic, antimicrobial effects and protection from diet related chronic diseases to its regular consumers. The millet polyphenols are a complex mixture of benzoic acid and cinnamic acid derivatives and exhibit enzyme inhibitory and anti-cataractogenic activities also. The non-starchy polysaccharides of the millet form bulk of its dietary fiber constituents and offer several health benefits including delayed nutrient absorption, increased faecal bulk and lowering of blood lipids. Regular consumption of finger millet as a food or even as snacks helps in managing diabetes and its complications by regulation of glucose homeostasis and prevention of dyslipideamia (Devi *et al.*, 2014). The mineral profile of the finger millet is the most important for the grain is valued. It is the best and cheap source for calcium. Several studies proving the benefits of calcium and iron from finger millet have reported. Majumdar *et al.* (2006) reported that ragi based convenience mixes had Ca and Fe contents of 35 to 330 mg and 1.2 to 9.6 mg, respectively. The continuous use of ragi based foods improved Ca and haemoglobin levels as reflected by dietary intervention studies in children.

Conclusion

Potential health benefits and nutritive value of finger millet grains were found comparable to major cereals and advantageous over other due to its mineral profile and dietary fibre content. Several processing technologies can be utilized to improve nutritional characteristics and sensory properties of finger millet and its products. Utilization of finger millet is gradually increasing due to the enhanced knowledge of consumers about the benefits of finger millet and also due to the technological advances being put forth by several R&D initiatives. Promoting utilization of millet grains in urban areas to open new markets for farmers to improve their income, developing highly improved products from millets is needed. Finger millet can be used in different food formulations for making value added products due to its well-balanced protein profile and gluten free properties. Although the consumption pattern of this millet is specific and continue to remain as such therefore its popularization in the broader range is essential and specific design of foods acceptable to the larger population can help in promoting the consumption of this millet.

References

Aliya SQ, Geervani P. (1981) An assessment of the protein quality and vitamin B content of commonly used fermented products of legumes and millets. Journal of Science and Food Agriculture. 32:837-842.

Anderson, J. W., Baird, P., Davis, R. H., Ferreri, S. Jr., Knudtson, M., Koraym, A., et al. (2009). Health benefits of dietary fiber. Nutr. Rev. 67, 188–205. doi: 10.1111/j.1753-4887.2009.00189.x

Antony U, Chandra TS. (1998) Antinutrient reduction and enhancement in protein, starch and mineral availability in fermented flour of finger millet (*Eleusine coracana*). Journal of Agricultural and Food Chemistry. 46:2578-2582.

Basappa SC, Somashekar D, Agarwal R, Suma K, Bharthi K. (1997) Nutritional composition of fermented ragi (Chhang) by phab and defined starter cultures as compared to unfermented ragi (*Eleusine coracana* G.). International Journal of Food Science and Nutrition; 48:313-319.

Chauhan Ekta Singh & Sarita (2018) Effects of Processing (Germination and Popping) on the Nutritional and Anti-nutritional Properties of Finger Millet (*Eleusine coracana*) Curr. Res. Nutr Food Sci Jour., Vol. 6(2) 566-572.

Chawla, R. and Patil, G.R. (2010). Soluble dietary fiber. Comprehensive Reviews in Food Science and Food Safety, 9, 178–196.

Chethan S, Malleshi NG (2007) Finger millet polyphenols: optimization of extraction and the effect of pH on their stability. Food Chem 105: 862-870.

Cho, S. S., Dreher, M. L. (2001). Edited, Handbook of Dietary Fiber. Marcel Dekker, Inc, New York, NY.

Dai, F. and Chau, C. (2016). Classification and regulatory perspectives of dietary fiber. J Food and Drug Analysis, 25, 37–42.

Davidek J, Velisek J, Pokorny J. (1990) Chemical changes during food processing. New York, Tokyo, Elsevier Amsterdam-Oxford., 448. 10.

Deosthale YG. (2002). The nutritive value of foods and the significance of some household processes. http://www.unu.edu, ,

Desai, A. D., Kulkarni, S. S., Sahoo, A. K., Ranveer, R. C., & Dandge, P. B. (2010). Effect of supplementation of malted ragi flour on the nutritional and sensorial quality characteristics of cake. Advance Journal of Food Science and Technology, 2(1), 67-71.

Devi PB, Vijayabharathi R, Sathyabama S., Malleshi NG & Priyadarisini VB (2014) Health benefits of finger millet (*Eleusine coracana* L.) polyphenols and dietary fiber: a review. Journal of Food Science & Technology ; 51(6):1021–1040

Geervani P, Vimala V, Uma Pradeep K, Rama Devi M. (1996) Effect of processing on protein digestibility biological value, and net protein utilization of millet and legume based infant mixes and biscuits. Plant Foods for Human Nutrition; 49:221-227.

Gopalan C, Rama SBV, Balasubramanian SC (2009) Nutritive value of Indian foods. National Institute of Nutrition, Indian Council of Medical Research, Hyderabad, India.

Gopalan, C., Ramasastri, B. V., & Balasubramanian, S. C. (1989). Nutritive value of indian foods: National Institute of Nutrition. Indian Council of Medical Research, Hyderabad, India Revised Ed.

Guha M, Malleshi NG (2006) A process for preparation of pre-cooked cereals and vegetables based foods suitable for use as instant soup mix or of similar type foods. Indian Patent 257/NF/06.

Gull A, Prasad K, Kumar P (2015) Effect of millet flours and carrot pomace on cooking qualities, color and texture of developed pasta. LWT-Food Sci.Technol 63: 470-474.

Guy, R. (2001) Extrusion cooking: technologies and applications, Woodhead Publishing, ISBN 978-185-5735-59-0, Cambridge, United Kingdom.

Hama, F., Icard-Vernière, C., Guyot, J.-P., Picq, C., Diawara, B. and Mouquet-Rivier, C. 2011. Changes in micro-and macronutrient composition of pearl millet and white sorghum during in field versus laboratory decortication. Journal of Cereal Science, 54, 425-433.

Harper JM, Jansen GR (1985) Production of Nutritious Foods in Developing Countries by Low-Cost Extrusion Technology. Food Reviews International; 1: 27-97.

HBT (2018) https://www.healthbenefitstimes.com/foxtail-millet/

Hejazi, S. N., & Orsat, V. (2016). Malting process optimization for protein digestibility enhancement in finger millet grain. Journal of Food Science and Technology, 53(4), 1929-1938.

Hotz C, Gibson RS. (2007). Traditional food-processing and preparation practices to enhance the bioavailability of micronutrients in plant-based diets. Journal of Nutrition; 37:1097–100.

Hulse JH, Laing EM, Pearson OE (1980) Sorghum and the millets: Their composition and nutritive value. Academic Press, London.

Ilango S and Antony U (2014). Assessment of the microbiological quality of koozh, a fermented millet beverage. African Journal of Microbiological Research, 8(3): 308-312, DOI: 10.5897/AJMR2013.6482.

Itagi HBN, Singh V, Indiramma AR, Prakash M (2013) Shelf stable multigrain halwa mixes: Preparation of halwa, their textural and sensory studies. J Food SciTechnol 50: 879-889.

Jyotsna R, Soumya C, Indrani D, Venkateswara RG (2011) Effect of replacement of wheat flour with finger millet flour (*Eleusinecorcana*) on the batter microscopy, rheology and quality characteristics of muffins. J Texture Studies 42: 478-489.

Kakade Subhash B., Hathan BS (2015) Finger Millet Processing: Review. International Journal of Agriculture Innovations and Research. 2015; 2319-1473.

Kalpana Platel, Sushma W. Eipeson, and Krishnapura Srinivasan (2010) Bioaccessible Mineral Content of Malted Finger Millet (*Eleusine coracana*), Wheat (*Triticum aestivum*), and Barley Hordeum vulgare. Journal of Agricultural and Food Chemistry; 58 (13), 8100-8103. DOI: 10.1021/jf100846e

Kamat SS (2008) A study on documentation and evaluation of indigenous method of preparation of papad with special reference to cereals and millets. University of Agricultural Sciences, Dharward, Karnataka.

Kik MC. (1957) The Nutritive Value of Rice and Its Byproducts. Bulletin 589, Agricultural Experiment Station. College of Agriculture and Home Economics. University of Arkansas, Fayetteville, USA,

Krishnan M, Prabhasankar P (2010) Studies on pasting, microstructure, sensory, and nutritional profile of pasta influenced by sprouted finger millet (*Eleucina coracana*) and green banana (*Musa paradisiaca*) flours. J Text Studies; 41: 825-841.

Krishnan R, Usha D, Sai MR, Malleshi NG (2011) Quality characteristics of biscuits prepared from finger millet seed coat based composite flour. Food Chem 129: 499-506.

Leszek M, van Zuilichem DJ (2011) Extrusion-Cooking and Related Techniques In: Extrusion-Cooking Techniques: Applications, Theory and Sustainability. p 1-24.

Liu J, Tang X, Zhang Y, Zhao W. Determination of the volatile composition in brown millets, milled millet and millet bran by gas chromatography / mass spectrometry. Molecules. 2012; 17:2271-82.

Mahadevappa VG, Raina PL (1978) Lipid profile and fatty acid composition of finger millet (*Eleusinecoracana*). Journal of Food Science and Technology, 15: 100-102.

Majumdar TK, Premavalli KS, Bawa A S (2006) Effect of puffing on calcium and iron contents of ragi varieties and their utilization. Journal of Food Science and Technology, 43(5):542-543.

Malleshi NG (2006) Decorticated finger millet (*Eleusine coracana*) and process for preparation of decorticated finger millet. United States Patent 2006/7029720 B2.

Malleshi NG (2007) Nutritional and technological features of ragi (finger millet) and processing for value addition. In: Krishne Gowda KT, Seetharam A (eds.) Food Uses of Small Millets and Avenues for Further Processing and Value Addition. Project Coordination Cell, All India Coordinated Small Millets Improvement Project, ICAR, UAS, GKVK, Bangalore, India.

Malleshi NG, Desikachar HSR (1981) Varietal differences in puffing quality of ragi (*Eleusine Coracana*). Journal of Food Science and Technology, 18: 30-32.

Malleshi NG, Desikachar HSR, Tharanathan RN (1986) Free sugars and non-starchy polysaccharides of finger millet (*Eleusine coracana*), pearl millet (*Pennisetumtyphoideum*), foxtail millet (Setariaitalica) and their malts. Food Chem 20: 253-261.

Malleshi NG, Desikachar HSR. (1986) Influence of malting conditions on quality of finger millet. Journal of Industrial Brewing.; 92:81

Mamiro PRS, Van Camp J, Mwikya SM, Huyghebaert A. (2001) *In vitro* extractability of calcium, iron, and zinc in finger millet and kidney beans during processing. Journal of Food Science; 66:1271–5.

Mbithi-Mwikya S, Van Camp J, Yiru Y, Huyghebaert A.(2000) Nutrient and antinutrient changes in finger millet (*Eleusine coracan*) during sprouting. LWT-Food Science and Technology. 2000; 33:9-14.

McDonough CM, Rooney LW, Earp CA (1986) Structural characteristics of Eleusine coracana (finger millet) using scanning and fluorescence microscopy. Food Microstructure 5: 247-256.

McDonough, C. M., Rooney, L. W., & Serna-Saldivar, S. O. (2000). The millets. Food Science And Technology-New York-Marcel Dekker-, 177-202.

Mugocha PT, Taylor JRN, Bester BH. (2000) Fermentation of a composite finger millet dairy beverage. World Journal of Microbiology and biotechnology. 16:341-344.

Nibedita M, Sukumar B (2003) Extrusion cooking technology employed to reduce the anti-nutritional factor tannin in sesame (*Sesamum indicum*) meal. Journal of Food Engineering 56: 201-202.

Nirmala M, Subba Rao MVSST and Muralikrishna G, (2000). Carbohydrates and their degrading enzymes from native and malted finger millet (Ragi, Eleusine coracana, Indaf-15). Food Chemistry, 69; 175-180.

Obilana AB, Manyasa E (2002) Millets. In: Belton PS, Taylor JRN (eds.) Pseudo cereals and less common cereals: Grain properties and utilization potential. Springer-Verlag: New York, pp: 177-217.

Onyango C, Noetzold H, Bley T and Henle T (2004). Proximate composition and digestibility of fermented and extruded uji from maize-finger millet blend. LWT-Food Sci. Technol, 37:827-832

Premavalli KS and Wadikar DD (2012)Major nutrients of finger-millet. In: 'Fingermillet: a valued cereal' M/s Nova Publishers., USA, 2012, pp: 27-52.

Premavalli KS, Majumdar TK, Madhura CV, Bawa AS (2003) Development of traditional products.Vragi based convenience mixes. Journal of Food Science and Technology, 40: 361-365.

Premavalli KS, Majumdar TK, Wadikar DD, Madhura CV, Swamy YSS, Vasudhish CR. (2006) Technical report on Standardization of Ragi based traditional food products and development of ragi; cereal/legume blends based functional foods (DFR/08/017/05). DRDO-DFRL

Purlis E. (2010) Browning development in bakery products – A review. Journal of Food Engineering; 99: 239–249.

Rajyalakshmi P, Geervani P. (1990). Studies on tribal foods of south India: effect of processing methods on the vitamin and *in vitro* protein digestibility of cereals/ millets and legumes. Journal of Food Science and Technology; 27:260-263.

Rao PU. (1994) Evaluation of protein quality of brown and white ragi (*Eleusine coracana*) before and after malting. Food Chemistry; 51:433-436.

Saha S, Gupta A, Singh SRK, Bharti N, Singh KP, et al. (2011) Compositional and varietal influence of finger millet flour on rheological properties of dough and quality of biscuit. LWT-Food SciTechnol 44: 616-621.

Senappa M. (1988) Sorghum and millets in east Africa with reference to their use in weaning foods In: Proceedings of workshop, Nairobi, Kenya, 12-16 October, 1987, Improving young child feeding in eastern and southern Africa: household level food technology. Alnwick D, Moses S, Schmidt O G eds. Ottawa, International Development Research Centre. 1988, 39-54

Shobana S, Malleshi NG (2007) Preparation and functional properties of decorticated finger millet (*Eleusine coracana*). J Food Eng 79: 529-538.

Shobana S, Sreerama YN, Malleshi NG (2009) Composition and enzyme inhibitory properties of finger millet (*Eleusine coracana* L.) seed coat phenolics: mode of inhibition of a-glucosidase and pancreatic amylase. Food Chem 115: 1268-1273.

Shukla K, Srivastava S (2011) Evaluation of finger millet incorporated noodles for nutritive value and glycemic index. J Food SciTechnol 51: 527-534.

Singh KP, Abhinav M, Mishra HN (2012) Fuzzy analysis of sensory attributes of bread prepared from millet-based composite flours. LWT-Food SciTechnol 48: 276-282.

Singh P, Raghuvanshi RS. (2012) Finger millet for food and nutritional security. African Journal of Food Science. 6:77-84.

Sneha Shigihalli, Usha Ravindra and P. Ravishankar (2018) Effect of Processing Methods on Phytic Acid Content in Selected White Finger Millet Varieties. Int. J.Curr.Microbiol.App. Sci; 7(2): 1829-1835

Stephen MM, John VC, Peter RSM, Wilfried O, Patrick K, et al. (2002) Evaluation of the nutritional characteristics of a finger millet based complementary food. J Agri Food Chem 50: 3030-3036.

Sudha ML, Vetrimani R, Rahim A (1998) Quality of vermicelli from finger millet (Eleusine coracana) and its blend with different milled wheat fractions. Food Res Int 31: 99-104.

Tharanathan RN, Mahadevamma S (2003) Grain legumes—a boon to human nutrition. Trends Food Sci Technol 14:507–518

United States Department of Agriculture / Human Nutrition Information Service (USDA/ HNIS). 1984. Composition of cereal grains and pasta. Agriculture Handbook No. 8-20. Washinton DC.

USDA (2018). United States National Research Council/ National Academy of Sciences. 1982.

Usha D., Malleshi NG. (2011) Changes in carbohydrates, proteins and lipids of finger millet after hydrothermal processing. LWT-Food Science and Technology. 44:1636–42.

Ushakumari SR (2009) Technological and Physico-chemical characteristics of hydrothermally-treated finger millet. University of Mysore, Mysore.

Varma V, Patel S. (2013) Value added products from nutria-cereals: Finger millet. Emiates Journal of Food and Agriculture.; 25:169-176.

Vasudish CR, Wadikar D.D., Semwal AD. and Sharma G.K. (2017) Millet Based Product Technologies And Their Commercialization Potential. Indian Food Industry-Mag; 36(4):38-43.

Vidyavati HG, Mushtari J, Vijayakumari J, Gokavi SS, Begum S, et al. (2004) Utilization of finger millet in the preparation of papad. Journal of Food Science & Technology; 41: 379-382.

Virupaksha, T.K.; Ramachandra G. and Nagaraju, D. (1975) Seed Proteins of Finger Millet and their Amino Acid Composition. J. Sci. Fd Agric. 1975, 26, 1237-1246

Wadikar DD, Kangane SS, Patki PE and Parate V (2014) Optimisation of a Multi-Millet Ready-to-Eat Extruded Snack with Digestibility and Nutritional Perspective. Indian Journal of Nutrition, 1 (1):104-110.

Wadikar DD, Premavalli KS, Satyanarayanaswamy YS and Bawa AS (2007) Lipid profile of finger millet (*Eleusine coracana*) varieties. Journal of Food Science & Technology, 44(1):78-81

Wadikar DD, Vasudhish CR, Premavalli KS, and Bawa AS. (2006) Effect of variety and processing on anti-nutrients in finger millet, Journal of Food Science & Technology, 43 (4): 370-373

Wankhede DB, Shehnaj A, Raghavendra RMR (1979) Carbohydrate composition of finger millet (Eleusine coracana) and foxtail millet (*Setaria italica*) Qualitas Plantarum. Plant Foods Hum Nutri 28: 293-303.

8

Proso Millet (*Panicum miliaceum*) Chemistry, Nutritional Attributes, Processing Innovations, Traditional and Modern Foods and Beverages Anti-nutrition and Health Benefits

Asna Urooj and Chandan Vinay Srigiripura

Department of Studies in Food Science and Nutrition, Manasagangothri University of Mysore, Mysuru – 570006, Karnataka, India

Introduction

Plants have served as an important food source in the human diet since times immemorial. In recent times, available plant food sources are facing the threat of limitation due to many reasons viz., shrinking of agricultural land, changes in traditional lifestyles, population growth, rapid urbanization, changes in climate and a variety of plant diseases (Strange and Scott, 2005). Technological, development in agriculture and food processing are associated with both benefits and drawbacks impacting the human health globally with people suffering from chronic hunger, malnutrition to over-nutrition and nutritional imbalances. The United Nation (UN) study reports that the world's food supply is under severe threat from loss of biodiversity and in the last two decades, approximately 20 percent of the earth's vegetated surface has become less productive (UN, 2019).

Plants, available in abundance, although are the cheapest food sources for human consumption, are still underutilized. Wheat (*Triticum aestivum*), rice (*Oryza sativa*), maize (*Zea mays*) and sugarcane (*Saccharum officinarum L)* are the four most widely cultivated crops in the world (Bennett, 2010). At, this juncture, it becomes necessary to explore the underutilized millet crops, including crops of tribal utility as long-term viable options for a sustaining the supply of food for humans.

Millets were the oldest crops known to humans but their importance and cultivation has reduced due to large scale cultivation of rice and wheat because of urbanization and industrialization. Millets are important for rain fed agriculture as they are drought and pest-resistant thus have a significant impact on developing countries. Millets are nutritionally rich, are considered as potential prebiotic and probiotic sources thus providing a number of health benefits to the consumers. Millets have the potential to improve nutrition quality when included as a part of the daily diet and are comparable with the staple cereals consumed by people around the globe (Saleh *et al.*, 2013).

Millets are classified as major and minor millets based on area grown and its grain size. The major millets include sorghum *(Sorghum vulgare*) and pearl millet (*Pennisetum glaucum*). Small millets are finger millet (*Eleusine coracana*), foxtail millet (*Setaria italica*), little millet (*Panicum sumatrense*), kodo millet (*Paspalum scrobiculatum)*, barnyard millet (*Echinochloa* s), proso millet (*Panicum miliaceum*) and brown top millet (*Brachiaria ramosa*) (Table 1).

Table 1: Millets of the world

Millet	Scientific name	Common names	Major areas of production	Use
Sorghum	*Sorghum bicolor*	Great millet, jowar, cholam, jola, jonna, durra, Egyptian millet, feterita, Guinea corn, jwari, juwar, milo, shallu, gaoliang, kaoliang, kafir corn, dura, dari, mtama, solam.	USA, Nigeria, Sudan, Mexico, Ethiopia, India, Argentina, China, Niger, Australia	Grown for food grain in Asia and Africa, for fodder in Americas
Pearl millet	*Pennisetum glaucum*	Bajra, cattail, bulrush, candlestick, sanyo, munga, seno	India, Western and Central Africa, Eastern and Southern Africa	Grown for food grain in Asia and Africa, for fodder in Americas
Finger millet	*Eleusine coracana*	Ragi, African,bird's foot, rapoko, Hunsa, wimbi, bulo, telebun, koracan, kurakkan	India, Ethiopia, Nepal, Uganda, Malawi, Burundi, Sri Lanka, Rwanda	Grown for food grain and beer making in Asia and Africa
Foxtail millet	*Setaria Italic*	Italian, German, Hungarian, Siberian, kangani, navane, thanahal	China, Myanmar, India, Eastern Europe	Grown for food grain and fodder
Proso millet	*Panicum milliaceum*	Common, hog, broom, samai, Russian, panivarigu, panic, maha meneri	Russia, USA, Ukraine, South Korea, Kazakhstan, France, Poland, Belarus, India, Iran	Grown for food grain and bird seed
Little millet	*Panicum sumatrense*	Blue panic, heenmeneri	India	Grown for food grain
Barnyard millet	*Echinochola crus-galli*	Japanese, sanwa, sawan, Korean, kweichou	India, Japan, China, Malaysia	Grown for food grain
Tef	*Eragrostis tef*	Abyssinian lovegrass	Ethiopia, Eritrea, Australia	Grown for food grain, and fodder
Fonio	Digitaria exilis	Fundi, hungry rice, acha	West Africa, Sudan, Ethiopia, Nigeria, Niger, Togo, Senagal, Mali	Grown for food grain in Africa

Source. ICAR 2018

Proso Millet

Proso millet (*Panicum miliaceum* L.) is one of the minor millets grown as a summer crop and the life cycle is between 60-100 days. Proso millet (PM) is also called as 'broom corn' millet, 'hog millet' 'common millet', 'barri', 'kashfi millet', 'chinese millet', 'panic millet' etc. PM was first domesticated in China between 8000-10000 BC and further distributed throughout the world. It is grown in China, India, Nepal, Africa, Russia, Ukraine, Belarus, Turkey, Romani and North America. PM is recognized as an underutilized crop as majority of the produce is used as a bird feed (Habiyaremye *et al.*, 2017; Das *et al.*, 2019). Proso millet belongs to the *'Graminae'* family, '*Panicoideae'* sub family and *'Paniceae'* tribe. It is a monocotyledon and the seeds comprise of pericarp, seed coat, aleurone layer and endosperm (Arendt and Zannini, 2013). The increase in prevalence of non-communicable diseases and knowledge on the role of food and food components in prevention and management of diseases and in the maintenance of health has led to the exploration of potential of millets as functional foods with good nutritional value.

Chemistry and Nutritional Attributes

Carbohydrates

The carbohydrate content of PM is 69.8 percent. The total starch and total sugars are 58 percent and 0.16 percent, respectively (Parameswaran *et al.*, 1991).The amylose and amylopectin content are 24.5 percent and 75.5 percent respectively. The size of small-spherical starch granule ranges from 1.25-12.5 microns and the size of the big-hexagonal granules ranges from 5-17.5 microns. Differences exist in the proportion of amylose content in PM (Muralikrishna *et al.*, 1982). Millets are classified based on the amylose content as waxy (low amylose) and non-waxy (high amylose) millets. Flours made from waxy and non-waxy PM contain 28 ± 0.7 and 2 ± 0.2 percent, respectively. Isolated starch from non-waxy PM contains 32.8 ± 0.7 percent of amylose, whereas the waxy variant has 2.8 ± 0.2 percent of amylose (Yang *et al.*, 2019). Proso millet (with husk) has a high total dietary fiber content of 20.04 *percent*. The soluble and insoluble fiber content are 13.1 percent and 17.83 percent, respectively. The husk contains 81.8 percent of the total dietary fiber. Dehusked grains contain 2.4 percent total dietary fiber and consist of 63.8 percent and 36.2 percent of insoluble and soluble fiber fractions (Geervani and Eggum, 1989).

The *in vitro* starch digestibility of starch isolated from PM ranges between 38.66 - 41.45 percent and is comparable to that of maize (42.82%) (Yanez *et al.*, 1991). The cooked PM starch has a Rapid Digestible Starch (RDS), Slowly Digestible Starch (SDS), Resistant Starch (RS) content of 37.2, 42.6 and 20.2

percent, respectively. The hydrolysis index is 70.9 which is comparable to fox tail millet (71) and higher than foxtail (68.9) and finger millet (66.4). The estimated glycemic index is 69.5 (Annor, 2015). The gelatinization temperature is between 80.3-81.8^{0}C which is higher than maize (75.5°C) (Yanez *et al.*, 1991). The solubility of isolated starches of PM is between 7.5 – 80 percent and the swelling power ranges between 13.66 – 21.51 g/g. The swelling power is inversely correlated with the amylose content (Wang *et al.*, 2019).

Protein

The protein content in the caryopsis of PM is between 11.3–12.7 percent, whereas the dehulled grains contain 16g of protein for every 100g of dry matter. The mean protein content in six varieties of PM from Sri Lanka is reported to be 14.4 ± 1.58 percent. The non-protein nitrogen is 17.3 ± 0.4 percent of the total nitrogen (Ravindran, 1992). The mean protein content in 8 varieties of PM from Russia is reported to be 13.13 ± 1.82 (% dry matter) (Kalinova and Moudry, 2006). Differences in the growing environment can lead to compositional variations. The amino acid profile as reported by Ravindran *(1992)* and Kalinova and Moudry (2006) are shown in table no 2. Glutamic acid, Leucine, serine, aspartic acid, proline, alanine, valine and phenylalanine are the major amino acids present in PM.

The biological value (BV) of PM is 52.35 ± 1.17 percent which is lesser than wheat (62.41 ± 0.22 %) (Geervani and Eggum, 1989; Sibian *et al.*, 2016). The lower BV can be attributed to the low lysine content. Lysine is the first limiting amino acid followed by methionine and cystine in PM. Among the soluble protein fractions in PM, prolamine fraction is predominantly present compared to albumin and globulin fractions. The prolamine fraction had lower lysine, arginine and glycine compared to albumin and globulin, thus contributing to lower protein quality indicators (Jones *et al.,* 1970). The Amino Acid Score (AAS) of PM is 79.27. Waxy PM has more AAS than non-waxy variety (80.64 vs 77.57). The *in vitro* protein digestibility is 68.4 – 84.6 percent (Ravindran, 1992; Gulati *et al.*, 2017).

Table 2: Amino acid composition of different cultivars of Proso millet

Amino acid / (g/100g)	PM (Sri Lanka)	PM (Russia)	PM (China)
Aspartic acid	6.7 ± 0.05	11.27	7.21 ± 0.97
Threonine	4.1 ± 0.14	4.08	6.65 ± 0.5
Serine	7.6 ± 0.11	4.91	8.82 ± 0.83
Glutamic acid	25.2 ± 0.09	24.05	32.55 ± 4.14
Proline	7.8 ± 0.05	NA	6.99 ± 0.63
Glycine	2.9 ± 0.05	4.48	1.86 ± 0.26
Alanine	11.7 ± 0.26	10.27	12.25 ± 1.56

Amino acid / (g/100g)	PM (Sri Lanka)	PM (Russia)	PM (China)
Cystine	1 ± 0.02	NA	NA
Valine	6.4 ± 0.26	6.0	4.17 ± 0.46
Methionine	4.1 ± 0.29	2.22	NA
Isoleucine	4.9 ± 0.12	5.34	3.9 ± 0.45
Leucine	14 ± 0.31	13	16.19 ± 2.08
Tyrosine	4.5 ± 0.57	4.25	NA
Phenylalanine	6.3 ± 0.05	6.01	NA
Histidine	2.4 ± 0.03	2.83	2.54 ± 0.23
Lysine	1.7 ± 0.06	3.3	1.26 ± 0.12
Arginine	4.1 ± 0.04	4.29	2.51 ± 0.52
PM – Proso millet			

Source: Ravindran 1992, Kalinova and Moudry 2006 and Shen 2018

Fat

The Indian variety TANU-145 contains 3.8 ± 0.2 percent, 4.3 ± 0.3 percent and 1.2 ± 0.0 percent fat in whole, brown and polished PM flour (Devisetti *et al.*, 2014). The mean fat content in nine varieties of dehusked waxy and non-waxy PM ranges are 2.87 ± 0.78 percent and 1.84 ± 0.51 percent, respectively. Linoleic acid was the predominant fatty acid found (61.74 ± 1.71%) followed by oleic acid (22.16 ± 2.05) and palmitic acid (12.09 ± 2.03). Stearic and linoleic acid were found to be 2.39 ± 0.38 and 1.61 ± 0.6 percent, respectively (Shen, 2018). Majority of the fat in the PM is in the free form ranging from 3.46 – 3.96 percent in the red PM and between 3.61-3.89 percent in the white PM grain (Lorenz and Hwang, 1986). Variations exist in the content of lipids among various cultivars of PM. The predominant fatty acid is polyunsaturated fatty acid. PM has higher fat content than brown rice 1.24 ± 0.08 percent, whole wheat 1.47 ± 0.05 percent (Longvah *et al.*, 2017).

Minerals

The total mineral content in PM ranges between 1.5 - 4.2 percent (Kalinova, 2007) The mineral content in the dehulled grain and caryopsis of PM, whole PM, rice and wheat are shown in table no 3. The dehulled grains are rich in potassium, phosphorus, zinc and magnesium. There are differences in the mineral content based on the region where the PM was grown. Although PM is a rich source of phosphorus, 67.3 percent of the phosphorus is present bound to the phytic acid, making it less available to absorption (Ravindran, 1991). PM has higher content of iron, sodium, zinc, copper and manganese compared to brown rice and whole wheat. Rest of the reported minerals are comparable with whole wheat and are higher than present in brown rice.

Table 3. Mineral content of Proso millet cultivars, brown rice and whole wheat

Mineral – (mg/100g)	Dehulled PM (Mean of 6 varieties from Sri Lanka)	Dehulled PM (Russian variety)	Whole PM grain (Mean of 3 Indian varieties)	Rice, Brown	Wheat, Whole
Calcium	30	20	18	10.93 ± 1.79	39.36 ± 5.65
Iron	11	4.68	48	1.02 ± 0.35	3.97 ± 0.78
Magnesium	120	110	253	93.91± 9.11	125 ± 14.8
Phosphorus	260	360	494	267 ± 64.9	315 ± 41.8
Potassium	370	200	499	199 ± 40.9	366 ± 59.6
Sodium	60	10	58	3.64 ± 0.08	2.5 ± 0.2
Zinc	11	2.6	7.8	1.68 ± 0.33	2.85 ± 0.65
Copper	4	0.6	1.6	0.37 ± 0.14	0.49 ± 0.12
Manganese	19	NA	3	1.7 ± 0.45	3.19 ± 0.59
Selenium (μg)	NA	NA	0.1	2.26 ± 0.43	47.76 ± 5.96

Source : Ravindran, 1991; Kalinova 2007; Vali *et al*., 2018 and Longvah *et al*., 2017

Vitamins

The vitamin content in dehulled PM, wheat and rice are shown in table 4. Riboflavin, pyridoxine and total folate content is higher than whole wheat and brown rice. Thiamine content in PM is comparable to wheat and higher than brown rice. Although varying, the α-tocopherol and niacin content are higher than whole wheat and brown rice.

Table 4: Mineral content of Proso millet cultivars, brown rice and whole wheat

Vitamin/Food	Proso millet	Wheat, whole	Rice, brown
Thiamine (mg)	0.4 - 0.8	0.46 ± 0.04	0.27 ± 0.02
Riboflavin (mg)	0.22 - 0.4	0.15 ± 0.04	0.06 ± 0.01
Niacin (mg)	1.5 - 3.7	2.68 ± 0.19	3.4 ± 0.12
Pyridoxine (mg)	0.5 - 0.8	0.26 ± 0.04	0.37 ± 0.05
α -Tocopherol (mg)	0.1 - 2.6	0.6 ± 0.3	0.62 ± 0.08
Pantothenic acid (mg)	0.85	1.08 ± 0.21	0.61 ± 0.04
Total folate (μg)	85	30.09 ± 3.79	11.51 ± 1.69

Source. Kalinova, 2007; Das *et al*., 2019 and Longvah *et al*., 2017

Anti-Nutritional Components In Proso Millet

Anti-Nutritional Factors (ANF) are phytochemicals found in many foods that are known to interfere with the digestibility, absorption and impair the utilization of several nutrients that are vital to health. ANFs are synthesized in plants as a part of the defense mechanism and are biologically active substances that can have deleterious effects in humans and animals when consumed in considerable quantities. Some of the ANFs found in foods are tannins, phytate, oxalates, saponins, lectins, cyanogenic glycosides, goitrogens, alkaloids and protease inhibitors (Gemede and Ratta, 2014).

Phytates and oxalates

Phytate is the salt of phytic acid and is the storage form of minerals, chiefly phosphorus. 75 percent of the phosphorus in kernels is present bound to phytates in plants. Phytate has been attributed to inhibit absorption of vital minerals in the gut leading to deficiencies. However, phytate is susceptible to degradation during processing such as germination. Dietary phytate have also been conferred with beneficial actions (Schlemmer *et al.*, 2019). Rajesh *et al.*(2009) have reported phytic acid content of 0.72 percent in whole PM flour, 0.83 percent in brown PM flour and 2.2 percent in polished PM flour (TNAU-145, India) (Devisetti *et al.*, 2014).The phytate content in the whole, dehusked and polished flour of PM from Assam was found to be 574.74 ± 12.92, 682.50 ± 12.99 and 194 ± 5.48 mg/100g respectively (Bora *et al.*, 2018). This indicates that the phytates are more concentrated in the bran layers than in the endosperm of the PM.

Oxalic acid present naturally in foods has the ability to form water soluble salts with minerals such as sodium, potassium, calcium and magnesium making them unavailable (Noonan, 1999). The total and water-soluble oxalate content in PM is 23 ± 0.4 mg/100g and 14 ± 1 mg/100g(dry weight basis), respectively. Foxtail millet and finger millet contain higher amount of soluble (15, 19 ± 4 mg/100g) and total oxalate content (29 ± 0.3 and 27 ± 0.6 mg/100g) (Ravindran, 1991).

Polyphenols

Polyphenols are natural chemicals present in plants which are byproducts of metabolism. Polyphenols are part of the defense system of the plant against ultraviolet rays and pathogen aggregation. Polyphenols are known to exert several beneficial effects on health, but are also known to exert antinutritional effects such as inhibition of non heme-iron. Tannins, a class of polyphenolic molecules are known to interact with protein as well as inactivation of enzymes (Pandey and Rizvi, 2009; Mennen *et al.*, 2005). The total phenolic compounds in whole, brown and polished flour of PM (TNAU-145 variety) is 1.2 ± 0.17, 0.4 ± 0.03 and 0.24 ± 0.02 mg/g of flour, respectively (Devisetti *et al.*, 2014). Bora *et al.* (2018) have reported 164.46 ± 1.89, 145.81 ± 0.94 and 69.37 ± 0.86 mg gallic acid equivalent (GAE)/100g of total polyphenolic content in PM flour from Assam. Ferulic acid, chlorogenic acid, caffeic acid and p-coumaric acid are the major phenolic acids found in PM (Zhang *et al.*, 2014).

The tannin content (catechin equivalent) in the 24 varieties of PM from Colorado, USA ranges between 0.055 to 0.178 percent. Ravindran reported tannin content in six varieties of PM from Sri Lanka, ranging from 0.5 to 0.69 percent and tannin content of 23 ± 0.4 mg/100g. (Ravindran, 1991). Lorenz reported tannin content in PM ranging from 0.1-0.14 percent among 24 varieties of PM from Colorado, USA (Lorenz, 1983).

Inhibition of Enzyme Activity

The seed extracts of 11 varieties of PM from Patencheru and Coimbatore did not exhibit any anti-tryptic or anti-chymotryptic activity (Chandrashekar *et al.*, 1982). Protein fractions isolated from PM are known to exhibit various levels of inhibition of carbohydrate digesting enzymes, *in vitro*.The IC_{50} values of hydrolysates from heat treated grains (100^{0}C) for inhibitory activity of alpha-amylase ranged between 0.12 – 8.21 and non-heated grains ranged between 0.77-6.32 mg/ml. Only the albumin fraction of the hydrolysate from heat treated grains exhibited inhibition as reported by an IC_{50} value of 1.37 ± 0.02 mg/ml. The peptide fractions of PM exhibited lower inhibitory activity compared to the hydrolysates. The IC_{50} values of peptides from heat treated grains (100^{0}C) for inhibitory activity of alpha-amylase was 0.24 ± 0.014 and non-heated grains ranged between 0.11-0.67 mg/ml. Peptides from grains heat treated at 65^{0}C did not exhibit any inhibitory activity. The fractions exhibited alpha glucosidase inhibition to a much lesser extent. The IC_{50} values of inhibitory activities of protein hydrolysates from non-heated grains, heat treated grains at 65^{0}C and 100^{0}C ranged between 0.12-1.46 mg/ml, 0.06-0.58 mg/ml and 0.35-0.89 mg/ml respectively. Whereas, the IC_{50} values of inhibitory activities of peptides from non-heated grains, heat treated grains at 65^{0}C and 100^{0}C ranged between 0.09-0.14 mg/ml, 0.05-0.22 mg/ml and 0.12-0.31 mg/ml respectively. This trend indicates the negative effect of heat processing on the carbohydrate inhibitory actions of extracts from PM (Karas *et al.*, 2019).

The alpha glucosidase inhibitory activity of phenolic extract of raw PM has been reported to be 29.47 ± 1.47 ug/ml (IC_{50}). The IC_{50} value of the extracts from the germinated PM was lower at 16.09 um/ml, whereas the IC_{50} value of extracts from steamed and microwaved PM were higher at 66.19 ug/ml and 84.62 ug/ml respectively, which is higher than the raw millet extract. The IC_{50} values of alpha amylase inhibitory activities of phenolic extracts from raw, germinated, steamed and microwaved PM were 34.15, 20.14, 94.37 and 84.14 ug/ml respectively (Pradeep and Sreerama, 2015). This provides evidence for the effect of processing on the phytochemical content and enzyme inhibitory activity of PM.

Processing

Processing of the millets involves application of technology to convert the millets in the raw form to edible form with emphasis on nutritional and sensory attributes. The major processing techniques are described as follows:

Decortication/Dehulling and milling

The dehulling process involves the removal of pericarp of the millets. The machinery used in the rice milling process is modified and used in this process. The traditional method using pestle or *denki* is still practiced for debranning of the small millets. The millets are dehulled using centrifugal sheller or scarified after treating with 0.2N hydrochloric acid followed by washing with water. The extent of decortication ranges between 12-30 percent of the outer surface. Decortication results in reduction of polyphenols, phytates, tannins, dietary fiber and minerals and improvement in the starch digestibility and mineral bio accessibility (Rathore *et al.*, 2016; Jaybhaye *et al.*, 2014).

Milling

The milling of small millets including PM is carried out using the technologies available for wheat and rice. The decorticated grains are further polished to obtain millet rice using the rice machinery and also subjected to pulverization using a hammer mill or plate mill. This process is required for obtaining semolina.The milled grains are also further processed to produce flakes (Malleshi, 1989). The dehulling of PM in a barley pearler followed by milling in a Quadrumat Jr miller has been demonstrated to yield PM flour of good quality with yield rates ranging between 65.6-87.3 percent (Lorenz *et al.*, 1980). Flaking of PM leads to reduction in protein, fat, fiber, minerals and phenolic compounds (Goudar and Satisha, 2016). Bora *et al.*(2018) have reported similar effect of processing on the phytochemical content of PM flour. The least phytate content was found in the polished PM flour (194/100mg) compared to dehusked (982.5 mg/100 g) and whole (574.74/100 g) PM flours. The total phenolic content (mg GAE/100 g) was reported to be 164.46 ± 1.89, 145.81 ± 0.94, and 69.37 ± 0.86 in the whole, dehusked and polished PM flour, respectively (Bora *et al.*, 2018). Roasting of the whole PM flour at 100^{0}C has been reported to increase the total phenolic and flavonoid content (Azad *et al.*, 2019).

Germination and Malting

Germination of cereals and legumes is an ancient practice that is known to improve nutritional and physico-chemical properties. When germination is induced in controlled conditions with regard to moisture and time, it is termed malting (Hubner and Arendt, 2013). The germination of PM grains for a period of 1-8 days in dark results in several changes in the nutrient profile. The total free amino acid content is reported to increase by 255 percent at the end of 7 days of germination. Significant changes in lysine and tryptophan is an important finding. Among the protein fractions, increase in the albumin and globulin (combined) by 35.6 percent and reduction in prolamin content by 36.3 percent has been reported. Changes in carbohydrate components include reduction in total carbohydrates (-17.5%), starch (-42.5%), amylose (-6.7%) and increase in the content of total sugars (+11.1%) (Parameswaran *et al.*, 1994). This reflects the enzymatic degradation of starch into sugars. The variety of PM with dark colored seeds is known to germinate more slowly and suffer lesser extent of damage due to imbibition (Khan *et al.*, 1997).

The high amylase activity in PM makes it suitable millet for malting. The ideal conditions to obtain optimal malt (highest activity of alpha and beta amylase) from PM include 5 days of germination with the moisture content at 44 percent and 22°C of temperature during the seeping and germinating stage (Zarnkow *et al.*, 2007).

Puffing and Parboiling

Puffing is a type of processing method used in creating snack products that can be in the ready to eat form. Rice is the most common puffed cereal. Puffing is carried out using several techniques such as dry heat, sand, gun puffing etc. Puffing improves the aroma and taste of the processed foods (Mishra *et al.*, 2014).Gun puffing of PM grains tempered to moisture content of 18 percent and at 140/160 psi of pressure is reported as the ideal conditions to obtain highly expanded and low density puffed PM grains. Puffing increases the protein and lipid content, but reduces the ash and dietary fiber of PM grains (Delost-Lewis *et al.*, 1992).

Parboiling is a processing technique used to increase the yield of cereals and millets undergoing decortication and to improve the milling efficiency. Parboiling involves soaking the grains in water, boiling in water or application of steam followed by drying. Parboiling of PM prior to preparation of porridge has been reported to increase the yield by 28 percent, decrease the total starch by 4.8 percent, rapidly digestible starch by 4.39 percent, increase the resistant starch by 1.2 percent and therefore decrease the estimated glycemic index by 2.5 percent compared to non-parboiled PM porridge. Parboiling results in increase of bound and free phenolics, but causes a reduction in *in vitro* protein digestibility, which can be secondary to interactions between phenolics and protein (Bora *et al.*, 2019).

Fermentation

Fermentation is a process that involves the action of microorganisms and the enzymes produced by them on the food components (chemical and nutritional constituents). Fermentation is known to improve the bio-accessibility, bioavailability, organoleptic property, and shelf life of foods (Nkhata *et al.*, 2018). Limited numbers of studies are available exploring the fermentation potential, conditions and products prepared from PM. The malting potential, optimum conditions and preparation of beer from PM has been evaluated. The conditions for the development of optimal malt from PM were 5 days of germination with a moisture content of 44 percent with the set temperature for steeping kept at 22^{0}C (Zarnkow *et al.*, 2007). Beer having comparable range of aroma compounds to the beer from barley has been developed using PM. The beer from PM was developed using *Saccro myces. bayanus* (Champagne-yeast) three fermenting yeasts *Saccro myces. cerevisiae* (Alt-, Kölsch- and Ale-yeast), and *Brettanomyces. bruxellensis* (Zarnkow *et al.*, 2010).

Traditional And Modern Food Products

Traditional Food Products

Bosa is a beer prepared from PM, traditionally prepared in the Egypt, Sudan and Turkey. With alcohol content ranges from 1-7 percent. PM is also the ingredient in the preparation of '*tongba', 'rakshi'* and *'marcha'*, which are fermented alcoholic drinks consumed in Nepal. Sur, Madua, Oshikundu and Koozh are fermented beverages prepared using PM in Himachal Pradesh, India, Arunachal Pradesh, India, Tamilnadu, India respectively. *'Kasha'* is a porridge prepared from PM after boiling in water or milk. In Korea, honey and eggs are added to the porridge and pancake called *'Noti'* is prepared in several regions. In Italy, 'Panissa' and 'Migliaccio' are among the traditional dishes prepared using PM (Arendt and Zannini, 2013; Das *et al.*, 2019).

Modern Food Products

Composite Flours and Baked Products

The PM flour alone is unsuitable for baking due to the absence of gluten-forming proteins. Thus, studies have been conducted using PM flour, alone or in combination with other millets with wheat flour in the preparation of baked products. Optimization of bread preparation using multiple millet flours along with wheat flour has been conducted. The composite flour included 9.1 percent barnyard, 10.1 percent finger-millet, 10.2 percent proso-millet and 69.6 percent wheat flour. The bread prepared from composite flour had lower loaf volume (139.2 vs 163.1 cm^3), higher firmness (2.5 vs 1.6 kg), darker crumb colour and higher acceptability than wheat bread (Singh *et al.,* 2012). Cakes with improved volume, colour and texture has been prepared with the use of heat (100^{0}C) and moisture (30%) treated PM flour as an ingredient (up to 75% of the total flour) (Fathi *et al.*, 2015).

Lorenz and Dilsaver (1980), have evaluated the baking properties and quality of bread prepared from composite flours using red and white PM flour and wheat flour. The mixing volume and bread volume reduced with increasing proportion of PM flour. The colour of the crumb as measured using Hunter crumb color score also decreased. The acceptability of the breads prepared using composite flour was satisfactory until a level of 15 percent. The study also included preparation and evaluation of noodles and cookies. Noodles were prepared replacing the wheat by PM flour at three proportions (20%, 40% and 60%). Noodles prepared from 20:80 PM and wheat flour had higher cooked weight and acceptability. Cookies were prepared from blends of PM and wheat, with PM replacing wheat flour at 5, 10, 20 and 30 percent. Improvements in the spread factor with increasing millet concentration were found. Except for unacceptable grittiness, all other sensory parameters were found to be satisfactory.

Schoenlechner *et al.* (2013) have further optimized the bread preparation using wheat and PM composite flour (50:50) with the addition of emulsifiers (DATEM and distilled monoglycerides) and enzymes (xylanase and transglutaminase). The bread from the composite flour had increased loaf volume, number of pores lower crumb firmness compared to wheat bread after the addition of emulsifiers. The addition of enzymes resulted in increased loaf volume and pore uniformity, but decreased the number of pores and elasticity.

Fermented Product

Pearl millet has been successfully used as an ingredient of a multi millet mix in the preparation of *idli* and *dosa,* which are traditional cereal-pulse, based fermented preparations. Germinated millet mix consisting of equal quantities of finger millet (*Eleusine coracana*), foxtail millet (*Setaria italica*), proso millet (*Panicum miliaceum*), little millet (*Panicum sumatrense*) and kodo millet (*Paspalum orbiculare*) flours were used along with black gram and rice powder in the ratio of 33:33:33. Water and salt (2%) were added and the batter was subjected to fermentation. Idli's prepared from the batter including millets had higher protein (15.1% vs 7.2 %), fat (5.2% vs 0.84%), calcium (128 mg vs 69 mg), ascorbic acid (3.1 mg vs 1.8) and iron (3.4 vs 1.96 mg) compared to the idli's prepared from rice-black gram batter. Similarly, the millet dosa had higher protein (15.8 % vs 6.6 %), fat (9.8% vs 1.6%), calcium (114 mg vs 74 mg), ascorbic acid (2.8 mg vs 2.2 mg) and iron (3.42 mg vs 1.8 mg) content compared to rice-black gram dosa. Reduction in tannin content by 66.1 percent in the millet idli and 45.3 percent in the millet dosa was found. But the millet-based preparations had lower sensory scores (Krishnamoorthy *et al.*, 2013).

Pasta

Cordelino *et al.* (2019) have explored the potential of PM as an ingredient for the preparation of pasta. Compared to cooked wheat pasta, PM based pasta from five varieties of PM had higher starch content and lower protein content. The RDS and SDS were lower than commercial gluten free pasta and comparable to wheat pasta. The millet pasta had significantly lower *in vitro* protein digestibility. Sensory analysis revealed higher uniformity, lower elasticity, higher grainy texture and lesser chewiness than wheat pasta.

Ready to Eat Flakes

The physicochemical and sensory aspects of ready to eat flakes prepared with various millets compared to commercial oat and rice flakes has been reported. The millets, including PM were gelatinized under steam and high pressure followed by flaking using a roller. The PM flakes had lower bulk density, higher expansion ratio, hardness and crispiness compared to oat and rice flakes. The nutritional evaluation found PM flakes to have lower crude fat, higher protein, higher dietary fiber, lower zinc and manganese compared to the oat and rice flakes. The PM flakes had higher sensory score than ragi flakes, but lower than that of little millet flakes (Takhellambam *et al.*, 2016).

Health Benefits

Millets are gaining popularity due to their nutrient profile, phytochemical content, ease of production and higher yield of productivity. The need for a whole food that can reduce the risk of development and aid in the management of non-communicable diseases that arise due to the chronic consumption of processed foods and refined foods. The health benefit of PM as contributed by the nutrient composition and phytochemical makeup is discussed further.

Nutrient profile

The nutritional profile of PM has some important considerations in daily nutrition. Each 100g of dehulled PM consumed provides 42.5 percent of the recommended dietary allowance (RDA) of folate, 27.5 - 65 percent of the RDA of iron, 35.2 – 74.4 percent of the RDA of magnesium, 42.3 – 82.3 percent of the RDA of phosphorus and 65 – 91.7 percent of the RDA of zinc of an Indian reference adult man (NIN, 2011). Dietary fiber is the indigestible part of the plant material that plays an important in the maintenance of health of humans. The whole PM is an excellent source of dietary fiber. Dietary fiber mainly constituting of insoluble and soluble fiber helps in the control of appetite, lower glycemia, favorably alter the lipid homeostasis, aid in the digestive health, support the normal microflora in the gut and reduce the risk of heart disease and colon cancer. Dietary fiber is also used as an ingredient to improve the physicochemical characteristics, cooking properties, texture and as a fat replacer in foods such as pasta, meat and meat products, breakfast cereals, baked foods, biscuits, ice-cream, beverages, yogurt etc (Dhingra *et al.*, 2012). Since majority of the fiber in PM is found in the husk, which is also the end product of milling, it can be used as a source of dietary fiber in preparation of processed foods.

Phytochemicals and their Benefits

Antioxidant Action

PM contains phytochemicals such as ferulic acid, chlorogenic acid, caffeic acid, coumaric acid etc which are antioxidants in nature. Consumption of polyphenols has been linked to reduced incidence of cardiovascular disease, protection against cancer, anti-diabetic effect, reduced risk of developing neurological diseases and anti-aging activity (Pandey and Rizvi, 2009). The antioxidant activity of extracts prepared from PM have been assessed *in vitro* and using cell lines. The peroxyl radical scavenging activity of PM is between 57.68 to 147.32 µmol of vitamin C per gm. The cellular antioxidant activity is reported between 2.51 to 4.38 µmol of quercetin equivalent/g (Zhang *et al.*, 2014). The 2, 2-diphenyl-1-picrylhydrazyl (DPPH) radical scavenging ability of the bound polyphenolic extracts of dehulled PM is reported between 1.5 – 5.1 µmol of ferulic acid equivalent/ g of dry matter and the reducing power is reported as 0.06 µmol of catechin equivalent/g of dry matter. The soluble extracts of dehulled PM extracts have a Trolox equivalent antioxidant activity between 0.2 – 1.4 µmol of trolox equivalent/g of dry matter and reducing power between 1.7 – 3.8 µmol of ascorbic acid equivalent/g of dry matter (Kumari *et al.*, 2017). Kim *et al.* (2010) have reported IC_{50} values ranging from 58 – 297 in the methanolic extracts of PM taken in various concentrations (10 to 20000 PPM), which was higher than that of ascorbic acid (1.1), alpha tocopherol (3.3 ± 0.1), BHA (5 ± 0.2) and butylated hydroxy toluene (56 ± 4.3).

Anti-Proliferative Action

The inhibition of cell proliferation, particularly the tumor cells is of medical importance as the incidence of cancer and deaths due to cancer is increasing. PM extracts have demonstrated antiproliferative action on the breast and liver cancer cells. Zhang *et al.*(2014) have reported EC_{50} values of PM extracts for the inhibition of growth of human breast cells ranging from 46.47 mg/ml to 104.1 mg/ml. The EC_{50} values against the growth of liver cells (HepG2) is reported between 51.37 to 92.38 mg/ml. Free extracts have higher inhibitory activities than bound extracts (Zhang *et al.*, 2014). Shen *et al.*(2018) have also reported antiproliferative activities of extracts (0-500 ug/ml) from Chinese variety of PM on MDA-MB-231-Breast Cell lines. A dose dependent action was observed.

Antihyperlipidemic Action

The whole and dehulled PM flour has demonstrated favorable alterations in lipid profile of wistar rats (n-48). Rats fed with high fat diet containing whole or dehulled PM flour in varying concentrations (10%, 20% and 40%) for a period of 28 days reduced the triglycerides, low density lipoprotein, total cholesterol and increased high density lipoprotein (HDL) in the serum compared to rats fed with high fat diet without PM flour (Bora (b)*et al.*, 2018). Nishizawa and Fudamoto (1995) have reported similar effects on feeding protein from PM in 24 male wistar rats. The diet consisting of 20 percent protein derived either from casein, soy protein or PM protein concentrate was fed to rats for a period of 21 days. Rats fed with PM protein had significantly higher HDL at the end of study period. These studies highlight the antihyperlipidemic potential of PM but warrants further studies. The PM protein also demonstrated hepatoprotective effect. In 15 male wistar rats, where liver injury was induced by D-galactosamine, addition of protein concentrate from PM in the diet (20% protein) for a period of 14 days showed significant reduction in the level of liver enzymes compared to rats fed with casein (Nishizawa *et al.*,2002).

Antihyperglycemic Action

In a study on genetically obese male mice with type 2 diabetes, mice fed with PM protein concentrate had significantly higher plasma insulin, HDL cholesterol and serum adiponectin compared to mice fed with casein protein-based diet for 21 days. In mice fed with high fat diet and PM protein concentrate, significant improvements in serum HDL cholesterol, serum adiponectin and reductions in plasma glucose and insulin was observed. Real time PCR analysis revealed upregulation of adiponectin gene expression and down regulation of TNF-alpha gene expression was observed (Park *et al.*, 2008). In a crossover study design, 12 healthy male subjects were fed with couscous, biscuit, extruded snacks or porridge containing 50 g of glycemic carbohydrates, prepared with decorticated PM or refined corn on separate days. They were subjected to oral glucose tolerance test. Although, the glycemic response as expressed by blood glucose area under curve (AUC) in mmol/min/L was found to be lower for millet biscuit (82.6 ± 13.7), extruded snack (189.3 ± 16.4) and porridge (40.1 ± 5.8) compared to corn-based biscuit (102 ± 10.3), extruded snack (198.7 ± 20.9) and porridge (52.2 ± 8.1), but statistically significant difference was not found. The millet couscous had higher blood glucose AUC (66.7 ± 11.6) than corn-based couscous (43.5 ± 5.8). These results highlight the fact that interactions between the food components vary based on the processing methods applied, which can alter their kinetics (McSweeney *et al.*, 2017).

Gluten Free Food

Gluten is a storage protein found in the endosperm of wheat, rye, barley and oats. Gluten has proinflammatory effect in people with celiac disease and can trigger intestinal symptoms in people with sensitivity to gluten without celiac disease. Although limited clinical evidence is available for the benefits of gluten free diet in non-celiac disease individuals, there has been a substantial increase in the proportion of people opting for gluten free diet (Niland and Cash, 2018). In this context, PM can be a part of the gluten free diet.

Conclusion

Proso millet has a nutritional value that is comparable with the common staple cereals such as wheat and rice. Proso millet can contribute considerably to the intake of several micronutrients, when used on a daily basis, but requires processing to reduce the content and interaction of anti-nutrients such as phytates and tannins. The phytochemical profile of PM and the limited *in vitro* and *in vivo* studies highlight the potential of PM and its components, chiefly the protein concentrate as a functional food/nutraceutical, but further research is required to establish the beneficial health actions. The value of proso millet can be further enhanced as a food ingredient or development of pure proso millet-based food products with acceptable sensory and physicochemical characteristics by the application of modern technology and research.

Abbreviations

PM – Proso millet

UN – United Nations

RDS - Rapid digestible starch

SDS - Slowly digestible starch

RS - Resistant starch

AAS – Amino acid score

ANF - Anti-nutritional factors

GAE - Gallic acid equivalent

USA – United States of America

IC_{50} - Half maximal inhibitory concentration

EC_{50} - Half maximal effective concentration

RDA - Recommended dietary allowance

NIN - National Institute of Nutrition

HDL - High density lipoprotein

AUC -Area under curve

References

Annor, G.A., Marcone, M., Corredig, M., Bertoft, E., and Seetharaman, K. (2015). Effects of the amount and type of fatty acids present in millets on their in vitro starch digestibility and expected glycemic index (eGI). J Cereal Sci. 64,76-81.

Arendt, E.K. and Zannini, E. (2013). Cereal grains for the food and beverage industries. Elsevier, 313-320.

Azad, K., Obyedul, M., Jeong, D.I., Adnan, M., Salitxay, T. Heo., J,W., Naznin, M.T., Lim, J.D., Cho, D.H., Park, B.J., and Park, C.H. (2019). Effect of Different Processing Methods on the Accumulation of the Phenolic Compounds and Antioxidant Profile of Broomcorn Millet (*Panicum miliaceum* L.) Flour. Foods. 8(7),230.

Bennett, B.C., (2010). Plants as food. Economic Botany. Encyclopedia of Life Support Systems (EOLSS). Eolss publishers. Oxford, UK.

Bora, P ., Das, P., Mohan, P., and Borthakur, A. (2018b). J Entomol Zool Stud. 6(3), 691-695.

Bora, P., Das, P., Bhattacharyya, R., and Saikia, A. (2018). Effect of processing on the phytochemical content and antioxidant capacity of proso millet (*Panicum miliaceum* L.) milled fractions. Int J Chem Stud. 6(4), 18-22.

Bora, P., Ragaee, S., and Marcone, M. (2019). Effect of parboiling on decortication yield of millet grains and phenolic acids and in vitro digestibility of selected millet products. Food Chemistry. 274, 718-725.

Chandrasekher, G., Raju, D.S. and Pattabiraman, T.N., (1982). Natural plant enzyme inhibitors. Protease inhibitors in millets. J Sci Food Agric. 33(5), 447-450.

Cordelino, I,G., Tyl C., Inamdar, L., Vickers, Z., Marti, A., and Ismail, B. (2019). Cooking quality, digestibility, and sensory properties of proso millet pasta as impacted by amylose content and prolamin profile. LWT – Food Science and Technology. 99, 1-7.

Das, S., Khound, R., Santra, M., and Santra, D,K. (2019). Beyond Bird Feed: Proso Millet for Human Health and Environment. Agriculture. 9(3), 64.

Delost-Lewis, K., Lorenz, K., and Tribelhorn, R. (1992). Puffing quality of experimental varieties of proso millets (*Panicum miliaceum*). Cereal Chem. 69(4), 359-365.

Devisetti, R., Yadahally, S,N., and Bhattacharya, S. (2014). Nutrients and antinutrients in foxtail and proso millet milled fractions: Evaluation of their flour functionality. LWT-Food Science and Technology. 59(2), 889-895.

Dhingra, D., Michael, M., Rajput, H., and Patil, R.T. (2012). Dietary fibre in foods: a review. Journal of Food Science and Technology. 49(3), 255-266.

Fathi, B., Aalami, M., Kashaninejad, M. and Sadeghi Mahoonak, A. (2016). Utilization of Heat Moisture Treated Proso Millet Flour in Production of Gluten Free Pound Cake. Journal of Food Quality. 39(6), 611-619.

Geervani, Eggum B.O.(1989). Nutrient composition and protein quality of minor millets. Plant Foods Hum Nutr. 39(2), 201-208.

Gemede, H.F., and Ratta, N. (2014). Antinutritional factors in plant foods: Potential health benefits and adverse effects. International Journal of Nutrition and Food Sciences. 3(4), 284-289.

Goudar, G., and Sathisha, G.J. (2016). Effect of extrusion and flaking on the retention of nutrients and phenolic compounds in millet grains. International Journal of Food Science and Nutrition. 1(4), 08-11.

Gulati., Li A., Holding, D., Santra, D., Zhang, Y., and Rose, D.J. (2017). Heating reduces proso millet protein digestibility via formation of hydrophobic aggregates. J Agric Food Chem. 65(9), 1952-1959.

Habiyaremye, C., Matanguihan, J.B., D'Alpoim Guedes, J., Ganjyal, G.M., Whiteman, M.R., Kidwell, K.K. and Murphy, K.M., (2017). Proso millet (*Panicum miliaceum* L.) and its potential for cultivation in the Pacific Northwest, US: a review. Front Plant Sci. 7, 1961.

Hübner, F, and Arendt, E.K. (2013). Germination of cereal grains as a way to improve the nutritional value: a review. Crit Rev Food Sci Nutr. 53(8), 853-861.

ICAR. (2018). https://www.google.com/search?q=the+story+of+millets and rlz=1C1CHBD _enIN814IN814 and oq=the+story+of+millets and aqs=chrome.69i57j69i60. 415j0j8 and sourceid =chrome and ie=UTF-8# accessed on 21.07.2019

Jaybhaye, R.V., Pardeshi, I.L., Vengaiah, C., and Srivastav. (2014). Processing and technology for millet based food products: a review. Journal of Ready to Eat Food. 1(2), 32-48.

Jones, R. W., Beckwith, A. C., Khoo, U., & Inglett, G. E. (1970). Protein composition of proso millet. Journal of agricultural and food chemistry, 18(1), 37-39.

Kalinova, J. (2007). Nutritionally important components of proso millet (*Panicum miliaceum* L.). Food. 1, 91-100.

Kalinova, J., and Moudry, J. 2006). Content and quality of protein in proso millet (*Panicum miliaceum* L.) varieties. Plant Foods for Human Nutrition. 61(1), 45-49.

Karas, M., Jakubczyk, A., Szymanowska, U., Jęderka, K., Lewicki, S., and Zlotek, U. (2019). Different Temperature Treatments of Millet Grains Affect the Biological Activity of Protein Hydrolyzates and Peptide Fractions. Nutrients. 11(3), 550-571.

Khan, M., Cavers, B., Kane, M., and Thompson K. (1997). Role of the pigmented seed coat of proso millet (*Panicum miliaceum* L.) in imbibition, germination and seed persistence. Seed Science Research. 7(1), 21-26.

Kim, J., Hyun, T.K., and Kim, M. (2010). Anti-oxidative activities of sorghum, foxtail millet and proso millet extracts. African Journal of Biotechnology. 9(18), 2683-2690.

Krishnamoorthy, S., Kunjithapatham, S., and Manickam, L. (2013). Traditional I ndian breakfast (Idli and Dosa) with enhanced nutritional content using millets. Nutrition and Dietetics. 70(3), 241-246.

Kumari, D., Madhujith, T., and Chandrasekara, A. (2017). Comparison of phenolic content and antioxidant activities of millet varieties grown in different locations in Sri Lanka. Food Science and Nutrition, 5(3), 474-485.

Longvah, T., An antan, I., Bhaskarachary, K., and Venkaiah, K. (2017). Indian food composition tables. Hyderabad: National Institute of Nutrition, Indian Council of Medical Research. 15-30.

Lorenz, K. and Hwang, Y.S. (1986). Lipids in proso millet (*Panicum miliaceum*) flours and brans. Cereal Chem. 387-390.

Lorenz, K. (1983). Tannins and phytate content in proso millets (*Panicum miliaceum*). Cereal Chem. 60, 424-426.

Lorenz, K., and Dilsaver, W. (1980). Rheological properties and food applications of proso millet flours. Cereal Chem. 21-24.

Lorenz, K., Dilsaver, W., and Bates, L. (1980). Proso millets. Milling characteristics, proximate compositions, nutritive value of flours. Cereal Chem. 57(1), 16-20.

Malleshi, N.G. (1989). Processing of small millets for food and industrial uses. Small millets in global agriculture. 325-339.

McSweeney, M,B., Ferenc, A., Smolkova, K., Lazier, A., Tucker, A., Seetharaman, K., Wright, A., Duizer, L,M., and Ramdath, D,D. (2017). Glycaemic response of proso millet-based (*Panicum miliaceum*) products. Int J Food Sci Nutr, 68(7), 873-880.

Mennen, L. I., Walker, R., Bennetau-Pelissero, C., & Scalbert, A. (2005). Risks and safety of polyphenol consumption. The American journal of clinical nutrition, 81(1), 326S-329S.

Mishra, G., Joshi, D.C., and Panda, B,K. (2014). Popping and puffing of cereal grains: a review. Journal of Grain Processing and Storage. 1(2), 34-46.

Muralikrishna, G., Paramahans, S.V., and Tharanathan, R.N., (1982). Carbohydrate Make-up of Minor Millets. Starch Stärke, 34(12), 397-401.

National Institute of Nutrition. (2011). Dietary guidelines for Indians – A manual. Hyderabad, India, 100.

Niland, B., and Cash, B.D. (2018). Health benefits and adverse effects of a gluten-free diet in non–celiac disease patients. Gastroenterology and Hepatology. 14(2), 82 – 91.

Nishizawa, N., and Fudamoto, Y. (1995). The elevation of plasma concentration of high-density lipoprotein cholesterol in mice fed with protein from proso millet. Biosci Biotechnol Biochem. 59(2), 333-335.

Nishizawa, N., Sato, D., Ito, Y., Nagasawa, T., Hatakeyama, Y., Choi, M.R., Choi, Y.Y., and Wei, Y.M. (2002). Effects of dietary protein of proso millet on liver injury induced by D-galactosamine in rats. Biosci Biotechnol Biochem. 66(1), 92-96.

Nkhata, S.G., Ayua, E., Kamau, E.H., and Shingiro, J.B. (2018). Fermentation and germination improve nutritional value of cereals and legumes through activation of endogenous enzymes. Food Science and Nutrition. 6(8), 2446-2458.

Pandey, K,B., and Rizvi, S.I. (2009). Plant polyphenols as dietary antioxidants in human health and disease. Oxid Med Cell Longev. 2(5), 270-278.

Parameswaran, K., Parvathy, S., and Sadasivam. (1991). Changes in the carbohydrates and nitrogenous components during germination of proso millet, Panicum miliaceum. Plant Foods for Human Nutrition. 97-102.

Park, K,O., Ito, Y., Nagasawa, T., Choi, M.R., and Nishizawa, N. (2008). Effects of dietary Korean proso-millet protein on plasma adiponectin, HDL cholesterol, insulin levels, and gene expression in obese type 2 diabetic mice. Biosci Biotechnol Biochem. 72(11), 2918-2925.

Pradeep, M., and Sreerama, Y.N. (2015). Impact of processing on the phenolic profiles of small millets: Evaluation of their antioxidant and enzyme inhibitory properties associated with hyperglycemia. Food Chemistry. 169, 455-463.

Rathore, S., Singh, K., and Kumar, V. (2016). Millet grain processing, utilization and its role in health promotion: A review. International Journal of Nutrition and Food Sciences. 5(5), 318-329.

Ravindran, G. (1991). Studies on millets: Proximate composition, mineral composition, and phytate and oxalate contents. Food Chemistry. 39(1), 99-107.

Ravindran, G. (1992). Seed protein of millets: amino acid composition, proteinase inhibitors and in-vitro protein digestibility. Food Chemistry. 44(1), 13-17.

Saleh, A,S., Zhang, Q., Chen, J., and Shen, Q. (2013). Millet grains: nutritional quality, processing, and potential health benefits. Compr Rev Food Sci Food Saf. 12(3), 281-295.

Schlemmer, U., Frolich, W., Prieto, R,M., and Grases, F. (2009). Phytate in foods and significance for humans: food sources, intake, processing, bioavailability, protective role and analysis. Mol Nutr Food Res. 53(S2), S330-S375.

Schoenlechner, R., Szatmari, M., Bagdi, A., and Tomoskozi, S. (2013). Optimisation of bread quality produced from wheat and proso millet (*Panicum miliaceum* L.) by adding emulsifiers, transglutaminase and xylanase. LWT-Food Science and Technology. 51(1), 361-366.

Shen, R., Ma, Y., Jiang, L., Dong, J., Zhu, Y., and Ren, G. (2018). Chemical composition, antioxidant, and antiproliferative activities of nine Chinese proso millet varieties. Food Agric Immunol, 29(1), 625-637.

Sibian, M,S., Saxena, D.C., and Riar, C.S. (2016). Nutritional and functional quality analysis and amino acid score evaluation of germinated wheat (*Triticum aestivum*) grain. International Journal of Food Science and Nutrition. 16-22.

Strange, R.N., and Scott, R. (2005). Plant disease: a threat to global food security. Annu. Rev. Phytopathol. 4383-116.

Takhellambam, R.D., Chimmad, B.V., and Prakasam, J.N. (2016). Ready-to-cook millet flakes based on minor millets for modern consumer. J Food Sci Technol, 53(2), 1312-1318.

United Nations (UN). https://www.un.org/sustainabledevelopment/blog/2019/05/nature-decline-unprecedented-report/ accessed on 21.7.2019.

Vali Pasha, K., Ratnavathi, C.V., Ajani, J., Raju, D., Manoj Kumar, S., and Beedu, S.R. (2018). Proximate, mineral composition and antioxidant activity of traditional small millets cultivated and consumed in Rayalaseema region of south India. J Sci Food Agric. 98(2), 652-660.

Wang, Y.S., Yu, Z., Chen, H.H., and Shang, M.S. (2019). Effect of Amylose Content on the Physicochemical Properties of Proso Millet Starch. Curr Top Nutraceutical Res. 17(1), 42-49.

Yanez, G.A., Walker, C.E., and Nelson, L.A. (1991). Some chemical and physical properties of proso millet (Panicum milliaceum) starch. Journal of Cereal Science. 13(3), 299-305.

Yang, Q., Zhang, W., Li, J., Gong, X., & Feng, B. (2019). Physicochemical properties of starches in proso (non-waxy and waxy) and foxtail millets (non-waxy and waxy). Molecules, 24(9), 1743.

Zarnkow, M., Faltermaier, A., Back, W., Gastl, M., and Arendt, E.K. (2010). Evaluation of different yeast strains on the quality of beer produced from malted proso millet (*Panicum miliaceum* L.). European Food Research and Technology. 231(2), 287-295.

Zarnkow, M., Kebler, M., Burberg, F., Back, W., Arendt, E.K., and Kreisz, S. (2007). The use of response surface methodology to optimise malting conditions of proso millet (*Panicum miliaceum* L.) as a raw material for gluten free foods. Journal of the Institute of Brewing. 113(3), 280-292.

Zhang, L., Liu, R., and Niu, W. (2014). Phytochemical and antiproliferative activity of proso millet. PloS one. 9(8),e104058.

9

Kodo Millet (*Paspalum scrobiculatum*) Chemistry, Nutritional Attributes Processing Innovations, Traditional and Modern Foods and Beverages Anti-nutrition and Health Benefits

Ajay Singh[1], Gopal Kumar Sharma[2], Pradyuman Kumar[3] and Kamalpreet Kaur[4]

[1]*Department of Food Technology, Mata Gujri College, Fatehgarh Sahib Punjab*

[2]*Defence Food Research Laboratory, Mysore, Karnatka*

[3]*Department of Food Engineering & Technology, SLIET, Longowal, Punjab*

[4]*Department of Microbiology, Mata Gujri College, Fatehgarh Sahib, Punjab*

Introduction

After long back of continuous steep turn down in cropping and consumption pattern of traditional grains, once again time is rolling back for them as is evident from available integrated meal or composite research formulations distributed all around the market. Today's food market is segmented majorly with them and also emphasize more of healthy consumptions owe to adopted governmental strategy as a promotional trends such as subsidies, mid day meal programmes, selection of grain of the year by FAO, 2014 and revised labeling policies etc. are the real time examples. Present lifestyle, ongoing food habits, hike in grain prices, adulteration practices, demand in functional foods and many more reasons lead to the time which emphasizes on its own for consumption of traditional grains. United Nations Food and Agriculture Organization (2015) reported that about 10.9 percent of world's population was under-nourished, 39 percent of adults $\geq$ 18 years of age were overweight and 13 percent among the obese category (Kumar and Tomer, 2018). Traditional grains as alternative medicines like pseudocereals and coarse cereals utilization are the only preference scale to overcome these troubles with their rich nutrient profile and functional attributes. Actually these are the nutritionally boosted grains with lots of capacity of nutritional returns to the consumers through least inputs at farming scale viz. less water requirement, less care, least pesticides and fertilizers distribution and not as much of long cropping cycle as like cereals (Vernta *et al.*, 2007). Furthermore, India succeeded all over the globe for the production for them. Existence of that much positive features even after their lead production, irony for these crops can be seen through their labeled recognition as "under-utilized grains" or restricted access as animal fodder, poultry feed or fish feed rather than for human use.

Pseudocereals group comprises of non grasses starchy seeds that are consumed majorly for their protein and minor nutrients. It includes amaranth, quinoa, buckwheat, chia seeds and breadnut. Similarly, kharif crops such as coarse cereals that have much value in day to day usage as food, fodder and fuel. These include sorghum, millets, barley and maize etc. Essential amino acid availability, unsaturated fatty acid profile, low glycaemic index (GI) status and being gluten free are the attributes that expresses them as "Nutri-cereals". Millets categorized as major millets (pearl millet, foxtail millet, proso or white millet and finger millet) and minor millets (barnyard millet, kodo millet, little millet, guinea millet and browntop millet). Sometimes teff and fonio is also placed alongwith millets. Among these, kodo millet (*Paspalum scrobiculatum*) gaining much importance these days in lieu of its exceptionally high fiber content and polyphenolic range.

Kodo is a perennial grass, length of which may vary from 120-150cm, to some extent bulbous at base. Kodo millet has its own significance as small millet that can flourish well in marginal soil where other crops may not survive. Its cultivation is restricted to Madhya Pradesh, Chattisgarh and Karnataka region of India (Singh, 1999). Bulk of this is consumed by producers at native ground thus ceases its recognition and contribution in economical development as well. Technologically speaking, enormous scope is still lying blank to quench majority of its functional and nutraceutical capacity for modern practices in development of value foods by means of enrichment, fortification and supplementation. As per the gathered information, global millet production met to a value of 27.83 million tons where maximum contribution was met by India (41.04%) followed by Nigeria (11.94%). Among Indian states, Madhya pradesh is the prominent state of India that occupies 33.4 percent of production area (Michaelraj and Shanmugan, 2013). In nutshell the production data for millet is contradictory and is rarely possible to estimate millet species wise production details. These contradictory differences in millet production data (2014) from FAOSTAT (28 million tons) and ICRISAT (56 million tons) is a dilemma concern (Taylor, 2019).

Kodo millet utilization restricted only to tribal population is long back dialogue. Now it retained its popularity and gaining much of importance and acceptance too with changing lifestyle due to advancement of urbanization.

Chemistry

It is believed that kodo is of African origin (3000 years) and domesticated in India, presently it is cultivated mostly in Deccan region and extend its cultivation practices in the foothills of Himalya also (Deshpande and Mohapatra, 2015). It flourish well in tropical and sub tropical land of African and Asian continent and yielded grains within a time span of 5-6 months in range of 250-1000 kg/ha. Whole seed is characterized by bran and husk that make up about 37 percent. As part of its taxonomical classification is concern, it belongs to kingdom: Plantae, subkingdom: Tracheobionta and is classified under division: Magnoliophyta, follow the class categorization as Liliopsida. Brief taxonomical hierarch is as follows;

Order : Cyperales

Family: Poaceae / Gramineae

Genus: *Paspalum* L.

Species: *Paspalum scrobiculatum* L.

Fig 1 *Paspalum scrobiculatum* L. (*Source*: USDA; Natural Resources Conservation Services)

Starch fractions for kodo millet shows mostly of polygonal appearance rarely of small spherical structures and possess molecular size in the range of 1.2 to 9.5μm. Scanning electron microscopy showed some indentations on to the starch granules and is due to the protein. Gelatinization temperatures showed by kodo millet starch molecules is 79.3°C (Kumari and Thayumanavan, 1998). It is stated in an investigation lead by Chandrasekara and Shahidi (2011) for kodo millet and reported that hull has marked efficiency to show total phenolic content 3 times high than that of whole grains with similar trend in its antioxidant potential in given order of hull > whole grain > dehulled grain. Soaking (96h) pronounced its better well efficiency to the extent that kodo millet extract retard the cell proliferation to 100 percent.

Nutritional Attributes

Kodo millet owe to the ease of access of complex carbohydrates, influence its consumers positively from being constipation, diabetic, obese and cardiovascular diseases. It is due to the majority of carbohydrates present as dietary fiber and is highest among (38%) major cereals (Sharma *et al.*, 2018). β-*glucan* is the major soluble dietary fiber present which impart all its favorable responses to the wellness of health. High polyphenol content (360mg/g) evidenced on its extraction with acetone somehow govern the high antioxidant capacity of the kodo millet grains (44%) and is far ahead in comparison with staple cereals i.e. wheat (30%), rice (24%) and barley (40%) (Sharma *et al.*, 2017). Both β-*glucan* (soluble fibre) and polyphenols are the constituents that maintain life long healthy capacity of our body and its parts. These phenolics and β-glucan*s* are the inhabitant of bran layers and endospermic cell walls respectively. Notable characteristics for both the constituents varied based on preliminary pre-processing of the grains, thus exerts meaningful effect on to their synthesis which in turn play antioxidative effects too.

Table 1: Comparative nutritive chart for staple cereal and kodo millet (per 100 g)

Grain / Nutrient	Kodo millet	Rice	Maize	Wheat
Energy (Kcal)	309	345	342	346
Protein (g)	8.3	6.8	11.1	12.1
Fat (g)	1.4	0.4	3.6	1.7
Fiber (g)	9	0.2	2.7	1.2
Calcium (mg)	27	10	10	48
Iron (mg)	0.5	3.2	2.3	4.9
Zinc (mg)	0.7	1.4	2.8	2.2
Thiamin (mg)	0.33	0.06	0.1	0.17
Riboflavin (mg)	0.09	0.06	0.1	0.17
Folacin (mg)	23.1	8	20	36.6

Source: Chapke and Tonapi., 2016

Today's preferred demand of functional food relies much on the selection of fibrous foods and foods with enough polyphenolic attributes by the consumers. Fortunately, highest dietary fibre has been reported in little millet (38%) and kodo millet (37%), respectively with the help of which, it maintains body glycemic index to be low thus, optimum to be used by diabetics (Saleh and Zhang, 2013).

Folic acid also known as vitamin B_9 is an essential micronutrient needed for brain development and DNA replication. Generally, females counteract its scarcity as of natural process occurring in them during menstrual blood loss owe to graffian follicle rupturing thus its recovery is questionable till the attainment of menopause. Furthermore, no one of human cell has the capacity to synthesize its own and thus it is an essential factor to be outsource for healthily body functioning. Few of microflora through fermentation is the major natural mode of its synthesis that can be achieved through various blend development in conjugation with probiotic strains (bacterial strains) in cultured dairy foods or drinks and alcoholic fermentations through yeasts etc. (Issoufou Amadou *et al.*, 2019). Kodo millet undergo primary and secondary transformations while in soaking cum germination and fermentation hours respectively that glimpses in its starch profiling owe to enzymatic cum microbial performance.

When talking of glycemic index status, experimentation also revealed that expected glycemic index (eGI) of cooked kodo millet starch was significantly lower than that of cooked rice flour even after refinement of kodo millet through decortications, protein and lipid fraction separation etc and hence advised to incorporate in routine diet to be healthy.

It is recognized that consumption of debranned kodo millet in the form of porridge is much of advantageous when illustrate competitively in respect to protein and lipid exhausted samples of kodo millet against rice and wheat. This phenomenon is better explained and follows the significant rise in glycemic index for debranned kodo millet porridge (42.5%) < protein fraction removed kodo millet porridge < lipid fraction removed kodo millet porridge < highest with removal of both (lipid and protein) fraction in same sample (62.5%) (Annor *et al.*, 2013). Interestingly, it is important to know about kodo millet that it is the grain with least phosphorous content among millets and also deficient in lysine amino acid but it achieves the status to show enormously high free radical quenching capacity of around 70 percent among millets when compared with other millets that falls within maximum free radical quenching capacity of around 15-53 percent. Types of processing and their decline approach to the human through thermal application, decortications and fractionation influences decrease in nutritional attributes in it. On the other hand kodo millet and all other millets in general are the store house to provide sulphur containing essential amino acids (cysteine and methionine) (Deshpande and Mohapatra, 2015).

Processing Innovations

Minimal Pre-processing

Soaking or Steeping

Steeping or soaking, germination or sprouting and pearling or debranning are among those pre-treatments applicable to millets, work to maintain the effective level of all the healthy concerns i.e. nutritious and functional ingredients. Progress of soaking and germination both enhances the total, free and bound polyphenols in the seeds. This is directly influenced with the activation of enzymes that synthesize the energy constituents for the growth of seedling side by side releases polyphenolic percent proportion with the breakage of complex structure of seeds (Sharma *et al.*, 2017).

Germination or Sprouting

Soaking followed by germination as another set of pre-treatment, synthesizes nutritional component so judiciously with subsequent reduction in anti-nutritional strength. Overnight water soaking followed by germination at 35°C for 38h as a separate treatment in lieu of pearling impact foremost in terms of reduction in β-glucan extracts from raw kodo millet sample (6.12%) to germinated samples (5.65%). This reduction is attributed to β-glucanase synthesis during germination stages that feeds on to it and lowers their level. Germination of kodo millet seeds, support the rise in crude protein content (1.24% to 1.65%) attributed to the synthesis of amino acids. Decrease in carbohydrates with the progress of germination is directly associated with raised soluble sugar level during germination stages (Saxena *et al.*, 2018).

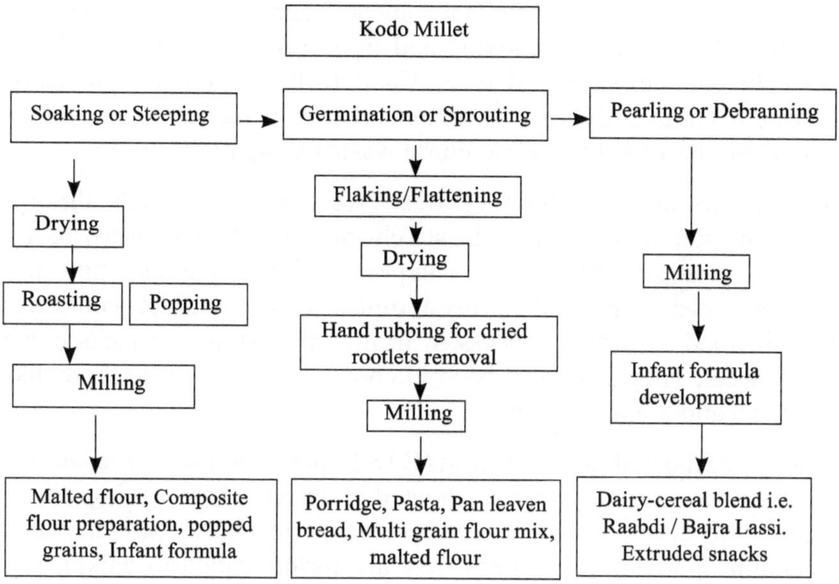

Fig 3: Kodo millet product (food and beverages) tree

Pearling or Debranning

Debranning is one of the considerable pre-processing treatments recognized for minimizing phenolic contents in grain, hence phytate and tannin ceases. In point of fact, anti-nutrients also graded under the class phenolics that are bran concentrated and are removed while pearling hours. Kodo millet is a dark toned grain but with the success of debranning, total phenolic level reduces to about 78 percent which signifies that phenol constituents are concentrated in testa and pericarp that deepens to aleurone layer (Chandrasekara *et al.*, 2012).

Hydrothermal and Dry Heat Treatments

Hydrothermal treatments generally applied through the way of stewing and parboiling. Both act through starch gelatinization, antinutrient minimization due to heat, open up the bonded chemical structures of nutrients for easy access and release of minerals require to maintain appropriate metabolism of body. Hydrothermally treated grains some time forms the basis to determine machinability characteristics, consistency judgment, ease in manual handling through rheological and viscosity parameters. All these stated parameters generally keep in mind while designing new baked products and also for infant food application for easy swallow consistency too. Parboiling has its own significance that is accepted worldwide and is applicable to any class of cereals, Pseudocereals or coarse cereals to carry out the necessary target functioning.

There is vast list of other treatments that somehow apply moistened heat such as baking, blanching, roasting, cooking and frying to lists of dishes enjoyed frequently in lieu of rice, wheat and maize. Catalytic hydrothermal liquefaction of food waste is another aspect of treatments that is of particular interest for energy valorization of food and agricultural waste (Maag *et al.*, 2018).

Flaking is an art of preparation of Ready-To-Eat (RTE) foods through moistened seeds; flaking gelatinizes the starch and works to inactivate *lipases*. There are three modes available that are in practice for flaking operation such as edge runner, roller runner and extrusion runner. Flaked products consumed as multi grain food or further process to frying or roasting to snacks either with salt or sweet. They also have been in choice as convenience foods like murmura, kheel and breakfast cereals.

Popping is one of the traditional dry heat (HTST operation) for the preparation of Ready-to-eat category of products. Popped grain served as evening snacks due to their extraordinary looks acquired by endospermic blast due to moisture and temperature gradient. Sometimes they are further rolled to chikki, bar, laddu and sattu form through sequential steps of seasoning and flavoring.

Fermentation

Fermentation can be preceded through spontaneous means without the addition of microbial strains or through specified culture activated previously. Fermented mass of carbohydrates add another advantage to the host but if preceded through probiotic strains, its healthful advantages become multiple. Probiotic foods help to repopulate the colonic beneficial bacteria that are reduced owe to disease, antibiotic and chemotherapeutic treatments. As per the gathered data, world's food consumers also opted for fermented foods with a prevalent fraction of 20-40 percent from being the total diet (Amadou *et al.*, 2011). An evidence of control of diarrhea among young children in Northern Ghana through fermented millet drink is an interesting intervention (Lei and Jakobsen, 2004). Fermentation can be channelized through controlled open and closed route. Controlled open fermentation of millet slurry (300 g/l) is allowed to ferment naturally for around 48h, dried and milled again to flour. Similarly, controlled close fermentation step forward in a similar manner but for autoclaved flour (adjusted moisture to 25%) substituted with selected probiotic strain (1.064×10^7spores/25 g of flour) for about 48h at 29°C. Both of the treatments account to about 15-37 percent and 20-40 percent reduction in phytic acid and tannins respectively (Singh *et al.*, 2017).

Prebiotic is an indigestible ingredient that stimulates the number and activity of probiotic strains. Oligofructose and inulin is the most deserving category of prebiotics that have plant origin and is associated with millet family. They have the special feature of not be hydrolyzed in upper gastrointestinal tract thus available for probiotic strains as selective feed in colon (Amadou *et al.*, 2019). Kodo millet is sufficiently enriched with these prebiotics and lead to formation of growth factors for long term survival of probiotic strains (Amadou *et al.*, 2013). Kodo millet is the abode of dietary fiber that is about 78.5 percent higher than our staple cereals viz. rice and wheat (Kumar *et al.*, 2018). This one is the only reason to be stated it as low glycemic food and succeeded to be the miracle selection by diabetic and obese patients.

Traditional and Modern Foods and Beverages

Foods

Fermented Foods

One of the major achievements among food innovation is healthy microbes through adaptation with selection of favorable bacterial and fungal strains that have the capability of transformation of various raw and complex substrates (useful or waste) to that of therapeutic health promoting, simplest wellness factors. Progress in this aspect has great potential for food industries to adapt well in any of geographical area as it has vast application scope to the fruits, vegetables, cereals, pseudo-cereals, coarse-cereals and milk as well. Entrepreneurs, that act swiftly and grab this novel initiative through commercial means are the real time players to turn the market towards functional, probiotic and nutraceutical side in alongwith achieved revenue benefits.

Idli is a fermented and steamed product obtained by coarse crushed mixing of rice with splitted black gram. It is consumed as staple diet in morning breakfast by Southern Indian society. Replacement of rice with 20, 40 and 60 percent proportion with kodo millet evaluated sensorially and found that 60 percent suits best in terms of flavor and taste, texture and color (Neelam *et al.*, 2013). Moreover on consumption Kodo millet based (60%) Idli had lower glycaemic index (GI 58.53) in comparison to control Idli (GI 67.11). High crude fiber in kodo millet (8.5 g/100 g) in comparison to white rice (0.2 mg/100 g) may be the possible reason to influence least glycemic response in kodo millet fortified idli instead of rice flour. Addition of this functionality in foods places an alternate for the patients that are in fear of being diabetic to enjoy diet in a habitual pattern.

Dosa is almost similar traditional fermented product described almost likely to Idli. It is a variety under the heading pancake usually prepared and enjoyed in Deccan India but its likeness is now days travel so long and enjoyed by every income group population of every age and most of the race. Dosa generally prepared from dhal of black gram and rice but in dry areas sorghum or millets; usually pearl millet and kodo millet is the maximum exploited grain for dosa preparation.

Nilamadana is prepared through fermentation practices, usually the food fermentation is carried out using some bacterial strains and is rarely seen that fungi is the main culture for the same purpose. Here the kodo millet grain as whole is allowed to ferment in presence of *Penicillium roqueforti; a* fungal strain. Sterile distilled water dissolved strains of *P.roqueforti* in 0.1 percent Tween-20 were inoculated in 30 g of whole kodo millet and inoculated for 5 days at 25°C. Surface growth on kodo millet grain of *p.roqueforti* imparted blue green colour to the fermented grains from where it get its name as "nilamadana". Analytical study for the fermented grains revealed that all the toxic metals reduce to the detection limit of 0.2mg/kg or lower through enzyme activation that feeds on to the toxic constituents. Metals that are metabolic regulators such as calcium, potassium and zinc rose to 42 percent, 27 percent and 15 percent respectively. Key effect of frequent consumption of "Nilamadana" relies in lipase inactivation that is responsible for the hydrolysis of triacylglycerols into glycerol and fatty acids and their absorption in small intestine. In absentia of lipase, triacylglyecerols absorption ceases and is the only way to control obesity. On the other hand, flavanoids reaches more than double their way as in comparison with unfermented grains (Dwivedi *et al.*, 2015).

Adai is another fermented combination of kodo millet, black gram dhal, green gram dhal and parboiled rice soaked in water in a ratio of 4:1:1:1. Beyond this formulation it also suggested from time to time by various researchers through alteration in blend development from some other grains with the main perspective to live healthy. Generally its cultivation is restricted to South Indian states thus it is mostly harvested and exploited in a better way by Deccan society under various food formulation practices. This is the main reason once again i.e. Adai is a traditional south Indian fermented snack flavored with chilli powder, asafoetida, cumin seeds, salt and curry leaves. The batter is cooked on a flat greased heated pan to caramel color and served rolled. Sometime sweet Adai is also constituted through Jaggery addition (Deshpande and Mohapatra, 2015).

Unfermented Foods

Extruded Snacks

In order to achieve functional characteristics kodo millet flour supplemented mass blended within defatted soy flour and water chest nut flour. Obtained extruded snacks through single screw extruded operation were on the liking preferences as judged through sensorial means with maximum crisp and minimum hard characteristics. They behave as meat analogue, rich in essential amino acid score with high bioavailability factors (Mohammad *et al.*, 2016).

The extrudates prepared from the blends of chick pea flour and kodo millet flour with differently treated horse gram flour (soaking, sprouting and pre-conditioning) was investigated (Sharmila and Athmaselvi, 2017). The twin screw extrudates with sprouted horse gram and overnight soaked horse gram flour has soothy taste and flavor thus are on liking preference by the panel but that made with flour blend containing dry untreated horse gram had overall least preference. Preference for the blend selection optimized to the level given as follows; where Horse gram-36 percent, Kodo millet-52 percent and Chick pea-12 percent selected as the best. These nutrient dense and calorie deficient extruded snacks add a new dimension for the small farmers as earning livelihood based on underutilized grains.

Papad prepared here with a combination of kodo millet flour and blackgram flour is the basis to set this mix in equal proportion followed with cumin seeds, salt and sodium bicarbonate to have appropriate sensory and textural likeness. Kodo millet flour previously blended with chilli powder and the whole blend is mixed with water so that stiff dough is made to roll out in circular shape and dried. Finally, flakes were spread on to the dry greased plate bed and sundried (Kamat and Yenagi, 2012).

Muruku is a popular snack dish consumed since long in Southern India, it is also known as chakli in other regions. Here, Kodo millet flour is the base ingredient to which chilli powder, sesame seeds, cumin seeds, asafoetida, butter and salt was combined appropriately and mixed to turn it to thick dough with the help of water mixing. Extrusion through hand extruder followed with deep frying in hot oil till it turns golden brown is the last course of action. Muruku is a long life stable snack that has the designation of consumption during evening tea refreshment and as enjoyment while traveling, picnic and celebrations.

Kolukattai is an Indian traditional salty and sweet snack segment available in Southern India. Salty kolukattai prepared from kodo millet flour by mixing it with water to make paste. Paste is seasoned using chopped onion, chopped green chillies, coriander leaves, asafoetida, mustard seeds and blackgram dhal and boiled with continuous stirring to make dough of thick consistency. Dough mold to oval shape and steamed using specialized perforated double or triple partitioned utensils for steam ease. Similarly, Sweet kolukattai is prepared where dough filled with a mixture of jagerry, shredded coconut and cardamom powder and then steam cooked (Deshpande *et al.*, 2015).

Pan leavened bread is sometime a part of routine practice of blend preparation by Indian residents through blending of kodo millet flour within wheat flour as basic feedstuff for the preparation of pan leavened bread by some families to have multi grain flour in winters. They feel it to be a good practice to have millet based diets in winters that help to stay fit and fine throughout the year (Amadou *et al*, 2019).

Porridge is a successful effort that developed with the use of kodo millet (decorticated, protein removed, lipid removed and protein-lipid removed in same sample), initiated by Annor and associated coworkers (2013). Moreover, analytical characterization direct its frequent consumption in the form of treated porridge samples and advocates healthful consumption as it influences positively in maintenance of glycemic index to least level and sluggish *in-vitro* starch digestibility that can't be done through rice and wheat porridge.

Composite Foods

Composite flour blend is an incorporation of multiple flour through varied proportion has their own significance if applied onto the ranges of appropriate consistency behavior when working on it and also for the desired nutritional target. Kodo and barnyard millet flour is applied in equal proportion with wheat flour (30% level) as best and defatted soy flour (10%). Analytical quantification for physical, chemical, functional and nutritional parameters decision revealed that the optimized blend has long shelf life only because of exceptional antioxidant rich profile of kodo and barnyard millet. It is generally advised that for chapatti making that WAC (Water absorption capacity) of 68 percent is suited best, chapatti flour with less WAC of 60 percent or less are otherwise hard and stiff (Vijayakumar and Mohankumar, 2009).

Ogi *is a* blended porridge of coarse cereals (kodo millet) and cereals (maize) produced traditionally by Africans. Fermented batter of mixed cereal millet flour used to feed childrens and adults as weaning and geriatric diet respectively. Sometimes, this fermented mass is dried and crushed to powder for storage ease that enjoy as breakfast cereal. Interestingly, it swells on hydration and cooking hence impact good to consumer with enhanced mouthfeel and palatibilty effect. Fermentation gives an extra advantage for its particular selection by childrens and aged people that shifted complex substances in to simpler components with easy in access by digestive enzymes of body thus assimilate needed energy (Vernta *et al.*, 2007).

Thatuvadai *is a* traditionally prepared Deccan Indian snack blended in kodo millet flour in alongwith small quantity of roasted Bengal gram dhal. The prepared dough of little thick consistency is adjusted for their taste and flavor by inclusion of dried powder of salt and chilli addition of curry leaves in which butter improves the flavor significantly. Finally, the dough is rolled out on polythene sheet for drying and then deep fried in hot oil (Deshpande and Mohapatra, 2015).

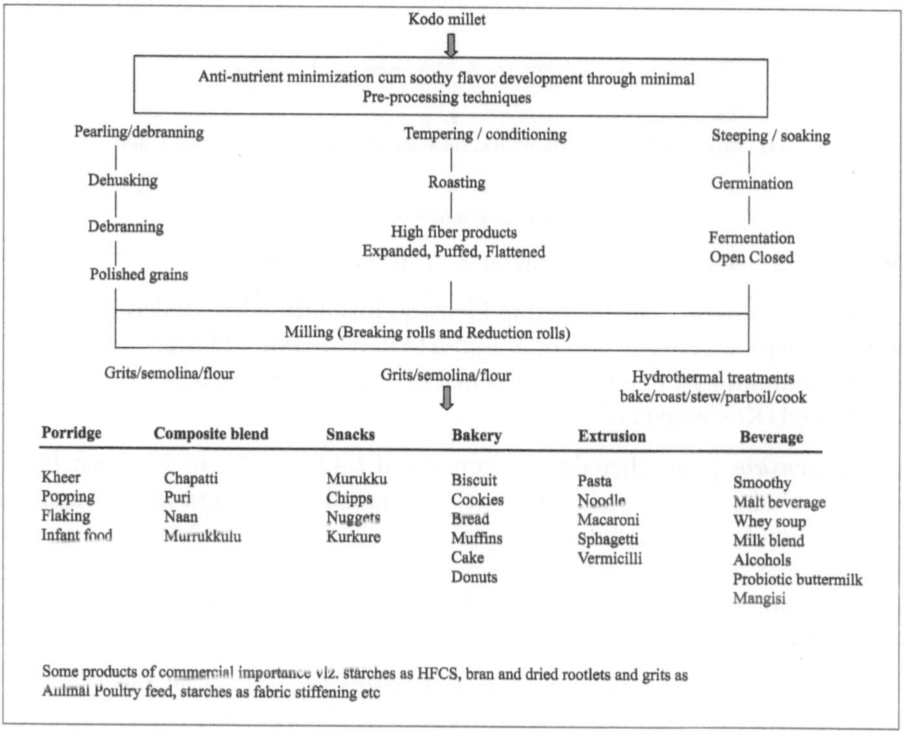

Fig 3: Kodo millet product (food and beverages) tree

Dairy Based Blends

Instant kheer mix and instant kutki mix are the Indian traditional milk-cereal composite blended dessert utilized by northern Indian community. It is the symbol of joy and is utilized during festival, celebrations, sweet-dish after meal and also in some religious ceremonies. North Indian state dairy co-coperative societies took an initiative of kheer production from last many years in Punjab Haryana and Chandigarh, which are served chilled. Sensory selection for this product emphasized researchers to develop instant kheer and kutki mix based on kodo millet so that enjoyment is boosted with today's technological upliftment of time management for working families.

Water soaking of grains i.e. kodo millet (2 hrs)

↓

Drying for 4- 5 hrs (8-10% moisture)

↓

Crush them to grits using electrical grinder

↓

Sautéing of grits for 5-10 minute (with 5% fat)

↓

Mixing of soatening grits with cane sugar powder and dairy whitener

↓

Packing and storage

Fig 4: Process flow diagram for the preparation of instant kheer mix

These instant mixes transformed to kheer so instantly when boiled with water or milk in a ratio of about 600ml of liquid to that of 100g of instant mix of kodo millet. Continuous stirring is maintained till the desired thickened consistency is achieved (Kashyap *et al.*, 2018).

Kodo payasam is another dessert consumed traditionally in southern India, it is like a substitute to kheer similarly consumed in northern India. Payasam consumption is also related to the symbol of celebration. Its basic preparatory steps involve seeds soaking and germination exposure that raises its nutrients bio-availability features in alongwith development of soothy taste among prepared stuffs (Rao, 2017).

Cook together kodo millet, saffron and milk on slow heat until soften to mashing

↓

Add sugar and stir gently to cook payasam

↓

Sidewise roast dry fruits in ghee till the appearance of golden colour

↓

Mixing of roasted dry fruits to the cooked payasam

↓

It can be served hot or cold

Fig 5: Process flow diagram for the preparation of kodo millet payasam

Beverages

Alcoholic Beverages

Boza, recognition of origin of beer is directed to boza (Arici and Daglioglu, 2018). It comes into existence around 8000-9000 years back in Islamic countries. It is an alcoholic drink that has some medicinal properties. Generally, any of millet variety can be utilized for its production. Now times, South Africa republic has Bouza or Boza among the important beverage of industrial produce. Crushed, millet grains are boiled with water and added sugar; allowed to ferment with lactic acid bacteria (LAB). Based on sweet and sour category and achieved LAB acidity are the real time drink with medicinal benefits. Being on fermentation side, health benefits is understood to all, low glycaemic index, gluten free grain and repopulation of healthy bacteria in GI tract has some other pronounced effect for wellness of life.

Mangisi is traditional alcoholic drink of African rural inhabitants that used to enjoy it during marriages and other cheer hours. It is prepared by spontaneous fermentation process, where malted kodo millet grains were mashed and milled followed with dilution in water. This water boiled and strained mass is now allowed to cool and rest for few days for spontaneous fermentation on its own. Sometime but rarely malt added adventitiously from previous batch with main perspective of efficient fermentation to raise alcohol percentage. Concluding, the produce comes-up with high alcohol value that the local inhabitant demands (Taylor and Duodu, 2019).

Malting is another category of early processing of millets that achieve high nutritional status these days that is likely to hit target in commercial importance for the production of various supplementary, complimentary, weaning foods and brewing. Malted substrate often dried sometimes and evident at the same time as amylase rich food because of high amylase activity occurred during imbibitions hours. No doubt among millets, finger millet is most frequently employed for this action but kodo millet has almost matching properties but is rarely employed for the production of green and caramelized malt. Later processes shift this mass into alcoholic liquid as beer through ladder approaches of wort preparation, fermentation and maturation.

Non-alcoholic Beverages

Dhanyamla is a liquid combination of water, rice, pulses and kodarva (*Paspalum scrobiculatum*) that has significant contribution in Ayurveda. Present combination is allowed to mild heat treatment and substituting for fermentation naturally. It is purgative and hot in its potency. Its healthily effects on the body can be observed in terms of relieves fatigue, cure pain of urinary bladder, increases appetite and hunger, aggravate pitta and mitigate vata and kapha. Furthermore in ayurveda, it is practiced as bodily application to allow toxin eradication from the body through sweating and relaxed in arthritic problems (Sekar, 2007)

Antinutrients

Antinutrients are the plants secondary metabolites, intended for their protection purpose. In actual they are polyphenolics that possess antioxidative, anticarcinogenic and antibacterial properties. Due to their complex cage like structure they chelate some of divalent cations with them and also hinder protein digestion as a result of which stunted and malnourished characters appears in consumers. Millets are much prominent with phytates and tannins that are generally bran centric and just wise attention through home scale pre-processing operations viz. soaking, germination, fermentation, debranning and heating retards their effect hence bioavailability of all the available nutrients rises. As per the thorough investigation lead by Taylor and Emmambux (2008) no antinutrient among kodo millet has been reported.

Phytates

Phytates or phytic acid (*myo*-inositol 1, 2, 3, 4, 5, 6, hexakisphosphate) is the primary source of inositol and storage phosphorus in plant seeds contributing ~70 percent of total phosphorus. Monogastric animals are reluctant to quench phosphorous from these seeds as they are lacking in dephosphorylation enzyme i.e. endogenous *phytases;* furthermore, these sequestrate other essential divalent cation (e.g. Ca^{2+}, Zn^{2+}, Mn^{2+}, Mg^{2+}, $Fe^{2+/3+}$), proteins and amino acids too with them (Singh and Kumar, 2018).

As of associated trouble, phytate prominent feeds supplemented with inorganic phosphate causes increase in faecal phosphate level and subsequent eutrophication of waterways and hence environmental issues. Pre-processing home scale treatments are the only cheapest mode to enjoy millet seeds fully with enhanced soothy flavor characteristics. As evidenced, phytates goes on decreasing with the advancement of germination time mainly because of hydrolytic activity of phytase that increases during germination, reduces phytate phosphorus into inositol monophosphate (Kumar *et al.*, 2018).

Goitrogens

As like millets in common, one more anti-nutritional factor that it contains is goitrogen. General perception about goitrogens is that moderate consumption do not influence to human body about its side effects but excessive consumption as like staple cereal causes interference with thyroid hormone production and ceases its level; all due to goitrogen (Leder *et al.*, 2004). If thyroid level diminishes its carrying capacity to iodine also decreases. In nutshell deficiency of iodine is a significant health problem which leads to development of enlarged thyroid gland known as goitre. Symptoms of goitre are dry skin, anxiety, depression and slow thinking. Sudan is the consumer of millet as primary energy source, owe to this goitre patients are maximally available in Sudan than anywhere else around the globe. People with thyroid problem thus control on to their millet consumption (Dunn *et al.*, 2002).

Kodo Poisoning

One more threat associated with kodo grains to be poisonous for both cattle and human is the manifestation by fungus. The poisoning of grain is associated with the fungus growth of *Aspergillus flavus*, *A. tamarii* and *Phomopsis paspalli.* They causes a fungal disease *Paspalum ergot* symptomatically appears through nervousness, lack of muscular coordination, staggering gait, spasms depression the severity of which leads to death of animal. Appearance of the same fungal poisoning in humans appears same as like allergic symptoms viz. nausea, vomiting, delirium, depression, intoxication, and unconsciousness. That all happens is due to the formation of a poisonous toxic compound cyclopiazonic acid (Antony *et al.,* 2003).

Health Benefits

Kodo millet and degenerative diseases (Diabetes, obesity, blood cholesterol level and cardiovascular disease and cancer)

India is the capital of diabetics; we are on top for the diabetic count all around the world. Helplessness or compulsion of sugar restricted diet is the only mean to be healthy. It is not applicable to avoid only the refined sugar but even for our basic staple cereals as well because they possess high glycemic index, least fiber value and many other factors also. It is also stated and proved through research studies by Rajasekaran and Chandra (2005) observed that animals consuming a feed with 55 percent kodo millet meal resulted in a reduction of 42 percent hyperglycemia, 27 percent of cholesterol and increased levels of enzymatic antioxidants (Glutathione, vitamin E and C) and non-enzymatic (glutathione reductase).

Kodo Millet and Celiac

Celiac is a genetically determined chronic inflammation to the intestine triggers on ingestion of wheat and interbreed species such as rye, barley, oat or triticale. Similar observations were also seen in case of gluten intolerant and wheat allergy but are not as similar as like celiac. Celiac patient has to strict to gluten free diet throughout its life as remedy to cure (Jnawali *et al.*, 2016). This lifelong strict adherence to gluten free diet in country like India where every meal weather for poor or rich, children to aged people and for men and women is revolving around wheat as staple cereal is tedious to manage. Millet as gluten free and competitive prolamine potential with much of nutritive substitutes is the demanding option now days to be fit and fine. It is already evidenced from the gathered data that prevalence of celiac in Indian community is 1 in 4 (Makharia *et al.*, 2011).

Kodo Millet and Ageing

Ageing is the natural phenomenon, advances with time due to cross linking of collagen fibers. Symptoms of which can be predicted easily through skin, eyes, hairs externally and heart efficiency, immunity, tumorous growth of cells and cancer as degenerative symptoms internally. Achievement of healthy state and to cure from these uneasy conditions can be through healthy meal provided with antioxidant constituents. The phenolics, flavanoids and proanthocyanidins are the compounds that possess healthful features in the body and ultimately quench free oxygen radical, target hit the carcinogens thus minimize the destructive consequences to body and masks all the ageing symptoms (externally and internally). Selection of antioxidants is also care taken by keeping in mind just not with the amount of available phenolics but also through extent of hydroxylation of these phenolic rings (Hegde and Chandra, 2005). No doubt, many more characteristics are also taken care of but major attribute is to hydroxyl phenolics rings. It is the only possible reason that much of our cosmetics and pharmaceutical oral doses anyhow boosted with natural antioxidants. That is why dark circle, loosening of skin and depigmentation are its possible side effects.

Table 6: Kodo millet and its role as alternative medicine

Disease	Prevention factor	Reference
Celiac disease	As of gluten free status	Nambiar *et al.*, 2011
Obesity and Diabetes	Control release of glucose and interference of fiber in their digestion is main factor	Bisoi *et al.*, 2012
Cardiovascular disease	Elevated levels of adinopectin benefit heart activity	Devi *et al.*, 2014
GI disturbances	Dietary fiber's water absorption and retention capacity in intestine	Sharma *et al.*, 2017
Micronutrient deficiency	Rich in divalent cation, Vitamin B complex enrich	Gupta *et al.*, 2012
Cancer	Lipid peroxidation ceases due to phenolics in millets	Chandrasekara and Shahidi, 2011
Ageing and Immunity	Antioxidative effect and collagen strengthening	Kaur *et al.*, 2019
Wound healing	*In-vivo* study for dermal effect of finger millet and kodo millet flour	Hegde *et al.*, 2005
Blood pressure	HDL profile, polyphenols and fiber mutually benefit from atherosclerosis	Verma and patel, 2017

Conclusion

In nutshell, Kodo millet utilization restricted only to tribal population is long back dialogue. Recent investigation and available full nutrient potential in them is the real time need to adapt with these wildly grown ignored crops. High fiber, loaded antioxidants and unsaturated fatty acids put them more advantageous side to be with them in ongoing hectic lifestyle of stress, laziness and biased eating habits. No doubt, government has also initiated through various promotion means in the same trend and adapted it for various mid-day meal phenomenon, subsidized channels for rural farmers, rural youth for self-entrepreneurship and even composite new product development are the appreciable initiatives. Processing innovations suggested herein and complimentary development with cereals, pseudocereals and by fermentation tactics give it more healthy vibes to run upon, lifetime. Now it retained its popularity and gaining much of importance even acceptance too by everyone owe to changing lifestyle issues due to advancement of urbanization. Thus, adoption of kodo millet throughout lifetime addresses more longevity and healthy days.

References

Amadou, I. (2019). Millet Based Fermented Beverages Processing. In Fermented Beverages (pp. 433-472). Woodhead Publishing.

Amadou, I., Gounga, M. E., and Le, G. W. (2013). Millets: Nutritional composition, some health benefits and processing-A review. Emirates Journal of Food and Agriculture, 501-508.

Amadou, I., LE, G. W., SHI, Y. H., Gbadamosi, O. S., Kamara, M. T., & Jin, S. U. N. (2011). Optimized lactobacillus plantarum lp6 solid-state fermentation and proteolytic hydrolysis improve some nutritional attributes of soybean protein meal. Journal of Food Biochemistry, 35(6), 1686-1694.

Annor, G. A., Marcone, M., Bertoft, E. and Seetharaman, K. (2013). In vitro starch digestibility and expected glycaemic index of kodo millet (*Paspalum scrobiculatum*) as affected by starch–protein–lipid interactions. Cereal Chemistry, 90(3), 211-217.

Bisoi, P. C., Sahoo, G., Mishra, S. K., Das, C. and Das, K. L. (2012). Hypoglycaemic effects of insoluble fiber rich fraction of different cereals and millets. Journal of Food Processing & Technology, 3(11), 1-11.

Chandrasekara, A. and Shahidi, F. (2011). Bioactivities and antiradical properties of millet grains and hulls. Journal of Agricultural and Food Chemistry, 59(17), 9563-9571.

Chandrasekara, A., Naczk, M. and Shahidi, F. (2012). Effect of processing on the antioxidant activity of millet grains. Food Chemistry, 133(1), 1-9.

Chapke, R. R. and Tonapi, V. A. (2016). Best practices for sorghum cultivation and importance of value-addition.

Deshpande, S. S., Mohapatra, D., Tripathi, M. K. and Sadvatha, R. H. (2015). Kodo millet-nutritional value and utilization in Indian foods. Journal of Grain Processing and Storage, 2(2), 16-23.

Devi, P. B., Vijayabharathi, R., Sathyabama, S., Malleshi, N. G., and Priyadarisini, V. B. (2014). Health benefits of finger millet (*Eleusine coracana* L.) polyphenols and dietary fiber: a review. Journal of Food Science and Technology, 51(6), 1021-1040.

Dwivedi, M., Vasantha, K. Y., Sreerama, Y. N., Haware, D. J., Singh, R. P., and Sattur, A. P. (2015). Nilamadana, new fungal fermented cereal based food. Journal of Functional Foods, 15, 217-224.

Gupta, N., Srivastava, A. K., & Pandey, V. N. (2012). Biodiversity and nutraceutical quality of some Indian millets. Proceedings of the National Academy of Sciences, India Section B: Biological Sciences, 82(2), 265-273.

Hegde, P. S. and Chandra, T. S. (2005). ESR spectroscopic study reveals higher free radical quenching potential in kodo millet (*Paspalum scrobiculatum*) compared to other millets, Food chemistry, 92, 177-182.

Hegde, P. S., Anitha, B. and Chandra, T. S. (2005). *In vivo* effect of whole grain flour of finger millet (*Eleusine coracana*) and kodo millet (*Paspalum scrobiculatum*) on rat dermal wound healing. Indian Journal of Experimental Biology, 43(3), 254-258.

Jnawali, P., Kumar, V., and Tanwar, B. (2016). Celiac disease: Overview and considerations for development of gluten-free foods. Food Science and Human Wellness, 5(4), 169-176.

Kashyap, A., Mehra, M. and Kashyap, Y. (2018). Study the physico-chemical attributes of kheer mix. Development of Food Science and Technology, 2856-2857.

Kaur, P., Purewal, S. S., Sandhu, K. S., Kaur, M., & Salar, R. K. (2019). Millets: A cereal grain with potent antioxidants and health benefits. Journal of Food Measurement and Characterization, 13(1), 793-806.

Kumar, A., Tomer, V., Kaur, A., Kumar, V. and Gupta, K. (2018). Millets: a solution to agrarian and nutritional challenges. Agriculture & Food Security, 7(1), 31.

Kumari, S. K., and Thayumanavan, B. (1998). Characterization of starches of proso, foxtail, barnyard, kodo, and little millets. Plant Foods for Human Nutrition, 53(1), 47-56.

Lei, V., and Jakobsen, M. (2004). Microbiological characterization and probiotic potential of koko and koko sour water, African spontaneously fermented millet porridge and drink. Journal of Applied Microbiology, 96(2), 384-397.

Makharia, G. K., Verma, A. K., Amarchand, R., Bhatnagar, S., Das, P., Goswami, A., and Anand, K. (2011). Prevalence of celiac disease in the northern part of India: a community based study. Journal of gastroenterology and hepatology, 26(5), 894-900.

Michaelraj, P. S. J., and Shanmugam, A. (2013). A study on millets based cultivation and consumption in India. Int J Market Financ Serv Manag Res, 2(4), 49-58.

Mohammad, A. Mohan, S. and Devendra, K.V. (2016). Textural Properties of Kodo (*Paspalum scrobiculatum* L.) based Soy Fortified Ready to Eat Extruded Snacks using Response Surface Methodology. Indian Journal of Science and Technology, 9(4), 1-6.

Nambiar, V. S., Dhaduk, J. J., Sareen, N., Shahu, T., and Desai, R. (2011). Potential functional implications of pearl millet (*Pennisetum glaucum*) in health and disease. Journal of Applied Pharmaceutical Science, 1(10), 62.

Neelam, Y., Kanchan, C., Alka, S., and Alka, G. (2013). Evaluation of hypoglycaemic properties of kodo millet based food products in healthy subjects. IOSR Journal of Pharmacy, 3(2), 14-20.

Plucknett, D. L., Smith, N. J. H., Williams, J. T., and Anishetty, N. M. (1983). Crop germplasm conservation and developing countries. Science, 220(4593), 163-169.

Saleh, A. S., Zhang, Q., Chen, J., and Shen, Q. (2013). Millet grains: nutritional quality, processing, and potential health benefits. Comprehensive Reviews in Food Science and Food Safety, 12(3), 281-295.

Shahidi, F., and Chandrasekara, A. (2013). Millet grain phenolics and their role in disease risk reduction and health promotion: A review. Journal of Functional Foods, 5(2), 570-581.

Sharma, S., Saxena, D. C., and Riar, C. S. (2017). Using combined optimization, GC–MS and analytical technique to analyze the germination effect on phenolics, dietary fibers, minerals and GABA contents of Kodo millet (*Paspalum scrobiculatum*). Food Chemistry, 233,20-28.

Sharma, S., Saxena, D. C., and Riar, C. S. (2018). Changes in the GABA and polyphenols contents of foxtail millet on germination and their relationship with *in vitro* antioxidant activity. Food chemistry, 245, 863-870.

Sharma, S., Saxena, D. C., and Riar, C. S. (2018). Characteristics of β-glucan extracted from raw and germinated foxtail (*Setaria italica*) and kodo (*Paspalum scrobiculatum*) millets. International Journal of Biological Macromolecules, 118, 141-148.

Sharma, S., Sharma, N., Handa, S., and Pathania, S. (2017). Evaluation of health potential of nutritionally enriched Kodo millet (*Eleusine coracana*) grown in Himachal Pradesh, India. Food Chemistry, 214, 162-168.

Sharmila, B. and Athmaselvi, K. A. (2017). Development of Ready to Eat Extruded Snacks from Blend of Under-Utilized Legumes and Millets. Journal of Pharmaceutical Sciences and Research, 9(6), 947.

Singh, A., Gupta, S., Kaur, R., and Gupta, H. R. (2017). Process optimization for anti-nutrient minimization of millets. Asian Journal of Dairy & Food Research, 36(4).

Singh, Chhida. (1999). Modern Techniques of Raising Field Crops. Oxford and IBH Publishing Co. Pvt. Ltd., New Delhi, 158–161.

Taylor, J. R. (2019). Sorghum and Millets: Taxonomy, History, Distribution, and Production. In Sorghum and Millets (pp. 1-21). AACC International Press.

Verma, V., and Patel, S. (2013). Value added products from nutri-cereals: Finger millet (*Eleusine coracana*). Emirates Journal of Food and Agriculture, 169-176.

Vernta, R. C. (2007). Nutri-cereals: value-addition of coarse cereals and millets. Postharvest Management and Value Addition, 272.

Vijayakumar, P. T., and Mohankumar, J. B. (2009). Formulation and characterization of millet flour blend incorporated composite flour. International Journal of Agriculture Sciences, 1(2), 46.

10

Teff (*Eragrostis tef*): Chemistry, Nutritional Attributes, Processing Innovations, Traditional and Modern Foods and Beverages

Pal Murugan Muthaiah, Athulya Dev, Santosh Pal and Janifer Raj Xavier and Anil Dutt Semwal

Defence Food Research Laboratory, Siddarthanagar, Mysore-570011, India

Introduction

Teff (*Eragrostis tef*) popularly known as Williams love grass or super grain or annual bunch grass originated in Ethiopia, South Africa has recently gained popularity with widespread cultivation and consumption in India, USA, South Africa and Eritrea (Arendt and Zannini, 2013). It generates around 500 million USA dollars as income to local farmers of Ethiopia. It is the second most important cash crop (Minten *et al*., 2013). As for the reasons portrayed for the recent fame and increased rate of acceptability for consumption of teff is due to its efficiency against celiac disease where availability of no gluten acts as a major essence for making it an essential functional food for gluten intolerant population. Teff is domesticated from Ethiopia highlands and major staple food for Ethiopian population whereas the ones cultivated and harvested from different regions of Australia, South Africa and USA is mainly used as forage crop for animal feed. The ideal highlands selected for teff cultivations should be middle elevations between 1800 and 2200 m above sea level. There should be adequate rainfall for the high yield of teff after the desired cultivation variety. Teff is a tiny, round and brown colored grain resembling millet. It is the smallest grain in the world. The major way of consumption of teff in Ethiopia in *Injera* with wot (stew). In 2013, teff had the largest share of cultivated area *i.e.,* 28 percent while the next largest being maize (20.3%). Market prize of teff is always huge folds larger than other cereals which defines the limited popularity of teff amongst other cereals. Teff has the largest share of total value of cereal production. The commercial surplus of teff is equal to the total of commercial surplus of all the other cereals altogether as maize, sorghum and wheat.

Origin

One of the known major catastrophically cultivated cereals cultivated known as teff, is also known as annual bunch grass, a species of love grass native to the Horn of Africa which is domesticated from highlands of Ethiopia, South Africa. Teff, belongs to Family: Poaceae, subfamily: Eragrostoidae, Tribe: Eragrosteae, Genus: Eragrostis (Bultosa and Taylor, 2003). It is now marginally grown in Australia, Germany, India, Netherlands, Spain and USA particularly in Idaho, California, Texas and Nevada. Due to its small size, a handful is enough to sow a large area which makes it particularly suited to seminomadic lifestyle. The name teff originated from *teff,* an Amharic word which means "lost". In Ethiopia, it is grown during rainy season (July – November). It is also termed as "emergency crop" as it is planted late in season. Teff germination occurs in 3 to 12 days after sowing. Temperature ranging from 15 to 35°C supports optimum germination while germination does not occur below 10°C. Teff is sown manually on firm and humid soil. Teff is harvested in 2 to 6 months when leaves and stem turn yellow. In Ethiopia, harvest is usually done with sickles during November to January. In 1996, US National Research Council characterized teff as having the "potential to improve nutrition, boost food security, development of rural areas and aid in sustainable land care."

It is a self-pollinated C4 plant which fixes carbon efficiently even in drought and higher temperatures thus making it a catastrophically grown crop. It is an intermediate between tropical and temperate grass. Teff cooks faster than other millets and quinoa due to their tiny size thus reducing fuel consumption. Teff is traditionally consumed in forms of Injera (spongy flatbread)- national dish of Ethiopia, atmit (gruels), unleavened bread (kitta), porridge, traditional alcoholic beverages (tella and arake) in Ethiopia but recent studies imply the trials for fortifying bakery food products with teff to replicate the nutritional component composition of the processed foods such as pasta, bread, gel like food even other western food products are also included in list. In family of cereals teff is one of the smallest known grains with an average length of 1 mm and weighing average weight of 1000 kernels of 12 teff varieties are 0.264 g. Mostly it is small and uniform in size. Color of teff varies from milky white, light brown and dark brown. Teff grows well with adequate amount of rainfall. The Ethiopian varieties of teff are mainly segregated into three varieties as Red (Quey), White (Nech) and mixed (Sergegna).

Fig. 1. Teff Grains (*Source*: http://plant.usda.gov)

The white variety is further diversified into main two variants as very white (Magna) and white (Nech) which superficially is according to the wholesalers where they have least distinction as per personal perception. This diversification is imprecise and subjective which can be confused. The white variety is expensive growing in Ethiopian highlands whereas red variety on the other hand is slowly gaining a huge popularity due to the high nutritional value which today the health consciousness of the people are more attracting towards such cereal grains. The quality of teff is judged by factors as physical appearance, impurity level, aroma, texture as well as nutritional quality. The value of commercial surplus of teff is equal in summation of the commercial value of altogether wheat, maize and sorghum. Teff is priced twice that of maize. The white teff is preferred more than red teff as the price of white teff is more than red teff which gives farmers more incentive to grow white teff than red teff. The white teff has more shelf life as compared to red teff.

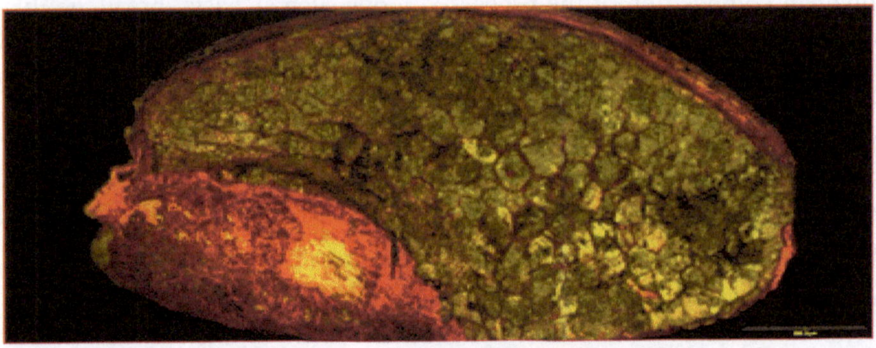

Fig 2: Teff grain: Longitudinal section with germ and endosperm (SEM image) (*Source*: Gebremariam *et al* ., 2014)

The "teff improvement project" released teffsa, the first teff variety in Ethiopia during March 2017. Further development areas to be carved on further:

a) Improving productivity of teff
b) Overcoming lodging malady as caused by the nitrogen and phosphorous in fertilizer
c) Developing climate smart and appropriate crop and soil management options
d) Developing tolerance to abiotic stress such as drought and soil acidity
e) Developing suitable pre- and post- harvest practices suitable for small and commercial farmer
f) Development of recipes and value added products using teff
g) Developing crop protection measures against diseases, insect pests and weeds
h) Improving or strengthening socio-economic condition through regular agricultural extension services

Nutritional Attributes

With the strong evident statistical data given in Ethiopian teff production and consumption indices it is witnessed that recent interest of people towards ample nutritionally enriched food products that can be used to combat gluten intolerance, celiac disease as well as mal absorption resulting as a result of celiac disease. Teff is a complete package of protein, carbohydrate as well as essential amino acids along with essential vitamins and minerals required by the body. According to the National Research of USA, teff can replenish our body with required amount of vitamins, minerals and essential amino acids. The amino acid of teff is equal to the protein content of egg but the lysine concentration of teff is less than that of egg and wheat. Teff consist of complex carbohydrate with high amount of slowly digestible starch. Teff has high amount of essential fatty acids, fiber, minerals as Ca and Fe and phytochemicals as polyphenols and phytates.

Table 1: Nutritional composition of teff

Nutritional components	White teff	Brown Teff
Moisture (%)	10.4	11.1
Protein (%)	11.1	10.5
Fat (%)	2.4	2.7
Carbohydrate (%)	73.6	73.1
Ash (%)	2.5	3.1
Crude Fiber	3.0	3.1
Energy (Kcal)	339	336

Carbohydrates

According to molecular size and degree of polymerization carbohydrates are classified into different classes as sugars, oligosaccharides, starch (amylose and amylopectin) and non-starch polysaccharide. In teff grain, the complex carbohydrate is known to be 80 percent, starch is 73 percent and as analyzed in 13 varieties, 20-26 percent is starch which is similar to sorghum.

Fig 3: Individual teff starch granules, where pg = polygonal granules

The rate of carbohydrates digested and absorbed by small intestine indicates the glycemic index which affects blood glucose level of the body. Due to rapid digestion and absorption of carbohydrates there is increase in level of blood glucose in a short period of time as well as increased metabolic perturbation which leads to type-2 diabetes and cardiovascular diseases. As stated, slowly digested carbohydrates are preferred over fast digesting carbohydrates.

Glycemic index measures how carbohydrate containing food can raise the blood glucose level (Fardetet *et al.*, 2006). Susceptibility of starch to form α-amylose, protein and its lipid content determines structure, encapsulation, crystal structure, degree of gelatinization, portion of damaged granules and starch granule retrogradation as well as glycemic index (Bultosa and Taylor, 2003). The structural analysis of starch granules by SEM (Scanning Electron Microscope) revealed that the teff granule to be size of 2-6 um, wheat 20-35 um, sorghum 20 um and maize 20 um. The least size can be observed in teff grain which is due to less surface area which can increase enzymatic attack which can decrease the *in vitro* starch digestibility. The glycemic index of teff is 74, white wheat 100, sorghum 72 and oats 71. "Hagberg Falling number", abbreviated to HFN of a grain or ground grain is determined according to Hagberg method which gives the measure of the alpha-amylase enzyme activity. Alpha-amylase degrades the starch to sugars(maltose and glucose). The falling number obtained relates to the amount of undigested sugar in starch. The traditional teff flour has falling number less than 250 gives an attractive product.

Protein

Teff contribute about two third of the daily protein requirement of Ethiopian population (Satheesh and Fanta, 2018). Crude protein content of teff varies from 8-11 percent and studies conducted on teffs fractional protein composition revealed that glutelin (45%), albumin (37%) and prolamins (12%) are major protein fractions

Table 2: Amino acid profile of Teff grain

Amino Acid	g/100g
Glutamic acid	3.349
Leucine	1.068
Aspartic acid	0.82
Alanine	0.747
Phenylalanine	0.698
Valine	0.686
Proline	0.664
Serine	0.622
Arginine	0.517
Threonine	0.51
Isoleucine	0.501
Glycine	0.477
Tyrosine	0.458
Methionine	0.428

Amino Acid	g/100g
Lysine	0.376
Histidine	0.301
Cystine	0.236
Tryptophan	0.139

Adebowale *et al.*(2011) reported that prolamins are one of the major protein constituents in teff . The amino acid profile as reported in Table 2 showed that glutamic acid, leucine, aspartic acid, alanine, phenylalanine, valine, proline, serine, arginine, threonine, isoleucine are the major amino acid in teff . Higher concentration of lysine is reported in teff while other amino acid such as glycine, tyrosine, methionine, lysine, histidine, cystine and tryptophan were also found in considerable quantities in teff. Teff is evident to be gluten free (Hopman *et al.*, 2008). Gluten is analyzed by detecting the presence of T-cell stimulatory epitopes which are seen to be absent which confirms no gluten. Teff is thus proven to be the best valuable ingredient for functional food destined for celiac patients who are gluten intolerant.

Fat

Cereals are not the best source of fat, consumption in high quantities can provide necessary fatty acids (Michaelsen *et al.*, 2011, Simoponlos *et al.*, 2011). These fatty acids help in development of growth and long term health. For instance, increased intake of ω-3 fatty acid (α-linolenic acid) which can decrease the biological markers of cardiovascular diseases and cancer, it can also combat inflammatory and auto-immuno diseases. The crude fat content of teff is less of maize and sorghum but is higher than wheat and rice (El-alfy *et al.*, 2012) Rice, sorghum and wheat have negligible amount of linoleic acid and α-linoleic acid and decortification of refining futher reduce the conccentration. But teff which is consumed as whole grain can lead to the least reduction of crude fat also increases the amount of polyunsaturated fatty acids as compared to other cereal grains. According to the Codex Standard (infant formula) the ratio of linoleic acid and α-linolenic acid is 5:15 for infants and analyzed concentration of linoleic acid: α-linolenic acid is 9:1 which can make teff to be considered favorable and good source of fatty acid as soybean.

Fiber

American Association of Cereal Chemists (AACC) defines dietary fiber as edible parts of or analogous carbohydrates resistant to digestion and absorption in human small intestine with complete or partial breakdown as well as absorption in large intestine with proven physiological effects to benefit health" (Deveries *et al.,* 2003; Cumminget *et al.,* 2009). The physiological effects may include (Champ *et al.*, 2003) faecal bulking (laxation), decreased glucose level or the decreased plasma LDL cholesterol. The crude fiber, dietary fiber (total and soluble) of teff is higher than other cereal grains such as as wheat, rice, sorghum and maize with the reason being the consumption without decortication and refining, due to smaller grain including bran and thus more fiber.

Minerals

The red teff has more concentration of iron and calcium than mixed or white teff grains whereas studies indicated the higher concentration of copper in white teff than red or mixed teff varieties (Abebe *et al.,* 2007; Mengesha *et al.,* 1966). Teff has higher concentration of iron, calcium, copper and zinc in comparison to sorghum and wheat (Hallberg and Rasmussen, 1981). When dilute hydrochloric acid is used to wash grains than deionized water, teff grain's iron content decreases from 39.7 mg/100gm to 3.5 mg/100gm. It leads to a huge impact and loss of acid labile intrinsic iron (Areda *et al.,* 1993). These studies could show that the mineral content of teff washed with dilute hydrochloric acid shows 50 percent less iron content than the ones washed with deionized water (Mangesha *et al.,* 1966). Washing grains decreased iron content and mineral contamination may be attributed to their smaller size and higher exposure to soil. When the traditional and manual methods of threshing of teff grains could be analyzed, it could be concluded that 30-38 percent increased amount of iron could be seen in traditionally threshed teff grains due to soil contamination. In manually threshed grains, the iron content is 16 mg/100gm which is still higher concentration compared to some other commonly consumed cereals.

Phytochemicals

The vast bulk of mineral absorption occurs in the small intestine. Bioavailability of minerals highly depends on host and dietary factors. Among dietary factors the phytochemicals can hinder the absorption of minerals in the body that constitutes major mineral absorption inhibitors called anti-nutritional factors as shown by effects including anti-diabetic, anti-cancers and antioxidative made antinutritional obscelete (Hurrell and Egli, 2010).

Phytates

These phytochemicals are commonly present in legumes and cereals. It is a primary form of phosphorous storage in seed which is around 60-90 percent concentration which in dry weight of cereal constitutes 1.5%. (Schlemmer *et al.,* 2009).

Polyphenols

Polyphenols are the secondary metabolites of plants that have defense mechanism against pathogens and ultraviolet radiations. These can protect the cell constituent against oxidative stress. (Monach *et al.,* 2004) The total phenolic content of cereals flour estimated by Folin-Ciocalten's method revealed the presence of the highest amount of GAE (mg/100gm) in red sorghum flour which is 1607 mg/100g GAE, barley 310mg/100g GAE, wheat 143 mg/100g GAE and teff has 140 mg/100g GAE (Hurrelli and Egli, 2010) Polyphenols can hamper iron absorption from plant based food. Polyphenols can form complexes with iron if they possess functional groups of catechol (orthodihydroxy benzene) and galloyl (trihydroxybenzene). Teff has a rich amount of phenolic acid mainly constitutes ferulic acid, vanillic, coumaric and cinnamic but the galloyl and catechol is negligible in amount so it does not affect iron bioavailability. It is also reported that wheat products enrichment with 30 percent teff could increase antioxidant activity from 1.4 to 2.4mM trolox equivalent anti-oxidant capacity (TEAC)/100g (Alaunyte *et al.,* 2012).

Processing Innovations

Gluten or similar compounds such as hordein in barley and secalin in rye from flour of wheat, barley, rye, oat and spelt are not suitable for a large number of people, babies in the first months of their lives. Hypersensitivity to gluten results in celiac disease. Rice, corn, tapioca, soy, buckwheat, arrowroot, potatoes and chestnuts are known crops which yield gluten free flour, this can be utilized for the production for various food products. Another source for gluten-free flour is *Eragrostis tef* (also called teff). For making gluten free product as teff flour it involves the consequent processing of teff seeds.

Teff Seeds

The processing methodologies involves in the milling of teff seed to flour includes harvesting, drying, transport, storage, grinding, mixing and packaging. Adequate amount of precautions need to be taken to prevent any kind of mixing of teff flour with non-gluten free crop or seed or flour. The equipment are also taken care that they are not utilized prior for any kind of non-gluten free products.

Harvesting

Teff is cultivated during rainy season of a year between July to November as teff needs adequate rainfall. According to the desired varieties grown the final stage of grains maturity varies. These seeds have the closest resemblance to millets as being the smallest grain with an average of approximate length of 1 mm. Colour varies from ivory or very white to dark brown or black.

Drying

The consequent stage after harvesting of teff is drying which is conducted to prevent any decay that can extend its shelf life. The grain with final moisture content of 12 percent is suitable for storage. Drying is preceding stage of storage which is carried on for few days after harvest before storage.

Transport

The transportation and handling of teff grains is the biggest challenge due to constraints of size being small which makes a huge challenge to protect the grains from any foreign materials which can impact gradually on quality of grains.

Storage

Falling number of teff flour plays a major role in baking and teff flour with falling number more than 250 is an excellent material for making bakery products (Roosjen, 2007). Harvesting the teff before the physiological maturity / proper ripening resulted in low level of falling number. Harvest of grains after ripening increases the falling number (Hidetoshi, 2001; Lukow *et al.,* 1995). During after-ripening of teff grain in cold areas the falling number go from an average of 230 immediately after harvesting to 260 after four to five weeks and make elevate to 330 within two to three months after harvesting. In warm areas the after- ripening effect is different, as starting at 230 which is immediately after harvesting and later may increase to 420. Teff flour with falling number in the range of 400 - 550 results in a dough or batter with high baking qualities can be used for manufacturing wide range of baked food products such as bread, pastry, cookies, pizza, pasta, noodles, etc. (Roosjen, 2007).

Grinding

This involves the great concern of protection of teff from any kind of foreign materials. The huge possibility of contamination may arise from the chances of cross contamination. The grinding is not carried out that fine as more fine the grain size is faster is the baking rate. Flour should be fine enough to pass sieve with pore seize of at most 150μ and at least 100 microns as sated according to the standard protocols. Usually the milling of teff grains is done using pin mil with integrated cooling is used, so that the flour is not burnt during grinding. Teff flour are related to the fact that due to the fine grinding, a relatively large surface is available for the absorption of water or a different liquid used for the preparation of a dough after grinding.

Traditional and Modern Food and Beverages

Teff bread

The gold standard baking test of teff bread was carried out as follows, where the teff flour was finely ground in a pin mill until minimally 70 percent of teff flour passed a sieve with a pore size of 100 microns.

Method of preparation: Dry ingredients are mixed thoroughly and combined with water and yeast and rested for 10-15 min. Batter made from the mixture either by manual beating or machine. The batter has been beaten for two minutes in lowest acceleration. The beaten batter scooped into two cake tins of 450 g and allowed to raise till edge of cake tin. After raising, it is baked in oven at 235 °C for 20 minutes approximately. After baking the bread was cooled and ready for consumption.

Table 3: Formulation of teff bread

S.No.	Ingredients	Weight percentage (%)
1.	Teff flour	100
2.	Citric acid	0.2
3.	Egg white powder	4.5
4.	Water	110
5.	Yeast	6

Fig 4: Teff Bread

Teff beer

Beer is commonly produced from barley after malting should not be consumed by celiac suffering patients. Beer prepared with gluten free ingredients can be consumed by gluten intolerant population. This type of food must not contain a particle of any material derived from wheat, barley or rye (FAO/WHO,1994). Celiac disease is a condition of the gastrointestinal tract due to inflammation caused by gluten present in cereals such as wheat, barley and rye (Green *et al.,* 2005). Beer usually contains appreciable amount of hordein (barley gluten prolamin). Malting and brewing of teff have not been extensively investigated except for the study conducted by Zarnkow *et al.* (2008) on optimization of malting conditions of teff. Optimal conditions of malting include germination for 4 days with 48 percent steeping at a temperature of 24°C (Bultosa *et al.*, 2007; Zarnkow *et al.*, 2008). Studies also indicated that teff is starchy suitable for brewing. The protein content of teff is comparable to wheat which indicates the teff malting causes an increase content of simple sugars and free amino nitrogen (FAN). In developing countries the commercial processing of these locally grown grains in to value added food and beverage product is an important driver for economic development.

Injera

Injera or Biddena is a sourdough raised flatbread with slightly spongy texture, traditionally made out of teff flour. It is the national dish of Ethiopia, Djibouti and Somalia. It is central to the dining process in those cultures as bread is the most fundamental component. Teff flour, warm water, self-rising flour, salt and yeast are used as major ingredients used in the preparation of injera.

Fig 5: Injera – Traditional Ethiopian cuisine

Method of preparation: Injera is prepared by mixing teff flour with water to make dough followed by fermentation and baking. Fermentation is triggered with addition of *ersho*, which is abundtant in aerobic Bacillus species and several yeasts (in order or abundance): *Candida milleri, Rhodotorula mucilaginosa, Kluyveromyces marxianus, Pichia maganishii* and *Debaromyces hansenii*. Ersho is a clear, yellow liquid that accumulates on the surface of teff flour batter fermented for 2-3 days. The starter culture is added in ratio of 1:1.6 (w/v) (Yetneberk *et al.,* 2004; Baye *et al.,* 2014). The fermentation generally continues 2–3 days (depending on weather conditions), after which the dough is thinned into a batter before baking on an open platter. Shape of injera compares with French crepe and Indian dosa which resembles flatbreads cooked in circle and shows the closest resemblance to South Indian *appam*. The bottom surface is smooth while the top is porous, the top helps to easily scoop in sauces and dishes. Injera became more common in United States during a big spike in Ethiopian immigration in 1980s and 1990s, largely because of Refugee Act passed in 1980. Teff is now being produced in United States by Teff Company in Idaho.

Use of teff in gluten free bakery product development

Various studies conducted on the use of teff flour for manufacture of gluten free bakery products are summarized in Table -4. The studies clearly showed that teff flour rich in nutrients can be a good alternative and can also improve the taste and flavor of gluten free bakery products. The studies also highlighted that great potential of combining teff with other flour sourdoughs for preparation of gluten free breads with enhanced sensory properties and nutritional attributes.

Table 4: Studies implying the application of teff in gluten-free bakery products

Product	Study	Conclusion	References
Cookies	Effect of 25 and 50 percent replacement with teff flour on the physico-chemical and sensory properties of gluten free sugar and peanut butter cookies.	Control and 25 percent teff cookies was preferred over the 50 percent in case of sugar cookies. No such preference was observed in case of peanut butter cookies.	Kenney *et al.* (2011)
Muffins	Effect of substitution of rice flour (control) with teff flour at 25 percent, 50 percent, 75 percent and 100 percent on the physical, textural, and sensory characteristics of gluten free muffins.	Demonstrated production of gluten free muffins with higher nutrition (protein (27%), iron (20.95%), calcium (25%) and fiber (2.21%)).	Assefa *et al.* (2011)
Pasta	Determined formulations for the production of pasta using oat, teff flour and egg.	Study showed that fiber and mineral content, with higher level of acceptance	Hager (2012)
Injera	Analyzed the effect of grain teff flour substitution with flaxseed on nutritional and functional properties of injera.	Substitution upto 9 percent flaxseed-with injera showed good proximate, nutritional and energy contents high in dietary fiber, alpha linolenic acid, lignans, proteins and TP of antioxidant nature.	Grima *et al.* (2013)

Product	Study	Conclusion	References
Bread	Produced sourdoughs from buckwheat, oat, quinoa, sorghum, teff and wheat flour using the hetero-fermentative lactic acid bacteria *Lactobacillus plantarum* FST 1.7 and added to a basic bread formulation of flour from the same grain type (20% addition level).	Sourdough addition increased the cell volume significantly in sorghum (+61 %), teff (+92 %) and wheat sourdough breads (+78 %). Increase in crumb porosity was reported.	Wolter *et al.* (2014)
Injera	Evaluated effect of teff variety and their blending with moringa on nutritional and sensory acceptability using two factor factorial design	Mixed teff with 10 percent moringa contained 45.74, 172.70, 5.85 and 8.55 mg/100 g Fe, Ca, Zn and beta-carotene contents respectively which were higher than the control	Tarekegh (2015)
Bread	Studied the influence of the addition of teff flour (5, 10 and 20%) and (buckwheat or rice) based sourdoughs using *Lactobacillus helveticus*) on the sensory quality and consumer preference of gluten-free (GF) breads.	20 percent teff enhanced sensory properties while 10 percent teff improved flavor score of breads made using teff and rice sourdough.	Campo (2016)
Cookies	Developed Teff-oat composites using gluten free teff flour containing essential amino acids with oat products containing beta-glucan	The elastic properties of teff-OBC (oat bran concentrate) and teff-WOF (whole oat flour) doughs were slightly higher than teff dough. Differences were found in shape and textural properties of the doughs and cookies. Overall, the teff-oat cookies were acceptable in color, flavor, and texture.	Inglett *et al.* (2016)

Anti-nutrients

Group of components present in wide variety of plant kingdom that hinders the uptake of nutrients by the body are known as antinutrients. Anti-nutritional factors in food reduce nutritional value of food interfering with digestibility and absorption of nutrients. In case of teff, mostly examined anti-nutritional factors are tannic and phytic acids.

Phytic acids

The capability of phytic acid to form chelating bivalent with minerals as Ca, Fe, Mg, Zn and Cu may even affect starch, protein and enzymes. The food preparation methods as steeping, germination and fermentation can reduce the phytate levels as during these processing methodologies to release endogenous and exogenous phytases which can reduce the phytate level thus reducing the responsible antinutritional factors. Teff grains have a high amount of phytates. Phytate also hinders the absorption of zinc and iron in the body. Any mineral which forms complex when it comes in contact with phytates as phytate-mineral and phytate-mineral peptide complexes in gastrointestinal tract can be reduced in absorption in mineral form which cannot be reciprocated to the free mineral form from the complexes formed (Sandstrom *et al.*, 1997; Manary *et al.*, 2002). Phytate get complexed with zinc secreted endogenously reducing the bioavailability of these minerals. These phytates which can reduce the bioavailability of minerals are referred to as antinutritional factors. The phytates in teff can be reduced with the help of individual or combination of various processing methodologies as soaking, fermentation, germination and various other methods. Phytate can be degraded by endogenous phytases, which can be produced during various processing methods. Even exogenous common enzymes can also reduce the content of phytates in teff grains.

Health Benefits of Phytate Consumption

- It can prevent crystallization inhibitors of calcium salts that can lead to the formation of kidney stone. Thus, phytates can reduce calcium salts from biological fluid.
- Phytates can reduce blood glucose levels and beneficial for diabetic patient.
- The phytates are evident to be anti- cancerous in nature.

The safe limit of phytate intake the human body is still unknown and under research amongst many recently conducted studies.

Tannins

The case of teff study concluded the tannin amount to be 0.111 mg/gm. Condensed tannins bind to protein and forms complexes that inhibit normal digestion in animals. Salivary mucin and histatins readily binds with tannins and precipitate them hindering absorption of proline. Salivary proline rich proteins are sometimes used to inactivate tannins.

Health Benefits

Teff is majorly a seed popularly accepted as a grain with versatile valuable applications. It can implicate any other grains nutritional benefits without any deduction due to decortication and refining. Teff was always one of the staple diets of Ethiopia, South Africa and recently spread to other countries due to the awareness about health benefits of teff as replica of a wholesome grain. Teff is not new but it is spreading beyond boundaries considering health conscious consumers acceptability rate.

Protein: Teff's mother country, Ethiopia as it is famous for its long distance runners. Ethiopians depending on teff as their staple diet get about 2/3rd of their daily consumption of protein from this small sized grain. In 150 g of teff flour consumed by a human body it can attain 25 percent of our daily value for protein. With the rich profile of essential amino acids and lysine present it can as a wholesome act as a natural daily diet for proper nourishment and development of all age group consumers.

Dietary fiber: As teff is consumed without decortications and maximum retention of bran present in the seed which can act as the best source of dietary fiber. Teff being a good source of dietary fiber including resistant starch present in seed and recent reports indicate role of resistant starch in managing blood sugar, weight and colon health. The presence of dietary fiber helps in the least absorption of toxins from the food consumed.

Gluten-free: Analysis of protein profile showed no traces of gluten in the seeds of teff. The gluten intolerance is a wide range problem faced by a huge population including the ones with celiac disease where malabsorption is a major side effect of celiac disease. The teff composition constitutes an ample amount of all varieties of nutrients that can substitute the ones responsible for malabsorption.

Magnesium: Teff shows a positive response amongst the consumers in regards of muscular, nervous and cardiovascular systems. It contains 69 percent of daily value required by the body. Magnesium is a major co factor required by our body which can help us to avoid migraines, heart attacks and diabetes.

Zinc: An ally to an individual's immune system and genetic code. It helps us fight off disease and builds proteins. It can provide 10 percent of daily value of daily intake for human body.

Calcium: Teff is rich in calcium and provides 123 mg of calcium from a single time consumption of teff. The calcium obtained from teff is five times higher than the whole wheat. It is a supporting pillar for healthy growth of teeth and bones.

Vitamin B_6: Teff can access 10 percent daily value of vitamin B_6 required by body which can help to prevent cardiovascular diseases, can boost the immunity, muscular and nervous system function. A supplementary vitamin which has a crucial role in various metabolic processes of an individual is vitamin B_6.

Vitamin C: Most grains do not boost vitamin C but teff is a good source of vitamin C which is crucial for our immune system, body tissues and skeletal system.

Iron: Inadequate iron intake is common among low and middle income nations particularly amongst infants, young children and pregnant women. Food fortification and nutritional supplements are one of the best medicines available and effective strategies to prevent iron deficiencies. By assuming an average daily intake of 200 g of teff enriched bread, it is possible to cover between 42 and 81 percent and 72 and 138 percent of daily intake requirements for iron in women and men respectively. As the iron bioavailability is widely going to vary depending on processing of teff as during injera preparation the fermentation of teff releases phytases which can reduce the level of phytates in seed increasing iron molar ratio. However, Bokhari *et al.*(2012) showed that consumption of 30 percent teff enriched wheat breads can help maintain serum iron levels in pregnant women. Thus, reduction in phytate level can help in increasing bioavailability of iron in teff.

Celiac disease: Worldwide 0.6-1.0 percent of the population is affected by celiac disease (CD). CD is caused by aberrant T-cell responses to glutens and gluten like proteins found in wheat, barley, rye and possibly oats. The symptoms include diarrhoea, abdominal pain and disturbances in nutrient absorption caused by histological alterations of the small bowel. Extra-intestinal complications such as osteoporosis, infertility and cancer have also been reported. The only available diet for those with CD are a strict gluten free diet which may consequently lead to inadequate intake of iron, calcium, folate and vitamin B12 and fiber in with those with CD. Teff contains a good amount of minerals, fiber and phytochemicals.

Diabetes: The global incidence of diabetes is increasing alarmingly and has become a major public health concern. Among macronutrients the type of carbohydrate consumed and its digestibility play a crucial role in glucose levels and thus can prevent the risk of diabetes. Teff contains relatively high amount of dietary fiber in comparison to cereals and decrease fasting blood glucose level and thus contribute in management of diabetes. High levels of phytic acid and polyphenolic content in teff with the ally of anti-oxidative property is likely to prevent diabetes.

Conclusion

Nutritional attributes of teff indicated that it being a staple diet of regions could also replenish the consumer with required amount of nutrients. The starch of teff is slowly digested does not elevate the blood sugar with low level of glycemic index. The amino acid profile shows the amino acid composition to be higher than other cereals composition except for lysine concentration. There is no gluten estimated in teff after analysis which marks it preferable for gluten intolerant people as well as people suffering from celiac disease. The polyunsaturated fats which can easily be digested and broken down with the digestion cycle proceedings are higher in teff than other cereal which retains the essential fatty acids in the body. The linoleic acid and α-linoleic acid ratio of teff is favorable when compared with standards from Codex. It is rich in iron and calcium compared to cereals, it also consists of an ample amount of other minerals. Higher levels of dietary fiber in tef prevents toxic component retention throughout the digestion cycle which can ultimately combat colon cancer.

References

Abebe, W., and Ronda, F. (2014). Rheological and textural properties of tef [*Eragrostis tef* (Zucc.) Trotter] grain flour gels. Journal of Cereal Science, 60(1), 122-130.

Abebe, W., Ronda, F., Villanueva, M., and Collar, C. (2015). Effect of tef [*Eragrostis tef* (Zucc.) Trotter] grain flour addition on viscoelastic properties and stickiness of wheat dough matrices and bread loaf volume. European Food Research and Technology, 241(4), 469-478.

Abebe, Y., Bogale, A., Hambidge, K. M., Stoecker, B. J., Bailey, K., and Gibson, R. S. (2007). Phytate, zinc, iron and calcium content of selected raw and prepared foods consumed in rural Sidama, Southern Ethiopia, and implications for bioavailability. Journal of Food Composition and Analysis, 20(3-4), 161-168.

Adebowale, A. R. A., Emmambux, M. N., Beukes, M., and Taylor, J. R. (2011). Fractionation and characterization of teff proteins. Journal of Cereal Science, 54(3), 380-386.

Adish, A. A., Esrey, S. A., Gyorkos, T. W., and Johns, T. (1999). Risk factors for iron deficiency anaemia in preschool children in northern Ethiopia. Public health nutrition, 2(3), 243-252.

Agren, G., & Gibson, R. (1968). Food Composition Table for Use in Ethiopia I. Ethiopian Nutrition Institute (ENI). Addis Ababa, Ethiopia.

Alaedini, A., and Green, P. H. (2005). Narrative review: celiac disease: understanding a complex autoimmune disorder. Annals of Internal Medicine, 142(4), 289-298.

Alaunyte, I., Stojceska, V., Plunkett, A., Ainsworth, P., & Derbyshire, E. (2012). Improving the quality of nutrient-rich Teff (*Eragrostis tef*) breads by combination of enzymes in straight dough and sourdough breadmaking. Journal of Cereal Science, 55(1), 22-30.

Alvarez-Jubete, L., Arendt, E. K., and Gallagher, E. (2010). Nutritive value of pseudocereals and their increasing use as functional gluten-free ingredients. Trends in Food Science & Technology, 21(2), 106-113.

Ambaw, A. (2013). Determination of iron fractions of laboratory threshed and field threshed teff injera. Addis Ababa University.

Areda, A., Ketema, S., Ingram, J., and Davis, R. H. (1993). The iron content of tef [*Eragrostis tef* (Zucc.) trotter] and its controversy. SINET: An Ethiopian Journal of Science (Ethiopia).

Arendt, E. K., Ryan, L. A., and Dal Bello, F. (2007). Impact of sourdough on the texture of bread. Food Microbiology, 24(2), 165-174.

Arendt-Elke, K., and Zannini, E. (2013). Amaranth, Cereal Grains for the Food and Beverage Industries. A volume in Woodhead Publishing Series in Food Science. Technology and Nutrition, 248, 439-473.

Assefa, K., Yu, J. K., Zeid, M., Belay, G., Tefera, H., Sorrells, M. E. (2011) "Breeding tef [*Eragrostis tef* (Zucc.) trotter]: conventional and molecular approaches," Plant Breed, 130(1). 1-9.

Banthia, N., Bindiganavile, V., Jones, J., and Novak, J. (2012). Fiber-reinforced concrete in precast concrete applications: Research leads to innovative products. PCI journal, 57(3).

Baye, K. (2014). Teff: nutrient composition and health benefits (Vol. 67). Intl Food Policy Res Inst.

Baye, K., Mouquet-Rivier, C., Icard-Vernière, C., Rochette, I., and Guyot, J. P. (2013). Influence of flour blend composition on fermentation kinetics and phytate hydrolysis of sourdough used to make injera. Food Chemistry, 138(1), 430-436.

Bokhari, F, E Derbyshire, W Li, CS Brennan, and V Stojceska. (2012). "A Study to Establish Whether Food Based Approaches Can Improve Serum Iron Levels in Child Bearing Aged Women." Journal of Human Nutrition and Dietetics. 25 (1): 95-100.

Bouzoubaâ, N., and Lachemi, M. (2001). Self-compacting concrete incorporating high volumes of class F fly ash: Preliminary results. Cement and Concrete Research, 31(3), 413-420.

Bultosa Beri, G. (2006). Physico-chemical and functional properties of grain tef [*Eragrostis tef* (Zucc.) Trotter] starch (Doctoral dissertation, University of Pretoria).

Bultosa, G. (2007). Physicochemical characteristics of grain and flour in 13 tef [*Eragrostis tef* (Zucc.) Trotter] grain varieties. Journal of Applied Science and Research, 3(12), 2042–2051

Bultosa, G., and Taylor, J. R. (2003). Chemical and physical characterisation of grain Tef [*Eragrostis tef* (Zucc.) Trotter] starch granule composition. Starch Stärke, 55(7), 304-312.

Campo Eva, Lis del Arco, Leyre Urtasun, Rosa Oria, Ana Ferrer-Mairal (2016) Impact of sourdough on sensory properties and consumers' preference of gluten-free breads enriched with teff flour, Journal of Cereal Science, 67, 75-82, ISSN 0733-5210, https://doi.org/10.1016/j.jcs.2015.09.010.

Champ, M., Langkilde, A. M., Brouns, F., Kettlitz, B., and Collet, Y. L. B. (2003). Advances in dietary fibre characterisation. 1. Definition of dietary fibre, physiological relevance, health benefits and analytical aspects. Nutrition Research Reviews, 16(1), 71-82.

DeVries, J. (2003). On defining dietary fibre. Proceedings of the Nutrition Society, 62(1), 37-43. doi:10.1079/PNS2002234

Dinku, A. (2005). The need for standardization of aggregates for concrete production in Ethiopian construction industry.

do Nascimento, K. D. O., Paes, S. D. N. D., de Oliveira, I. R., Reis, I. P., and Augusta, I. M. (2018). Teff: suitability for different food applications and as a raw material of gluten-free, a literature review. Journal of Food and Nutrition Research, 6(2), 74-81.

Fardet, Anthony, Fanny Leenhardt, Delphine Lioger, Augustin Scalbert, and Christian Rémésy. 2006. "Parameters Controlling the Glycaemic Response to Breads." Nutrition Research Reviews. 19 (01): 18-25.

Gallagher, E., Gormley, T. R., and Arendt, E. K. (2004). Recent advances in the formulation of gluten-free cereal-based products. Trends in Food Science & Technology, 15(3-4), 143-152.

Gebeyehu, D., Phhnfl, P., and Gebeye, D. (n.d.). Ethiopian Environment and Forest in Ethiopia: Highlights on Major Achievements and Contributions Getachew Dc_Jegn Seyount Keiemwork.

Gebremariam, M. M., Zarnkow, M., and Becker, T. (2013). Thermal stability of starch degrading enzymes of teff (*Eragrostis tef*) malt during isothermal mashing. Process Biochemistry, 48(12), 1928-1932.

Gebremariam, M. M., Zarnkow, M., and Becker, T. (2014). Teff (*Eragrostis tef*) as a raw material for malting, brewing and manufacturing of gluten-free foods and beverages: a review. Journal of Food Science and Technology, 51(11), 2881-2895.

Girma, T., Bultosa, G., and Bussa, N. (2013). Effect of grain Tef (Eragrostis tef (Zucc.) Trotter) flour substitutions with flaxseed on mineral content, antioxidant activity, phytic acid content and microbial quality of injera. Science, Technology and Arts Research Journal, 2(3), 51-58.

Green, P. H. (2005). The many faces of celiac disease: clinical presentation of celiac disease in the adult population. Gastroenterology, 128(4), S74-S78.

Hager, A. S., Lauck, F., Zannini, E., & Arendt, E. K. (2012). Development of gluten-free fresh egg pasta based on oat and teff flour. European Food Research and Technology, 235(5), 861-871.

Hallberg, L, and E Björn-Rasmussen. 1981. "Measurement of Iron Absorption from Meals Contaminated with Iron." The American Journal of Clinical Nutrition. 34 (12): 2808-2815.

Haynes, H. H., and Rail, R. D. (1976). Concrete for Ocean Thermal Energy, Conversion Structures. Civil Engineering Lab (Navy) Port Hueneme Calif.

Hidetoshi I (2001) The relationship between grain ripening and falling number in winter wheat seeds. Jpn J Crop Sci 70:373–378

Hopman, E., Dekking, L., Blokland, M. L., Wuisman, M., Zuijderduin, W., Koning, F., and Schweizer, J. (2008). Tef in the diet of celiac patients in The Netherlands. Scandinavian Journal of Gastroenterology, 43(3), 277-282.

Hurrell, Richard, and Ines Egli. 2010. "Iron Bioavailability and Dietary Reference Values." The American Journal of Clinical Nutrition. 91 (5): 1461S-1467S.

Inglett, G. E., Chen, D., and Liu, S. X. (2016). Physical properties of gluten free sugar cookies containing teff and functional oat products. J Food Res, 5(3), 72.

Kalagi, M. G. R., Patil, D. R., Rao, S., and Nayak, M. N. Natural Fiber Reinforced Polymer Composite Materials for Wind Turbine Blades. Applications International Journal of Scientific Development and Research (IJSDR) www. ijsdr. org ISSN, 2455-2631.

Kenney, E. S., Butler, C., Moore, C., Bhaduri, S., Ghatak, R., and Navder, K. P. (2011). The effect of substituting teff flour in gluten-free sugar cookies and peanut butter cookies. Journal of the American Dietetic Association, 111(9), A63.

Kosmatka, S. H., Kerkhoff, B., and Panarese, W. C. (2002). Design and control of concrete mixtures (Vol. 5420, pp. 60077-1083). Skokie, IL: Portland Cement Association.

Kotásková, E., Sumczynski, D., Mlček, J., and Valášek, P. (2016). Determination of free and bound phenolics using HPLC-DAD, antioxidant activity and in vitro digestibility of Eragrostis tef. Journal of Food Composition and Analysis, 46, 15-21.

Lukow OM, White NDG, Sinha RN (1995) Influence of ambient storage conditions on the bread making quality of two HRS wheats. J Stored Prod Res 31:279–289

Manach C, Scalbert A, Morand C, Rémésy C, Jiménez L. Polyphenols: food sources and bioavailability. Am J Clin Nutr. 2004 May;79(5):727-47. doi: 10.1093/ajcn/79.5.727. PMID: 15113710.

Manary MJ, Krebs NF, Gibson RS, Broadhead RL and Hambidge KM (2002) Community-based dietary phytate reduction and its effect on iron status in Malawian children. Annals of Tropical Paediatrics 22, 133–136.

Mengesha, M. H. (1966). Chemical composition of teff (Eragrostis tef) compared with that of wheat, barley and grain sorghum. Economic Botany, 20(3), 268-273.

Michaelsen, K. F., Dewey, K. G., Perez Exposito, A. B., Nurhasan, M., Lauritzen, L., & Roos, N. (2011). Food sources and intake of n-6 and n-3 fatty acids in low-income countries with emphasis on infants, young children (6–24 months), and pregnant and lactating women. Maternal & Child Nutrition, 7, 124-140.

Minelli, F. (2005). Plain and fiber reinforced concrete beams under shear loading: Structural behavior and design aspects. Università di Brescia.

Minten B, Tamru S, Engida E, Kuma T (2013). Ethiopia's Value Chains on the Move: The Case of Tef. Ethiopian Strategic Support Program, Working Paper 52. Addis Ababa, Ethiopia

Moreno Amador, M. D. L., Comino Montilla, I. M., and Sousa Martín, C. (2014). Alternative grains as potential raw material for gluten–free food development in the diet of celiac and gluten–sensitive patients.

Nanni, A. (1991). Pseudoductility of fiber reinforced concrete. Journal of Materials in Civil Engineering, 3(1), 78-90.

Roosjen J (2007) Processing of teff flour. European patent specification, publication number: WO 2005/025319 (24.03.2005 Gazette 2005/12), European patent office. (http://www.abs-africa.info/uploads/media/Teff-PatentEP_1_646_287_B1_01.pdf).

Sandstrom B (1997) Bioavailability of zinc. European Journal of Clinical Nutrition 51, S17–S19.

Satheesh, N., and Fanta, S. W. (2018). Review on structural, nutritional and anti-nutritional composition of Teff (Eragrostis tef) in comparison with Quinoa (*Chenopodium quinoa* Willd.). Cogent Food & Agriculture, 4(1), 1546942.

Schlemmer, U., W. Frølich, R.M. Prieto, and F. Grases. 2009. "Phytate in Foods and Significance for Humans: Food Sources, Intake, Processing, Bioavailability, Protective Role and Analysis." Molecular Nutrition & Food Research. 53 (S2): S330-S375.

Simopoulos, Artemis P. 2001. "N– 3 Fatty Acids and Human Health: Defining Strategies for Public Policy." Lipids. 36 (1): S83-S89

Suliburska, J., and Krejpcio, Z. (2014). Evaluation of the content and bioaccessibility of iron, zinc, calcium and magnesium from groats, rice, leguminous grains and nuts. Journal of Food Science and Technology, 51(3), 589-594.

Taha S. El-Alfy, Shahira M. Ezzat and Amani A. Sleem (2012) Chemical and biological study of the seeds of Eragrostis tef (Zucc.) Trotter, Natural Product Research, 26:7, 619-629, DOI: 10.1080/14786419.2010.538924

Tarekegh, G. (2015). Blending of Teff (Eragrostis teff) varietes with Moringa (Moringa Oleifera) leaf powder to improve the nuturitonal content of injera (Doctoral dissertation, Harmaya University).

Wolter, A., Hager, A. S., Zannini, E., Czerny, M., & Arendt, E. K. (2014). Impact of sourdough fermented with Lactobacillus plantarum FST 1.7 on baking and sensory properties of gluten-free breads. European food Research and Technology, 239(1), 1-12.

Wrigley, C. (2004). Encyclopedia of Grain Science. Elsevier.

Yetneberk, S., de Kock, H. L., Rooney, L. W., and Taylor, J. R. (2004). Effects of sorghum cultivar on injera quality. Cereal chemistry, 81(3), 314-321.

Zarnkow, M., Almaguer, C., Burberg, F., Back, W., Arendt, E. K., & Kreisz, S. (2008). The use of response surface methodology to optimise malting conditions of teff (Eragrostis tef (Zucc.) Trotter) as a raw material for gluten-free foods. Brewing Science, 61(5/6), 94–104.

Zegeye, A. (1997). Acceptability of injera with stewed chicken. Food Quality and Preference, 8(4), 293-295.

Zimmermann, M. B., and Hurrell, R. F. (2007). Nutritional iron deficiency. The Lancet, 370(9586), 511-520.

Zimmermann, M. B., Chassard, C., Rohner, F., N'goran, E. K., Nindjin, C., Dostal, A., and Hurrell, R. F. (2010). The effects of iron fortification on the gut microbiota in African children: a randomized controlled trial in Cote d'Ivoire. The American Journal of Clinical Nutrition, 92(6), 1406-1415.

[illegible] (20..). [illegible] for [illegible]

[illegible]

11

Fonio Millet: Chemistry, Nutritional Attributes, Processing Innovations Traditional and Modern Foods and Beverages, Anti-nutritional Factors and Health Benefits

Sakshi Sharma[1], Pal Murugan Muthaiah [1], Vasudevan[2], GK Sharma[1] and Anil Dutt Semwal[1]

[1]Grain Science and Technology Division, Defence Food Research Laboratory, Siddarthanagar, Mysore-570011, India
[2]Food Engineering and Packaging Division, Defence Food Research Laboratory, Siddarthanagar, Mysore-570011, India

Introduction

Fonio is an indigenous cereal crop of West Africa. Fonio grain belongs to genus *Digitaria* Haller that consists of 230 to 325 grass species, both annual and perennial. The fonio species is a part of Poaceae family, sub- family of Panicoideae, and the tribe of Paniceae. Fonio grains have a protective covering (husk) outside the kernel, similar to rice. The difference in the chemical composition in the different anatomical fractions among different cereal grain varies. The major anatomical sections of dehusked fonio are tegument, embryo or germ and endosperm. Fonio grains are extraordinarily tiny and are consumed as a whole grain. Handling and processing of fonio grains for end-product utilization is dependent upon the size, shape and structure of the fonio grains.

The major significant constituent of cereal crops are protein and carbohydrates. Other nutritional factors such as minerals, vitamins, flavonoids, etc. also constitute a significant place in providing essential nutrients in the diet. Fonio, due to its high protein content, is regarded to be one of the best tasting and nutritious cereal grain among African cereals (Jideani. 1990, NRC, 1996, Sadiq *et al*. 2015). The nutritional composition of the fonio grains have been extensively studied (Jideani and Akingbala, 1993; Jideani *et al*., 1996; Irving and Jideani, 1997;Serna Saldivar, 2003; Cruz *et al*., 2011; Emmambux and Taylor, 2013).

Fonio undergoes mechanical as well as thermal processing such as parboiling, pre cooking and roasting before consumption to reduce their cooking time and develop a characteristic flavor. Due to advances in the technology and improvement in processing innovations, fonio is being utilized to develop bread, porridge, couscous etc, Fonio is considered a potential source of nutraceuticals such as antioxidants essential amino acids, high dietary fiber and thus proves to be beneficial for cardiovascular diseases help in control of diabetes and weight loss. Fonio is safe for people suffering from gluten intolerance and digestive difficulties.

This chapter gives in-depth detail about the origin and botany, chemical composition and nutritional properties, processing, traditional and modern foods and beverages, health benefits of fonio.

Origin and Botany

Fonio is widely cultivated in dry savannas of West Africa particularly in Senegal, Mali, Burkina Faso, Guinea and Nigeria. Fonio is known to be one of Africa's oldest cereal crop and is deemed to be cultivated since 5000 BC (Small, 2015). Total yield in general varies from 1-1.4 tons/ha (Ndoye and Nwasike, 1993). Currently, the fonio crop is grown on about 500,000 ha in West Africa, and is consumed by 3–4 million people (Small, 2015). Unlike other cereal grains, the cultivation of fonio has not spread beyond its center of origin, West Africa yet (Taylor, 2019). Fonio is said to be cultivated around River Niger and in West Africa between the Senegal to Lake Chad. It is widely distributed in the regions of Mali, Ivory Coast, Benin, Burkina Faso and

Senegal and it is said to be staple food for mountain dependent peoples of Fouta-Djalon, Guinea.

Fonio plant grows about 45-50 cm long with panicle of up to 15 cm long. Fonio is the world's fastest maturing cereal crop. The early types of varieties can produce grain in only 6–8 weeks after sowing while some other varieties may take up to 6 months. Thus, fonio is sometimes referred to as "The Grain of life" (Arendt and Bello, 2011). This millet grows well in lowlands experiencing tropical climate possessing a rainfall ranging within 600 to 1200 mm with temperature ranging from 25 to 30°C. It grows well in sandy, loamy, stony, shallow and even on poor and infertile soils with an exception of clayey soils. It even shows markable growth in acidic soils with high content of aluminum that are lethal to other crops.

There are two species of Fonio that are known White Fonio (*Digitaria exilis*) and Black Fonio (*D. iburua*). The Fonio seeds in general vary from very white to fawn yellow or purplish in colour. Black fonio is said to be 'blacker' because it produces flowering and fruiting heads of reddish or dark brown in colour, although both species produce white and dark seeds (Small, 2015). A comparison of white and black fonio and their structural differences has been presented (Table 1 and Fig 1).

Morphologically, it is an ascending, free-tillering annual cereal grass. The stem is slender and kneed which grows up to a height of 800 mm. Arrangement of leaves will be simple and alternate with a glabrous leaf blade which is linear to lanceolate in shape. When it comes to inflorescence, it has a terminal digitate panicle, which possess 2-5 slender spike like recemes. Here the spikelets are stalked and narrowly ellipsoid which is surrounded by lemma, palea and glumes. Fruit beared is a grain type, which is a minute caryopsis with a shape of oblong to globose-ellipsoid and the colour ranges between white to pale brown or purplish.

Fonio is known by many names in Africa such as *fundi, findi*, hungry rice and Asian millet (National Research Council, 1996). White and black fonio are also known by the names Acha and Ibura respectively. It has been known by several English names such as acha, acha grass, white acha, fonio millet, fundi millet, hungry rice, hungry koos and hungry millet; petit mil, fonio blanc and millet digitaire in French; alongside Ivory coast as Fini. Pom and pohin; Podgi at Benin; entaya, foyo, fira and fodio at Niger and by several other names in different regions.

Table 1. Comparison of White v/s Black Fonio

S.No.	Parameter	White Fonio	Black Fonio
1.	Size	White fonio is generally up to 75 cm (2.5 feet) tall	Black fonio is taller,sometimes reaching 1.4 m (4.5 feet)
2.	Cultivation	White fonio is the most widely used of the two species, grown from Senegal to Chad, particularly on the upland plateau of central Nigeria.	Cultivation of black fonio is largely restricted to the Jos-Bauchi Plateau of Nigeria and to northern regions of Togo and in countries outside of Africa.
3.	Other Names	Acha	Ibura

Chemical Composition and Nutritional Properties

Fonio grain has a protective covering called husk (glume) over it. Fonio grain with the outer protective husk is known as paddy fonio. Paddy fonio length ranges between 1.5 to 1.8 mm and width is about 0.9 mm (Irving and Jideani, 1997; Cruz *et al.*, 2011). Paddy fonio is dehusked to get edible whole grains of fonio; husks comprises of about 23 percent of the weight of paddy fonio (Cruz *et al.*, 2011). Fonio grains are small in size with variable length of 1-1.55 mm (Small, 2015) and 0.75 mm in width (Irving and Jideani, 1997). The thousand kernel weight of fonio millet ranges from 0.4 to 0.6 g (Arendt and Bello, 2011; Small, 2015). The external structure of dehusked fonio caryopsis, called tegument, is majorily composed of pericarp and aleurone layer. The endosperm is composed of starch, aleurone layer as well as protein bodies. The carbohydrate content is approximately 75 percent (db) (Emmambux and Taylor, 2013). The starchy endosperm contains about 27 percent of amylose content (Jideani *et al.*, 1996). Aleurone layer consists of lipid droplets besides protein bodies (Irving and Jideani, 1997). Contrary to wheat, the protein bodies in fonio grains are abundant at the periphery and decrease gradually toward the center (Cruz *et al.*, 2011). Germ portion of fonio grain is separated from endosperm by scutellum. Germ portion of fonio grain is about 1/3rd of total caryopsis (Irving and Jideani, 1997). Embryo and scutellum cells are rich in lipid and protein bodies and contain globoids that are phytic acid inclusions (Ballogou *et al.*, 2013).

Carbohydrates

Carbohydrates are the major constituent of the fonio grain. The carbohydrate content of fonio varies from 67.1 to 91 percent (Irving and Jideani, 1997; Cruz *et al.*, 2011; Temple and Bassa, 1991; Jideani and Akingbala, 1993; Serna Saldivar, 2003; Fliedel *et al.*, 2004; Chukwu and Abdul- Kadir, 2008). Major carbohyrates found in fonio grain are monosaccharides, oligosaccharides and polysaccharides such as starch and cellulose. The predominant carbohydrate present in fonio grain is starch, the content of which is found to be about 68 percent (Cruz *et al.*, 2011). The amylose content ranged between 22.1 to 28 percent among fonio grains (Jideani and Akingbala, 1993; Jideani *et al.*, 1996). Starch granules of fonio grains are polygonal shaped and about 6 to 8μm in size (Jideani *et al.*, 1996). Starch hydrolyses to yield soluble sugars,which are present in cereal grain in minute concentrations. The amount of soluble sugar found in fonio grain is nearly 1 percent (Cruz *et al.*, 2011).The soluble sugars identified in general in cereal grains are glucose, saccharose, stachyose, fructose and raffinose. Saccharose content of fonio grain was reported to vary from 0.7 to 0.8 percent (Cruz *et al.*, 2011). Lasekan (1994) reported pentosan content in fonio grain of about 33g/ Kg. Crude fiber content for fonio grains were reported in the range of 0.41- 11.3 percent (Jideani and Akingbala, 1993; Serna Saldivar, 2003). A large variation in the amount of crude fiber in fonio grains could be ascribed to environmental and genetic factors, agronomical conditions and geographical location and different analytical methods (Ballogou Vénérande *et al.*, 2015).

Protein

Fonio grains had protein content in the range of 5.1 percent to 11 percent (Cruz *et al.*, 2011; Fliedel *et al.*, 2004). The individual osborne fractions for the fonio grains were explored by Jideani *et al.* (1994b). The highest fraction was observed to be of glutelin (14%), followed by prolamin (5.5%), albumin (3.5%) and globulin (1.8%). Amino acid profile of fonio grains were explored by various researchers, and found that prolamin and glutenin fractions as storage protein whereas albumin and prolamin as metabolic proteins. Essential amino acids present in fonio were histidine, methionine, phenylalanine, leucine, isoleucine, lysine, threonine, tryptophan and valine (Ballogou Vénérande *et al.*, 2015). Fonio, similar to other cereal grains is found to be limited in the amino acid lysine, but rich in methionine and cysteine (Jideani, 1990; Lumen *et al.*, 1993; Vietmeyer *et al.*, 1996). Comparison of amino acid profile between black and white fonio was investigated by Mbaye and Nwasike (1993). Black fonio contained higher amount of all amino acids except for methionine which was found to be relatively higher in white fonio (Table 2). Aspartic acid was not detected in black fonio in their study.

Table 2. Amino Acid Profile of Black and White Fonio

S.No.	Amino Acid	White Fonio (mg/100 g/grain)	Black Fonio (mg/100 g/grain)
1.	Glutamic Acid	1591	2871
2.	Leucine	772	1395
3.	Alanine	709	1170
4.	Proline	559	862
5.	Phenylalanine	402	803
6.	Serine	402	673
7.	Valine	457	614
8.	Isoleucine	315	508
9.	Tyrosine	284	450
10.	Threonine	315	389
11.	Arginine	300	389
12.	Methionine	441	355
13.	Glycine	252	331
14.	Cysteine	221	261
15.	Histidine	165	261
16.	Lysine	205	225
17.	Tryptophan	111	215
18.	Aspartic Acid	512	-

Source: Mbaye and Nwasike (1993)

Lipids

Lipid content for fonio grains varied from 1.3 percent to 5.2 percent (Temple and Bassa, 1991; Jideani and Akingbala, 1993; Irving and Jideani, 1997; Serna Saldivar, 2003; Fliedel *et al*., 2004; Chukwu and Abdul- Kadir, 2008; Cruz *et al*., 2011). Fatty acid profile of fonio grain revealed more than 75 percent of the unsaturated fatty acid content, with dominant fatty acids being linoleic acid about 45.7 percent and oleic acid about 30.6 percent. Other fatty acids such as palmitic acid, stearic acid, behenic acid, linolenic acid, arachidic acid, palmitoleic acid and myristic acid were present in fonio grain in minor quantities (Table 3).

Table 3: Fatty Acid Profile of Fonio Grain

S. No.	Fatty Acid (%)	Fonio
1.	C14:0 myristic acid	0,1
2.	C16:0 palmitic acid	16,8
3.	C16:1 palmitoleic acid	0,3
4.	C18:0 stearic acid	4,1
5.	C18:1 oleic acid	30,6
6.	C18:2 linoleic acid	45,7
7.	C18:3 linolenic acid	0,8
8.	C20:0 arachidic acid	1,1
9.	C22:0 behenic acid	0,4

Source: Ballogou *et al*., 2015; Fliedel *et al*., 2004

Minerals

The ash content of fonio grains vary from 1-6 percent (Jideani and Akingbala, 1993; Fliedel *et al.*, 2004). Major macro elements present in fonio grains are calcium, phosphorous, potassium, sodium, magnesium and sulphur. Micronutrients present in fonio grains are iron, copper, manganese and zinc. Fonio grains were reported to be richer in sulphur content than other cereals (Fliedel *et al.*, 2004).

Vitamins

Fonio grains were found to be a source of B-complex vitamins, chiefly thiaminni cotinic acid and riboflavin. The concentration of nicotinic acid (3 mg/100g) has been found to be relatively more than the other vitamins in fonio grains (Serna Saldivar, 2003).

Table 4: Vitamins and Minerals in Fonio Grains

S. No.	Nutrient	Concentration
	Macronutrients	
1.	Calcium (%)	0.0067- 0.03
2.	Phosphorous (%)	0.09- 0.25
3.	Potassium (%)	0.02- 0.26
4.	Sodium (%)	0.005- 0.03
5.	Magnesium (%)	0.07- 0.85
6.	Sulphur (%)	0.16
	Micronutrients	
7.	Iron (ppm)	36- 133.6
8.	Copper (ppm)	1.5- 15
9.	Manganese (ppm)	21.6- 30
10.	Zinc (ppm)	30- 42.3
	Vitamins	
11.	Thiamine (mg/ 100g)	0.3-0.48
12.	Riboflavin (mg/ 100g)	0.005- 0.1
13.	Nicotinic Acid (mg/ 100g)	3

Source: Ballogou *et al.*, 2015

Polyphenols and Other Bioactive Compounds

A wide variation in total phenols have been observed in both white and black fonio. Chukwurah *et al.* (2016) observed higher content of free phenolics in the black fonio as compared to white fonio. In fonio grains, like other cereal grains, phenolic acids are present in both free as well as bound form. The major pehnolic acids found in fonio are ferulic acid, vanillic acid, chlorogenic acid, p- coumaric acid, gallic acid, protocatechuic acid, caffeic acid, syringic acid, sinapic acid, salicylic acid, etc. The predominant phenolic acid present in fonio was protocatechuic acid. The cooking process has been found to significantly reduce the total free phenolic acids in fonio grains. Aprofound decreasewas observed in two phenolic acids namely,protocatechuic acid (47%) and vanillic acid (42%). There could be variable effect of thermal treatment on the phenolic profile of different cereal grains. The thermal processing is considered to cause depolymerization of oligomeric and polymeric structures, concomitant polymerization reactions, and formation of complex of soluble phenolics and macromolecules found in the food matrix. These tend to reduce the solubility of phenols. Therefore, thermal treatment during cooking reduced the soluble fractions in fonio, and, with increase in the bound fraction, predominance of occurrence of polymerization reactions might have been induced. However, the reduction of soluble phenolics might also be associated to oxidation enhanced by longer cooking time which in turn led to effective extraction of bound phenolics acids by opening of cell matrix in alkaline conditions (Massaretto *et al.*, 2011; N'dri *et al.*, 2013).

The comparative analysis between white and black fonio extracts shows higher antioxidant activity of black fonio than white fonio.

Annongu *et al.* (2019) analyzed both white and black fonio to have similar total flavonoid content of about 137 mg/100 g. Flavonoids like apigenin (150 mg/kg) and luteolin (350 mg/kg) are present in white fonio. C-glycosyl flavones such as orientin and vitexin have been detected in white fonio. Proanthocyanidins can be detected in lower levels in both white (11 mg/100 g) as well as black (9.7 mg/100 g) fonio (Annongu *et al.*, 2019; Boncompagni *et al.*, 2019; Zhu, 2020). Sartelet *et al.* (1996) investigated the potential role of fonio flavonioids (500 mg/kg of the edible whole cereal grains) to the goitrogenic processes. The presence of apigenin (A = 150 mg/kg) and of luteolin (L1 = 350 mg/kg) was detected, of which 10 percent of A and 80 percent of L1 are present in free form, whereas the remaining 90 percent of A and 20 percent of L1 are bound as o- glycosylflavones. Both the apigenin and luteolin aglycones possess strong anti-thyroid peroxidise (TPO) activities which significantly reduces the hormomogenic capacity of this enzyme.

Other bioactive compounds present in fonio grains are alkaloids (42 mg/100 g), saponins (1.8 mg/100 g), anthraquinones, terpenoids, steroid,phytosterols, cardenolides, cardiacglycosides,and phlobatanninsin low quantities (Egbebi and Muhammad, 2016;Annongu *et al.*, 2019; Zhu, 2020; Adams and Yakubu, 2020).

Table 5. Comparison of Nutritional Compositon between White and Black Fonio

S.No.	Nutrient	White Fonio	Black Fonio
1.	Moisture (%)	7	7
2.	Carbohydrate (%)	79.72	76.91
3.	Protein (%)	7.11	8.75
4.	Fat (%)	3	4
5.	Ash (%)	2.13	2.31
6.	Fibers (%)	0.79	1.03
7.	Energy (Kcal/ 100g)	374.32	378.64

Source: Sadiq *et al.*, 2015

Flavor Compounds

The flavour profiling of compounds present in cooked and malted white fonio have been carried out (Lasekan and Feijao-Teixeira, 2004; Lasekan *et al.*, 2001). Lasekan *et al.* (2001) identified and quantified 23 volatile compounds in cooked fonio. They divided volatiles into three main groups compounds formed by starch degradation, compounds formed by reaction between protein and starch and compounds resulting from lipid oxidation. The volatile compounds found in the cooked fonio were aldehydes, hetero-cyclics, hydrocarbons, alcohols, ester, terpene and two benzene compounds. Major volatiles found in cooked fonio were ethylbenzoate, 2, 5-dimethylpyrazine, 2-pentylfuran, 2, 6- dimethylpyrazine, trisiloxane, and 1,4-cyclohexadiene. They identified 10 compounds as potential contributors to cooked fonio flavor namely, pentanal, hexanal, hexanol, nonanal, benzaldehyde, 2-mcthylpyrazinc, 2,5-dimethylpyrazine, 2, 6-dimethypyrazine, 2-acetylpyrroline and 2-pentylfuran were identified as potential contributors to the flavour of cooked acha.Ethylacetate,1-hydroxy-2-propanone, pyrazine, 2,3-dimethylpyrazine, hexanal and 2,5-dimethylpyrazine have been identified as the major aroma compounds in the malted white fonio(Lasekan *et a*l., 2004).

Comparison of Nutritive Value of Fonio Grains with other Cereal Grains

Fonio grains comprises of higher amounts of two sulphur containing amino acids (methionine and cysteine) which are otherwise deficient in other major cereals (Vietmeyer *et al.*, 1996). Fonio grains contains almost twice as methionine as egg protein (Al-Khayri *et al.*, 2019).The calorific value, fat and protein content of fonio grain are comparable to other major cereal grains (Obilana, 2003; Oyenuga, 1968; Temple and Bassa, 1991). Higher ash content has been reported for fonio grain as compared to other major cereals. The fiber content of fonio is similar to other major and small millets, but more than wheat, maize and sorghum. Higher concentration of major and microelements have been reported in fonio grains as compared to other cereal grains (Al-Khayri *et al.*, 2019).

As compared to sorghum and pearl millet, the total phenolic content present in fonio grain was lower. This comparitively low total phenolic content provides an increased organoleptic acceptability to fonio based products. The *in vitro* antioxidant activity of white fonio grains have been found to be comparable to sorghum and pearl millet (N'Dri *et al.*, 2013).

A comparison about major nutrients in different grains has been presented in Table 6.

Table 6. Major nutrients of Fonio in comparison with other cereals of Africa

Nutrient	Fonio	Wheat	Sorghum	Pearl Millet	Finger Millet	Teff
Carbohydrate (%)	75	71	73	70	74	73
Protein(%)	9	7.8	7.9	11	7.3	9.6
Fat(%)	1.8	1.1	2.8	4.8	1.3	2
Fiber(%)	3	2	2.3	2.3	3.6	3
Ash(%)	2.9	1.6	1.6	1.9	2.6	2.9
Energy(%)	1411	1105	1142	1483	1403	1411

Source: Obilana, 2003

Antinutritional Factors

Phytates

Phytate is the most commonly occurring antinutritional factor present in cereals. These phytaes inhibits the absorption of micronutrients like iron, zinc, etc. (Hurrell, 2003). The average phytate content of fonio have been observed to be 514 mg/ 100g in paddy fonio. Phytate content have been found to significantly reduce during processing and cooking of fonio grains.

A significant decrease of phytate content has been observed in mid wet fonio due to processing treatments such as dehulling, milling and washing. Dehulling and milling reduces major concentrations of phytates from fonio as bran layers are removed which contains majority of phytates (Doesthale *et al.*, 1979; Chen *et al.*, 1998; Bryant *et al.*, 2005; Mitchikpe, 2007). The washing of fonio takes a long time, and thus, phytates have been observed to reduce due to leaching phenomenon, also observed during soaking process (Mahgoub and Elhag, 1998; Egli *et al.*, 2002; Hurrell, 2003). Despite these processing treatments, the remaining amount of phytates still present may act as inhibitory for micronutrients such as iron. Koreissi-Dembe′le *et al.* (2013) have emphasized that complete degradation of phytate content from fonio in order to assure iron absorption to atleast 18 percent. The molar ratios of phytate to iron and phytate to zinc of processed fonio products depicts poor iron availability, however, adequate zinc bioavailability.

Processing of Fonio

Fonio grains undergo complex processing before making it edible for consumption. Knowledge of physico chemical and functional properties of fonio as well as wide deviation in these properties among different varieties is required for mechanizing the processing of fonio grains into various fonio-based products. Generally, fonio grain is processed mechanically (dehusking and milling) and thermally (pre-cooking, parboiling and roasting) (Ballogou *et al.*, 2015). Processing of fonio paddy grains to whitened fonio grains is achieved in two stages. After threshing, raw fonio or paddy fonio is dehusked in first stage to remove outer husk and obtain dehusked fonio grains. In the second stage outer bran layer (pericarp and germ) are removed from the dehusked fonio grains to obtain whitened fonio grains. Dehusking and milling of fonio grain is a laborious task involving 4 to 5 successive poundings alternated with winnowing. The productivity of the whole process is very low and the obtained fonio grains have to be washed for removal of dirt and sand adding additional steps, time and effort in the processing operations. Therefore, mechanizing the processing operations of fonio grain is essential to improve production and quality of fonio grain (Cruz, 2004). A flowchart depicting fonio grain processing has been presented in Fig1. A dehuller has been done previously by small grains like fonio millets (Smith, 1996).

Various research projects have been developed with an aim to improve processing of fonio grains and design machinery for mechanization of post-harvest operations carried out during processing of grain. For instance, a project focusing on this issue was carried out by CIRAD in collaboration with national institutes of Mali (IER), Guinea (IRAG) and Burkina Faso (IRSAT) (Cruz, 2004).

Parboiling and pre-cooking both includes heat and moisture to treat the fonio grains. Paddy fonio or raw fonio are subjected to the parboiling process. Fonio paddy are hydrated in hot water till the moisture content of the grains reach about 32-34 percent (Cruz *et al.*, 2009, 2011; Ballogou *et al.*, 2013). Thereafter, paddy grains are steamed for about 10 to 30 minutes, and then dried in shade or sun. The grains are dehusked and milled to produce parboiled fonio grains.

Knowledge of parboiling process variables is important for processing of grains, particularly soaking or hydration step during parboiling. This requires a thorough study of water absorption characteristics of fonio grains. A high water absorption capacity of food grain is considered desirable for processing operations like wet milling as mechanical stress required for wet milling reduces when grains absorb saturation moisture (Tunde-akintunde, 2010; Ituen *et al.*, 1986). Extraction of important components such as starch is also facilitated in grains with higher water absorption. Protein and carbohydrates in cereal grains mainly absorb water. The proteins have a higher water absorption capacity compared to carbohydrates. Therefore, grains having higher protein content will have high water absorption capacity and will tend to absorb water at a faster rate, at the same temperature (Tunde-akintunde, 2010).

After dehusking and milling, the cooking process of fonio also is a long and technical errand (Cruz, 2004). The flowchart describing the individual steps of cooking fonio is presented in Fig 3. As a solution to this, a new product had been developed by small scale- enterprises in the mid-nineties, i.e. the pre-cooked fonio.

Dehusked and milled grains undergo pre-cooking and roasting treatments. For pre- cooking treatment, dehusked and milled fonio grains were washed for removal of dust and sand particles. After draining, the grains are steam cooked for 30 to 48 min and subsequently sun dried to obtain pre-cooked grains (Rivier and Cruz, 2007; Ballogou *et al.*, 2013).

For roasting, dehusked and milled fonio grains are first washed to remove dust and sand particles and sun dried. After completion of drying, the dried grains are roasted in a pot at around 300°C for approximately 30 to 35 min and cooled at room temperature to obtain roasted fonio grains.

Traditional and Modern Food Products from Fonio

Various attempts have been made to produce bakery products such as crackers, biscuits, cake, sugar cookie, biscuits, breads, etc. utilizing fonio flour (McWatters *et al.*, 2003; Jideani *et al.*, 2007; Jideani *et al.*, 2008; Agu *et al.*, 2014; Ayo and Gidado, 2017; Ayo and Ayo, 2018; Olagunju *et al.*, 2018; Ayo *et al.*, 2019). Ayo and Nkama (2003) demonstrated substitution of wheat flour with acha flour for producing acceptable and good quality biscuits. Ayo and Nkama (2004) evaluated the effect of acha grain flour on quality characteristics of the acha– wheat composite bread and found that up to 30 percent acha grain flour could be utilized in the production of composite bread without any significant effect. According to the, the high content of methionine and cysteine in the acha grain could improve the protein quality of the bread. Quality of bread produced from fonio has also been studied(Igyor. 2005, Jideani *et al.*, 2007, Jideani *et al.*, 2008). Acha is a highly nutritious grain having attractive flavor, which is made into porridge and couscous, or is ground and mixed with other flours to make breads (Agu *et al.*, 2015; Mbaeyi- Mwaoha and Uchendu, 2016).

In some places, fonio is considered an exotic crop, and is reserved particularly for chiefs, royalty, special occasions and diabetics. Several food preparations,

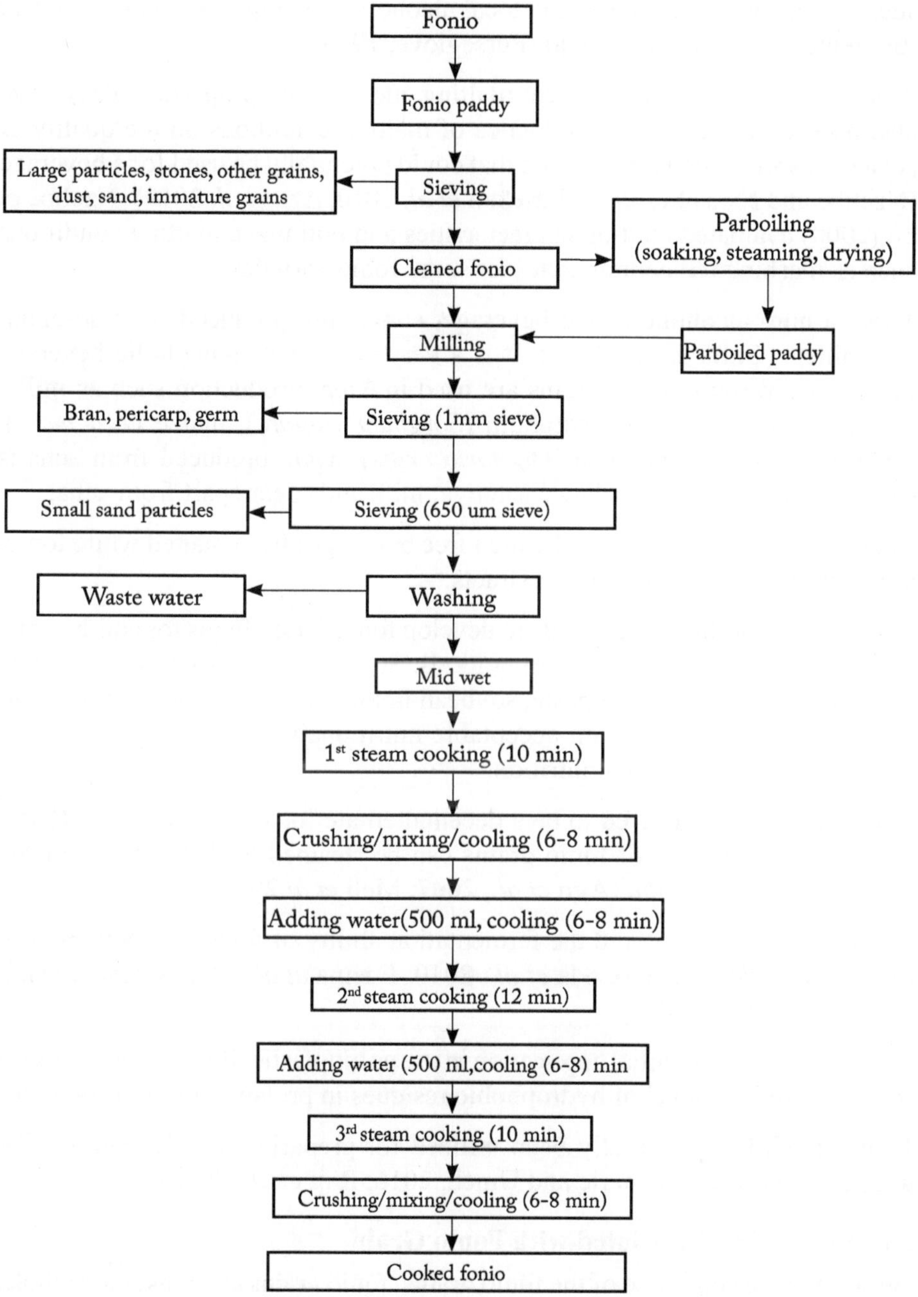

Fig. 1: Flowchart depicting fonio processing and cooking of fonio
Source: Koreissi-Dembele *et al.*, 2013

such as porridges and `tuwo' and local alcoholic beverages, such as `pito' and `burukutu' are made from fonio (Purseglove, 1975).

Acha and iboru possess excellent malting and brewing properties. They have also been examined for the influence of malting conditions on the quality of acha malt and roasted malt extract that could potentially be used for a beverage (Nzelibe and Nwasike, 1995; Lasekan *et al.* 2010, Ayo *et al.* 2019). Nzelibe *et al.* (2000) evaluated malting characteristics and optimized malting conditions such as mashing temperature, etc. for some fonio varieties.

Kunu, a non- alcoholic cereal beverage, commonly produced from sorghum and millets (Gaffa *et al.*, 2002). It is a very popular non-alcoholic beverage in Nigeria. Various cereal grains are used in *Kunu* production such as millet (*Pennisetum typhoideum*), sorghum (*Sorghum vulgare*), maize (*Zea mays*), rice (*Oryza sativa*) and acha (*Digitaria exilis*). *Kunu* produced from acha is called *Kunun acha* and includes tsamiya, milk and sugar apart from acha.

Badejo *et al.* (2017) developed gluten free beverages from malted white fonio, along woth tigernut and ginger extracts.

Various attempts have been made to develop fonio-based foods for children and well as fonio- based infant weaning foods. Sosanya *et al.* (2018) mixed white fonio flour with cashewnut paste, soybean flour, crayfish powder and mineral-vitamin mix to develop well acceptable nutritionally rich food products for addressing the issue of malnutrition.

Fonio has been discovered to be a decentalternate for semolina (NRC, 1996). Coarse fonio particles of fonio grains can be steamed to develop granulated products such as *Dambu* (Agu *et al.*, 2007; Meli *et al.*2013).

Many researchers explored the fermentation ability of fonio grains inorder to produce nutritious food (Coda *et al.*, 2010; Edema *et al.* 2013; Chinma *et al.*, 2016; Ishola *et al.*, 2018).

Fonio grains find potential application in use as bioplastic films and coatings for foods due to high level of hydrophobic residues in prolamin protein fraction.

Fonio starch has potentially been utilized for preparing noodles and can be added into porridge (Umerie and Umeh, 2016; Raji *et al.*, 2018).

Health Benefits Associated with Fonio Grain

Because of the small size of the fonio grains, fonio grains are consumed whole. Consumption of whole fonio grain makes it an exceptional source of dietary fiber and nutraceuticals with involved benefits suitable for health conscious people as well as for people with obesity and diseases such as diabetes. Fonio grains are considered potential source of nutraceuticals, namely antioxidant phenolics and cholesterol lowering waxes (Jideani and Jideani, 2011). The polyphenols present in fonio grains impart antioxidant and free radical scavenging activity(Chukwurah *et al.*, 2016; Glew *et al.*, 2013; N'Dri *et al.*, 2013).

Fonio grains have been shown to have antidiabetic potential and glycemic control capacity (Jideani, 1999; Balde *et al.*, 2008; Jideani and Jideani, 2011; Alegbejo *et al.*, 2011; Ibrahim and Saidu, 2017; Adams and Yakubu, 2020). Ibrahim and Saidu (2017) examined *in vivo* glycemic property of processed white fonio grains have been examined in diabetic Wistar rats. The dehulled samples showed higher glycemic response than the hulled samples which might be attributed to the dietary fiber present in the hulls.

Fonio grain is a rich source of vitamins, dietary fiber, phenolics and other bioactive compounds. These compounds have been confirmed to have antidiabetic and glycemic control potential. Fibers bind with amylases, thus decreasing the interaction between amylases and starch. The phenolic compounds can regulate insulin signalling pathways by interacting with gut microbiota, in turn, regulating the glycemic response (Zhang *et al.*, 2020).

Fonio grains show potential antithyroid property due to the presence of polyphenols, flavonoids, phytosterols, and n-3 fatty acids. *In vitro* antithyroid property of fonio grain flavonoids have been identified. Flavonoids like apigenin and luteolin displayed antthyroid peroxidase (TPO) activity and decreased the activity of the cyclic adenosine monophosphate phosphodiesterase (Sartelet *et al.*, 1996).

Fonio grains consist of higher amount of sulphur amino acid, which are believed to be essential for proper function of heart and nerve transmission (CIRAD, 2006).

Fonio grains are gluten free and thus are advantageous for people affected from celiac diseases (Saturni *et al.*, 2010).

Utilization of Fonio other than as Food

Fonio has been utilized as an important religious and cultural crop, particularly in Lamba and Tamberma communities. Fonio has an important religious and cultural value in weddings, women initiation ceremonies and traditional baptism of newborn child (Koroch *et al.*, 2013; Adoukonou- Sagbadja *et al.*, 2006). In Akposso and Akébou communities, some rituals are done before the sowing, harvesting and consumption of the newly harvested fonio. A traditional festival named *Ovazu* is also held in these two communities every year for celebrating the multi-secular cultural associations with the crop of fonio (Adoukonou- Sagbadja *et al*., 2006).

Fonio has significant role in traditional medicine and pharmaceutical uses (Ogaji and Famoriyo, 2010; Jideani, 2012). Muazu *et al*.(2010) has explored utlilization of Fonio starch as a glidant in folic acid tablet formulation and reported it to serve as an alternative glidant for pharmaceutical granulations. Fonio starch was hydrolysed for production of nanocrystals that were subsequently carboxymethylated. Both native as well as modified fonio starch can be utilized as carriers for controlled release of a drug naproxen having low solubility (Odeniyi *et al*., 2019).

In traditional medicine, fonio has been utilized for meteorism constipation as a diuretic for treatment or elimination of blood clots for treatment of diarrhea, loss of appetite, dysentery, chickenpox, stomach ache and asthma (Neuwinger, 2000; Koroch *et al*., 2013; Adoukonou- Sagbadja *et al*., 2006).

Fonio grains are used as a valuable feed for the ruminant livestocks such as sheep, cattle and goat (Adoukonou- Sagbadja *et al*., 2006). Fonio is easily digested and is highly palatable (Vodouhe *et al*., 2007). It is a rich source of potassium, phosphorous and trace minerals which are suitable to formulate specific nutrient rich feed for pigs and poultry (Clottey, 2006). Fonio straw ash is used as potash for cooking. Chopped fonio straw is mixed with clay to build walls of houses (Vodouhe *et al*., 2004).

Future Prospects

Harvesting and post- harvesting operations of fonio is a tedious process and thus, restricts the utilization of this indigenous African grain on a larger scale (Adoukonou-Sagbadja *et al*. 2006). Diversity in agronomic, physico-chemical, and nutritional properties of fonio is minimal. The significant issues associated with production of fonio grains are lower productivity, smaller size of fonio grains, lower pest resistance and higher susceptibility to diseases and lodging. Considering the highly nutritious nature and associated health benefits of fonio grains, there is a dire need for exploration of modern breeding and farming techniques and mechanized systems for developing improved fonio varieties with increased productivity and enhanced nutritional value. Globally, fonio grains trade and distribution for local as well as export purposes shall also be focused on forgetting world wide popularity.

Both white and black fonio varieties should be explored for variation in nutrient value and phytochemical content. Diversity in content of dietary fibre and the phytochemicals in cereals are determined genetically, however, few environmental aspects are also known to play a significant role. Selecting varieties with higher content of phytochemicals and bioactive components through breeding programmes may aid in producing new types of grains with improved health benefits (Shewry, 2009). Biochemical assays such as *in vitro* bile acid binding methodology can be utilized for evaluation of health effects of fonio grains, which, in turn can be further utilized in screening of food fractions for potential health benefits in animal and human studies programmes. The researchers shall also take into account the acceptance of fonio based food products by the consumers and focus on the sensory attributes, along with nutritional benefits of fonio grains. Fonio grains has a great potential to be developed in to wide range of food and beverage products by incorporating different ingredients and processing innovations. However, the sensory and nutritional attributes of the fonio grains is not to be compromised and newer technological aids should be involved keeping in view the perception of products in association to consumer acceptability for successful development of fonio- based food products.

Conclusion

The physicochemical and nutritional properties, processing, traditional and modern food products along with health benefits associated with the fonio grain has been summarized in this chapter. Fonio grain is a rich source of both macronutrients (protein, lipids and starch) and micronutriens (zinc, iron, vitamin B, etc.). Fonio grains undergo various processing operations before consumption, including dehulling, milling, parboiling, roasting, etc. to develop products such as precooked and mid wet fonio and various other fonio based gluten free products like beverages, breads, puddings, crackers, breakfast cereals and weaning foods. Several health benefits such as antidiabetic potential, glycemic control capacity, antithyroid property associated with fonio grain have been highlighted. Overall, fonio grain can be considered nutritionally at par with other cereal grains and can be utilized widely for food and non- food applications.

References

Adams, D. M., and Yakubu, M. T. (2020). Aqueous extract of Digitaria exilis grains ameliorate diabetes in streptozotocin-induced diabetic male Wistar rats. Journal of Ethnopharmacology, 249, 112383.

Adoukonou-Sagbadja, H., Dansi, A., Vodouhè, R. and Akpagana, K. (2006). Indigenous knowledge and traditional conservation of fonio millet (*Digitaria exilis, Digitaria iburua*) in Togo. Biodiversity and Conservation, 15(8), 2379-2395.

Agu, H. O., Ayo, J. A. and Jideani, A. I. O. (2015).Evaluation of the quality of malted acha-soy breakfast cereal flour.African Journal of Food, Agriculture, Nutrition and Development, 15(5), 10542-10558.

Agu, H. O., Ayo, J. A., Paul, A. M., and Folorunsho, F. (2007). Quality characteristics of biscuits made from wheat and African breadfruit (*Treculia africana*).

Agu, H. O., Ezeh, G. C., and Jideani, A. I. O. (2014). Quality assessment of acha-based biscuit improved with Bambara nut and unripe plantain. African Journal of Food Science, 8(5), 278-286.

Al-Khayri, J. M., Jain, S. M., and Johnson, D. V. (Eds.). (2019). Advances in Plant Breeding Strategies: Nut and Beverage Crops: 4 (Vol. 4). Springer Nature.

Annongu, A. A., Edoh, J. H., Ojo, K. R. M., Olugbenga, G. A., Olaniyan, A. J., Olajumoke, B. B. Z., and Kevin, E. E. (2019). Phytochemicals and nutrients constituents of varieties of hungry rice and their significance in nutrition. Phytochemicals and Nutrients Constituents of Varieties of Hungry Rice and their Significance in Nutrition, 41(1), 10-10.

Arendt, E., and Dal Bello, F. (Eds.). (2011). Gluten-free cereal products and beverages. Elsevier.

Ayo, J. A., and Nkama, I. (2003). Effect of acha (D. exilis) flour on the physico chemical and sensory qualities of biscuits. Nutrition and Food Science, 33, 125-130.

Ayo, J. A., Agbatutu, A., and Iribom, G. K. (2019).Effect of germination duration on the chemical composition, functional and pasting properties of acha flour and its potential for biscuit production.Nigerian Annals of Pure and Applied Sciences, 2, 9-17.

Ayo, J. A., and Ayo, V. A. (2018). Proximate composition, physical and sensory quality of acha-moringa seed flour blend biscuits. Asian Food Science Journal, 1-7.

Ayo, J. A., and Gidado, F. E. (2017). Physicochemical, phytochemical and sensory evaluation of acha-carrot flours blend biscuit. Current Journal of Applied Science and Technology, 1-15.

Badejo, A. A., Olawoyin, B., Salawu, S. O., Fasuhanmi, O. S., Boligon, A. A. and Enujiugha, V.N. (2017).Antioxidative potentials and chromatographic analysis of beverages fromblends of gluten-free acha (Digitaria exilis) and tigernut (Cyperus esculentus) extracts. Food Measure, 11, 2094–2101.

Balde, C. P., Hereijgers, B. P., Bitter, J. H., and Jong, K. P. D. (2008). Sodium alanate nanoparticles– linking size to hydrogen storage properties. Journal of the American Chemical Society, 130(21), 6761-6765.

Ballogou, V. Y., Sagbo, F. S., Soumanou, M. M., Manful, J. T., Toukourou, F. and Hounhouigan, J. D. (2015). Effect of processing method on physico-chemical and functional properties of two fonio (Digitaria exilis) landraces. Journal of Food Science and Technology, 52(3), 1570-1577.

Ballogou, V. Y., Soumanou, M. M., Toukourou, F., and Hounhouigan, J. D. (2013). Structure and nutritional composition of fonio (Digitaria exilis) grains: a review. International Research Journal of Biological Sciences, 2(1), 73-79.

Boncompagni, E., Nielsen, E., Sanogo, M. D., Coghetto, V., Temporiti, M. E., Daglia, M., and Doria, E. (2019). Anti-nutritional metabolites in six traditional African cereals. Journal of Food and Nutrition Research, 58(2).

Bryant, R. J., Dorsch, J. A., Peterson, K. L., Rutger, J. N., and Raboy, V. (2005). Phosphorus and mineral concentrations in whole grain and milled low phytic acid (lpa) 1-1 Rice.Cereal Chemistry, 82(5), 517-522.

Chen, H., Siebenmorgen, T. J., and Griffin, K. (1998).Quality characteristics of long-grain rice milled in two commercial systems.Cereal Chemistry, 75(4), 560-565.

Chinma, C. E., Anuonye, J. C., Ocheme, O. B., Abdullahi, S., Oni, S., Yakubu, C. M., and Azeez, S. O. (2016). Effect of acha and bambara nut sourdough flour addition on the quality of bread. LWT, 70, 223-228.

Chukwu, O., and Abdul-kadir, A. J. (2008).Proximate chemical composition of acha (*Digitaria exilis* and *Digitaria iburua*) grains.Journal of food Technology, 6(5), 214-216.

Chukwurah, P. N., Uyoh, E. A., Usen, I. N., Ekerette, E. E., and Ogbonna, N. C. (2016). Assessment of intra and inter species variation in antioxidant composition and activity in marginalized fonio millet (*Digitaria* spp.). Journal of Cereals and Oilseeds, 7(1), 1-6.

Clottey, V. A., Avornyo, F., Addo-Kwafo, A., and Agyare, W. A. (2006). The potential of fonio (*Digitaria exilis, Stapf*) as feed for monogastrics. Livest Res Rural Dev, 18.

Coda, R., Rizzello, C. G., and Gobbetti, M. (2010). Use of sourdough fermentation and pseudo-cereals and leguminous flours for the making of a functional bread enriched of γ-aminobutyric acid (GABA). International Journal of Food Microbiology, 137(2-3), 236-245.

Cruz, J. F. (2004). Fonio: a small grain with potential. Leisa-Leusde-, 20, 16-17.

Cruz, J. F., Beavogui, F., and Dramé, D. (2011).Fonio, an African cereal [Le fonio, une céréale africaine].Wageningen: CTA.

Cruz, J. F., Rivier, M., and Fliedel, G. (2009). D8. Etude de l'étuvage du fonio et essais expérimentaux: projet n°. 015403 FONIO. Amélioration de la qualité et de la compétitivité de la filière fonio en Afrique de l'Ouest.

de Lumen, B. O., Thompson, S., and Odegard, W. J. (1993). Sulfur amino acid-rich proteins in acha (*Digitaria exilis*), a promising underutilized African cereal.Journal of Agricultural and Food Chemistry, 41(7), 1045-1047.

Doesthale, Y. G., Devara, S., Rao, S., and Belavady, B. (1979). Effect of milling on mineral and trace element composition of raw and parboiled rice. Journal of the Science of Food and Agriculture, 30(1), 40-46.

Dury, S., Meuriot, V., Fliedel, G., Blancher, S., Guindo, F. B., Drame, D., Bricas, N., Diakite, L.,and Cruz, J. F. (2007). The retail market prices of fonio reveal the demand for quality characteristics in Bamako, Mali.

Egbebi, O. A., and Muhammad, A. A. (2016). Microbiological analysis of ready-to-eat suya meat sold in Owo, Ondo State. Int. J. Innov. Biochem. Microbiol. Res, 4, 11-15.

Egli, I., Davidsson, L., Juillerat, M. A., Barclay, D., and Hurrell, R. F. (2002).The influence of soaking and germination on the phytase activity and phytic acid content of grains and seeds potentially useful for complementary feedin. Journal of Food Science, 67(9), 3484-3488.

Emmambux, M. N., and Taylor, J. R. (2013). Morphology, physical, chemical, and functional properties of starches from cereals, legumes, and tubers cultivated in Africa: A review. Starch-Stärke, 65(9-10), 715-729.

Fliedel, G., Ouattara, M., Grabulos, J., Drame, D., and Cruz, J. F. (2004). Effet du blanchiment mécanique sur la qualité technologique, culinaire et nutritionnelle du fonio, céréale d'Afrique de l'Ouest.

Gaffa, T., Jideani, I. A., and Nkama, I. (2002). Traditional production, consumption and storage of Kunu–a non alcoholic cereal beverage. Plant Foods for Human Nutrition, 57(1), 73-81.

Glew, R. H., Laabes, E. P., Presley, J. M., Schulze, J., Andrews, R., Wang, Y. C., Chang, Y. C.,and Chuang, L. T. (2013). Fatty acid, amino acid, mineral and antioxidant contents of acha (*Digitaria exilis*) grown on the Jos Plateau, Nigeria. International Journal of Nutrition and Metabolism, 5(1), 1.

Hallberg, L., Brune, M., and Rossander, L. (1989). Iron absorption in man: ascorbic acid and dose-dependent inhibition by phytate. The American Journal of Clinical Nutrition, 49(1), 140-144.

Hurrell, R.F. (2003). Influence of vegetable protein sources on trace element and mineral bioavailability. The Journal of Nutrition, 133(9), 2973S-2977S.

Igyor, M. A. (2005). Substitution of wheat flour with "acha"(Digitaria exilis) for bread making. Botswana Journal of Technology, 14(2), 51-57.

Irving, D. W., and Jideani, I. A. (1997).Microstructure and composition of Digitaria exilis Stapf (acha): a potential crop. Cereal Chemistry, 74(3), 224-228.

Ituen, E. U. U., Mittal, J. P., and Adeoti, J. S. (1986).Water Absorption in Cereal Grains and Its Effect on Their Rupture Stress 1.Journal of Food Process Engineering, 8(3), 147-158.

Jideani, A. I., and Akingbala, J. O. (1993).Some physicochemical properties of acha (Digitaria exilis Stapf) and iburu (*Digitaria iburua* Stapf) grains.Journal of the Science of Food and Agriculture, 63(3), 369-374.

Jideani, A. I., Apenten, R. K. O., and Muller, H. G. (1994b).The effect of cooking on proteins from acha (*Digitaria exilis*) and durum wheat. Journal of the Science of Food and Agriculture, 65(4), 465-476.

Jideani, I. A. (1990).Acha, Digitaria exilis, the neglected cereal.Agriculture International (UK), 42(5), 132-143.

Jideani, I. A. (1999).Traditional and possible technological uses of *Digitaria exilis* (acha) and *Digitaria iburua* (iburu): A review.Plant Foods for Human Nutrition, 54(4), 363-374.

Jideani, I. A. (2012). Digitaria exilis (acha/fonio), *Digitaria iburua* (iburu/fonio) and Eluesine coracana (tamba/finger millet) Non-conventional cereal grains with potentials. Scientific Research and Essays, 7(45), 3834-3843.

Jideani, I. A., and Jideani, V. A. (2011). Developments on the cereal grains *Digitaria exilis* (acha) and *Digitaria iburua* (iburu). Journal of Food Science and Technology, 48(3), 251-259.

Jideani, I. A., Owusu, R. K., and Muller, H. G. (1994a).Proteins of acha (*Digitaria exilis* Stapf): Solubility fractionation, gel filtration, and electrophoresis of protein fractions.Food Chemistry, 51(1), 51-59.

Koreissi-Dembélé, Y., Fanou-Fogny, N., Hulshof, P. J., and Brouwer, I. D. (2013). Fonio (Digitaria exilis) landraces in Mali: Nutrient and phytate content, genetic diversity and effect of processing. Journal of Food Composition and Analysis, 29(2), 134-143.

Lasekan, D. O. (1994). Chemical composition of acha (*Digitaria exilis*) flour. J. Food Sci. Agric, 14, 177-179.

Lasekan, O. O., Teixeira, J. F., and Salva, T. J. G. (2001). Volatile flavour compounds of cooked acha (*Digitaria exilis* Stapf). Food Chemistry, 75(3), 333-337.

Lasekan, O., Terezinha Salva, J., and Abbas, K. (2010). Effect of malting conditions and quality characteristics of malt and roasted malt extract from 'acha'grains (*Digitaria exilis* Stapf). Journal of the Science of Food and Agriculture, 90(5), 850-860.

Mahgoub, S. E., and Elhag, S. A. (1998).Effect of milling, soaking, malting, heat-treatment and fermentation on phytate level of four Sudanese sorghum cultivars.Food Chemistry, 61(1-2), 77-80.

Massaretto, I. L., Alves, M. F. M., de Mira, N. V. M., Carmona, A. K., and Marquez, U. M. L. (2011). Phenolic compounds in raw and cooked rice (Oryza sativa L.) and their inhibitory effect on the activity of angiotensin I-converting enzyme. Journal of Cereal Science, 54(2), 236-240.

Mbaeyi-Nwaoha, I. E., and Uchendu, N. O. (2016). Production and evaluation of breakfast cereals from blends of acha and fermented soybean paste (okara). Journal of food Science and Technology, 53(1), 50-70.

Mbaye, N., and Nwasike, C. (1993). Fonio millet (*Digitaria exilis*) in West Africa. Riley KW, Gupta SC, Seethaam A. and JN.

McWatters, K. H., Ouedraogo, J. B., Resurreccion, A. V., Hung, Y. C., and Phillips, R. D. (2003). Physical and sensory characteristics of sugar cookies containing mixtures of wheat, fonio (Digitaria exilis) and cowpea (*Vigna unguiculata*) flours.International Journal of Food Science and Technology, 38(4), 403-410.

Meli, C. F., Njintang, N. Y., Noumi, G. B., Bernard, C., Relkin, P., Armand, M., and Mbofung, C. M. (2013). Processing, physicochemical, nutritional, and organoleptical properties of dackere, an African cereal/tuber food semolina. Food and Bioprocess Technology, 6(3), 699-709.

Mitchikpe, C. E. S. (2007). Towards a food-based Approach to Improve Iron and Zinc Status of Rural Beninese Children: Enhancing Mineral Bioavailability from Sorghum-based Food.

Muazu, J., Musa, H., and Bhatia, P. G. (2010).Evaluation of the glidant property of fonio starch. Research Journal of Applied Sciences, Engineering and Technology, 2(2), 149-152.

N'Dri, D., Mazzeo, T., Zaupa, M., Ferracane, R., Fogliano, V., and Pellegrini, N. (2013).Effect of cooking on the total antioxidant capacity and phenolic profile of some whole-meal African Cereals. Journal of the Science of Food and Agriculture, 93(1), 29-36.

National Research Council. (1996). Lost Crops of Africa: volume I: grains. National Academies Press.

Ndoye, M., and Nwasike, C. C. (1993). Fonio millet (*Digitaria exilis* Stapf) in West Africa. Advances in Small Millets. Oxford and IBH, New Delhi, India, 85-94.

Neuwinger, H. D. (2000). African Traditional Medicine: a Dictionary of Plant use and Applications. With Supplement: Search System for Diseases. Medpharm.

Nzelibe, H. C., and Nwasike, C. C. (1995). The brewing potential of "acha"(Digitaria exilis) malt compared with pearl millet (*Pennisetum typhoides*) malts and sorghum (Sorghum bicolor) malts. Journal of the Institute of Brewing, 101(5), 345-350.

Nzelibe, H. C., Obaleye, S., and Onyenekwe, P. C. (2000). Malting characteristics of different varieties of fonio millet (*Digitaria exilis*). European Food Research and Technology, 211(2), 126-129.

Obilana, A. B. (2003). Overview: importance of millets in Africa. World (all cultivated millet species), 38, 28.

Odeniyi, M. A., Adepoju, A. O., and Jaiyeoba, K. T. (2019). Native and modified *Digitaria exilis* starch nanoparticles as a carrier system for the controlled release of naproxen. Starch-Stärke, 71(9-10), 1900067.

Ogaji I, Famoriyo P (2010). Binding effects of *Digitaria exilis* Stapf (*Gramineae*) starch on some physicochemical properties of paracetamol tablets in a wet granulation process. Int. J. Pharm. Sci. 2(2):5610569

Olagunju, A. I., Omoba, O. S., Enujiugha, V. N., and Aluko, R. E. (2018). Development of value-added nutritious crackers with high antidiabetic properties from blends of Acha (Digitaria exilis) and blanched Pigeon pea (*Cajanus cajan*). Food Science and Nutrition, 6(7), 1791-1802.

Oyenuga, V. A. (1968). Nigeria's food and feeding-stuffs.

Purseglove, J. W. (1975). Pandanaceae. Tropical Crops Monocotyledons. Vols. 1 and 2. ELBS edn.

Raji, A.O., Nassam, H. O., Abolaji, G., Ayorinsw, J. O. and Raji, M.O. (2018). Nutritional properties of fonio starch-defatted moringa seed flour blends prepared at different ratios. Croatian Journal of Food Science and Technology, 10(2), 145-156.

Rivier, M., and Cruz, J. F. (2007). D7. Étude de la précuisson du fonio au sein de petites entreprises de transformation à Bamako (Mali) et à Ouagadougou (*Burkina Faso*): projet n°. 015403 FONIO. Amélioration de la qualité et de la compétitivité de la filière fonio en Afrique de l'Ouest.

Sadiq, I. Z., Maiwada, S. A., Dauda, D., Jamilu, Y. M., and Madungurum, M. A. (2015). Comparative nutritional analysis of black fonio (*Digitaria iburua*) and white fonio (*Digitaria exili*).International Research Journal of Biological Sciences, 4(6), 4-9.

Sartelet, H., Serghat, S., Lobstein, A., Ingenbleek, Y., Anton, R., Petitfrere, E., Augie-Augie, L. Martiny, and Haye, B. (1996). Flavonoids extracted from fonio millet (*Digitaria exilis*) reveal potent antithyroid properties. Nutrition, 12(2), 100-106.

Saturni, L., Ferretti, G., and Bacchetti, T. (2010). The gluten-free diet: safety and nutritional quality. Nutrients, 2(1), 16-34.

Serna Saldivar S.O. (2003).Cereals/Dietary Importance. In: Caballero B, Trugo L and Finglas (Eds) Encyclopedia of Food Sciences and Nutrition: Academic Press, London, UK, 1027-1033.

Shewry, P. R. (2009). The Health Grain Programme Opens new Opportunities for Improving Wheat for Nutrition and Health.

Small, E. (2015). 47. Teff and Fonio–Africa's sustainable cereals. Biodiversity, 16(1), 27-41.

Smith, I. F. (1996). Sorghum and Millets: Proceedings of the Symposia. Vienna: International Association for the Cereal Science and Technology. 171- 197.

Taylor, J. R. (2019). Sorghum and Millets: Taxonomy, History, Distribution, and Production. In. Sorghum and Millets (pp. 1-21). AACC International Press.

Temple, V. J., and Bassa, J. D. (1991).Proximate chemical composition of Acha (*Digitaria exilis*) grain.Journal of the Science of Food and Agriculture, 56(4), 561-563.

Tunde-Akintunde, T. Y. (2010). Water absorption characteristics of Nigerian acha (*Digitaria exilis*). International Journal of Food Engineering, 6(5).

Umerie, S. C., and Umeh, S. O. (2016). Isolation and characterization of *Digitaria exilis* seed starch as raw material for noodles making. International Journal of Agriculture and Biosciences, 5(6), 366-369.

Vietmeyer N.D., Borlaugh N.E., Axtell J., Burton G.W., Harlan J.R. and Rachie K.O., (Ed) Fonio (Acha). Lost crop in Africa Chap. 3., BOSTID Publication, 58–75 (1996).

Vodouhe, R., Atta-Krah, K., Achigan-Dako, G. E., Eyog-Matig, O., and Avohou, H. (2007). Plant genetic resources and food security in West and Central Africa: Regional Conference, 26-30 April 2004.

Zhu, F. (2020).Fonio grains: Physicochemical properties, nutritional potential, and food applications.Comprehensive Reviews in Food Science and Food Safety, 19(6), 3365-3389.

12

Role of Millets as Traditional and Alternative Medicine: A Way Forward with Enhanced Health Attributes

Ajay Singh[*1], *Gopal Kumar Sharma*[2], *Abhimannyu Kalne*[3] *and Narender Raju Panjagari*[4]

[1]*Department of Food Technology, Mata Gujri College, Fatehgarh Sahib Punjab, India*
[2]*Department of Food Technology, Defence Food Research Lab, Mysuru Karnatka-570 044, India*
[3]*Department of Agricultural Processing and Food Engineering Indira Gandhi Krishi Vishwavidyalaya, Raipur, Chattisgarh, India*
[4]*Dairy Technology Division, National Dairy Research Institute, Karnal Haryana, India*

Introduction

Since our existence, we practiced to be fit and fine with the food as a single option traditionally to cure ourselves also which we come to know about their selection with long term experiences. It is a type of diet therapy where food works like medicine. Presently, we practice the same through medicinal diet therapy where selected foods and food supplements are suggested to be taken with medicine for longevity promotion and fit (Zou, 2016). Ongoing life style and approach of being in high status involves restricted physical activities because of deskbound and easy access in every aspect. Long term association with these types of habits somehow affects our health that proves fatal and makes us sick. No doubt, involvement of physical workout is on top most priority to have healthy sustainable life but healthy eating can never be ignored to get rid of risks viz. diabetes, obesity, ceased immunity, elevated blood pressure, cancer and other heart inefficiency. So once again diet therapy comes into action in a traditional adopted pattern as a repeat. Today, market is adapted with the trends of series of formulations with much of nutraceutical and functionality in them whether it is a matter of concern for beverage, bakery, frozen dessert or some other means. Furthermore multigrain, value added, composite dairy cereal blend, probiotic and prebiotic formulations are the functional food categories that are even on liking preferences by consumers to combat with the disease associated with sedentary life style. As a basic rule stated 2000 years back that "Grains to nourish, fruit to assist, livestock to benefit and vegetables to supplement". If we follow the statement and keep them all balanced to supply energy (Wu and Liang, 2018) no harm will be there in terms of health. Basically we are blind consumers of cereals for our daily energy demand and satiety feeling without knowing the values of healthy eating viz, pseudocereals, coarse cereals and by-products to make ourselves fit (Berti *et al.*, 2005). Cereals are embedded with high glycemic index and less fiber and saturated fats in them moreover they are gluten prominent thus are on least selection by us all. On the other hand pseudocereals and coarse cereals with gluten free status, prone with high fiber, least glycemic index and unsaturated fatty acid profile boosted with polyphenols, antioxidants and also with micronutrients viz. minerals and vitamins thus target hit by food technologists, scientists and policy makers for economic upliftment of nation with healthy manpower of its residents. Among these 'Millets' are classified under coarse cereals which shows all the above stated positive traits along with its capacity to grow with least inputs (least care and attention, less water supply and acid tolerant soil grown capacity) thus grab the status as "Nutricereals" (Singh *et al.*, 2015).

Sorghum; the major millet and other minor millets (kodo, kutki, barnyard and foxtail millet etc.) have great potential to contribute in national food security mission and to achieve sustainable livelihood specifically in dryland ecosystem. Other beneficial traits associated with them are the nutritional capacity, low glycemic index, high fiber, antioxidant richness and foremost the gluten free status. No doubt value addition, composite blend development, fermented probiotic based formulation are the enhanced viability aspect adopted for these crops with commercial vision and are the only vehicle to set well being in today's anxiety. Latest advances in food researches direct the farmers to re-adopt the traditionally grown crops, aware people about their benefits by way of ample of food and pharmaceutical scope for betterment of all.

As per the Global Nutrition Report 2016 rank based categorization of stunted population, India ranks 114 out of 132 countries even Bangladesh and Nepal are for better and step ahead marginally higher than India. Similarly, prevalence of anaemia in women of reproductive age, India ranks 170 out of 185 countries at a prevalent rate of 48.1 precent, compared with the U.S. which is the best at 11.9 percent. These findings indicate that though India ranks first in milk, millets, fruits and second in terms of green leafy vegetables production but side wise we also won the world's diabetic race as topper and acquired the status as "Diabetic capital" (Rao, 2015). Ignorance, negligency, lack of knowledge about selection of food and its eating pattern the mainstream reasons to get struck with chronic diseases like blood pressure, obesity, diabetes, cardiovascular diseases, cancer, etc. So implementation of better governmental policies through promotion of best farming practices in agriculture, subsidized raw material and implements approaches, wise selection of eatables, mid day meal actions are the skillful efforts that a nation can try to become healthy and prosperous. Introduction of millets is also among those action plans to get rid of malnutrition, anaemia and high expeises on food is a problem.

Millets: Botanical Description and Chemical Composition

Millets refers to a group of annual grasses mainly found in arid and semiarid regions that are categorized under five genera: Penissetum, Eleusine, Setaria, Panicum and Paspalum. Millet grass produces coarse grains that are harvested, dried milled and consumed as animal fodder and human food both. The seeds of above stated genera are of paramount importance in Africa, Asia, China and Russia Federation. These miracle seeds are gaining importance because of their potential to provide nutritional benefits. Millets proved their importance in nutrition content as they are approximate 3 to 5 times superior to wheat and rice in terms of minerals, proteins and vitamins. The popularity of millets also due to organic crops as there is no need of pesticides and fertilizers to flourish well as are climate resilient crop thus suited best to low economic farmers to earn their livelihood and long term survival. Non-glutinous, non-acid forming, rain fed grown crop with maturity in about 65 days only are the rich attributes for the selection by marginal farmers (Saleh *et al.*, 2013).

The term millet includes a number of small-grained cereal grasses, tiny in size, round in shape and can be white, gray, yellow or red shade. Based on the grain size, millets have been classified as major millets which include sorghum/great millet (*Sorghum bicolor* (L.)) and pearl millet/ spiked millet/ bajra (*Pennisetum glaucum* (L.) R. Br.) and small millets which include finger millet/ragi (*Eleusine coracana* (L.) Gaertn), foxtail millet/ italian millet/ kakum (*Setaria italica* (L.)(, kodo millet/ kodon (*Paspalum scrobiculatum* (L.)), proso millet/ common millet/ cheena (*Panicum miliaceum* (L.)), barnyard millet/ sanwa (*Echinochloa frumentacea* (*L.*)) and little millet/ kutki (*Panicum sumatrense* Roth). Now India is on top in production with production capacity of 11 million tons for millets (Taylor, 2019).

If we look in depth for the nutritional comparison with that of staple cereals i.e. wheat rice and maize of any continent, we find that millets are much ahead of them in each and every aspect. Specifically in case of micronutrients if we talk about B-complex vitamins that majorly works as co-enzymes, participates in about 300 enzymatic bodily reactions and we have to depend for it on medicinal pills is abundantly available in millets. Finger millet has 30 times more calcium than rice (Masih *et al.*, 2019). It also compete with that of calcium in milk availability and wise combination of milk and millet as value added composite dairy cereal blend is really an innovative option for growing childrens. Similarly proso millet and pearl millet is the store house for iron that is so least associated with rice and wheat. Furthermore, kodo millet is exceptionally high for the fibre and polyphenolics content in them among millets and staple cereals too that performs against various cognitive, degenerative and chronic diseases. Only sorghum and pearl millet are lacking sufficient fibre in them that are a step low than rice, wheat and maize.

Table 1: Categorization of millets with their common names

Sr No.	Millet category	Scientific name	Common name	
			English	Hindi
1	Major millets	*Pennisetum glaucum*	Pearl millet	Bajra
2		*Setaria italic*	Foxtail millet	Kangni
3		*Panicum miliaceum*	Proso millet	Barre
4		*Eleusine coracana*	Finger millet	Madua
5	Minor millets	*Echinochloa* spp	Barnyard millet	Sanwak
6		*Paspalum scrobiculatum*	Kodo millet	Kodra
7		*Panicum sumatrense*	Little millet	Kutki
8		*Brachiaria deflexa*	Guinea millet	--
9		*Urochloa ramose*	Browntop millet	--
10		*Fonio* (White & Black)	Hungary rice	--

Food habits: Drift from Animal Fodder to Functional Foods

Vast scope of utilization (feed, fodder, pharmaceuticals, biofertilizer and biofuels) emphasizes government to reforms policies favoring millets with advancement of time by the way of millet promotion, subsidies, mid day meal operation and much more. Now, restricted access of millet consumption is again rolling on to its use for human purpose than to animal fodder just (Singh and Kumar, 2018).

Some ethnic beverages or socio-economic cultural food practices or some Geographic indications have their own importance that are followed by our ancestors with nutri-cereals in them but diminishes in present era. Their health benefits are sufficient to avoid health risks associated with long list of marked adapted sodas, cola drinks, soft drinks, sherbets and others. Rarely, those traditional concepts are even practiced today by some tribal communities. Refreshing beverages and thirst quenching drink are on top among them like kodo ka jaanr; a traditional fermented drink of Sikkim based on finger millet has its own peculiarity to provide energy and nutrients to their regulars (Hati *et al.*, 2019). Sikkim is the abode of creations with millets where two more innovations viz. Tchang or Jhar and Rokshi as fermented drinks are also crowned to it. Chhaang or hot beer is a step ahead for Sikkimise, where yeast based fermentation to get beer also in use these days too. Above discussed all these fermented drinks have soothy flavours and fantastic aromas as practiced within bamboo cases. Drifted time changes the technologies and consumers behavior too that use to select high CO_2 purged, as today's choice, they are the dense calorie sweetening agents and nothing besides this.

i. Worldwide scenario

No doubt they are the crops of Asia and Africa as they are stamped to be originated and adopted by people in these continents and maintain that symbol with about 97 precent contribution of production from developing countries.

Table 2: Top 10 millet producing countries

Country	Production (Tones)
India	10,910,000
Nigeria	5.000,000
Niger	2,955,000
China	1,620,000
Mali	1,152,331
Burkina Faso	1,109,000
Sudan	1,090,000
Ethiopia	807,056
Chad	582,000
Senegal	572,155
World	29,870,058

Source. Sarita and Ekta, 2016

ii. Indian statistics and contribution

India is the leading country in millet production. We were habitual of millet as our staple crop in and around 1970's but by the year 2000 ignorance of millet reaches to the extent that staple crop recognition is reached to wheat and rice. No doubt production practices improve with time but per capita consumption ceases so drastically (50-75%). As a result of which the generated data in 2005 states that millet instead of its intended staple diet source is preferred to use only for various industrial applications of alcohol production and livestock fodder (Basavaraj *et al.*, 2013).

The largest millet consuming Indian states are also reported distinctively with type of variety selected for production and consumption. Population from different Indian states is adapted to plough different types of millets e.g. Gujarat is in lead with pearl millet, Karnataka leading with finger millet and Maharashtra ahead with sorghum for consumption (Varadhrajan *et al.*, 2013). On the other hand Kerala, Orissa, West Bengal and Tamil Nadu are the states with their choice of cereal eating confined only to rice. A study reported by National Nutrition Monitoring Bureau in 2006 for urban residents revealed that only 2 precent of their total calories were contributed by the millets (Konapur *et al.*, 2014).

Millets: A way forward to develop nutritionally secured nation

Small millets are not small in nutrition rather some of them are wonders in terms of seeds with nutraceutical features. This is the reason why one should involve millets in routine diet to harness benefits of millets. Environment protection is the additional benefit that is directly related to least chemical dosing and least water requirement while cropping of millets. So these wonder seeds will write so many success stories of right eating and healthy living for their nation's residents if adopted both in farming and consumption. This is also one of the reason why government reframed their policies and amended time to time or provides flexibility to farmers, traders and consumers too for selection of these types of healthy seeds for farming, trading and utilization respectively. Furthermore, mid day meal activities, subsidized availability, free seed sample distribution campaign, research institute establishment and contract farming are some of the initiatives and lively examples initiated by central and state government aimed with rise in average life expectancy of the nation's people.

Recent adopted trends by developed countries are the role model like where production of gluten free products in each and every segment of prepared food is the novel adoption. Due to this initiative, they arise as significant industries and hence valued their commercial viability. Also, biofortification phenomenon adoption where level of minerals and vitamins are prominently rises for better metabolic functional impact on to the body of consumers is the modest activity explored herewith.

So here authors are in the opinion of come forward with millets and put a drift in ongoing trend to be healthy because mostly poor countries are on its production side, struggling with least yield owe to limited resources thus production ceases while stable nations are still far away in concern to the millet.

i. Competitive Characterization with other Staple Cereals

Most appropriately suited characteristic of millets consumption is its gluten free trait. It explores the new hope for the patients who are suffering in any of these three ailments viz. gluten intolerant, wheat allergy or celiac disease. We Indians are really in need of gluten free options because all around the globe we are the only consumers of wheat in multiple forms from breakfast to dinner, religious occasion to other family functions, gifts, elderly and children preferred diets as well. Not just this, we also ranked top in diabetic patients and for obese population too. So adoption of millet is the need of time for all of us that direct to roll back to traditional practices of growing and consuming millets to fit and stay healthy with longer life.

Status of millet as gluten free grain, high fiber and low glycemic index traits are the next apprehension demanded by today's end user of any age, race, sex and of any location to get rid from chronic diseases related to them. Wheat as our staple diet is energy concentrated grain with gluten, similarly rice both with high glycemic index (Vaidya *et al.*, 2014). Least fiber characteristics in them directly influence our body and long term continuous adaptation on to them make us seek with lists of disturbed functionality of body organs. This is one of the reason why millets are for most preferred choice as like practiced ancestrally in multi grain preparation, millet based pan leavened bread and porridge preparation to be strong and fine even in the era where medical facilities are negligible. Much in minerals and vitamins in millets as in comparison with wheat and rice work to maintain body metabolism and somehow supplement our immune system to work efficiently for long.

Table 3: Comparative characterization of treated millet flour against staple cereal flour

Parameters	Raw FMF (mg/g)	GFMF (mg/g)	PFMF (mg/g)	Wheat Flour (mg/g)	Rice Flour (mg/g)
Protein	12.57 ± 1.22^a	10.92 ± 1.34^b	12.02 ± 1.34^a	11.8 ± 0.57^c	6.8 ± 0.87^d
Fat	3.22 ± 0.41^a	2.75 ± 0.59^b	3.37 ± 0.59^a	1.5 ± 0.045^c	0.6 ± 0.21^d
Ash	3.14 ± 0.36^a	3.35 ± 0.73^a	3.10 ± 0.73^a	1.5 ± 0.08^b	1.5 ± 0.05^b
Carbohydrates	71.14 ± 3.5^a	72.91 ± 2.45^a	71.96 ± 2.45^a	74.98 ± 6.98^b	81.21 ± 7.80^c
Crude Fiber	10.21 ± 1.67^a	10.03 ± 1.45^a	6.93 ± 1.45^b	1.2 ± 0.023^c	0.2 ± 0.010^d
Moisture	9.93 ± 1.89^a	10.07 ± 0.98^a	9.55 ± 0.98^a	10.22 ± 0.90^a	9.89 ± 1.23^a

FMF: Foxtail millet flour, GFMF: Germinated foxtail millet flour, PFMF: Pearled foxtail millet flour *Source*: Singh and Hathan, 2014

ii. Possible ways to Encourage Millet Consumption

Enhanced demand is the only key to popularize it by users and it will only be initiated if promoted for its positive features. Involvement of little pre-processing strategy to the harvested seeds is the innovative idea that enhances nutrient bioavailability and ceases anti-nutrients. Innovative technological pre-processing techniques such as debranning or pearling, soaking or steeping, sprouting or germination, hydrothermal treatment and fermentation (open and closed) are some of the adopted phenomenon by tribal community and even by house hold ladies of villages to enjoy the taste of millets without knowing the enhanced positive features of bio-availability of nutrients. It is enormously highlighted in some of the researches that soaking, germination and fermentation elevate the divalent cations and fiber level of millets with enhanced soothy taste and significant reduction in the level of anti-nutrients.

Simplified practices and vast scope of innovative formulation application fortified with millets like popping, puffing, extrusion, baking, and multigrain combination labeled with enhanced health attributes are the paradigm to encourage millet adoption by farmers, traders and consumers. Some of the latest innovation in breakfast flakes and infant formula are the technological innovation of the field practiced through long researches and developmental work (Raju *et al.*, 2015).

Enhanced consumption of millets is somehow directly or indirectly in dealt with the beneficial prospects of least labor input in fields, nutritional superiority, employment opportunities in pre-processing of harvested grains and environment betterment with least pesticides, fertilizers and water uses. Actually, coarse cereals are the environment resilient crops that adapt to any climate and even in damp soils thus contain full scope for marginal farmers to earn livelihood. Environment safety is another aspect, evident from least eutrophication and biomagnification values as utilized with negligible of pesticides and fertilizers. These are the crops depending seasonal rains and hence valued water recyclability. Lastly but not the least same crop is the source for feed for animals (green fodder), fish (powdered seed and chaff), poultry (husk and hull) and human (kernel).

iii. Functional Millet Foods in Indian Market

Currently many companies launched millet based products in Indian market. The demand for millet and value added products of millets are increasing as consumers becoming aware with the benefits of millets. Cleaned and graded millet is also available in attractive packing in market. On other hand value added products such as ragi choco balls, millet biscuits, Millet upma rawa, millet idli rawa, millet flour, instant idli mixes, dosa mix also available in supermarkets as well on online marketing platform. Though now days, general awareness for health benefits of millet is increased, still there is huge potential to exploit it commercially by way of value addition; dairy-cereal blends, cereal-millet composite preparation and also with dairy and fruit industries effluents to have functional drinks, bakery, extruded and heat desiccated product in alliance with millets.

Table 4: Millets; A view of traditional to the modern drift in commercial aspect

Sr No.	Traditional concept	Novel trends	Emerging trends
1	Flour grades	Instant porridges and instant foods	Gluten free foods
2	Dumpling, porridges and gruels	Expanded snack foods	Biofortified grains and their products
3	Popcorn	Ready-to-eat breakfast cereals	Phytochemical rich foods
4	Pan leavened breads	Pasta	Protein biopolymers
5	Noodles	Bioethanol production	Cellulosic bioethanol
6	Couscous	Starches and syrups	Fractionated starches
7	Malt, alcoholic beverages and animal feed	Forage and ground cover	Gluten free beers

Source. Taylor, 2019

iv. Medication through millet: Traditional Concept to Present Reforms

I. Diabetes and Millets: - That is the history since long when no one in Indian continent is found with diabetes but with the transformation of time, eating habits and lifestyles shortfalls its life expectancy rate with an average to about 60 years. In an estimate in regard to the global diabetic count, it is expected that it get double originally from 171 million in 2000 to 366 million in 2030 with special threat to India with highest number (Kaveeshwar and Cornwall, 2014). We are prosperous enough in nature and nature producing organic and inorganic produces but lack of awareness, fraudulent practices and biased governmental policies now we are diseased enough and competing with rest of countries to achieve peak in this regard (Ezzati *et al.*, 2002). Present data of obesity, diabetes, cardiovascular disease, mortality rate of neonatal infant, anaemic girls, under developed children and many more conferred this recognition of India in present status while traditionally consumption of millets as staple cereal is the logic to save us from being diabetic, obese, malnourished, undernourish and anaemic etc. Various epidemiological evidences show that diet rich with various plant sources especially of coarse cereal classes are protective in nature against lists of degenerative diseases viz. age related disorders, cardiovascular functionality, cancer and diabetes (Kaur *et al.*, 2014). Millet itself recognized as functional food owe to the resistant starches, high dietary fibre and antioxidants presence. Reduction in tumor formation, lowering of blood pressure and supplying of gastrointestinal bulk thus delayed in satiety feeling are the pre-eminent characters that make us all feel to fit well.

Diabetes is a metabolic syndrome judged by imbalance in carbohydrate, protein and lipid metabolism (Eckel *et al.*, 2005). The term "Refining" is the drastic factor and if applied in "eatables" e.g. flour, sugar, oil and then long term strictness to the same diet, diabetes will confirm. Adoption of practices as we usually care take of traditionally means coarse grinding or grits consumption as porridge, millet consumption and lacking of sieving stages now days evidenced so easily that fibers are the factors that interferes the unnecessary absorption and hence maintains the glycemic index to normal. Thus once again adoption of whole grain consumption habits, diet shift pattern to multi grain with millets or dairy millet composite blend are the natural means to stay fit and fine in special perspective to diabetes and obese (Park and Kim, 2011). A comparative study for finger millet consumption in competitive to wheat and rice induced lower plasma glucose level. That all occurs due to high fibre value and might be due to anti-nutritional factors that interfere their digestion. It is also evaluated that finger millet polyphenol is the major ant-diabetic constituent (Bijwan *et al.*, 2019).

Delayed wound healing responses is also influenced with decrease in antioxidants in body as reported when studied on to the rats in experimentation. However the same is when fed with finger millet based diet shown to have fast wound healing because of raised antioxidant level (Shahidi and Chandrasekara, 2013). Barnyard millet (heat treated and dehulled) influence positively by lowering in glycemic index. In addition to all this, millet and other coarse cereals on extrusion exhausted with all the readily digestible sugars and slow digested sugars will remain thus also shown to possess ant-diabetic characteristics.

Nambiar *et al.* (2011) reviewed health benefits of pearl millet. It is reported as grain that retains its alkaline properties even after cooking and is also gluten free grain (Leder, 2004). It is also rich in potassium, B-vitamins, phosphorous, iron, copper, magnesium, manganese and zinc (NIN, 2003). Pearl millet flour has reported with maltose and D-ribose as predominant sugars and low levels of fructose and glucose (Oshodi *et al.*, 1999). Diet consisting of pearl millet flour is thus considered to be of prime importance in management of diabetes mellitus (Ballolli, 2017). have reported that pearl millet has the lowest GI (55) as compared to combined mix flour of whole and dehusked green gram, Jowar and Ragi. Such low glycemic index foods are helpful to manage diabetes by improving metabolic control of blood pressure. Thus pearl millet based food products / traditional recipes need to promote for the diabetic patients.

II. Obesity and millets: -Obesity is another commonly observed result of sedentary lifestyle with changing food habits. About 250 million people are added in obesity class daily to the 7.2 billion who already live on the Earth. Yet more than 66 precent of the world population are currently malnourished and starving (Bhat *et al.*, 2019). In the same time, 21st century's rich people disease is obesity with more than half of the world's adult population either overweight or obese (Baker *et al.*, 2016). How more than 66 precent malnourished people are obese as per the sayings above, but it really occurs just because of fluctuating policy for the governments, biased and eating unknowingly and selection of eatables as taste demands not through healthy perspective. This is the major drawback that can be seen in malnourished and obese or overweight population when compare with traditional data of enjoy healthy eating habits.

Among millets, foxtail (*Setaria italica*) and kodo (*Paspalum scrobiculatum*) varieties are the home for dietary fibre thus are on likeness preference for the patients suffering from various ailments like diabetes, constipation, obesity and cardiovascular diseases. They are embedded with high level, ranged in from 22 to 38 precent respectively. Owe to this complex carbohydrate; influence slow release of glucose into blood streams to prevent abrupt rise in glycemic index. These are the natural remedies for betterment of living stay on this planet as it experience avoid the person to be diabetic owe to its effect to counter glucose diffusion in body cells. Previously focused literature survey also recognize the competitive potential of foxtail millet insoluble fiber on to the commercially available insoluble fibers obtained from soy. β-glucan is the major fraction with these stated characteristic insoluble dietary fibre of β-(1, 3–1,4)-D-glucan monomer unit that contribute in health upliftment with reduction in serum cholesterol level and blood lipid level as well.

One of the similar investigation suggested by Aoe *et al.* (2014) favor our saying that diets with β-glucan fortification in white rice works against the onset of obesity and its related disease. Also it is not contradictory to state that all the clinical studies comes to the common approach of multiple health impact that almost increases with application of minimal pre-processing stages to the harvested kernels like germination that positively impact the bioavailability of minerals (divalent cation) i.e. Zn^{2+}, Mg^{2+} and Fe^{2+} similarly for vitamins alongwith elevated protein and hence experience the achieved bio-fortification. On the other hand dense calorie compounds such as fatty acids, carbohydrates decreases significantly with progress of germination thus suits better to have healthy days ahead.

III. Anaemia and millets: - One of the recent posts from Bajpai (2018) confirms India's status to be home of 194.6 million under nourished people, highest in world and it occupies 112th rank among 141 nations in reference to the child development index.

Similarly US national institute of health reported that 1.6 billion people of are anaemic, Major cause of being anaemic is iron deficiency. Iron deficiency reflexes person's to be with ceases physical performance and other psychological behavioral changes. India alone contributes about 50 precent of maternal deaths all around the globe due to anaemia. A study conducted by Scott *et al.* (2018) on to the school going childrens of Maharashtra revealed that consumption of Iron bio-fortified pearl millet improves iron status and some measures of cognitive behavioral and physical performances in Indian adolescents.

We also get relaxed from being anaemic if we are supplied with millets in our meals. Researchers call upon it as ''biofortification" to combat micronutrient deficiency; focused specially in lieu of iron deficiency anaemia. It was also suggested time to time (Julia *et al.*, 2017) that benefits of bio-fortified crops will be of great interest for the patients who are iron deficient at baseline and for those who consumed biofortified crops in a frequent manner. In a nut shell effective intervention to improve iron status primarily can be achieve in this very way for a nation where about 70 precent population is engaged in farming that in due course relaxed us with elevated serum ferritin and total body iron level.

Infants, pregnant and lactating women are on the major risk of iron deficiency and hence anaemic for developing countries. Other ailments associated with iron deficient body is impairment in cognitive performance, reduced work efficiency, ceased immunity in alongwith various pregnancy complications viz. low birth weight babies, poor learning capacity and reduced psychometer skills etc. Iron itself effect 15 precent of world's population as iron deficient people where 35 precent of women and 43 precent of young childrens are concerned (Oladiji *et al.*, 2007). Drastically, haemoglobin depletion and reduced working efficiency of iron dependent intra-cellular enzymes meant for maintenance of metabolic well being is the big issue here to address.

IV. Cardiovascular disease and millets: - Heart is the core organ of body that possess all the necessary actions required throughout whether we are in sleep or in awakened mode. Pollution problems, smoking, alcohol drinking and high calorie diet consumption in regular way causes severe threat alarming to death of person. Present outcomes through animal studies demonstrated that rats who adapted to strict diet of native and treated starch from barnyard millet influences them with least blood glucose, serum cholesterol as in comparison with diet based on rice and other minor millets (Saleh *et al*, 2013). In addition, serum triglycerides were reduced significantly for finger millet and proso millet groups of hyper-lipidemic rats than those fed with white rice and sorghum groups. Moreover, concentrations of serum total, HDL, and low density lipoprotein (LDL) cholesterols were also elevated significantly in sorghum and white rice consumption than of finger millet and proso millet groups. Hence it may concluded that millets work effectively against cardiovascular disease, blood pressure problems and obesity (Saleh *et al.*, 2013).

In nutshell, millets are fat rich such as pearl millet containing about 5 mg/100g of fat as stated in National Institute of Nutrition report (2003). High fat in pearl millet is directly attributed to high germ proportion that by way of lipase action develops high free fatty acids and is probable reason to rancidity after milling within week if allowed (Lai and Varriano, 1980). Most advantageous term here is the unsaturated fatty acid profile presence with prevalence of about 75 precent. Among unsaturation, ω-3 linolenic acid alone contribute around 4 precent of the total fatty acid of it and it is assumed by researchers that intake of HDL is somehow more efficient in replacement of LDL proportion from body fat to get rid of all ailments associated with fat consumption (Vanisha *et al.*, 2012).

V. Cancer and millets:- Millets are the store house of plants secondary metabolites that possess quite benefits to the human health. Certain species in this category are rich in bioactive phytochemical that prominently headed in finger millet (high in reducing capacity), foxtail millet, proso millet and kodo millet (high in polyphenolics). Obtainable bioactive phytochemicals in them includes; phenol, tannins, alkaloids, flavonoids and saponins. Each of them possess its specialty to improve health status *viz.* metabolic activities are maintained through alkaloids, flavonoids are the defense mechanism of body hence possess certain biological properties *viz.* antibacterial, antitoxic, anti-inflammatory and somehow as antioxidants thus anti-tumorous as well. Similarly saponins are the anti-carcinogens and have hypolipidomic effect. All these together put cumulative medicinal effect on to the body and all.

They quench the reactive oxygen species and hence reduce their count in the body which ultimately impact positively from being cancerous, skin related disorders (dark circles, blemishes, loosening of skin) and heart diseases.

Antioxidants (Phenolic compounds) of millet have anticancer properties. Reported data from time to time with phenolic content of 51.4 mg/100g dry weight, 608.1 mg/100g and 761 mg/100g (Huang *et al.*, 1992) in pearl millet favors its properties to be directly linked to inhibit tumor development. These phenolics majorly are of testa and pericarp concentric thus consumption of whole grain is the only reason to combat against these associated risks.

Millet is among harsh condition adapted crop with lot in nutrition and nutraceutical for their consumers. To be on well being side and also to enjoy the advantages of healthy life, antioxidants and polyphenolics are the top most priorities to select to eat. Plant polyphenols are the secondary metabolites that evidenced themselves in various food researches, pharmaceutical, biotechnological and microbiological applications as potent anticarcinogens besides antimicrobial, antidiarrhoeal and anti-infalmmatory properties. Various in-vitro cytotoxic assays studied for phenolic extracts from millets reported antiproliferative activities against human breast cancer and liver cancer cells (Nidhi *et al.*, 2015). Also, it could be inferred that some other millet extract based *in-vitro* cytotoxic studies (Sreeramulu *et al.*, 2009) conferred them to be the abode of natural antioxidant thus relieves us from oxidative damage and cytotoxic effects too.

Conclusion

First of all, millets habitat to flourish well even in low rainfall and least attention and care while in field practices and even less attention while in storage time is advantageous. These are traditionally accepted, adapted as medicine not as grains owe to the least glycemic index character, ant-diabetic and cholesterol lowering impact to humans. So lots of available blended composition as; beverages, porridge, milk blended new products and baked stuffs compatibility gives it new horizons to run upon in the present time of competitive innovation with healthy cum nutritious approach. Evidenced clinical researches describes its full potential if consumed through a value chain on composite dairy foods with enhanced health attributes. Each millet grain is naturally provided with many of variable characters that are needed now days to have healthy disease-free longevity.

References

Aoe, S., Ikenaga, T., Noguchi, H., Kohashi, C., Kakumoto, K., & Kohda, N. (2014). Effect of cooked white rice with high β-glucan barley on appetite and energy intake in healthy Japanese subjects: a randomized controlled trial. Plant Foods for Human Nutrition, 69(4), 325-330.

Arnold, T. 26 Global Progress in the Scaling Up Nutrition (SUN) Movement. The Biology of The First 1,000 DAYS, 409.

Bajpai, A. (2018). Child rights in India: Law, policy, and practice. Oxford University Press.

Baker, R., Taylor, E., Essafi, S., Jarvis, J. D., and Odok, C. (2016). Engaging young people in the prevention of noncommunicable diseases. Bulletin of the World Health Organization, 94(7), 484.

Ballolli, U. (2017). Prevalence and management of adolescent obesity through millet based foods (Doctoral dissertation, UASD).

Basavaraj, G., Rao, P. P., Basu, K., Reddy, C. R., Kumar, A. A., Rao, P. S., and Reddy, B. V. S. (2013). Assessing viability of bio-ethanol production from sweet sorghum in India. Energy Policy, 56, 501-508.

Berti, C., Riso, P., Brusamolino, A., and Porrini, M. (2005). Effect on appetite control of minor cereal and pseudocereal products. British Journal of Nutrition, 94(5), 850-858.

Bhat, Z. F., Morton, J. D., Mason, S. L., and Bekhit, A. E. D. A. (2019). Pulsed electric field improved protein digestion of beef during *in-vitro* gastrointestinal simulation. LWT, 102, 45-51.

Bijlwan, M., Naik, B., Sharma, D., Singh, A., and Kumar, V. (2019). Recent developments in dough based bakery products: A mini review.

Eckel, R. H., Grundy, S. M., and Zimmet, P. Z. (2005). The Metabolic Syndrome. The Lancet, 365(9468), 1415-1428.

Ezzati, M., Lopez, A. D., Rodgers, A., Vander Hoorn, S., Murray, C. J., and Comparative Risk Assessment Collaborating Group. (2002). Selected major risk factors and global and regional burden of disease. The Lancet, 360(9343), 1347-1360.

Hati, S., Das, S., and Mandal, S. (2019). Technological Advancement of Functional Fermented Dairy Beverages. In Engineering Tools in the Beverage Industry (pp. 101-136). Woodhead Publishing.

Huang, M. T., Ho, C. T., Wang, Z. Y., Ferraro, T., Finnegan-Olive, T., Lou, Y. R., and Conney, A. H. (1992). Inhibitory effect of topical application of a green tea polyphenol fraction on tumor initiation and promotion in mouse skin. Carcinogenesis, 13(6), 947-954.

Julia, L.F., Jere, D.H. and Saurabh, M. (2017). Iron-biofortified staple food crops for improving iron status; A review of the current evidence. Current Opinion in Biotechnology. 44, 138-145.

Kaur, K. D., Jha, A., Sabikhi, L., and Singh, A. K. (2014). Significance of coarse cereals in health and nutrition: a review. Journal of Food Science and Technology, 51(8), 1429-1441.

Kaveeshwar, S. A., and Cornwall, J. (2014). The current state of diabetes mellitus in India. The Australasian Medical Journal, 7(1), 45.

Konapur, A., Gavaravarapu, S. R. M., Gupta, S., and Nair, K. M. (2014). Millets in meeting nutrition security: issues and way forward for India. Indian J Nutr Diet, 51(3), 306-321.

Lai, C. C., and Varriano-Marston, E. (1980). Lipid content and fatty acid composition of free and bound lipids in pearl millets. Cereal Chemistry, 57(4), 271-274.

Léder, I. (2004). Sorghum and millets. Cultivated Plants, Primarily as Food Sources, 1, 66-84.

Manoj Kumar, C. T., Raju, P. N., and Singh, A. K. (2015). Barley-pearl millet-milk based mineral fortified complementary food: moisture adsorption isotherms. Indian J. Dairy Sci, 68, 3.

Masih, J., Verbeke, W., Deutsch, J., Sharma, A., Sharma, A., Rajkumar, R., and Matharu, P. S. (2019). Big Data Study for Gluten-Free Foods in India and USA Using Online Reviews and Social Media.

Nambiar, V. S., and Matela, H. (2012). Potential functions of lemon grass (*Cymbopogon citratus*) in health and disease. International Journal of Pharmaceutical and Biological Archives, 3(5), 1035-1043.

Nambiar, V. S., Dhaduk, J. J., Sareen, N., Shahu, T., and Desai, R. (2011). Potential functional implications of pearl millet (*Pennisetum glaucum*) in health and disease. Journal of Applied Pharmaceutical Science, 1(10), 62.

Nidhi S, Meenu G, Anamika S. and Jayanthi A. (2015). Evaluation of antimicrobial and anticancer properties of finger millet (*Eleusine coracana*) and pearl millet (Pennisetum glaucum) extracts. The pharma Innovation Journal, 3(11); 82-86.

NIN (2003). Nutritive value of Indian Foods, Ed Gopalan and Deosthale, National Institute of Nutrition, Hyderabad, 2003

Oladiji, A.T., Jacob, T.O. and Yakubu, M.T. (2007). Anti-anaemic potentials of aqueous extracts of Sorghum Bicolor (L.) moench stem bark in rats. doi:10.1016/j.jep.2007.01.013.

Oshodi, HN Ogungbenle, MO Oladimeji, A. A. (1999). Chemical composition, nutritionally valuable minerals and functional properties of benniseed (*Sesamum radiatum*), pearl millet (Pennisetum typhoides) and quinoa (*Chenopodium quinoa*) flours. International Journal of Food Sciences and Nutrition, 50(5), 325-331.

Park, T., and Kim, Y. (2011). Phytochemicals as potential agents for prevention and treatment of obesity and metabolic diseases. Anti-Obes Drug Discov Dev Bentham, Dubai, 1, 150-185.

Rao, G. H. (2015). Non-traditional approaches to diagnosis and management of Type-2 diabetes mellitus: point of view. J Diabetes Metab, 6(489), 2.

Saleh, A. S., Zhang, Q., Chen, J., and Shen, Q. (2013). Millet grains: nutritional quality, processing, and potential health benefits. Comprehensive reviews in Food Science and Food Safety, 12(3), 281-295.

Sarita, E. S., and Singh, E. (2016). Potential of millets: nutrients composition and health benefits. Journal of Scientific and Innovative Research, 5(2), 46-50.

Scott, S. P., Murray-Kolb, L. E., Wenger, M. J., Udipi, S. A., Ghugre, P. S., Boy, E., and Haas, J. D. (2018). Cognitive performance in Indian school-going adolescents is positively affected by consumption of iron-biofortified pearl millet: A 6-month randomized controlled efficacy trial. The Journal of Nutrition, 148(9), 1462-1471.

Shahidi, F., and Chandrasekara, A. (2013). Millet grain phenolics and their role in disease risk reduction and health promotion: A review. Journal of Functional Foods, 5(2), 570-581.

Singh, A. K., Rehal, J., Kaur, A., and Jyot, G. (2015). Enhancement of attributes of cereals by germination and fermentation: a review. Critical Reviews in Food Science and Nutrition, 55(11), 1575-1589.

Singh, A., and Kumar, P. (2018). Gluten free approach in fat and sugar amended biscuits: A healthy concern for obese and diabetic individuals. Journal of Food Processing and Preservation, 42(3), e13546.

Sreeramulu D, Reddy CV, Raghunath M. (2009) Antioxidant activity of commonly consumed cereals, millets, pulses and legumes in India. Indian Journal of Biochemistry and Biophysics, 46; 112-5.

Taylor, John R. N. (2019). Sorghum and Millets, taxonomy, history, distribution and production. Sorghum and millets (2nd Ed.), https://doi.org/10.1016/B978-0-12-811527-5.00001-0.

Vaidya, R., Mohan, V., Bai, M. R., and Vasudevan, S. (2014). Glycemic Index of Indian Cereal Staple Foods and their Relationship to Diabetes and Metabolic Syndrome. In Wheat and Rice in Disease Prevention and Health (pp. 333-346). Academic Press.

Varadharajan, K. S., Thomas, T., and Kurpad, A. V. (2013). Poverty and the state of nutrition in India. Asia Pacific Journal of Clinical Nutrition, 22(3), 326-339.

Wu, Q., and Liang, X. (2018). Food therapy and medical diet therapy of traditional Chinese medicine. Clinical Nutrition Experimental, 18, 1-5.

Zou, P. (2016). Traditional Chinese medicine, food therapy, and hypertension control: a narrative review of Chinese literature. The American Journal of Chinese Medicine, 44(08), 1579-1594.

13

Millet Protein Hydrolysates and Peptides as Functional Foods

***Himani Agrawal*[1,2], *Vikas Dadwal*[1,2], *Rashim Kumari*[1,2] *and Mahesh Gupta*[1,2]**

[1]CSIR-Institute of Himalayan Bioresource Technology, Palampur-176061 Himachal Pradesh, India

[2]Academy of Scientific and Innovative Research (AcSIR), Ghaziabad-201002 India

Introduction

Millets like finger millet, foxtail millet, pearl millet, sorghum millet and other small millets are the ancient super grains and nutrition reservoirs for better health. In semi-arid regions, these are the important food and fodder crops, and primarily acquiring more importance in the world. Millets are adapted to a wide range of temperatures, moisture-regimes and input conditions supplying food and feed to millions of dryland farmers, particularly in the developing world (Devi *et al.*, 2014). Besides, millets form the essential raw material for the production of potable alcohol and starch in industrialized countries (Li *et al.*, 2008). For the present and future, millets are the superfoods and their short growing season from planted seeds to mature, ready to harvest plants in as little as 65 days make them commercially sound. When properly stored, whole millets will keep for two or more years (Devi *et al.*, 2014).

Millets are cultivated in tribal and rain-fed, low-fertile land and mountainous areas including Telangana, Tamil Nadu, Uttar Pradesh, Karnataka, Maharashtra, Rajasthan, Madhya Pradesh, Haryana, Gujarat, Chhattisgarh, and Andhra Pradesh. During the pre-Green Revolution era, in 1965-66, millets cultivation was 36.90 million hectares. However, the area under millet cultivation declined to 14.72 million hectares in 2016-17, because of change in dietary habits, consumption pattern, less demand, low yield, unavailability of millets, and conversion of irrigated area for cultivation of wheat and rice (Mohan, 2018). By seeking this transformation, the Government has also decided to declare 2018 as "National Year of Millets" (Swaminathan, 2018).

Millets are known as one of the oldest foods to humans but in favour of rice and wheat, they were rejected with industrialization and urbanization. As a gift of newly acquired lifestyles, with cardiovascular diseases, hypertension and diabetes, millets have reverted as a sustainable choice to live a healthy life without consuming loads of expansive medicines that have serious side-effects in the long run. In our inner ecosystem, millets also act as a prebiotic feeding micro-flora. Millets hydrate our colon to keep us from being constipated. Millets contain Niacin (Vitamin B_3) and magnesium which can help in lowering cholesterol and reduce the effects of heart attacks and migraines respectively. All millet varieties show high antioxidant activity and it also prevents cardiovascular disease by decreasing C-reactive protein and triglycerides. Millets are non-allergenic and gluten-free. Besides, free fatty acids like linoleic acids, oleic acid and palmitic acid are also present in millet grains in the free form (Bagdi *et al.*, 2011). Hence, millets are now being pronounced as " Adbhut Anaj/Miracle grains and nutria-cereals".

Millet demand is going to higher due to their importance as functional health foods and value addition for food and non-food uses. In recent years, millets grains and their ingredients are accepted in terms of functional food and nutraceuticals and receiving scientific attention. Functional foods are those food or dietary constituents that offer a health benefit beyond basic nutrition. Food can be made functional by applying any technological or biotechnological means to increase the concentration of, add, remove or modify a particular component as well as to improve its bioavailability, provided that component has been demonstrated to have a functional effect (Hasler, 2002).

Usually, a dietary protein considered as a source of essential amino acids and energy, required for the maintenance of physiological functions and normal growth. The human body requires sufficient intake of protein to maintain the body nitrogen balance (Dideriksen *et al.*, 2013). Excess amount of protein ingestion results in excess metabolism and excretion. Conversely, the body utilizes its own proteins as a source of nitrogen due to inadequate intake of dietary protein; hence, regular and adequate intake of dietary protein is necessary. In the human body, there are numerous key functions including provision of energy (4 kcal/g protein), cell signalling. repair of tissues, structural and enzymatic, performed by protein (Hoffer, 2016).

Recently, identification and characterization of bioactive peptides and hydrolysates from animal and plant sources has become an emerging area of interest. Protein hydrolysates have been defined as "mixtures of polypeptides, oligopeptides and amino acids that are manufactured from protein sources using partial hydrolysis" (Schaafsma, 2009). Bioactive peptides are specific inactive fragments in the sequence of the parent protein, released by the action of proteolytic enzymes and exert several biological effects upon activation. The size of these bioactive peptides is in the range of 2-20 amino acids with a molecular mass of less than 6000 Da. These bioactive peptides exert several biological functions depending on their amino acids composition and sequences. Bioactive peptides possess hypocholesterolemic, antithrombotic, antioxidative, antimicrobial, immunomodulatory, mineral binding and opiate-like activity. Over the last two decades, the interest has been grown in this area and novel bioactive peptides continually being discovered, as these short-chain peptides from hydrolysed proteins may be utilized more efficiently than free amino acids and also possess a higher nutritive value (Grimble *et al.*, 1986). It has appeared that, in human nutrition, protein hydrolysates have many uses such as sports nutrition products, weight-control, ingredients in energy drinks and (Frokjaer, 1994), also nutrition source for immuno-compromised and elderly patients (Nagodawithana *et al.*, 2008). Millet proteins have also contain several bioactive peptides. The present chapter deals with the bioactive peptides identified from millet protein hydrolysates. It describes a specific characteristic of millet derived hydrolysates and peptides. Furthermore, it represents an overview of hydrolysate, bioactive peptides, production and the structure-activity relationship of millet derived antioxidant peptides and their safety concerns.

Millets Protein and their Functionality

In comparison to rice and wheat, millets contain considerably higher protein content. The protein content in millets like barnyard millet (11.6%), foxtail millet (12.3%), proso millet (12.5%), bajra (11.6%), and jowar (10.4%), is equivalent to wheat (11.8%) and much higher than rice (6.8%). However, finger millet comprises lesser protein (7.3%) but it is rich in minerals and calcium in contrast to rice and wheat. Millets typically contain higher quantities of essential amino acids than rice and maize (Obilana and Manyasa, 2002). Relative percentages and distribution of total protein within the pearl millet comprised of endosperm (59.5%), germ (31.2%) and pericarp (9.4%). The higher germ to endosperm ratio of pearl millet results in higher percentages of the albumin and globulin protein in addition to higher protein content. From a nutritional perspective, this also favourably impacts the amino acid composition, particularly in terms of tryptophan and lysine content (Serna-Saldivar, 1995). Main storage protein of foxtail millet is prolamin which is more than 55 percent and rich in hydrophobic amino acids. Moreover, foxtail millet protein contains 19.89 percent branched chain amino acids, 10.20 percent aromatic amino acids, and 10.15 percent Pro of the total amino acids which signifies the special physiological functions and physiochemical properties of foxtail millet protein (Kumar and Parameswaran, 1998).

Oxidative Stress

Oxidation of biomolecules is a free radical-mediated process, plays a crucial role in all living microorganism. Hydroxyl (OH) and superoxide anion are free radical molecules, atoms and ions with unpaired electrons (Zhang *et al.*, 2011). Mitochondrial respiration and exogenous chemicals generate an excessive amount of free radicals may cause oxidative damage, cell death and tissue damage by oxidizing biomolecules (Kehrer, 1993). Furthermore, during the processing and storage of food products, fats and oils get oxidized which worsen the quality of nutritive values and lipid content (Rajapakse *et al.*, 2005). Upon consumption of these toxic products, several diseases can rise. Air pollutants and oxidants in tobacco can be absorbed to blood circulation or cause some harmful reactions in the skin and exert adverse effects. Moreover, generation of varieties of oxidants is also stimulated by ultraviolet radiation (Hruza and Pentland, 1993; Ichihashi *et al.*, 2003). Under normal condition, antioxidant defence system can remove reactive oxygen species either through enzymatic reaction (glutathione peroxidase and superoxide dismutase) or through non-enzymatic reaction (coenzymes cofactors, trace elements and vitamins) (Johansen *et al.*, 2005). Conversely, in a certain condition, the endogenous defence system fails to protect the body from reactive radicals, resulted in oxidative stress (Kaneto *et al.*, 1999). Oxidative stress is a condition of excessive or inadequate elimination of reactive oxygen species (ROS) and reactive nitrogen species (RNS), highly reactive molecules which play as a promoter rather than an initiator of chronic diseases (Rice-Evans and Diplock,

1993) (Fig 1). These harmful oxidation process can be delayed or inhibit using bio-absorbable antioxidants. In recent years, various synthetic antioxidants such as propyl gallate, butylated hydroxytoluene (BHT) and butylated hydroxyanisole (BHA) have been used to inhibit the free radical-mediated process. Though, these synthetic antioxidants are efficient and economical but also exhibit various hazardous and toxic effects (Ito *et al.*, 1985). Hence, in the area of biochemistry and human nutrition, natural antioxidants from food sources with no or little side effects currently have become an area of interest. Among the public, there has also been an increasing tendency to choose natural antioxidants over a synthetic one.

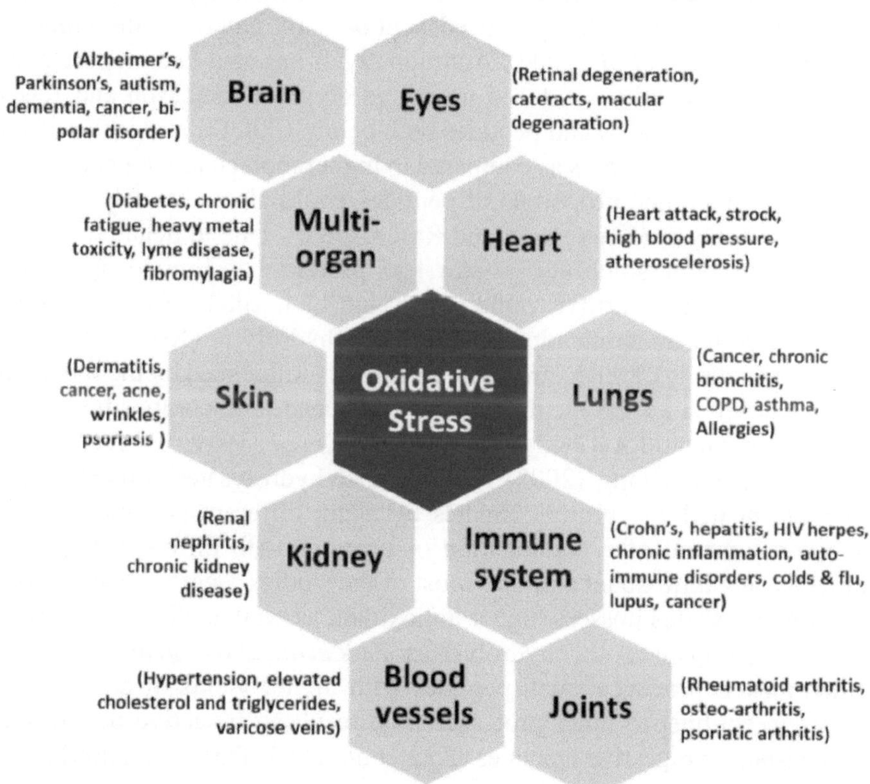

Fig. 1: Impact of free radical oxidative stress on human body

Preparation Technique for Bioactive Peptides and Hydrolysates

Bioactive peptides can be generated from precursor proteins using several techniques such as enzymatic hydrolysis (enzymes derived from plants and microorganisms) and microbial fermentation (Korhonen and Pihlanto, 2006). Fig 2 illustrates the different techniques used for the production of bioactive peptides. To enhance the nutritional and functional properties of proteins, enzymatic hydrolysis is the most widely accepted technique (de la Barca *et al.*, 2000; Moure *et al.*, 2005; Yim and Lee, 2000). For example, after enzymatic digestion, glycinin and α-conglycinin showed improved antioxidant activity (Saito *et al.*, 2003) because of more active amino acid R groups exposed upon hydrolysis, resulted in increased antioxidant activity (Matoba, 2002). Development of hydrophobicity is the additional advantage of hydrolysis since proteolysis unfolds the protein chains. Furthermore, hydrolysis also reduces allergenic agents (Moure *et al.* 2005). Solubility also increases due to increased free carboxyl and amino groups, improved by peptide bond cleavage. Hence, hydrophobicity can be decreased or increase upon hydrolysis, depending on the molecular weight of the produced peptides and nature of the precursor protein (de la Barca *et al.*, 2000). Additionally, it has also been reported that small bioactive peptides produced upon hydrolysis (Kristinsson and Rasco, 2000, Moure *et al.*, 2005) and peptides with below 1000 Da molecular weight showed much less bitterness as compared to higher molecular weight peptides. Conversely, functional properties of peptides could also adversely affect by extensive hydrolysis (Kristinsson and Rasco, 2000). Due to negative sensory attributes and poor solubility, alfalfa leaf protein has limited application in spite of the high nutritive value (Chen and Qiu, 2003). Such limitations can be overcome by application of enzymatic hydrolysis (Xie *et al.*, 2008). Antioxidant activity of the hydrolysates is directly influenced by the specificity of protease, which affects the amount, size, sequence and composition of free amino acid and peptides (Chen *et al.*, 1995; Jeon *et al.*, 1999; Wu *et al.*, 2003). Peñta-Ramos and Xiong (2002) reported that hydrolysates produced from native and heated soy protein isolates by using different enzymes resulted in the development of a combination of peptides with different degrees of hydrolysis and antioxidant activity. Most of the studies reported that alcalase-derived hydrolysates possess higher antioxidant as well as diverse biological activities as compare to other hydrolysates (Park *et al.*, 2001; Qian *et al.*, 2008). Furthermore, it produces small peptides with higher yields of antioxidative peptides in contrast to other proteases. The resulting bioactive peptides are more resistant to digestive enzymes (Kim *et al.*, 2001; Park *et al.*, 2001).

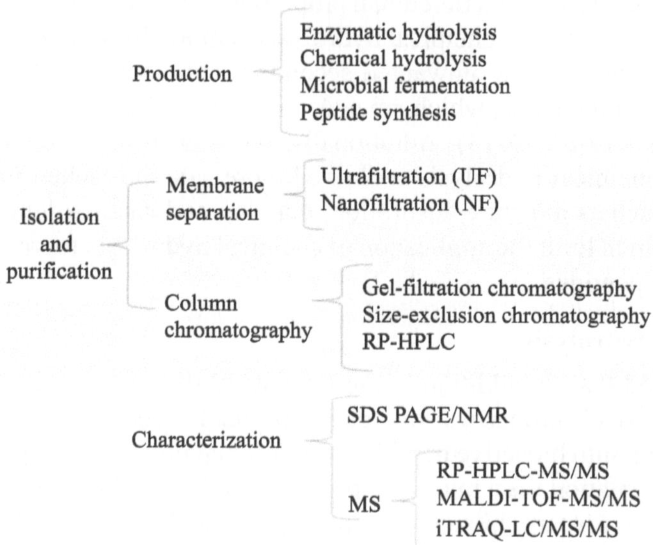

Fig. 2: Flow diagram of key technologies in the production, isolation, purification, and characterization of bioactive peptides.

Enzymatic Hydrolysis

Enzymatic hydrolysis is the key process for the production of protein hydrolysates and peptides (Wang *et al.*, 2014). Several factors including enzymc/substrate ratio, degree of hydrolysis (DH), hydrolysis time, enzyme specificity and amino acid sequence affect the composition and functional properties of bioactive peptides (Malumba *et al.*, 2008; Merz *et al.*, 2015, Zhuang *et al.*, 2013). It is now clear that different cleavage behaviours in the peptide chain via diverse proteases can yield various functional bioactive products, and the substrate hydrolysis rate tends to decrease with increasing hydrolysis time and eventually reaches a stable point (Tavano, 2013). Some frequently using proteases in the production of bioactive peptides are papain (EC 3.4.22.2), pepsin (EC 3.4.23.15), trypsin (EC 3.4.21.4), neutrase (EC 3.4.24.28), Flavorzyme (EC 3.4.15.1), thermolysin (EC 3.4.24.27) and alcalase (EC 3.4.21.1).

Chemical Hydrolysis

Another approach to improve the functional properties of food-derived proteins is chemical hydrolysis (Fountoulakis and Lahm, 1998). For the production of food protein-derived bioactive peptides; chemical reagents, alkaline hydrolysis and acid hydrolysis have been commonly applied. Due to high digestibility and the presence of an appropriate amount of amino acids bioactive peptides, have its functional properties (Zhu *et al.*, 2008). As compare to enzymatic hydrolysis, chemical hydrolysis is more difficult to be controlled and also produce several unwanted secondary compounds such as L-amino acids and D-amino acids, which are uneconomical and difficult to be removed (Samaraweera *et al.*, 2013). Additionally, the final product obtained in the process of chemical hydrolysis contained more salt and lacked in functional properties such as moisture adsorption, foaming and water solubility (Kim *et al.* 2004), which limit the application of chemical hydrolysis for the production of bioactive peptides.

Microbial Hydrolysis

It is supposed that, in addition to the long storability, fermentation enhances the nutraceutical properties of foods, probably because of the protein fragmentation into bioactive peptides by microbial proteases (Rajapakse *et al.*, 2005). Fermentation via a microbial protease is a favourable technique besides hydrolysis to prepare bioactive peptides, particularly from those protein sources that are discarded (Sheriff *et al.*, 2012). Fungi, lactic acid bacteria (LAB) and *Bacillus subtilis* accomplished the liberation of a large number of dipeptides, tripeptides, and other bioactive peptides during fermentation (Cao *et al.*, 2009; Singh *et al.*, 2014). These microorganisms are accepted as safe and cheap sources of proteases and do not cause any loss of essential amino acids (Singh *et al.*, 2014). Additionally, fermented products have functional properties and better health-related benefits, such as lower glucosinolate content (He *et al.*, 2012).

Bioactive Peptides and Hydrolysates

The biological activity and functionality of protein isolate, hydrolysate and its peptides are currently the focus of many types of research as they are widely distributed and safe and maybe use for the prevention of chronic diseases (Baydar *et al.*, 2007; Rajapakse *et al.*, 2005). Bioactive peptides are specific protein fragments, having a positive impact on body functionality and human health (Sa´nchez and Va´zquez, 2017). These specific peptides motifs remain inactive within the polypeptide chain of parent protein, thus released by proteolytic enzymes and exert several biological effects upon activation (Dziuba and Darewicz, 2007). Due to their positive health benefits, bioactive peptides are currently the focusing area and have also been considered as the new generation of biologically active regulators. Bioactive peptides are attributed to several biological activities such as antioxidant, antihypertensive, antithrombotic, antimicrobial, immunomodulatory, opioid and mineral binding activity (Sa´nchez and Va´zquez, 2017). Numerous peptides from protein sources have been found to exhibit antioxidant activity. Antioxidents peptides from foods contain 2-20 amino acid residues, with low molecular weight, easy absorption, high activity, low cost and considered to be safe and healthy compounds. As compared to enzymatic antioxidant they have various advantages including no hazardous immunoreaction and having more stability in a different situation. Beyond their antioxidant activity, they also possess functional and nutritional properties (Hattori *et al.*, 1998; Xie *et al.*, 2008). Due to the ability to exert antihypertensive effects by inhibiting ACE, attention towards bioactive peptides from various sources has been increased (Cavazos and Gonzalez de Mejia, 2013; Sa´nchez and Va´zquez, 2017). ACE catalyzes the conversion of the peptide hormone angiotensin I to angiotensin II, which is a blood vessel constrictor and leads to the development of high blood pressure or hypertension (De Leo *et al.*, 2009). Blood vessel dilator peptide, bradykinin is also degraded by ACE. Therefore, the ACE inhibitory activity of bioactive peptides contributes to antihypertensive properties. Studies suggested that protein hydrolysates from millets contain bioactive peptides with ACE inhibitory activity. Several bioactive peptides have been identified from plant sources such as wheat (Kumagai, 2010), rice (Selamassakul *et al.*, 2016), barley (Bamdad and Chen, 2013), corn (Zhu *et al.*, 2019), soy (Singh *et al.*, 2014) and amaranth (Silva-Sánchez *et al.*, 2008). In addition to plant sources, other sources include fish, eggs and meat. Millets grains has now also become an area of interest to identify bioactive peptides. Till now pearl millet, finger millet, green sorghum millet and foxtail millet is used to identify peptides with antioxidant and antihypertensive activity.

Pearl Millet (*Pennisetum glaucum*)

An antioxidant peptide from pearl millet using trypsin enzymatic digestion has been identified (Agrawal *et al.*, 2016) (Table 1). After digestion of protein, the hydrolysate was purified using a different chromatographic technique such as ultrafiltration membrane fractionation, gel-filtration chromatography and reverse phase-ultra fast liquid chromatography (RP-UFLC). The antioxidant activity was significantly increased after digestion and purification due to the generation of the smaller size of peptides and free amino acids during hydrolysis reaction (Kou *et al.*, 2013). The DPPH free radical scavenging activity of protein isolate from pearl millet was about 13.03 percent which increase after digestion (25.98 %) and purification (67.66 %). Similar activity was also observed with other radicals such as ABTS, metal chelating, hydroxyl radical and reducing power activity. The amino acid sequence of pearl millet antioxidant peptide was identified as SDRDLLGPNNQYLPK with a molecular mass of 1729.2 Da. It has also been reported that peptides with molecular weight less than 6000 Da exhibit string free radical scavenging activities (Sun *et al.*, 2004).

Finger Millet (*Eleusine coracana*)

Finger millet has been reported for its high nutritional value and a rich source of amino acids such as methionine, isoleucine, leucine and phenylalanine (Chandra *et al.*, 2016). Two antioxidant peptides from finger millet have been identified as TSSSLNMAVRGGLTR and STTVGLGISMRSASVR with a molecular mass of 1564.7991 Da and 1636.8567 Da respectively (Agrawal *et al.*, 2019) (Table 1). In silico analysis of the interaction of finger millet peptides and free radicals (DPPH and ABTS) demonstrated that peptides interact with free radical by forming strong hydrogen bonding and hydrophobic interaction. It has been reported that inhibitors interact with their target molecules through the force of electrostatic interaction, hydrogen bonds, hydrophobic and ven der Waal (Mirzaei *et al.*, 2018). Finger millet peptide TSSSLNMAVRGGLTR and STTVGLGISMRSASVR contain 46.66 percent and 50 percent ratio of hydrophobic amino acid which increase the peptide's solubility in lipid medium and accessibility to hydrophobic radical species (Siow and Gan, 2013). Selection of suitable enzyme and digestion condition are necessary to obtain protein hydrolysate and peptides with desirable bioactivity and functional properties. It has also been reported that as compare to pepsin, protein digested with trypsin enzyme possess a higher degree of hydrolysis, peptide content and antioxidant activity (Agrawal *et al.*, 2019).

Sorghum Millet (*Sorghum bicolor*)

ACE inhibitory peptide (Thr-Leu-Ser) from sweet sorghum protein was identified using alcalase hydrolysis. The peptide was purified using different chromatographic techniques such as ultrafiltration, ion exchange, gel filtration, and reverse-phase high-performance liquid chromatography (RP-HPLC). The molecular docking analysis showed that the ACE inhibitory effect of tripeptide was primarily attributed to the presence of Ser residue at C-terminal, which can efficiently interact with the S1 and S2 pockets of ACE. The IC-50 value of peptide was found to be 102.1 μM for ACE inhibition which revealed that the tripeptide from the sweet sorghum protein hydrolysate can be utilized as pharmaceuticals or functional food ingredients for the prevention of hypertension (Wu *et al.*, 2016).

In another study seven peptides (VPPSKLSP, VAITLTMK, GLLGKNFTSK, LDSCKDYVME, HQTSEFK, VSKSVLVK and TSVEIITSSK) with antioxidant activity were also identified from green tender sorghum. Peptides VAITLTMK and VSKSVLVK found to exhibit the strong DPPH radical scavenging capacity (74.19% and 77.64%, respectively) and peptides VPPSKLSP and VAITLTMK also possessed the highest hydrophobic ratio (62.5%). Interaction between peptides and radical species could also help to increase lipid inhibitory activity (Siow and Gan, 2013). Moreover, peptide sequence, having valine at the N-terminal reported to exhibit strong antioxidant activity in an oil system (Chen *et al.*, 1995). Hence, these peptides can be used as promising natural antioxidants in several applications such as nutraceuticals, food, cosmetics and medicinal industries (Agrawal *et al.*, 2017).

α-kafirin from a white sorghum variety has also been used to produce protein hydrolysate using chymotrypsin (Vasudeva *et al.*, 2006). Four fractions obtained from hydrolysate after Gel-filtration possessed significant *in vitro* ACE inhibitory properties. Two of the fractions inhibited the enzyme competitively and the other two fractions showed uncompetitive inhibition. Previously it has been reported that these fractions from the chymotryptic hydrolysates of a-kafirin may have similar peptide sequence as a tandem-repeated proline-rich peptide sequence of maize endosperm proteins, which has also been reported to have ACE inhibitory activity (Maruyama *et al.*, 1989).

Foxtail Millet (*Setaria italica*)

Three potential antioxidant peptides (SGYYMH, LGTFQN and LHALLL) were identified from foxtail millet (*Setaria italica*) meal fermented by *Lactobacillus paracasei* (Table 1). The synthesized peptides with the same sequence were also found to be effective for the inhibition of the *Escherichia coli* ATCC 8099 growth. This study demonstrated that fermented foxtail millet meal possesses potential natural antioxidants peptides along with antimicrobial activity (Amadou *et al.*, 2013).

Defatted foxtail millet protein hydrolysate digested with alcalase has also been reported to possess antioxidant activity based on DPPH, ABTS, reducing power, metal-chelating, and inhibition of linoleic acid autoxidation. After fractionation with Gel filtration (Sephadex G-25), the antioxidant activity was increased up to 85.71 percent in the linoleic acid oxidation system. Amino acid composition showed that fractionated foxtail millet protein hydrolysate, with higher antioxidant capacity, had the highest hydrophobicity (8.62 kJ/moL amino acid residue "AAR") and hydrophobic amino acids content (51.94%) which resulted in increased solubility in lipid medium, therefore enhancing the antioxidant activity (Mohamed *et al.*, 2012).

In another study, raw, extruded and fermented foxtail millet protein hydrolysate, digested with two enzymes pepsin and pancreatin, was used to determine blood pressure lowering effect in spontaneously hypertensive rats (SHRs) (Chen *et al.*, 2017). After 4 weeks of treatment systolic blood pressure (SBP) reduction of more than 20 mmHg was observed in 200mg/kg of body weight of raw and extruded foxtail millet protein hydrolysates without changing the heart rate; hence, peptides performed to preserve the normal cardiac responses resulted in reduced risk to develop arrhythmias and other related complications (Majumder *et al.*, 2015). Furthermore, serum ACE activity and level of plasma angiotensin II were also reduced in the SHRs. However, the ACE inhibition effect of each foxtail millet protein hydrolysates was different due to the presence of peptides with different structure, amino acid composition and molecular weights (Majumder *et al.*, 2015). It has been reported that antioxidants have beneficial effects on hypertension. It has also been confirmed that ACE inhibitors improve BP condition by decreasing lipid peroxidation and increasing antioxidant status. Moreover, the combination of antioxidant peptides also exerts a synergistic effect on the SHRs without having ACE inhibitory activity in foxtail millet protein hydrolysates. Extruded foxtail millet protein hydrolysate improved the oxidative status in SHRs which resulted in improve vascular status, explaining the remarkable reducing trends of SBP.

Table 1: Peptides identified from pearl millet, finger millet, green sorghum millet and foxtail millet

	Hydrolysis	**Peptide**	**Activity**	**Reference**
Pearl millet	Trypsin	SDRDLLGPNNQYLPK	Antioxidant	Agrawal *et al.*, 2016
Finger millet	Trypsin	TSSSLNMAVRGGLTR	Antioxidant	Agrawal *et al.*, 2019
		STTVGLGISMRSASVR.		
Sorghum millet	Alcalase	VPPSKLSP	Antioxidant	Agrawal *et al.* ,2017
		VAITLTMK		
		GLLGKNFTSK		
		LDSCKDYVME		
		HQTSEFK		
		VSKSVLVK		
		TSVEIITSSK		
	Alcalase	TLS	ACE inhibitory	Wu *et al.*, 2016
Foxtail millet	Lactobacillus paracasei fermentation	SGYYMH	Antioxidant	Amadou *et al.*, 2013
		LGTFQN		
		LHALLL		

Structure-antioxidant Activity Relationship

The food derived protein as a potential precursor of antioxidant peptides can be assessed by the structural analysis. Peptides possess potent antioxidant activity based on their amino acid sequence and composition. Hydrophobic amino acids such as methionine (M), phenylalanine (F), proline (P), isoleucine (I), leucine (L), Valine (V), Alanine (A) and glycine (G) reported for the potential antioxidant activity of peptides (Sila and Bougatef, 2016). As shown in fig 3 2 ratio of hydrophobic amino acid in pearl millet peptide (SDRDLLG-PNNQYLPK-40%), finger millet peptide (TSSSLNMAVRGGLTR-46.6% and STTVGLGISMRSASVR-50%), green sorghum millet peptide (VPPSKL-SP-62.5%, VAITLTMK-62.5%, GLLGKNFTSK-50%, LDSCKDYVME-30%, HQTSEFK-14.28%, VSKSVLVK-50%, and TSVEIITSSK-30%) and foxtail millet peptide (SGYYMH-33.33%, LGTFQN-50% and LHALLL-83.33%) showed a positive impact on their antioxidant activity. Previously, it has been reported that due to a single hydrogen atom in the side chain of glycine residue peptide backbone get high flexibility (Nimalaratne *et al.*, 2015) and by donating a proton it also neutralizes active free radical (Chen *et al.*, 2012). As shown in Fig 3 pearl millet peptide contain single glycine residue, finger millet peptide contains two glycine residue in both peptides, GLLGKNFTSK peptide of green sorghum millet only contain two glycine residue and peptides SGYYMH and LGTFQN from foxtail millet also contains single residue of glycine, which contribute to the antioxidant activity of peptides. Acidic (Aspartic acid and Glutamic acid) and basic (Arginine) amino acids in peptides have also been reported to possess metal ion chelating activity due to the presence of amino and carboxyl groups in their side chains (Giménez *et al.*, 2009). As shown in Fig 3 pearl millet contains two acidic and one basic amino acid residue in their sequence while both peptides of finger millet contain only two residues of basic amino acid. Green sorghum millet peptide LDSCKDYVME and HQTSEFK contain three and one acidic amino acid respectively. The presence of these acidic and basic amino acid residue might be favourable to antioxidant activities of the peptide. Millet peptides also contain repeating amino acids such as di- and tri-peptides which revealed faster absorption and higher biological action in contrast to their constituent and free amino acids (Siow and Gan, 2013). As represented in figure 2 pearl millet contain two di-peptides (LL and NN), finger millet contains one tri-peptide (SSS) and two di-peptide (GG and TT), green sorghum millet contains four di-peptide (PP, LL, II and SS) and foxtail millet contain one di-peptide (YY) and one tri-peptide (LLL).

The antihypertensive properties of bioactive peptides are determined by their ability to inhibit the angiotensin-1 converting enzyme (ACE). The existence and position of particular amino acids in peptides can also influence the ACE-inhibitory activity. Specifically, peptides containing the valine-alanine-proline epitope have been reported to possess strong ACE-inhibitory effect, and these have been reported to be produced during sourdough fermentation of wheat by lactic acid bacteria (Rizzello *et al.*, 2008). Processes such as lactic acid fermentation and malting, which induce significant proteolytic activity, are applied extensively to sorghum and millets. Furthermore, sorghum and millet proteins are relatively rich in amino acids such as proline, alanine and valine. This suggests that sorghum and millet foods produced using lactic acid fermentation and malting could potentially be important sources of bioactive peptides with health-promoting properties.

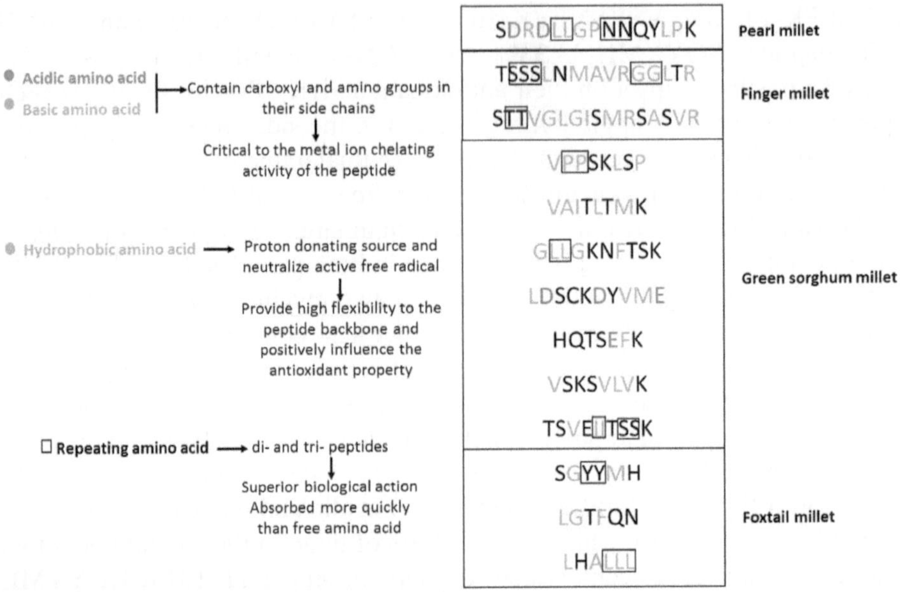

Fig. 3: Structure-antioxidant activity relationship of peptides from pearl millet, finger millet, green sorghum millet and foxtail mille

Health Effects

It has been reported that free radicals are involved in the initiation or progress of several degenerative diseases. For example, recent evidence suggests that oxidative stress may contribute to type II diabetes by increasing insulin resistance or impairing insulin secretion. Uncontrolled generation of free radicals worsens the development and progression of diabetes and its complications (Johansen *et al.*, 2005). It also has been suggested that oxidative stress plays a key role in the initiation or progress of cardiovascular complications (like atherosclerotic process and alteration in lipid metabolism) in patients with metabolic syndrome (Palmieri *et al.*, 2006). Therefore, due to the close relationship between oxidative stress and diseases, control of oxidative stress seems to be one of the crucial steps in slowing down the progress of these diseases or preventing complications. In this regard, a vast number of antioxidants are isolated and identified from natural sources to control oxidative stress. Beside various well-known natural antioxidants like Vitamin C, polyphenols, flavonoids and carotenoids, peptides with antioxidative properties are also the focus of recent research. For instance, ingestion of Douchi (fermented soybean food) extracts to rats had an effect of increasing SOD activities in liver and kidney, catalase (CAT) activity in liver, GSH-Px activity in the kidney as well as decreasing serum thiobarbituric acid reactive substances (TBARS) in liver and kidney. These results were attributed to the peptides and free amino acids components of Douchi extracts which act as an antioxidant (Wang *et al.*, 2008). In addition, it has been found that the radical-scavenging capacity of plasma rose while an elevated concentration of MDA in the aorta dropped following the intake of egg white hydrolysates (0.5 g/kg/day of egg white hydrolysates) by spontaneously hypertensive rats (SHR). It has been postulated that egg white hydrolysates may contribute to the prevention of oxidative stress by the increment of plasma radical-scavenging capacity and inhibition of lipid peroxidation (Manso *et al.*, 2008). Pena-Ramos and Xiong have reported that soy protein isolate and its hydrolysates decreased TBARS by 28–65 percent. In another study, it was shown that the ingestion of soy protein isolate (SPI) or soy peptide to male Wistar rats reduced paraquat (PQ)-induced oxidative stress. Such effects were not observed when the amino acid mixture was fed to rats. Additionally, soy protein isolates and soy peptide prevented the elevation of the serum TBARS concentration (Takenaka *et al.*, 2003). Soymilk-kefir has exhibited to have significant antimutagenic and antioxidant activities. It was proposed that fermented soymilk can be considered as a food with preventative roles in mutagenic and oxidative damage (Liu *et al.*, 2005).

Safety Parameters

Hydrolysis of protein is achieved by several techniques such as acid and alkaline hydrolysis, heat treatment and enzymatic incubation. Enzymatic proteolysis has several advantages over chemical processes such as varietal choices, high reaction velocity, specificity and mild process circumstances. With an accepted safe use in human nutrition, digestive proteolytic enzymes used are trypsin, chymotrypsin, pepsin and so on obtained from different microorganisms, animals and edible parts of plants (food grade). Enzymes obtained from animals and edible parts of plants generally cause no health risks. Enzymes obtained from microorganisms used in food fermentations are also safe to use only when the organism source is not pathogenic and also not producing toxic compounds remaining in the final product. Before using all individual strains should be tested for toxicological programs. *Escherichia coli, Bacillus, Saccharomyces cerevisiae* and *Aspergillus niger* strains are important as microorganism sources (Schaafsma, 2009). According to the Codex Alimentarius Commission 1992, it is not necessary to label the food grade enzyme when they are measured as processing aids. The most important safety issues regarding enzyme preparations are as source organism, accidental reactions (conversion of histidine in histamine) and skin irritations (occupational health problem possibly due to exposure to the enzyme). As compared to free amino acids and intact proteins, amino acids from protein hydrolysates especially from di- and tripeptides enter into the circulation and absorbed more quickly (Bilsborough and Mann, 2006) because membrane peptide transporters have more ability as compared to amino acids transporters over the brush border (Gambrell, 1998). Ideally, a high level of peptide transportation may cause consequences on the utilization of protein hydrolysate and metabolic effects but there are no significant adverse effects have been reported so far due to ingestion of protein hydrolysates. In food industries, processes used for the production of protein hydrolysates are common and usually use food-grade source materials, equipment and processing aids. Furthermore, proteins are naturally hydrolysed in the gastrointestinal tract after ingestion of food, resulting in the release of di-peptides, tri-peptides and free amino acids, which are transported across the intestine wall (Doorten *et al.*, 2009). Protein hydrolysates and peptides with low MW are widely used in the formulation of hypoallergenic foods for infants due to less allergenicity than their native protein (Høst and Halken 2004). Though, the confirmation of the nontoxic nature of the product is necessary before ingestion. Prediction of peptides toxicity can be done by using *in silico* process (ToxinPred available at http://www.imtech.res.in/raghava/ toxinpred/) (Gupta *et al.*, 2013) followed by *in vivo* toxicity assessment before human consumption. The *in vivo* toxicity and safety assessment of food products must be carried out following the guidelines suggested by international authorities. Large amounts of scientific evidence and tests are required for this and usually carried out on vertebrate cell lines, vertebrate models, and unicellular microbial species (Marques *et al.*, 2011).

Use as Functional Food Ingredients

The scientific literature suggested that food proteins contain a large number of the potentially bioactive peptide in their sequences, but do not reveal any bioactivity in its original primary structure. Bioactive peptides can be produced during protein hydrolysis, either in controlled proteolysis or in the normal digestive process *in vitro*. As components of the human diet, very little information is available about the quantities, bioavailability and bioactivity of the peptides, generating during the digestive process of foods. Hence, it is necessary to disrupt the structure of the proteins in order to measure and quantify the impact of bioactivity of these peptides, either as food component or as isolated, partial or totally purified and extract or concentrated bioactive peptides of interest for health. Currently, there are a lot of technologies are available for the extraction, hydrolysis, concentration and purification as well as the determination of the amino acid composition and sequence, but are expensive. The cost-benefit balance will have to be evaluated on a case-by-case basis to justify the investment, considering both economic and human health aspects. It is expected that the attempt of using total hydrolysate and/or specific fractions of hydrolysate will be given greater priority in formulating nutritional supplements and developing foods with specific physiological functionality. Only the United States of North America, Canada, some European and Eastern countries have developed and used functional products with application of bioactive peptides. As functional ingredients in different foods, millet protein hydrolysates have several applications due to their characteristic properties such as the ability to foaming and emulsification, protein solubility, water holding capacity, oil holding capacity, in addition to combining numerous bioactivities with benefits for health.

Unlike the European and Asian countries, North America, Australia, New Zealand and in South American countries, in general, little has been done with respect to the application of bioactive peptides and hydrolysates as food supplements and/or ingredients, in the formulation of specific functional foods or foods consumed by the general population. This gap as to effective utilization of our sources of bioactive peptides is due to the lack of greater integration between universities and research institutes and the food industry, for the development of integrated multidisciplinary projects, with funding shared between interested sectors: research, industry, and funding agencies for food research and technology. From the 1980s there was a profound transformation of the concepts of healthy nutrition for the human species with the discovery of new components in foods, considered non-nutrients, but which for being bioactive contributed to improving human health in two complementary aspects: 1) acting as modulators of metabolism and possibly acting on the different metabolic cycles, also acting on the human genome influencing its expression to improve the effectiveness, benefiting the health of individuals and/or populations; 2) reducing the risks of chronic non-communicable diseases, thus contributing to improving human health, well-being, and increase longevity. Foods with these characteristics came to be called "functional foods" and their importance has been recognized not only by the Eastern countries but also by all countries of Western culture (Urala and Lähteenmäki, 2007). In a simplified and concise way, functional food is one that contains one or more bioactive substances in concentrations suitable for the quantity ingested with the diet to be enough to exercise in the metabolism the functions.

Conclusion

Numerous efforts have been done to produce functional millet protein hydrolysates and peptides, some are successful and some not. There is a huge possibility for millet protein hydrolysates and peptides to be produced and sold as functional food ingredients. However, a standardized protocol to study the functional properties of these protein hydrolysates and peptides are needed, as well as the use of different endogenous enzymes to make functional millet protein hydrolysates and peptides. As millet peptides have been explored only in the area of antioxidant and antihypertensive. Further multifunctional effect of these peptides related to antidiabetic, anticancer, anti-inflammatory and anti-obesity also need to be examined. On the molecular level, with more basic research, the future of millet protein hydrolysates and peptides could be bright, especially in light of the cultivation problems facing millets. It is time for millets and their bioactive ingredients research to take a new, fresher, and more scientific direction, because the field has been more or less stagnant for the last few years or so.

References

Agrawal, H., Joshi, R., and Gupta, M. (2016). Isolation, purification and characterization of antioxidative peptide of pearl millet (*Pennisetum glaucum*) protein hydrolysate. Food Chemistry, 204, 365-372. https://doi.org/10.1016/j.foodchem.2016.02.127

Agrawal, H., Joshi, R., and Gupta, M. (2017). Isolation and characterisation of enzymatic hydrolysed peptides of green tender sorghum and their antioxidant activities. *LWT-Food Science and Technology,* 84, 608-616. https://doi.org/10.1016/j.lwt.2017.06.036

Agrawal, H., Joshi, R., and Gupta, M. (2019). Purification, identification and characterization of two novel antioxidant peptides from finger millet (*Eleusine coracana*) protein hydrolysate. *Food Research International*, 120, 697-707. https://doi.org/10.1016/j.foodres.2018.11.028

Amadou, I., Le, G. W., Amza, T., Sun, J., and Shi, Y. H. (2013). Purification and characterization of foxtail millet-derived peptides with antioxidant and antimicrobial activities. *Food Research International*, 51(1), 422-428. https://doi.org/10.1016/j.foodres.2012.12.045

Bagdi, A., Balázs, G., Schmidt, J., Szatmári, M., Schoenlechner, R., Berghofer, E. and Tömösközia, S. J. A. A. (2011). Protein characterization and nutrient composition of Hungarian proso millet varieties and the effect of decortication. Acta Alimentaria, 40(1), 128-141. https://doi.org/10.1556/aalim.40.2011.1.15

Bamdad, F., and Chen, L. (2013). Antioxidant capacities of fractionated barley hordein hydrolysates in relation to peptide structures. Molecular Nutrition & Food Research, 57(3), 493-503. https://doi.org/10.1002/mnfr.201200252

Baydar, N. G., Özkan, G., & Yaşar, S. (2007). Evaluation of the antiradical and antioxidant potential of grape extracts. *Food Control*, 18(9), 1131-1136. https://doi.org/10.1016/j.foodcont.2006.06.011

Bilsborough, S., and Mann, N. (2006). A review of issues of dietary protein intake in humans. International Journal of Sport Nutrition and Exercise Metabolism, 16(2), 129-152. https://doi.org/10.1123/ijsnem.16.2.129

Cao, X. H., Liao, Z. Y., Wang, C. L., Yang, W. Y., and Lu, M. F. (2009). Evaluation of a lipopeptide biosurfactant from Bacillus natto TK-1 as a potential source of anti-adhesive, antimicrobial and antitumor activities. Brazilian Journal of Microbiology, 40(2), 373-379. https://doi.org/10.1590/S1517-83822009000200030

Cavazos, A., and Gonzalez de Mejia, E. (2013). Identification of bioactive peptides from cereal storage proteins and their potential role in prevention of chronic diseases. Comprehensive Reviews in Food Science and Food Safety, 12(4), 364-380. https://doi.org/10.1111/1541-4337.12017

Chandra, D., Chandra, S., and Sharma, A. K. (2016). Review of Finger millet (Eleusine coracana (L.) Gaertn): a power house of health benefiting nutrients. Food Science and Human Wellness, 5(3), 149-155. https://doi.org/10.1016/j.fshw.2016.05.004

Chen, C., Chi, Y. J., Zhao, M. Y., and Xu, W. (2012). Influence of degree of hydrolysis on functional properties, antioxidant and ACE inhibitory activities of egg white protein hydrolysate. Food Science and Biotechnology, 21(1), 27-34. https://doi.org/10.1007/s10068-012-0004-6

Chen, H. M., Muramoto, K., and Yamauchi, F. (1995). Structural analysis of antioxidative peptides from Soybean. beta.-Conglycinin. Journal of Agricultural and Food Chemistry, 43(3), 574-578. https://doi.org/10.1021/jf00051a004

Chen, J., Duan, W., Ren, X., Wang, C., Pan, Z., Diao, X., and Shen, Q. (2017). Effect of foxtail millet protein hydrolysates on lowering blood pressure in spontaneously hypertensive rats. European Journal of Nutrition, 56(6), 2129-2138. https://doi.org/10.1007/s00394-016-1252-7

Chen, W., and Qiu, Y. (2003). Leaf protein's utilization status and its prospect. *Food Science*, 24(2), 158-161.

Codex Alimantarius Commission (1992). Inventory of processing aids. *Codex Alimantarius* 1A, 1–39.

De la Barca, A. C., Ruiz-Salazar, R. A., & Jara-Marini, M. E. (2000). Enzymatic hydrolysis and synthesis of soy protein to improve its amino acid composition and functional properties. Journal of Food Science, 65(2), 246-253. https://doi.org/10.1111/j.1365-2621.2000.tb15988.x

De Leo, F., Panarese, S., Gallerani, R., and Ceci, L. R. (2009). Angiotensin converting enzyme (ACE) inhibitory peptides: production and implementation of functional food. Current Pharmaceutical Design, 15(31), 3622-3643. https://doi.org/10.2174/138161209789271834

Devi, P. B., Vijayabharathi, R., Sathyabama, S., Malleshi, N. G., and Priyadarisini, V. B. (2014). Health benefits of finger millet (*Eleusine coracana* L.) polyphenols and dietary fiber: a review. Journal of Food Science and Technology, 51(6), 1021-1040. https://doi.org/10.1007/s13197-011-0584-9

Dideriksen, K., Reitelseder, S., & Holm, L. (2013). Influence of amino acids, dietary protein, and physical activity on muscle mass development in humans. *Nutrients*, 5(3), 852-876. https://doi.org/10.3390/nu5030852

Doorten, A. P. S., Vd Wiel, J. A. G., and Jonker, D. (2009). Safety evaluation of an IPP tripeptide-containing milk protein hydrolysate. Food and Chemical Toxicology, 47(1), 55-61. https://doi.org/10.1016/j.fct.2008.10.001

Dziuba, M., and Darewicz, M. (2007). Food proteins as precursors of bioactive peptides—classification into families. Food Science and Technology International, 13(6), 393-404. https://doi.org/10.1177/1082013208085933

Fountoulakis, M., and Lahm, H. W. (1998). Hydrolysis and amino acid composition analysis of proteins. Journal of Chromatography A, 826(2), 109-134. https://doi.org/10.1016/S0021-9673(98)00721-3

Frøkjaer, S. (1994). Use of hydrosylates for protein supplementation. *Food Technology (Chicago)*, 48(10), 86-88.

Gambrell RC. 1998. Amino Acids and Proteins for the Athlete: The Anabolic Edge. Medical Science Sports Exercise, 1462-1463.

Giménez, B., Alemán, A., Montero, P., and Gómez-Guillén, M. C. (2009). Antioxidant and functional properties of gelatin hydrolysates obtained from skin of sole and squid. Food Chemistry, 114(3), 976-983. https://doi.org/10.1016/j.foodchem.2008.10.050

Grimble, G. K., Keohane, P. P., Higgins, B. E., Kaminski, M. V., and Silk, D. B. A. (1986). Effect of peptide chain length on amino acid and nitrogen absorption from two lactalbumin hydrolysates in the normal human jejunum. Clinical Science, 71(1), 65-69. https://doi.org/10.1042/cs0710065

Gupta, S., Kapoor, P., Chaudhary, K., Gautam, A., Kumar, R., Raghava, G. P., and Open Source Drug Discovery Consortium. (2013). *In silico* approach for predicting toxicity of peptides and proteins. *PloS one*, 8(9), e73957. https://doi.org/10.1371/journal.pone.0073957

Hasler, C. M. (2002). Functional foods: benefits, concerns and challenges—a position paper from the American Council on Science and Health. The Journal of Nutrition, 132(12), 3772-3781. https://doi.org/10.1093/jn/132.12.3772

Hattori, M., Yamaji-Tsukamoto, K., Kumagai, H., Feng, Y., and Takahashi, K. (1998). Antioxidative activity of soluble elastin peptides. Journal of Agricultural and Food Chemistry, 46(6), 2167-2170. https://doi.org/10.1021/jf970484r

He, R., Ju, X., Yuan, J., Wang, L., Girgih, A. T., and Aluko, R. E. (2012). Antioxidant activities of rapeseed peptides produced by solid state fermentation. Food Research International, 49(1), 432-438. https://doi.org/10.1016/j.foodres.2012.08.023

Hoffer, L. J. (2016). Human protein and amino acid requirements. Journal of Parenteral and Enteral Nutrition, 40(4), 460-474. https://doi.org/10.1177/0148607115624084

Høst, A., and Halken, S. (2004). Hypoallergenic formulas–when, to whom and how long: after more than 15 years we know the right indication!. Allergy, 59, 45-52. https://doi.org/10.1111/j.1398-9995.2004.00574.x

Hruza, L. L., and Pentland, A. P. (1993). Mechanisms of UV-induced inflammation. Journal of Investigative Dermatology, 100(1), S35-S41. https://doi.org/10.1038/jid.1993.21

Ichihashi, M., Ueda, M., Budiyanto, A., Bito, T., Oka, M., Fukunaga, M., & Horikawa, T. (2003). UV-induced skin damage. *Toxicology*, 189(1-2), 21-39. https://doi.org/10.1016/S0300-483X(03)00150-1

Ito, N., Fukushima, S., and Tsuda, H. (1985). Carcinogenicity and modification of the carcinogenic response by BHA, BHT, and other antioxidants. *CRC Critical Reviews in Toxicology,* 15(2), 109-150. https://doi.org/10.3109/10408448509029322

Jeon, Y. J., Byun, H. G., & Kim, S. K. (1999). Improvement of functional properties of cod frame protein hydrolysates using ultrafiltration membranes. Process Biochemistry, 35(5), 471-478. https://doi.org/10.1016/S0032-9592(99)00098-9

Johansen, J. S., Harris, A. K., Rychly, D. J., and Ergul, A. (2005). Oxidative stress and the use of antioxidants in diabetes: linking basic science to clinical practice. *Cardiovascular Diabetology*, 4(1), 1-11. https://doi.org/10.1186/1475-2840-4-5

Kaneto, H., Kajimoto, Y., Miyagawa, J. I., Matsuoka, T. A., Fujitani, Y., Umayahara, Y., ... and Hori, M. (1999). Beneficial effects of antioxidants in diabetes: possible protection of pancreatic beta-cells against glucose toxicity. Diabetes, 48(12), 2398-2406. https://doi.org/10.2337/diabetes.48.12.2398

Kehrer, J. P. (1993). Free radicals as mediators of tissue injury and disease. *Critical Reviews in Toxicology,* 23(1), 21-48. https://doi.org/10.3109/10408449309104073

Kim, J. M., Whang, J. H., Kim, K. M., Koh, J. H., and Suh, H. J. (2004). Preparation of corn gluten hydrolysate with angiotensin I converting enzyme inhibitory activity and its solubility and moisture sorption. Process Biochemistry, 39(8), 989-994. https://doi.org/10.1016/S0032-9592(03)00205-X

Kim, S. K., Kim, Y. T., Byun, H. G., Park, P. J., and Ito, H. (2001). Purification and characterization of antioxidative peptides from bovine skin. *BMB Reports,* 34(3), 219-224.

Korhonen, H., and Pihlanto, A. (2006). Bioactive peptides: production and functionality. *International Dairy Journal*, 16(9), 945-960. https://doi.org/10.1016/j.idairyj.2005.10.012

Kou, X., Gao, J., Xue, Z., Zhang, Z., Wang, H., and Wang, X. (2013). Purification and identification of antioxidant peptides from chickpea (*Cicer arietinum* L.) albumin hydrolysates. LWT-Food Science and Technology, 50(2), 591-598. https://doi.org/10.1016/j.lwt.2012.08.002

Kristinsson, H. G., and Rasco, B. A. (2000). Biochemical and functional properties of Atlantic salmon (Salmo salar) muscle proteins hydrolyzed with various alkaline proteases. Journal of Agricultural and Food Chemistry, 48(3), 657-666. https://doi.org/10.1021/jf990447v

Kumagai, H. (2010). Wheat proteins and peptides. Bioactive proteins and peptides as functional foods and nutraceuticals, 29, 289. DOI:10.1002/9780813811048

Kumar, K. K., and Parameswaran, K. P. (1998). Characterisation of storage protein from selected varieties of foxtail millet (*Setaria italica* (L) Beauv). Journal of the Science of Food and Agriculture, 77(4), 535-542. https://doi.org/10.1002/(SICI)1097-0010(199808)77:4<535::AID-JSFA77>3.0.CO;2-G

Lee, J. H., and Yim, M. H. (2000). Functional properties of fractionated soy protein isolates by proteases from Meju. Food Science and Biotechnology, 9(4), 253-257.

Li, J., Chen, Z., Guan, X., Liu, J., Zhang, M., and Xu, B. (2008). Optimization of germination conditions to enhance hydroxyl radical inhibition by water-soluble protein from stress millet. *Journal of Cereal Science*, 48(3), 619-624. https://doi.org/10.1016/j.jcs.2008.01.008

Liu, J. R., Chen, M. J., and Lin, C. W. (2005). Antimutagenic and antioxidant properties of milk– kefir and soymilk– kefir. Journal of Agricultural and Food Chemistry, 53(7), 2467-2474. https://doi.org/10.1021/jf048934k

Majumder, K., Chakrabarti, S., Morton, J. S., Panahi, S., Kaufman, S., Davidge, S. T., and Wu, J. (2015). Egg-derived ACE-inhibitory peptides IQW and LKP reduce blood pressure in spontaneously hypertensive rats. Journal of Functional Foods, 13, 50-60. https://doi.org/10.1016/j.jff.2014.12.028

Malumba, P., Vanderghem, C., Deroanne, C., and Béra, F. (2008). Influence of drying temperature on the solubility, the purity of isolates and the electrophoretic patterns of corn proteins. Food Chemistry, 111(3), 564-572. https://doi.org/10.1016/j.foodchem.2008.04.030

Manso, M. A., Miguel, M., Even, J., Hernández, R., Aleixandre, A., and López-Fandiño, R. (2008). Effect of the long-term intake of an egg white hydrolysate on the oxidative status and blood lipid profile of spontaneously hypertensive rats. Food Chemistry, 109(2), 361-367. https://doi.org/10.1016/j.foodchem.2007.12.049

Maruyama, S., Miyoshi, S., Kaneko, T., & Tanaka, H. (1989). Angiotensin I-converting enzyme inhibitory activities of synthetic peptides related to the tandem repeated sequence of a maize endosperm protein. Agricultural and Biological Chemistry, 53(4), 1077-1081. https://doi.org/10.1080/00021369.1989.10869427

Matoba, T., and Kenkyu, D. T. (2002). How does the radical-scavenging activity of soy protein food change during heating. Daizu Tanpakushitsu Kenkyu, 5, 47-50.

Marques, A., Lourenço, H. M., Nunes, M. L., Roseiro, C., Santos, C., Barranco, A., and Cencic, A. (2011). New tools to assess toxicity, bioaccessibility and uptake of chemical contaminants in meat and seafood. Food Research International, 44(2), 510-522. https://doi.org/10.1016/j.foodres.2010.12.003

Merz, M., Eisele, T., Claaßen, W., Appel, D., Rabe, S., Stressler, T., and Fischer, L. (2015). Continuous long-term hydrolysis of wheat gluten using a principally food-grade enzyme membrane reactor system. Biochemical Engineering Journal, 99, 114-123. https://doi.org/10.1016/j.bej.2015.03.019

Mirzaei, M., Mirdamadi, S., Ehsani, M. R., & Aminlari, M. (2018). Production of antioxidant and ACE-inhibitory peptides from Kluyveromyces marxianus protein hydrolysates: Purification and molecular docking. Journal of Food and Drug Analysis, 26(2), 696-705. https://doi.org/10.1016/j.jfda.2017.07.008

Mohamed, T. K., Issoufou, A., and Zhou, H. (2012). Antioxidant activity of fractionated foxtail millet protein hydrolysate. International Food Research Journal, 19(1).

Mohan C. 2018. 2018 National Year of Millets. Krishi Jagran/New Delhi.

Moure, A., Domínguez, H., and Parajó, J. C. (2005). Fractionation and enzymatic hydrolysis of soluble protein present in waste liquors from soy processing. Journal of Agricultural and Food Chemistry, 53(19), 7600-7608. https://doi.org/10.1021/jf0505325

Nagodawithana, T. W., Nelles, L., and Trivedi, N. B. (2008). Protein hydrolysates as hypoallergenic, flavors and palatants for companion animals. In Protein hydrolysates in biotechnology (pp. 191-207). Springer, Dordrecht. https://doi.org/10.1007/978-1-4020-6674-0_11

Nimalaratne, C., Bandara, N., and Wu, J. (2015). Purification and characterization of antioxidant peptides from enzymatically hydrolyzed chicken egg white. Food Chemistry, 188, 467-472. https://doi.org/10.1016/j.foodchem.2015.05.014

Obilana, A. B., and Manyasa, E. (2002). Millets. In Pseudocereals and less common cereals (pp. 177-217). Springer, Berlin, Heidelberg. https://doi.org/10.1007/978-3-662-09544-7_6

Palmieri, V. O., Grattagliano, I., Portincasa, P., and Palasciano, G. (2006). Systemic oxidative alterations are associated with visceral adiposity and liver steatosis in patients with metabolic syndrome. The Journal of Nutrition, 136(12), 3022-3026. https://doi.org/10.1093/jn/136.12.3022

Park, P. J., Jung, W. K., Nam, K. S., Shahidi, F., and Kim, S. K. (2001). Purification and characterization of antioxidative peptides from protein hydrolysate of lecithin-free egg yolk. Journal of the American Oil Chemists' Society, 78(6), 651-656. https://doi.org/10.1007/s11746-001-0321-0

Peñta-Ramos, E. A., and Xiong, Y. L. (2002). Antioxidant activity of soy protein hydrolysates in a liposomal system. Journal of Food Science, 67(8), 2952-2956. https://doi.org/10.1111/j.1365-2621.2002.tb08844.x

Qian, Z. J., Jung, W. K., and Kim, S. K. (2008). Free radical scavenging activity of a novel antioxidative peptide purified from hydrolysate of bullfrog skin, Rana catesbeiana Shaw. Bioresource Technology, 99(6), 1690-1698. https://doi.org/10.1016/j.biortech.2007.04.005

Rajapakse, N., Mendis, E., Jung, W. K., Je, J. Y., and Kim, S. K. (2005). Purification of a radical scavenging peptide from fermented mussel sauce and its antioxidant properties. Food Research International, 38(2), 175-182. https://doi.org/10.1016/j.foodres.2004.10.002

Rice-Evans, C. A., and Diplock, A. T. (1993). Current status of antioxidant therapy. Free Radical Biology and Medicine, 15(1), 77-96. https://doi.org/10.1016/0891-5849(93)90127-G

Rizzello, C. G., Cassone, A., Di Cagno, R., and Gobbetti, M. (2008). Synthesis of angiotensin I-converting enzyme (ACE)-inhibitory peptides and γ-aminobutyric acid (GABA) during sourdough fermentation by selected lactic acid bacteria. Journal of Agricultural and Food Chemistry, 56(16), 6936-6943. https://doi.org/10.1021/jf800512u

Saito, K., Jin, D. H., Ogawa, T., Muramoto, K., Hatakeyama, E., Yasuhara, T., and Nokihara, K. (2003). Antioxidative properties of tripeptide libraries prepared by the combinatorial chemistry. Journal of Agricultural and Food Chemistry, 51(12), 3668-3674. https://doi.org/10.1021/jf021191n

Samaraweera, H., Moon, S. H., Lee, E., Yang, S. H., Suh, J. W., and Ahn, D. U. (2013). Chemical hydrolysis of phosvitin and the functional properties of the hydrolysates. International Journal of Engineering Research & Technology, 2(11), 3373.

Sánchez, A., and Vázquez, A. (2017). Bioactive peptides: A review. Food Quality and Safety, 1 (1), 29–46. https://doi.org/10.1093/fqsafe/fyx006

Schaafsma, G. (2009). Safety of protein hydrolysates, fractions thereof and bioactive peptides in human nutrition. European Journal of Clinical Nutrition, 63(10), 1161-1168. https://doi.org/10.1038/ejcn.2009.56

Selamassakul, O., Laohakunjit, N., Kerdchoechuen, O., and Ratanakhanokchai, K. (2016). A novel multi-biofunctional protein from brown rice hydrolysed by endo/endo-exoproteases. Food & Function, 7(6), 2635-2644. https://doi.org/10.1039/C5FO01344E

Serna-Saldivar, S. (1995). Structure and chemistry of sorghum and millets. Sorghum and millets: chemistry and technology, 69-124.

Sheriff, M., Zainab, K. A., Laminu, H. G., and Maisaratu, A. (2012). Hydrolysis of gelatinized maize, millet and sorghum starch by amylases of Aspergillus niger. Bioscience Research, 9(2), 92-98.

Sila, A., and Bougatef, A. (2016). Antioxidant peptides from marine by-products: Isolation, identification and application in food systems. A review. Journal of Functional Foods, 21, 10-26. https://doi.org/10.1016/j.jff.2015.11.007

Silva-Sánchez, C., De La Rosa, A. B., León-Galván, M. F., De Lumen, B. O., de León-Rodríguez, A., & De Mejía, E. G. (2008). Bioactive peptides in amaranth (Amaranthus hypochondriacus) seed. Journal of Agricultural and Food Chemistry, 56(4), 1233-1240. https://doi.org/10.1021/jf072911z

Singh, B. P., Vij, S., and Hati, S. (2014). Functional significance of bioactive peptides derived from soybean. Peptides, 54, 171-179. https://doi.org/10.1016/j.peptides.2014.01.022

Siow, H. L., & Gan, C. Y. (2013). Extraction of antioxidative and antihypertensive bioactive peptides from Parkia speciosa seeds. Food Chemistry, 141(4), 3435-3442. https://doi.org/10.1016/j.foodchem.2013.06.030

Sun, J., He, H., and Xie, B. J. (2004). Novel antioxidant peptides from fermented mushroom Ganoderma lucidum. Journal of Agricultural and Food Chemistry, 52(21), 6646-6652. https://doi.org/10.1021/jf0495136

Swaminathan MS. 2018. Shaping our agriculture future. Science Reporter.

Takenaka, A., Annaka, H., Kimura, Y., Aoki, H., and Igarashi, K. (2003). Reduction of paraquat-induced oxidative stress in rats by dietary soy peptide. Bioscience, Biotechnology, and Biochemistry, 67(2), 278-283. https://doi.org/10.1271/bbb.67.278

Tavano, O. L. (2013). Protein hydrolysis using proteases: An important tool for food biotechnology. Journal of Molecular Catalysis B: Enzymatic, 90, 1-11. https://doi.org/10.1016/j.molcatb.2013.01.011

Urala, N., and Lähteenmäki, L. (2007). Consumers' changing attitudes towards functional foods. Food Quality and Preference, 18(1), 1-12. https://doi.org/10.1016/j.foodqual.2005.06.007

Vasudeva, K., Sajeeda, N., Chandrashekar, A., and Rajini, P. S. (2006). Chymotryptic hydrolysates of a-kafirin, the storage protein of sorghum (*Sorghum bicolor*) exhibited angiotensin converting enzyme inhibitory activity. Food Chemistry, 100(1), 306-311. https://doi.org/10.1016/j.foodchem.2005.10.004

Wang, D., Wang, L. J., Zhu, F. X., Zhu, J. Y., Chen, X. D., Zou, L., and Saito, M. (2008). In vitro and in vivo studies on the antioxidant activities of the aqueous extracts of Douchi (a traditional Chinese salt-fermented soybean food). Food Chemistry, 107(4), 1421-1428. https://doi.org/10.1016/j.foodchem.2007.09.072

Wang, X. J., Zheng, X. Q., Kopparapu, N. K., Cong, W. S., Deng, Y. P., Sun, X. J., and Liu, X. L. (2014). Purification and evaluation of a novel antioxidant peptide from corn protein hydrolysate. Process Biochemistry, 49(9), 1562-1569. https://doi.org/10.1016/j.procbio.2014.05.014

Wu, H. C., Chen, H. M., and Shiau, C. Y. (2003). Free amino acids and peptides as related to antioxidant properties in protein hydrolysates of mackerel (*Scomber austriasicus*). Food Research International, 36(9-10), 949-957. https://doi.org/10.1016/S0963-9969(03)00104-2

Wu, Q., Du, J., Jia, J., and Kuang, C. (2016). Production of ACE inhibitory peptides from sweet sorghum grain protein using alcalase: Hydrolysis kinetic, purification and molecular docking study. Food Chemistry, 199, 140-149. https://doi.org/10.1016/j.foodchem.2015.12.012

Xie, Z., Huang, J., Xu, X., and Jin, Z. (2008). Antioxidant activity of peptides isolated from alfalfa leaf protein hydrolysate. Food Chemistry, 111(2), 370-376. https://doi.org/10.1016/j.foodchem.2008.03.078

Zhang, T., Li, Y., Miao, M., and Jiang, B. (2011). Purification and characterisation of a new antioxidant peptide from chickpea (*Cicer arietium* L.) protein hydrolysates. Food Chemistry, 128(1), 28-33. https://doi.org/10.1016/j.foodchem.2011.02.072

Zhu, B., He, H., and Hou, T. (2019). A comprehensive review of corn protein derived bioactive peptides: Production, characterization, bioactivities, and transport pathways. Comprehensive Reviews in Food Science and Food Safety, 18(1), 329-345. https://doi.org/10.1111/1541-4337.12411

Zhu, L., Chen, J., Tang, X., and Xiong, Y. L. (2008). Reducing, radical scavenging, and chelation properties of in vitro digests of alcalase-treated zein hydrolysate. Journal of Agricultural and Food Chemistry, 56(8), 2714-2721. https://doi.org/10.1021/jf703697e

Zhuang, H., Tang, N., Dong, S. T., Sun, B., and Liu, J. B. (2013). Optimisation of antioxidant peptide preparation from corn gluten meal. Journal of the Science of Food and Agriculture, 93(13), 3264-3270. https://doi.org/10.1002/jsfa.6170

14

Pseudo Cereals as Super Foods: An Overview

***Rahul Vashishth*[1]*, Charu Goel*[2]*, Anil Dutt Semwal*[2] and *Gopal Kumar Sharma*[2]**

[1]*Vignan's Foundation for Science, Technology and Research, Guntur-522213*
[2]*Defence Food Research Laboratory, Mysore-570011*

Introduction

In recent years, pseudocereals and alternative cereal substitutes are gaining attention from agriculture, processing and consumer point of views. The justification for such changes can be found in the recent thought of re-extending the range of starch rich seeds available in human diet, because from world-wide perspective cereals consumption is largely restricted to the cereals wheat, rice and maize (Venskutonis and Kraujalis, 2013). In upcoming years, the increase in population can lead to an unfavorable pressure on agriculture and nutrition needs.

Pseudocereals are the underutilized grains which are described as the ill-defined group of species that belongs to the non-grassy family of grains but, like cereals, they are dry starchy seeds with composition very much similar to that of cereals. The word "pseudo" means "virtual" hence, due to their similar composition as that of cereals, these grains were named "Pseudocereals". Cereals are monocotyledonous species, as classified by taxonomists on the grounds of their characteristics that they share in grass family, *Poaceae*. Pseudocereals are mono and di cotyledons non-grassy plants having perisperm enfolding embryo (Fletcher, 2016). The worldwide recognized group of pseudocereals includes dicotyledonous plants such as Buckwheat (*Fagopyrom esculentum* and *Fagopyrum tartaricum*) placed in *Polygonaceae*, Quinoa (*Chenopodium quinoa wild*) placed in the *Chenopodiaceae* family and grain Amaranth (*Amaranthus spp.*) placed in the *Amaranthaceae* family. These grains are grouped under a single umbrella of pseudocereals because the fruits obtained from these plants renders similar function and composition as that of cereals. Pseudocereals can be used as a partial or complete substitution of cereals in some framework. The major similarities between pseudocereals and cereals relates to their storage materials, starch and protein, as they consist of approximately similar relative proportions. Pseudocereal fruit can be harvested, dried, pounded and ground into meal or flour for their further used as a major ingredient or can be used along with other cereal grains.

Major pseudocereals well known worldwide are buckwheat, amaranth and quinoa. Amaranth and quinoa were widely grown in Mesoamerica during Hispanic period. They were grown throughout Latin America for thousands of year and expected to be cultivated as early as 4000BC (De la cruz Torres, 2008). In New World times, they were given equal importance with maize and beans. Cultivation of these grains is increasing steadily and expected to be grown in more than 70 countries, including Denmark, England, France, India, Sweden and United States (FAOSTAT-2018). Moreover, buckwheat has been cultivated in China for about 3000 years. It is native to Asia and later expanded to Central and Eastern Europe with Turkish and Mongolian intervention during late Middle Ages (Wijngaard and Arendt, 2006). Two buckwheat species majorly cultivated for food consumption in Asia, *Fagopyrum esculentum* (common buckwheat) and *Fagopyrum tartaricum* (tartary buckwheat) (Mazza and Oomah, 2003).

Pseudocereals are environment friendly crop with great resistance potential against adverse and changing environment conditions, especially suitable for high altitude regions and possess good adaptability to different soil conditions. These forgotten and underutilized grains can be used in various ways; grabbing tremendous commercial attention due to their beneficial and balanced nutritional importance in view of recent nutraceutical regulations. Buckwheat is a rich source of rutin, flavonoids (orientin, vitexin, quercetin, isovitexin etc.) and bioactive compounds such as fagopyritol and fagopyrins with various proven medical benefits. It also consists of substantial quantity of resistant starch and complex carbohydrates and thus can be effectively exploited for the production of low-glycemic foods and improve lipid metabolism by restricting oxidation of low density lipoprotein (LDL). Quinoa and amaranth are known for their balanced amino acid composition and good protein quality. Protein quality of quinoa is found to be similar to that of milk casein. Amaranth grains are a good source of calcium and lysine and can be used as complementary protein with other cereal grains. Pseudocereals are largely devoid of prolamins which renders them suitable for the production of food products consumed by individuals suffering from celiac disease. Pseudocereals based food products are in great demand due to the increased awareness among consumers regarding nutritional benefits of these grains. Almost all the parts from these plants can be utilized for various food or industrial applications. The grains can be dried and ground to make flour and used as whole or as partial substitution in developing variety of food products. They can be introduced into health beneficiary diets as they are loaded with nutritionally rich constituents. With their continuous consumption, some "nutritional related modern civilization diseases" such as indigestion, obesity, diabetes can be controlled. Pseudocereals are effective nutritional supplements to conventional cereals (particularly products made from wheat and maize) due to their beneficial effect on health.

Structure and Physical Properties

Pseudocereals grains specifically amaranth and quinoa are an appreciable source of protein with balanced amino acid profile, vitamins, mineral and polyunsaturated fatty acid due to their higher bran fraction (seed coat surrounding starch-rich perisperm and a campylotropus embryo) in comparison with cereals (Bressani, 1994). Buckwheat has high starch content due to its greater grain to endosperm ratio.

Pseudocereals (buckwheat, amaranth and quinoa) kernels are structurally divided into three main parts i.e. endosperm, embryo and seed coat. The endosperm is primarily the storage zone of grain consisting of starch as the major part; protein and fat embedded into it and called as starch-protein matrix. The embryo is the germ part that is an oil storage zone of the grain, with high protein and mineral content. The seed coat, also known as pericarp or bran, made up of mainly cellulose and hemicellulose with some traces of protein and lignin (Baltensperger, 2003). The relative content of these three portions in entire grains will vary according to the cultivar, environment conditions and soil quality.

Amaranth is an annual, broad leave plant with embryo of amaranth encloses the starch-rich endosperm in a ring formation. The probable variation in height of plant ranges from 0.5- 3 mm. The flower can achieve the length of 90 cm and varies in color. It consists of rather larger embryo and accounts about 25% of the grain weight (Bressani, 1994). The outer structure of amaranth grain is smooth, thin, shiny and slightly flattened with lens like shape. The grain size varies from 1.0 to 1.7 mm in length and 0.9 to 1.3 mm of width and an average weight of 0.6 to 1.0 mg which is quite lesser than other cereals (wheat and maize) and other counterpart like buckwheat; due to very small size of amaranth seeds and higher grain to hull ratio (Reguera and Haros, 2017). The color of seed coat varies from white, yellow, pink brown to black which depends on the type of cultivar. The amaranth grain can be consumed with the seed coat and not necessarily be removed during processing (unlike the quinoa and buckwheat grain).

Buckwheat is a single seed enclosed by indehiscent pericarp that is tightly rounded around seed. It is three edged seed with acute angle fold, 4-9 mm length gives it a pyramid like shape with a round base (Izydorczyk *et al*, 2014). The buckwheat bran can be thin or thick depending on variety which renders it greater or lesser adhesion to pericarp or endosperm with varying color from silver-gray to black. Single buckwheat grain can weight on an average between 10 to 20 mg (Karpati and Banyai, 1980). The surface appearance of grain varies from shiny to dull depending on the cultivar and environment conditions. Buckwheat grain generally consists of broad cotyledons, central larger embryo, rich in protein, dividing the soft white endosperm into two equal parts. The surrounding testa membrane is generally yellow-green in color (Baltensperger, 2003). The thick hull of buckwheat is generally being removed during processing. The dehulled buckwheat endosperm or groats vary slightly in their color, shape and size depending on the endogenous factors such as species, compact starch content and processing conditions. Tartary buckwheat generally results in significantly smaller and rounder groats in comparison with other common buckwheat (Izdorczyk *et al.*, 2014).

The fruit of quinoa is an achene with varying color from yellow to magenta which is dependent on the quantity of betacyanins pigments (Mabry *et al.*, 1963). The fruit is generally held in five-lobed perianth that become fragile during drying and can be easily removed by mere rubbing between the palms. Grains can be cylindrical, conical or ellipsoidal in shape depending on the cultivar (Vega-Galvez *et al.*, 2010). The diameter for each grain ranges from 1.0 to 2.6 mm. They are very small and light weight having 1000 kernel weight ranging from 1.9 to 4.3 g (Koziol, 1993; Vilche *et al.*, 2003). Quinoa is a complex structure grain with seeds covered by dry, two layered pericarp, and outer layer often is papillose (Prego *et al.*, 1998). The seed coat present just beneath the pericarp. The color of seed coat can vary from transparent, white, and brown to black sometimes (Saturni *et al.*, 2010). If the grain is cut through its major axis the embryo, the central part containing white color starch perisperm can be seen coiled around by two cotyledon and radicle. The interesting fact about quinoa structure is that in this the hypocotyl-radicle axis is surrounded by one or two layered living endosperm tissue which is not true for the common cereal grains. Starch is majorly stored in non-living, thin walled pericarp. The embryo from quinoa have good amount of stored lipid bodies, protein bodies and mineral content.

Physical Properties

The physical properties of the grain such as shape, size, surface characteristics, surface area, volume and weight are essential parameters required in design and development of grain handling equipment at different stages of harvesting and processing (Vilche *et al.*, 2003; Abalone *et al.*, 2004). Physical characteristic of grain affects certain other engineering properties of grain which are necessary for designing the unit operations, equipment and storage conditions (Vashishth *et al.*, 2017). They are divided into two main categories: mechanical and aerodynamical properties. The mechanical properties of grain involve the angle of repose, coefficient of friction etc. and aerodynamical properties of grains involves terminal velocity and drag force. Both the mechanical and aerodynamical properties are significantly affected by change in physical properties of grain. The gravimetric properties of grain effects both mechanical and aerodynamical properties which is important while designing the equipment related to aeration, drying and transportation (Vashishth *et al.*, 2017). The bulk density of the grain determines the storage condition and transports system required for the grain; true density and porosity dictates the resistance that will be shown by the grain against the air flow during aeration or drying process. Pseudocereals are small and light weight grains which may not withstand the conventional processing condition of cereals and hence suitable processing conditions for pseudocereals should be designed before processing to produce better grain quality and its end products (Abolane *et al.*, 2004; Babic *et al.*, 2011).

Chemical Composition of Pseudocereals

Pseudocereals are rich source of protein, fiber, minerals, insoluble polysaccharides and polyunsaturated fatty acid; thereby gaining immense popularity as a nutritional grain (Schoenlechner *et al.*, 2008; Alemaychu, 2015). The chemical composition of common pseudocereals is depicted in Table 1. It is becoming an important part of human diet due to its nutritional quality and proven beneficial nutraceutical properties. The high percentage of resistance starch (RS) in comparison with cereal grain and presence of unique bioactive component such as rutin, fagopyritol and phytosterols enhance their importance for being utilized as functional ingredient in food preparations (Reguera and Haros, 2017).

Table 1: Proximate composition of common pseudocereals (g/100g db)

S.No.	Parameter	Buckwheat	Quinoa	Amaranth
1.	Carbohydrates	58.5-73.7	64.1-70.4	55.1-67.3
2.	Dietary Fibre	6.7-29.5	3.8-19.7	6.7-20.6
3.	Protein	9.0-15.2	11.0-16.5	13.5-16.5
4.	Lipid	1.3-3.4	4.1-9.7	5.6-8.8
5.	Ash	1.4-3.9	1.3-3.8	1.5-3.3

Source: Ruales and Nair, 1993; Sanz-Penella *et al*, 2013; Reguera and Haros, 2017; USDA, 2018

Carbohydrate content

Carbohydrate frames the principle constituent in pseudocereals composition contributing around 40-70 percent of the dry matter (Table 1). They are considered as the major source of energy in human diet as they are inexpensive and present in abundance. They are present in plant in two major from as structural and storage carbohydrates. Structural carbohydrates are those which are utilized by plant for developing their structure and cell walls whereas the storage carbohydrates are the stored reservoir of energy for plants. Carbohydrates can be classified on the basis of their degree of polymerization into mono, di, oligo or polysaccharides. In foods, there are two categories of carbohydrates: available and non-available carbohydrates. The available carbohydrates are those which can easily be utilized by body during digestion whereas unavailable carbohydrates are not digested by human and hence, not available for absorption (Shelton and Lee, 2000).

Buckwheat carbohydrate content ranges from 67-74 percent of which, major portion is contributed by starch ranging from 50-75 percent (Steadman *et al.*, 2000). The starch granules content in buckwheat ranges from 20-45 percent.

Buckwheat starch are generally polygonal or round in shape with varying size of 2-14 μm and mean diameter of 5 μm, possessing a spherical shape. They tend to produce detectable flat areas, which are formed due to flat and tight packing of starch in endosperm (Acquistucci and Fornal, 1997; Soral-Smietana *et al.*, 1984; Wolter *et al.*, 2013). Starch quality can be depicted on the basis of amylose and amylopectin content which provides the functional properties to the starch (Hager *et al.*, 2012). The amylose content of buckwheat starch ranges from 15 to 52 percent and at times present at a concentration of 1:1 with amylopectin. The degree of polymerization for starch varies from 12 to 45 percent glucose units (Mazza, 1993) with molecular weight of 94,900 indicating long chain amylopectin exhibiting high viscosity. The gelatinization temperature for buckwheat starch ranges from 68 to 81°C with mean temperature of 74°C (Li *et al.*, 1997; Praznik *et al.*, 1999; Qian and Kuhn, 1999b). The low amylose and high amylopectin content with long chain tends to produces starch gel with lower strength. Certain amylose in buckwheat tends to form complexes with lipid molecule and affects the functional properties of starch such as swelling power and water solubility. The buckwheat starch tends to produce higher viscosity than that of cereal starches and similar to that of root starches (Arendt and Zannini, 2013). Higher viscosity of buckwheat starches can be ascribed to inter and intra molecular hydrogen bonding, higher swelling and gelling properties (Yoshimito *et al.*, 2004). Along with the higher peak and set back viscosity. The buckwheat starch tends to have greater water holding capacity as compared to cereal starches which can be attributed to the smaller starch granule size of buckwheat starch and greater surface area. Buckwheat starch uniquely has higher resistance starch content (35%) (Skrabanja and Kraft, 1998; Skrabanja *et al.*, 2001). The resistant starch content of grain tends to vary during processing and found dependent on starch granule size, amylose content, amylose-lipid interaction and starch-protein interaction. The enzyme present in human digestive system cannot digest this starch and hence considered as a dietary fiber. The higher content of resistant starch, retrograded starch and dietary fiber reduces the rate at which the starch will be digested and hence can be utilized in the production of low GI food.

Amaranth is an achene grain having carbohydrate content of 58 to 71 percent with 65 to 75 percent of starch (Venskutonis and Kraujalis, 2013). Starch is the major portion of the carbohydrates present in amaranth seeds. Starch granules of amaranth ranges from 0.5-3.0 μm, polygonal in shape with A-type diffraction pattern and crystallinity of 45.5 percent (Uriyapongson and Rayas-Duarte, 1994; Radosaljevic *et al.*, 1998). Amaranth seed contains 4-5 percent of dietary fiber, two to three folds' higher sucrose than wheat grain and higher non-starch polysaccharides (Venskutonis and Kraujalis, 2013). The starch content in amaranth is mainly aggregated in perisperm with cumulative granule diameter of 80 μm (Wilhelm *et al.*, 1998; Berghofer and Schonelechner, 2002). The amylose content in starch varies from 7 - 34.3 percent which tends to vary with variety of grain. The starch content also found to vary significantly among species of amaranth ranging from 2 to 37.6 percent (Wu and Corke, 1999). Amaranth starches found quite low in amylose content ranging from 0.1 to 1.1 percent and high amylopectin content ranging from 89 to 99.9 percent resulting in weak starch gel and categorized as waxy starch (Stone and Lorenz, 1984; Perez *et al.*, 1993; Marcone, 2005). The gelatinization temperature of amaranth starch ranges from 62 to 68°C (Saunders and Becker, 1984). Amaranth starch has been considered as waxy starch due to their low amylose content. Starch has been found quite stable against the freeze - thaw cycle and retrogradation during storage after processing (Yanez *et al.*, 1986; Uriyapongson *et al.*, 1994; Paredes-Lopez *et al.*, 1994); which is attributed to its small size of starch granule which exposes greater surface area during both freezing and thaw cycle resulting in higher stability (Becker *et al.*, 1981; Lehman, 1988). The *in-vitro* starch digestibility for amaranth starch is relatively higher than other pseudocereals which can be attributed to the small size of starch granule which renders them more susceptible to digestive enzymes (Kong *et al.*, 2012). The role of amaranth in production of low GI food is quite debatable as the research findings on ability of grain to maintain the glucose level are quite diverse but it can be effectively used for controlling the blood cholesterol level. Some of the researchers found the positive effect of amaranth on the rate of glucose release and reduced cholesterol levels in albino rats whereas some of them found increased glucose levels in rats during consumption of amaranth rich diet. Production of amaranth and wheat based white bread resulted in GI of 91.7 for 50:50 replacement level and GI of 136.2 for a 25:75 replacement level. Hence, amaranth can be efficiently used for controlling the blood cholesterol levels but incorporation of amaranth in diet to maintain the blood glucose level should be avoided (Chaturvedi *et al.*, 1997; Caselato-Sousa and Amaya-Furfan, 2012).

Quinoa is an achene grain having carbohydrate content of 61 to 74 percent with 58 to 64 percent of starch (Rapo-Carrasco *et al.*, 2003; Repo-Carrasco-Valencia and Arana, 2017;), making major portion of the carbohydrates in quinoa. Starch granules from quinoa are polygonal in shape with A-type diffraction pattern. Quinoa seed contains 2-3 percent of dietary fiber, sucrose (2.79 %) and traces of flatulence causing oligosaccharide such as stachyose, verbacose and raffinose (Quian and Kuhn, 1999b; Iglesias-Puig *et al.*, 2015). The starch content in quinoa is mainly aggregated in perisperm and some are present in bran of the grain with individual grain diameter of 1.0 to 2.5 μm and aggregated granule diameter of 6.4 to 32 μm with oval shape granule(Atwell *et al.*, 1983). The amylose content in starch varies from 11 to 12.2 percent which tends to vary with variety of grain. The quinoa starch being rich in amylopectin content tends to gelatinized faster at a lower temperature in comparison with other cereal starches. The gelatinization temperature of quinoa starch ranges from 57 to 71°C (Taylor and Parker, 2002). Even though the starch is low in amylose content, these starches been found quite stable against the freeze and thaw cycle, and retrogradation during storage after processing. Similar to that of amaranth starch the stability of quinoa starch against freeze thaw cycle and processing can be attributed to the small size of starch granule which exposes greater surface area during both freezing and thaw cycle resulting in higher stability (Ahamed *et al.*, 1996). I*n-vitro* starch digestibility for quinoa said to be relatively higher than other pseudocereals which can be attributed to the small size of starch granules and higher amylopectin content which renders them more susceptible to digestive enzymes. Higher concentration of dietary fiber in quinoa were reported (14-27.6%) in comparison with other grains. The resistant starch content of quinoa can be increased by 19 percent by modification (Linsberger-Martin *et al.*, 2012). Quinoa as a whole grain can be efficiently used for the production of low GI food as it has tendency to regulate the blood glucose level. It can also be effectively used for controlling the blood cholesterol level due to its high soluble and insoluble dietary fiber content.

Protein

Pseudocereals are considered as good source of nutritionally balanced protein, higher biological value and protein efficiency ratio (Table 1 & 2).

Buckwheat is one of the good source of protein ranging from 12 to 15 percent, localized in protein bodies with mean diameter of 1-10 μm (Wei *et al.*, 2003; Alvarez-Jubete *et al.*, 2010). On the basis of solubility, they are classified into four major groups Globulin, Albumin, Glutelins and Prolamins. Protein content in buckwheat varies depending on variety, environment conditions and fertilizers used during the production (Christa and Soral-Smietana, 2008). The protein content is unevenly distributed throughout the grain with highest concentration present in embryo (56%) followed by outer fraction (40%) and minimum amount is present in the inner fraction of the grain (1-2%). The hull contains around 4 percent of total protein of the grain (Morita *et al.*, 2006; Christa and Soral-Smietana, 2008). Protein quality of buckwheat is regarded better than cereal protein because of its higher biological value of 92.3 percent in comparison with 81.5 percent of dried and defatted milk, and 62.5 percent of wheat protein (Rayas-Duarte *et al.*, 1998). The major portion of endosperm protein is made up of salt soluble globulin (40%) which is majorly divided into three distinct fractions 13S, 8S and 2S, which contributes 30 percent of the total protein. Albumin fraction of buckwheat protein ranges from 18-

25 percent total protein with molecular weight of 8kD-12 kD which again depends on the variety of the grain. The prolamin content is either absent or present in low concentration (Maksimovic, *et al.*, 1966; Radovic *et al.*, 1996; Radovic *et al.*, 1999).

Buckwheat protein is well known for its nutritionally balanced amino acid profile. The presence of anti-nutritional factors such as proteases, phytic acid and saponins; the protein quality tends to reduce but still the biological value for buckwheat protein remains as high as 90 percent which is higher than that of cereals. Due to the presence of antinutritional factors the actual protein digestibility get reduced to 80 percent as the protolytic enzymes could not hydrolyses the protein (Bonafaccia and Kreft, 1994). The buckwheat protein majorly consists of similar or higher concentration of glutamic acid, aspartic acid and arginine followed by lysine. The concentration of lysine in buckwheat can be as high as twice the concentration of lysine present in wheat whereas glutamine and proline concentration is found less than in wheat protein. High concentration of lysine in buckwheat makes it suitable as a supplementary grain with other cereals (wheat and rice). Threonine and methionine are the first and second limiting amino acid hence, it is advisable to mix buckwheat with other cereal grain to prepare food with a balanced amino acid profile. Prolamin is the major amino acid causing irritation in people suffering from gluten intolerance or celiac disease, hence the individual suffering from celiac disease can incorporate buckwheat in their diet as buckwheat contains nil or marginal concentration of prolamins (Pomeranz and Robins, 1972; Wei *et al.*, 2003; Krkoskava and Mrazora, 2005; Gua and Yao, 2006; Sano *et al.*, 2014).

Table 2: Essential amino acid composition of common pseudocereals (g/100g protein)

S.No.	Amino acid	Buckwheat	Quinoa	Amaranth	Reference Pattern
1	Threonine	1.9-4.04	2.3-4.7	3.3-3.5	3.4
2	Leucine	2.8-6.12	3.9-6.5	5.4-6.4	6.6
3	Isoleucine	2.6-3.50	2.3-7.4	3.6-4.2	2.8
4	Lysine	4.9-6.70	6.5-9.8	4.8-5.6	5.8
5	Phenylalanine + Tyrosine	3.5-7.42	3.0-4.7	7.0-7.3	6.3
6	Tryptophan	1.2-1.06	0.00-1.3	1.1-1.8	1.1
7	Valine	3.4-4.97	2.9-6.0	4.2-4.6	3.5
8	Methionine+Cysteine	1.0-4.10	0.9-3.1	4.5-4.7	2.5

Source: Taylor and Parker, 2002; Berghofer and Schoenlechner, 2002; Caselato-Sousa and Amaya-Farfan, 2012; Venskutonis and Kraujalis, 2013; D'Amico *et al.*, 2017; FAO/USDA/WHO-2018.

Amaranth shows higher protein content with better protein quality in comparison with quinoa and buckwheat. Major portion of this protein will be stored in germ and outer seed coat (65%) and remaining stored in perisperm containing 35 percent of protein (Saunders and Becker, 1984). Protein content and protein quality depends on the cultivar and growing conditions. Cereals has alcohol soluble prolamin as their major storage protein but in pseudocereals, the major storage proteins are globulin and albumins. The amaranth protein contains approximately 40 percent albumins, 20 percent globulins, 25-30 percent glutelins and only 2-3 percent prolamins. The prolamins content in amaranth proteins has been controversial as some researchers have found very low concentration of prolamin ranging between 0.48 to 0.79 percent and 1.2 percent. Higher globulin concentration in amaranth exhibits excellent functional properties along with nutritional value. The functional properties of amaranth globulin isolates have been reported to be better than soya globulins isolates, pointing towards amaranth protein quality and stability at isoelectric point (Marcon and Kakuda, 1999). Amaranth has outstanding amino acid profile with significantly high concentrations of essential amino acids. The major functional proteins present in amaranth are 11S globulin, 7S globulin, α-amylase inhibitors, trypsin inhibitors, antimicrobial proteins, non-specific lipid transfer protein, superoxide dismulase, glucosc-1-phosphate, glucosyl transferase, enyltransferase, polyamide oxidase, granule bound starch synthase and acetolactate synthase (Montoya-Rodriguez *et al.*, 2015).

The amino acid content of amaranth has been reported to be very much closer to the ideal amino acid composition mentioned by FAO/WHO for human diet. Amaranth protein is rich in lysine, arginine along with higher concentration of methionine, tryptophan and sulfur containing amino acid (Palombini *et al.*, 2013). The lysine concentration of amaranth can be 2-2.5 times of lysine present in cereals. Literature regarding limiting amino acid in amaranth are debatable; as literature reporting leucine (on the basis of chemical score), threonine (on the basis of protein efficiency ratio) or valine as limiting amino acid at times are present (Montoya-Rodriguez *et al.*, 2015). The protein digestibility of raw amaranth ranged from 74-86 percent which tends to increase during processing (Gamel *et al.*, 2004; Escudero *et al.*, 2004). A high protein efficiency ratio (PER) and biological value for amaranth protein was also determined in literature which ranged from 1.2-1.5 and 75-87 percent respectively (Pederson *et al.*, 1987; Bressani *et al.*, 1989; Abreu *et al.*, 1994). The increase in the protein digestibility can be attributed to the unfolding of protein structure and protein-carbohydrate complex, and reduction in anti-nutritional factor such as phytic acid, proteases and other complex molecule. On the basis of higher protein digestibility, protein efficiency ratio and amino acid profile, amaranth can be efficiently utilized along with other cereals or as whole in the human balanced nutritional diet.

Quinoa is well known as complete food due to its balanced nutritional and anti-nutritional constituents. The nutritional quality is majorly attributed to its high protein quality of 8-22 percent (Hager *et al.*, 2012; Rosero *et al.*, 2013). Protein content and quality of quinoa varies with the genotype, environment conditions and the fertilizers used (Kakabouki *et al.*, 2014). The maturity level of grain also tends to affect the final protein content of quinoa seeds. The major part of the quinoa proteins is stored in embryo (38%) in order to provide nutrition for the seed growth and development, whereas the endosperm (7%) consists of marginal level of protein (Ando *et al.*, 2002; D'Amico *et al.*, 2015). In quinoa, major proteins were albumin and globulin. Both albumin and globulin in combination contributes 60 to 80 percent of the total protein depending on the cultivar and environment condition. Even though the percent level of albumin and globulin have been well established but still the ratio of globulin and albumin yet not been established as different researches has come up with different conclusion. In two research studies revealed, albumin as a major protein fraction but other three studies depicted globulin as the major protein fraction. These changing fractions of albumin can be justified on the basis of second albumin fraction present in quinoa. The protein fraction of quinoa seeds is made of medium sized protein with average molecular weight of 10 to 70 kD and few weak bands of 70 to 100 kD (Drzewiecki *et al.*, 2003; Abugoch, 2009; Hager, *et al.*, 2012). Quinoa protein frame work is majorly designed with two major protein subunits i.e. 11S and 2S-polypeptide with relatively varying solubility in acidic medium. 11S tends to precipitate out at pH 5 whereas 2S subunit is still soluble (Brinegar, 1997).

The amino acid content of quinoa had been reported as well balanced and similar to that of milk protein casein. In comparison with the levels of amino acid requirement for pre-school children given by FAO/WHO quinoa can be a perfect ingredient as it has all the essential amino acid and not deficient even in single amino acid. Quinoa protein is rich in lysine, arginine along with higher concentration of methionine, tryptophan and sulfur containing amino acid (Guzman-Maldonado, 1998). The lysine concentration of amaranth can be 2 times of lysine present in cereals. Isoleucine and valine are the first and second limiting amino acid respectively. The protein digestibility of raw quinoa ranges from 76-80.5 percent which can be improved up to 83 to 84 percent during processing (Ranhotra *et al.*, 1993; Gonzalez *et al.*, 2012; Rosero *et al.*, 2013). The increase in the protein digestibility can be attributed to the unfolding of protein structure, protein-carbohydrate complex and reduction in anti-nutritional factor such as phytic acid, proteases and other complex molecule. The quality of quinoa protein was found variety dependent as the protein efficiency ratio (PER) of quinoa protein was reported in range of 1.95-

2.33, while some authors reported PER of 2.7 to 3.1 as well. Chauhan *et al.*, (1992) reported an interesting aspect of quinoa protein. They observed the uniform distribution of amino acid in quinoa which does not change during fractionation of grain. The hull, dehulled grain, bran and flour were reported to possess similar amino acid composition as that of grain. On the basis of higher protein digestibility, protein efficiency ratio and balanced amino acid profile, quinoa can be efficiently utilized along with other cereals for achieving a balance nutritional diet.

Lipid

In recent years, the consumption of lipid in common human diet has increased up to 30 percent (NCHS, 2014). The increase in lipid percentage in diet is complemented by the increased awareness of consumer regarding balanced fat which has led to the consumption of fat rich in polyunsaturated fatty acid (PUFA) and lower in saturated fatty acids (SFA). Major upsurge in fat based diet is reported in Mediterranean diet which involves consumption of olive oil, stated to be rich in PUFA and tends to produce beneficial health effect by reducing cholesterol levels (Hu, 2003; Widmer *et al.*, 2015). The PUFAs are not synthesized or derived from other sources in human body and hence they need to be supplied through diet in human. The oils from the grains such as buckwheat, amaranth and quinoa found to be rich in PUFA such as oleic, linoleic and linolenic acids which offers exceptional quality to these unconventional grains oils and enhances the possible changes for these all to bc introduccd in human dict.

In general majority of buckwheat oil is reported to be present in embryo (8.2%) followed by testa (2%), pericarp (0.5%) and finally endosperm (0.4%). The oil content in buckwheat (2.4 to 3.2%) is mostly higher than other common cereals but was found lower than grains like amaranth and quinoa seeds (Biacs *et al.*, 2002). The lipid concentration in buckwheat tends to vary with the cultivar, species, seeding time and environment conditions as the higher oil concentrations were reported for plants grown at higher altitudes (Venskutonis and Kraujalis, 2013). Buckwheat fat has higher neutral lipid ranging from 81-85 percent of total lipids, which is more than double the amount of neutral lipids present in wheat and rye (Ikeda, 2002). It exhibits polar lipid content of 15.4 to 18.8 percent of total lipids with five major lipids: digalactosyl diglycerides (DGD), phosphatidyl choline (PC), phosphatidyl etholamine (PE), lysophosphatidyl etholamine (LPE) and phosphatidyl serine (PS) in which DGD and PC possesses maximum lipid content of 43.3 percent and 17.9-25.9 percent of entire lipids, respectively (Obara and Miyata,1969; Lakshminarayana, *et al.*, 1984; Gamel *et al.*, 2007). Fatty acids namely oleic, linoleic, linolenic, arachidic, palmitic, stearic, lignoceric and behenic acids contribute to more than 93 percent of total fatty acids present in buckwheat oil with oleic, linoleic and linolenic contributing 880 mg/g of total fatty acid content (Horbowicz and Obendrof, 1992). The SFA content of buckwheat ranges from 16-25 percent of total fatty acid with the leading SFAs namely palmitic and stearic acid (Table 3). The ratio among saturated and unsaturated fatty acid (UFA) is regarded as an indicator for nutritional and functional quality of lipid, buckwheat has S/U value ranging from 2.6 – 3.4. Therefore, buckwheat is considered as good quality fat. It contains an average tocopherol content of 50 mg/kg of oil which provides the antioxidant potential and oxidation stability to the buckwheat oil (Kim *et al.*, 2001).

Table 3: Fatty acid composition of common pseudocereals (g/100g fat)

S.No.	Fatty acid	Buckwheat	Quinoa	Amaranth
1	Myristic	0.3-0.5	0.1-0.2	0.12-0.29
2	Palmitic	15.6-19.7	9.7-11.4	14.09-25.91
3	Palmitoleic	0.1-0.2	0.1-0.2	0.33-0.60
4	Stearic	2.0-3.0	0.6-0.8	3.17-4.62
5	Oleic	35.2-37.1	24.5-25.6	20.18-36.22
6	Linoleic	34.8-39.0	50.2-52.8	24.84-55.90
7	Linolenic	0.0-1.93	3.9-7.00	0.2-1.29
8	Arachadonic	1.6-1.8	0.4-0.7	0.11-0.35
9	Eicosenoic	2.0-2.3	--	0.15-0.65
10	Behenic	0.8-1.1	--	0.00-0.35

Source: Ruales and Nair, 1993b; Koziol, 1993; Jahaniaval *et al.*, (2000); Vidueiros (2015).

In amaranth oil content is widely held in germ and seed coat contributing around 7.4 percent of entire lipids with trivial share existing in perisperm (2.3% of entire fat) (Bestchart *et al*, 1981; Becker, 1994). Amaranth possesses fat content of 5.8 to 19.3 percent db which is quite higher than that of buckwheat, quinoa and cereal grains (Valencia-Chamorro *et al.*, 2017). The fat obtained from amaranth is generally yellow in color which is comparable to the color of corn oil. The fat content in buckwheat tends to vary with the cultivar, species, seeding time and environment conditions as the higher oil content has been observed for plants grown at higher altitudes. Neutral lipid content in amaranth expected around 90 percent with rest contributed by polar lipids. The major polar lipids in amaranth are monogalactosyl diglycerides (MGD), phosphatidyl choline (PC), digalactosyl diglycerides (DGD), phosphatidyl etholamine (PE) andphosphotidyl inositol (PI), in which PC and MGD are the major lipids with highest lipid content of 1.8 to 2.6 percent and 1.3 to

15.6 percent of total lipids, respectively (Laxminarayan *et al.*, 1984). Fatty acids profile for amaranth has been observed pretty comparable to the cotton seeds and corn with UFA content of 77.1 percent of entire lipid and degree of unsaturation of 74-77 percent (Jahaniaval *et al.*, 2000; Martirosyan *et al.*, 2007). The major UFAs present in amaranth are oleic (19-35%) and linoleic (25-62%) of total fatty acid (table 4). The saturated to unsaturated fatty acid ratio for amaranth grain ranges from 2.5 to 3.7 percent (Dodak *et al.*, 1997; Jahaniaval *et al.*, 2000). Tocopherol is regarded as essential constituent of lipid fraction in amaranth. The tocopherol content in amaranth tends to vary from 0.78-2.95 mg/100g seed for α tocopherol, 0.71-6.74 mg/100g seed for β tocopherol and 0.11-2.05 mg/100g seed for δ-tocopherol (Bruni *et al.*, 2011; Valencia-Chamorro *et al.*, 2017). The tocopherol content in amaranth depends heavily on the variety of grain, accession, growing condition and maturity stage during harvesting of grain. Amaranth tends to possess good concentration of squalene which a is bioactive component and used widely in pharmaceutical industries. The squalene content in amaranth is not affected in the course of processing and found stable during thermal processing which render amaranth as a significantly important source of squalene.

Quinoa grains possess major lipid in embryo and perisperm with lipid concentration of 10.2 and 5-8 percent respectively (Prego *et al.*, 1998). The lipid present in embryo is around 49 percent of total lipid content whereas perisperm contributes around 46 percent to the total lipid content. Quinoa possess an average fat content of 4-9.7 percent which is pretty high than buckwheat and cereal grains but lower in comparison with amaranth. The colour of amaranth fat generally lies between colorless to yellowish. Lipid content of buckwheat tends to differ according to cultivar, species, seeding time and environment conditions. Among three major pseudocereals quinoa possess highest lipid content on average followed by amaranth and buckwheat. Quinoa contains lower neutral lipids than amaranth and buckwheat. The entire neutral as well as polar lipid content for quinoa is around 55.9 percent and 25.2 percent respectively, of whole lipid content of grain (Przybylski *et al.*, 1994). The polar lipid content in quinoa contains six major lipids namely lysophosphatidyl etholamine (LPE), phosphatidyl choline (PC), phosphatidyl inositol (PI), phosphatidyl etholamine (PE), and phosphatidyl serine (PS) and lysophosphatidyl choline (LPC) in which LPE and PEare the major fraction with highest lipid faction of 43.2 and 18.5 percent of total polar lipids, respectively (Przybylski *et al.*, 1994). Fatty acids namely oleic, linoleic, linolenic, arachidic, palmitic, stearic, lignoceric and behenic acids contributes to more than 93% of total fatty acids present in quinoa oil with oleic, linoleic and linolenic contributing major proportion of total fatty acid content (Ruales and Nair, 1993a). The SFA content for quinoa ranges from

11-17 percent of overall fatty acid with predominant SFAs such as palmitic contributing about 8.5 percent to total fatty acid content followed by other major SFA myristic, stearic, behenic and lignoceric acid but in inconsequential quantities (Vidueiros *et al.*, 2015). The quinoa contains pretty high UFA content contributes around 82.7 to 85 percent of total fatty acid concentration with oleic, linoleic and linolenic acids as the predominant UFAs (Table 4). The occurrence of high linoleic and linolenic acid concentration claims it as a valuable source of essential fatty acids (Vidueiros *et al.*, 2015). The ratio of SFA to UFA for quinoa ranges from 3.7 to 4.9, which is higher than previously reported ratio for amaranth and buckwheat (Ruales and Nair, 1993b). Quinoa contains an average tocopherol content of 17.5μg/kg of α-tocopherol and 47.2 μg/kg of γ-tocopherol on dry weight bases. The availability of tocopherol at such concentration provides the antioxidant potential and oxidation stability to quinoa oil (Repo-Carrasco *et al.*, 2003; Alvarez-Jubete *et al.*, 2010).

Table 4: Mineral profile of common pseudocereals (mg/100g)

S.No.	Mineral	Buckwheat	Quinoa	Amaranth
1	Calcium	14.8-22.1	32.7-110	159-204
2	Copper	0.37-0.71	0.77-5.1	0.69
3	Iron	2.1-4.7	2.1-9.0	7.6-9.2
4	Magnesium	173.6-267.60	173.6-500	248-279
5	Manganese	1.00-1.64	1.00-4.0	3.33-3.66
6	Phosphorus	260-490	260-535	530-557
7	Potassium	402.3-565.50	120-900	470-508
8	Zinc	1.0-2.9	0.8-2.3	1.6-4.2

Source: Ruales and Nair, 1993a; Bruni *et al*, 2001; Alvarez-Jubete *et al.*, 2009; Caselato-Sousa and Amaya-Farfan, 2012; Sanz-Penella *et al.*, 2013; .

Mineral Profile

Pseudocereals are considered as an efficient dietary source of essential minerals and trace metals like Iron (Fe), Magnesium (Mg), Manganese (Mn) and calcium (Ca) (Table 4). The distribution of these minerals in kernel is quite analogous to that of cereal grains as the major share of minerals are concentrated in seed coat, hull and aleurone layer of kernels. The availability of minerals in amaranth tends to differ according to the color of plant as the dark colored plant were reported to be rich in Fe, Mg and Ca content than that of white seeds genotype. Whereas, no such effect was observed in minerals like potassium (K), sodium (Na) and phosphorus (P) (Makus, 1990; Stalknecht and Schulz-Schaeffer, 1993). Fe, Mg and Cu content of grain tends to vary significantly among different varieties. Quinoa is considered as mineral rich with occurrence of higher minerals concentration than popular cereal grains.

Phosphorus, sodium, potassium and magnesium content of quinoa grain is majorly localized in embryo whereas as pericarp consist of higher concentration of Ca and P, existing as bound P and Ca in protein bodies (Chauhan *et al.*, 1992; Coulter and Lorenz, 1990; Kent and Evers, 1994). Akin to amaranth the availability of mineral in quinoa tends to vary with cultivar, species, climatic condition and maturity level of grain during harvesting. Like other pseudocereals, buckwheat is also considered as rich source of dietary minerals and trace metal which plays a significant part in growth and development. Significant share of minerals in buckwheat is stored in embryo and aleurone layer with Mg, Zn and Co as the major minerals. Fe, Mn, Co, Mo, Ni and Al are generally localized in both hull and seed coat of grain which tends to reduces during dehulling and milling process (Kuhn *et al.*, 1996). Phosphorus is mainly concentrates in phytates and protein bodies. Buckwheat is a poor source of calcium. On the ranking of Recommended Daily Allowances (RDA), pseudocereals can be classified as vital source of minerals like magnesium, phosphorus and iron. The concentration of iron in quinoa reported to be in range of 8-9 mg/100g, buckwheat (3-4 mg/100g) and amaranth (8.02-9.33 mg/100g) (Mosha and Svanberget, 1990; Longvah *et al.*, 2017).

Vitamins

The vitamin profile for pseudocereals is generally falls comparable to cereals with quinoa been exception (Table 5). Buckwheat is regarded as a vital source of water soluble vitamins mainly vitamin B1 (Thiamin), B2 (Riboflavin), B3 (Niacin), B6 (Pyridoxine) and Vitamin C (Ascorbic acid). Among buckwheat varieties, tartary buckwheat said to possess higher concentration of B and C vitamins than common buckwheat (Ikeda *et al.*, 2002; Ahmed *et al.*, 2014). Vitamin E, also exist in buckwheat which provides the shielding effect against oxidative spoilage to the buckwheat oil. Major tocopherol present in buckwheat is α-tocopherol with absence of tocotrienols. The availability of vitamin in buckwheat can be improved by fermentation or sprouting the grains. During sprouting vitamin C concentration can be enhance upto 0.25 mg/g in buckwheat (Lintschinger *et al.*, 1997; Kim *et al.*, 2004; Ahmed *et al.*, 2014). The occurrence of various tocopherol in buckwheat varies with variety of grain. Various forms of tocopherol such as α, β and ϒ exhibit different antioxidant potential hence the change in different tocopherol concentration results in change in antioxidant potential of grain (Przyvylski *et al.*, 1998; Zielinski *et al.*, 2001; Kim *et al.*, 2002).

Amaranth is considered as a better source of vitamin than buckwheat among pseudocereals. Amaranth is regarded as a vital source of B complex vitamins such as riboflavin and thiamin, vitamin C, folic acid along with fat soluble vitamin E (Caseato-Sousa and Amaya-farfin, 2012). It contains all forms of tocopherol namely α, β and δ-tocopherol and α and γ-tocotrienols. The α and γ-tocopherol are reported to be present at higher quantity 47.53 mg/kg dry weight and 17-26 mg/kg dry weight respectively in contrast to δ-tocopherol (USDA, 2010; Ruales and Nair, 1993; Alvarez-Jubete *et al.*, 2009; Tang and Tsao, 2017).

Table 5: Vitamin concentration of pseudocereal kernels.

S.No.	Vitamin	Buckwheat	Quinoa	Amaranth
1	β-Carotene	0.21	0.39	Nd
2	Thiamine	0.46	0.29-0.40	0.07-0.10
3	Riboflavin	0.14	0.20-0.39	0.19-0.23
4	Niacin	1.80	1.24-160	1.17-145
5	Pantothemic acid	1.05	Nd	Nd
6	Pyridoxine	0.73	0.49	Nd
7	Folates (μg)	0.025	0.08-0.18	0.05-0.10
8	Vitamin C	5.00	0.18-16.4	3.36-7.24
9	Tocopherol	4.5-12.9	8.7	5.5
10	α-tocopherol	0.30-24.8	2.6-5.37	0.085
11	β-tocopherol	0.00-54.6	Nd	Nd
12	□-tocopherol	Nd	5.3	5.14
13	δ-tocopherol	0.8	Nd	0.24
14	α-tocotrienols	Nd	Nd	Nd
15	β-tocotrienols	Nd	0.3	Nd
16	□-tocotrienols	0.00-0.10	Nd	Nd

Source: Reguera and Haros, 2017.

Quinoa is considered as the best source of vitamins among pseudocereals with its higher concentration of B complex vitamins and folic acid concentrations. Quinoa also contains significant concentration of vitamin E. The occurrence of carotenoids, vitamin-E, riboflavin and folic acid in quinoa is pretty high than wheat and other common cereals. According to RDA, quinoa can be classified as vibrant source of riboflavin, thiamin, folic acid, niacin and vitamin E as it can fulfill the daily requirement of certain minerals and vitamins for children of age between 1-3 years. The concentration of vitamins in quinoa tends to vary with cultivar, species, environment conditions and maturity of seeds during harvesting (Gordillo-Bastidas *et al*, 2013; Tang and Tsao, 2017).

Bioactive Component

The existence of bioactive compounds in these pseudocereals is well documented in research. The major bioactive compounds present in pseudocereals are polyphenols, phytosterols, resistant starches and other dietary component possessing some functional values in human health. Polyphenols like vallic, *p*-hydroxybenzoic and ferulic acid along with flavonoids like catechin, kaempferol and quarcetin contributes towards bioactive compounds available in these grains (Fig. 1 & Table 6).

Table 6: Bioactive components in common pseudocereals grain (mg/kg)

S.No.	Bioactive compound	Buckwheat	Quinoa	Amaranth
1	Caffeic acid	82.7-90.7	2.16-8.7	6.41-6.61
2	Ferulic acid	11.31-12.34	76-298	120-620.0
3	Vanillic acid	5.3-5.47	14-146.6	15.5-69.5
4	p-Coumaric acid	12.84-15.11	1.1-95.0	1.2-17.4
5	p-Hydroxybenzoic acid	110	15.84-38.80	8.5-20.90
6	Catechin	Nd	3.89-4.62	Nd
7	Kaempferol	Nd	0.25-542	22.4-59.7

Source: Brady *et al.*, 2007; Hirose *et al.*, 2010; Repo-Carrasco-Valencia *et al.*, (2011); Pasko *et al.*, 2011; Miranda *et al.*, 2014; Tang *et al.*, 2015

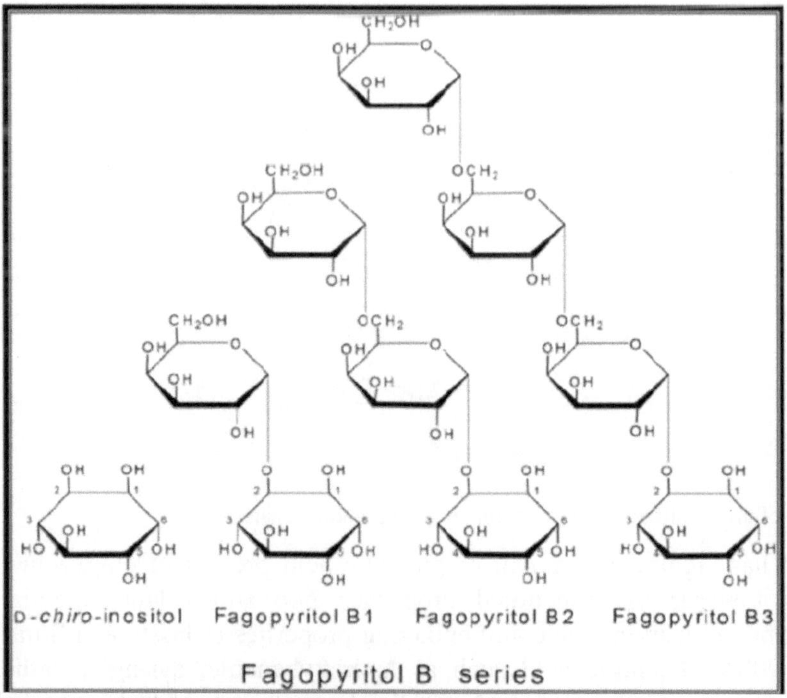

Fig. 1. Fagopyritol B series formed during linkage between D-chiro-inositol and galactose molecule. Adapted from Horbowicz *et al.* (1992).

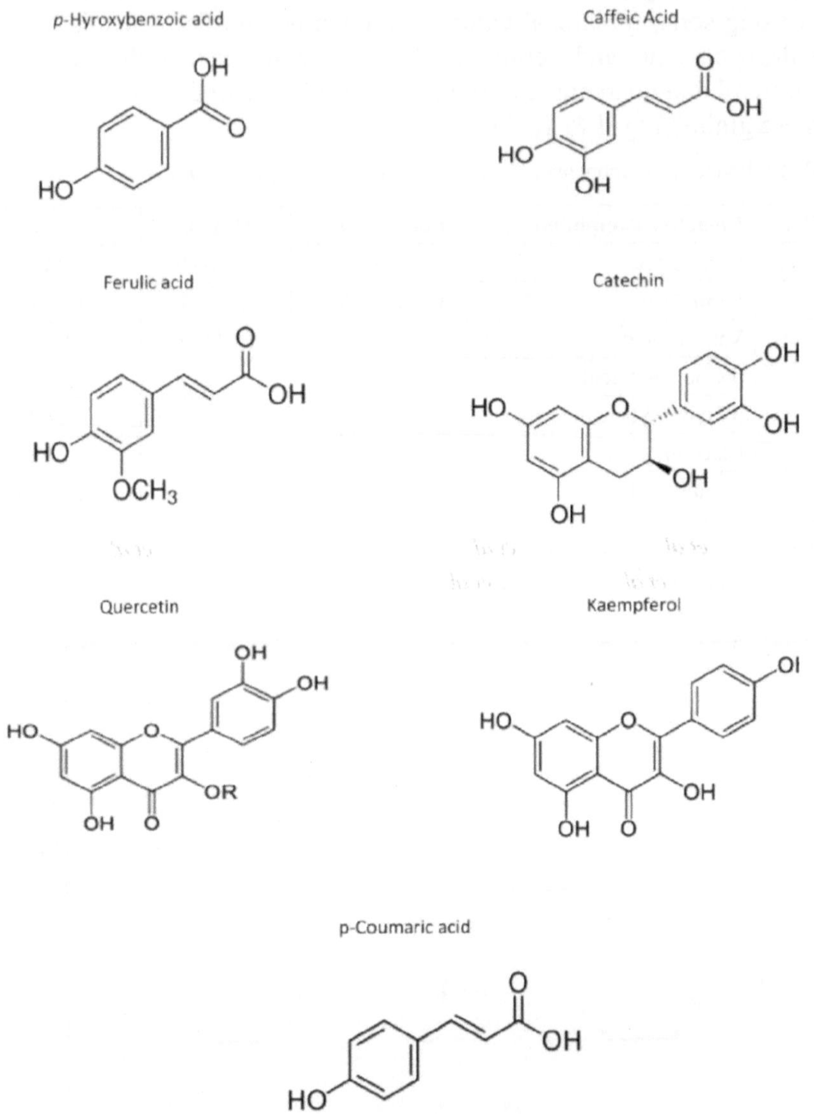

Fig. 2: Some of the identified bioactive component in pseudo cereals

Buckwheat has significant amount of phenolic acid present in grain which contributes towards the functional properties like antioxidant potential, antimutagenic, anticarcinogenic and antiaging properties (Christa and Soral-Smietana, 2008). Phenolic acid such as hydroxybenzoic, syringic, vallic, *p*-hydroxybenzoic, p-coumaric and ferulic possess antioxidant potential and majorly present in bran portion of buckwheat (Steadman *et al.*, 2000;

Steadman *et al.*, 2001b). Buckwheat contains high concentration of para-hydroxybenzoic acid (11 mg/100g) and caffeic acid (8.5 mg/100g) and low level of folic acid (1.2 mg/100g) in contrast with other cereal grain(Mattila *et al.*, 2005). The phenolic content in buckwheat husk and powder was observed as 30 and 15mg/100g respectively (Guo *et al.*, 2012). The primary antioxidant in buckwheat is rutin which is unique to it among pseudocereals (Gallardo *et al.*, 2006; Zhang *et al.*, 2012; Kreft *et al.*, 2006). Buckwheat bran expected to possess 2-7 folds greater antioxidant potential than that of wheat, barley and other cereal grains. Existence of dietary fiber and resistant starch results in the slow release of glucose in blood on buckwheat consumption which renders buckwheat as low GI food. Buckwheat can efficiently be utilized for controlling the blood cholesterol levels because of the existence of bioactive component like fagopyrins, known for regulating cholesterol synthesis level in human. The availability of bioactive content in buckwheat can possibly be improved during sprouting and fermentation (Negi *et al.*, 2018).

Akin to buckwheat, amaranth also has significant antioxidant potential. The phenolic content of amaranth falls in range of 2-4 mg gallic acid/g. whereas some researchers reported lower levels of phenolic acid (0.9 to 1.1 mg gallic acid/g). This difference in phenolics availability can be accord to different species and growing conditions (Guzman-Maldonado and Paredes-Lopez, 1998). Rutin, nicotiflorin and isoquercetin are the major polyphenolic acid existing in amaranth. Rutin is the predominant phenolic acid in amaranth followed by nicotiflorin. Other phenolic acids available in amaranth are p-hydroxybenzoic, vanillic and syringic acid with concentration of 0.17-0.22 mg/100g, 0.15-0.18 mg/100g and 0-0.08 mg/100g respectively (Barba de la Rosea *et al.*, 2009). Dark colored amaranth was found rich in betalains, class of red and yellow pigment. These pigments generally replace anthocyanin in some species. The antioxidant capacityof betalain availablein amaranth was found greater than ascorbic acid, rutin and catechin, indicating the possible use of these pigments as strong natural antioxidant (Cai *et al.*, 2003). Studies also expressed that antioxidant capacity of this pigment (betalain) also depends on their chemical structure as the betalain pigment with more number of functional groups attached to it shows high antioxidant capacity. The other non-nutritional bioactive compounds like fiber and phytic acid at an optimum concentration helps in controlling the higher cholesterol levels and helps in maintaining the better gut health.

Quinoa in general has better antioxidant potential than amaranth but inferior than buckwheat. The average phenol content in quinoa ranges from 1.59-374 mg gallic acid/100g and percentage soluble phenolic acid varied from 7 to 61% (Miranda *et al.*, 2010; Rapo-Carrasco-Valenscia *et al.*, 2010; Rapo-Carrasco-

Valenscia *et al.*, 2011). This vast range in concentration of in-soluble and soluble phenolic acid is accorded to different quinoa varieties ranging from white to dark brown. Quinoa is also regarded as significant source of flavonoid with concentration ranging from 36.2 to 144.3 mg/100g with glycoside of flavanols, kaempferol and quarcetin (Hirose *et al.*, 2010; Repo-Carrasco-Valencia *et al.*, 2010). Amaranth also contains significant concentration of squalene ranging from 3.39 to 5.84 percent of total lipid content. Squalene is the triterpens with significant functional value as it possesses the potentialto control the blood cholesterol and functionas potential antioxidant. Quinoa also contains certain levels of saponins which exhibits constructive effect on human health. Saponins are steroids and triterpenoids glycoside which present in pericarp and gives bitter taste to quinoa. Varieties having higher concentration of saponins tend to taste bitter with low threshold value of 100 mg/100g quinoa seeds (Koziol, 1991; Taylor and Parker, 2002). Saponins can efficiently be used to produce anticarcinogenic, anti-inflammatory, immune modulating and hypo-cholesteromic effects.

Antinutritional Factors

Phytic Acid

Phytic acid is stated to be present in whole grain such as cereals, legumes and pseudocereals. Phytates in plant are produced as a part of defense mechanism against insect it is in storage form of phosphorus with majority of the phosphorus bound to these phytates. Chemical form of phytic acid involves 1,3,4,5,6-hexakisphosphate, InsP_6 and phytates. Role of phytic acid towards human health on consumption is bit controversial as some reports suggest that phytic acid on consumption produce beneficial health impacts such as the hydrolyzed product from phytate tends to exhibit anticarcinogenic properties and even help in reducing cardiovascular disease (Haros *et al.*, 2000). Further, it is well known to reduce the bioavailability of minerals and protein by forming insoluble complex with these components and rendering them unavailable for absorption in human gut. Phytates are negatively charged molecule which effectively attracts the positive charge mineral such as Ca, Mg, Zn, Cu and Iron and form the complex resulting in reduced bioavailability of mineral present in pseudocereals which results in further threat for the developing countries fighting against mineral deficiency syndromes. Saponin content of pseudocereals are quit high with phytate content of 35 to 38 μmol/g, 4.8 to 2.1 μmol/g and 9.3 to 20.3 μmol/g for buckwheat, amaranth and quinoa seedsrespectively (Bohn *et al.*, 2008; Sanz-Penella *et al.*, 2013; Garcia-Mantrana *et al.*, 2014; Iglesias-Puig *et al.*, 2015). Majority of these phytates are concentrated in embryonic tissues and outer layer of grain. Although pseudocereals are good source of minerals but at contrary the existence of significant concentration of phytate limits the absorption of minerals during digestion. Hence the reduction of phytate during processing place major role while utilizing pseudocereals in diet. Reduction in phytate content can be attained by using various method such as thermal processing, soaking, roasting and enzyme based processing.

Saponins

Saponins are the triterpenoids consisting linear arrangement of one hexose or pentose glycoside unit joined to sapogenin aglycone, which can possibly a steroid or triterpenoid aglycone. These saponins have the tendency to form complex with protein, lipids and mineral such as Fe and Zn resulting in haemolytic effect. They can enhance the membrane permeability resulting in upsurge consumption of food and increased drug assimilation (Vega-Galvez *et al.*, 2010). However, as previously explained saponin do possess health promoting anti-carcinogenic, anti-inflammatory and hypocholesterolemic effects. Existence of saponin content in quinoa varies with species of grain rendering some of the grains bitter in taste. Two major saponin recognized are saponin-A relating to β-D-glucopyranosyl-[β-D-glucopyranosyl-(1-3)-α-L-arabino-pyranosyl-(1-3)]-3-β-23 dihydroxy-12-en-28-oatc-30 methyl ester, and saponin-B relating to β-D-glucopyranosyl-[β-D-glucopyranosyl-(1-3)-α-L-arabino-pyranosyl-(1-3)]-3-β-23 dihydroxyolean-12-en-28-oate (Ruales and Nair, 1993b). Reduction in saponin content of grain can be attained by soaking, scrubbing and washing the quinoa seeds to reduce the bitter taste. Using the above methods saponin B can be removed completely along with 50% reduction in saponin A. Relatively lower concentration of saponins were reported in buckwheat and amaranth and hence products produce using amaranth or buckwheat will not produce significant toxicity or hazard to consumer (Reguero and Haros, 2017)

Protease Inhibitors

Protease inhibitors are defined as the proteins or other food constitutes which has the capacity to bind to the proteolytic enzyme and inhibit their activity by forming very stable complex. Buckwheat reported to have higher level of proteolytic inhibitors. The lower protein digestibility of buckwheat is also accorded to the existence of other antinutritional factor namely tannins and protease inhibitors. But, the researchers reported fairly low concentration of protease in quinoa and amaranth. Therefore, it can be assumed that they may not affect the activity of proteolytic enzymes and protein digestibility (Bodroza-Solarov *et al.*, 2008; Vega-Galvez *et al.*, 2010).

Milling Process

In general, the physical methods are always preferred over chemical methods to produce flour from pseudocereals and this is attributed to the cost effective and ecofriendly nature of this milling. Physical milling involves high shear application because of dry matter milling which tends to produce heat and causes the temperature to rise a bit. The increase in temperature tends to bring about the changes in certain properties of by-products obtained during milling.

Objective of milling comprises the generation of intermediate products which can be used for manufacturing the foodstuffs. The milling processes are generally categorized in dry and wet milling methods. Both processes involve the use of water but the volume of water used in dry milling is significantly lower than wet milling. Dry milling involves separation of anatomical part of grain i.e. endosperm, germ and pericarp, whereas the wet milling involves the separation of purest component of grain such as starch, protein, dietary fiber and oil.

Dry Milling Method

A general process of dry milling can be elaborated as grinding, sifting, separation and regrinding [Fig 3]. These processing steps will be repeated again and again till a particular component of grain is not obtained. Before starting the milling process of pseudocereals, the grains are cleaned to separate any traces of foreign material using different unit operation and equipment.

Quinoa and amaranth can be milled using two generalized methods of dry milling. First process involves conditioning of grain to required moisture content (tempering). Tempering causes toughening of pericarp and germ so that, they are not separated/broken down during milling. The conditioned grains are milled and separation of various fractions is carried out by sieving through screen with varying diameter 0.8 mm to 0.5 mm. The starch particles being larger in size will not pass through the first sieve of 0.8 mm, further separation will take place on seed coat fraction with particle size larger than 0.5mm followed by germ fraction of particle size smaller than 0.5mm hence, it will be collected at last. The dry milled fraction of amaranth will have protein fraction with protein content of 20.4-24.3 percent and lipid content (11.0-13.2%), and flour fraction with starch content of 73.8 percent and protein content of 6.5 percent (Chauhan *et al.*, 1992; Farfan *et al.*, 1978; Mckevith, 2004; Miroshnichenko *et al.*, 2009).

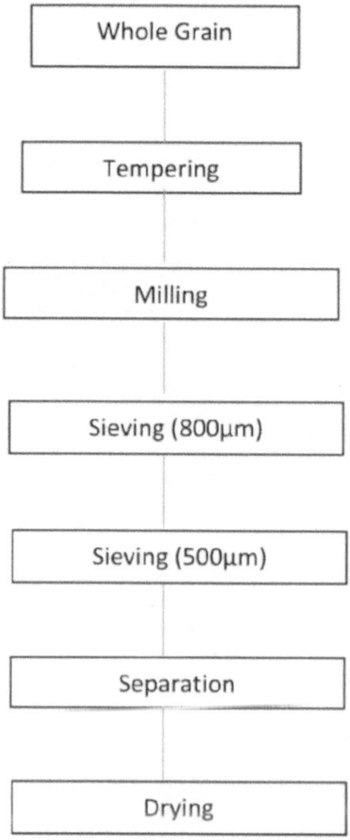

Fig. 3: General diagram for dry milling of pseudo cereals.

The dry milling for buckwheat varies slightly from amaranth and quinoa. After cleaning, the buckwheat grain is precooked by boiling in sufficient volume of water at 90-100°C for 30 min and then carefully dried to the desired moisture content (Tosi *et al.*, 2000). During precooking starch in buckwheat absorbs water and expands which produces break in husk. On further cooling/drying the starch becomes hard and husk becomes fragile.

This results in easy removal of husk from kernel during milling. The milling groats can be achieved using combination of pearler and hammer mill.

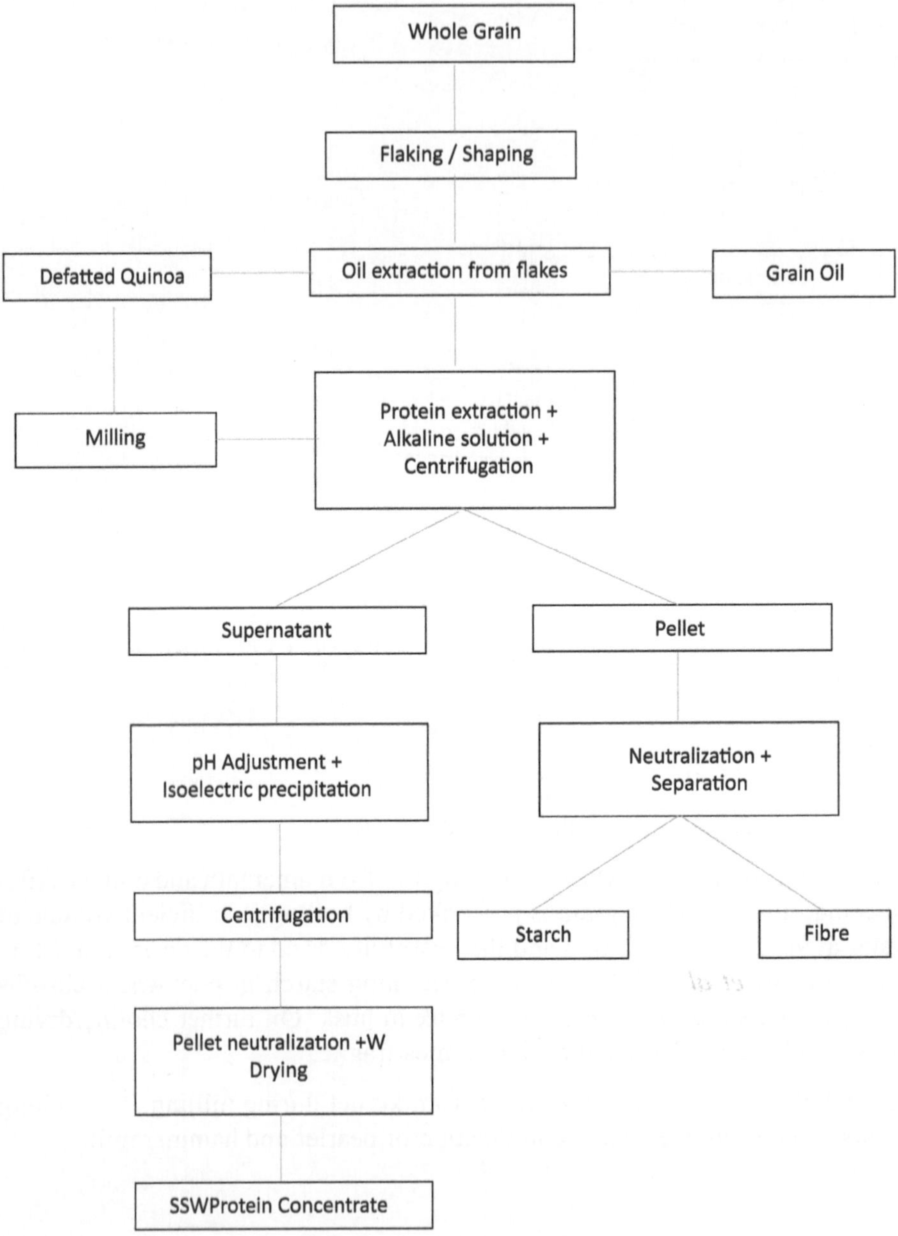

Fig. 4: Schematic diagram of quinoa wet milling process (Scanlin, 2010)

Wet Milling

The wet milling process of pseudocereals implicates the utilization of excess quantity of water to prepare slurry in which constituents of grain will be dissolved and dispersed/suspended (Fig. 4). Initially the grain will be cleaned to remove any foreign material followed by steeping the grains in steeping tank where they are soaked in weak alkaline or aqueous sulphur dioxide solution (SO_2) at desired temperature for predetermined time depending on kind of pseudocereal and method is used. During steeping, the grain will start disintegrating and some part of it will start dissolving in the steeping solution. These dissolved particles will be separated and dried separately. The grain tends to get very soft during steeping and this can be easily disintegrated by slightly rubbing between the fingers.

First milling operation is performed at this stage and the resulting suspension is milled again to promote reduction in particle size of endosperm while leaving the fibrous fraction intact. The suspension is filtered using sieve of decreasing mesh size until it reaches to last mesh which permits the movement of starch and protein (Fig. 4). Protein being smaller and lighter fraction will be separated using centrifugation and further dried to desired moisture content. The sediment starch particles will be washed, dried and stored for further use (Wronkowska and Haros, 2014).

Yield and degree of starch recovery during milling

$$\text{Starch Yield} = \frac{\text{(Weight of the starch obtianed during milling)}}{\text{(Weight of the dry grain)}} \text{X } 100$$

$$\text{Degree of Starch Recovery} = \frac{\text{(Weight of the starch obtianed during milling)}}{\text{(Weight of dry starch present in grain)}} \text{X } 100$$

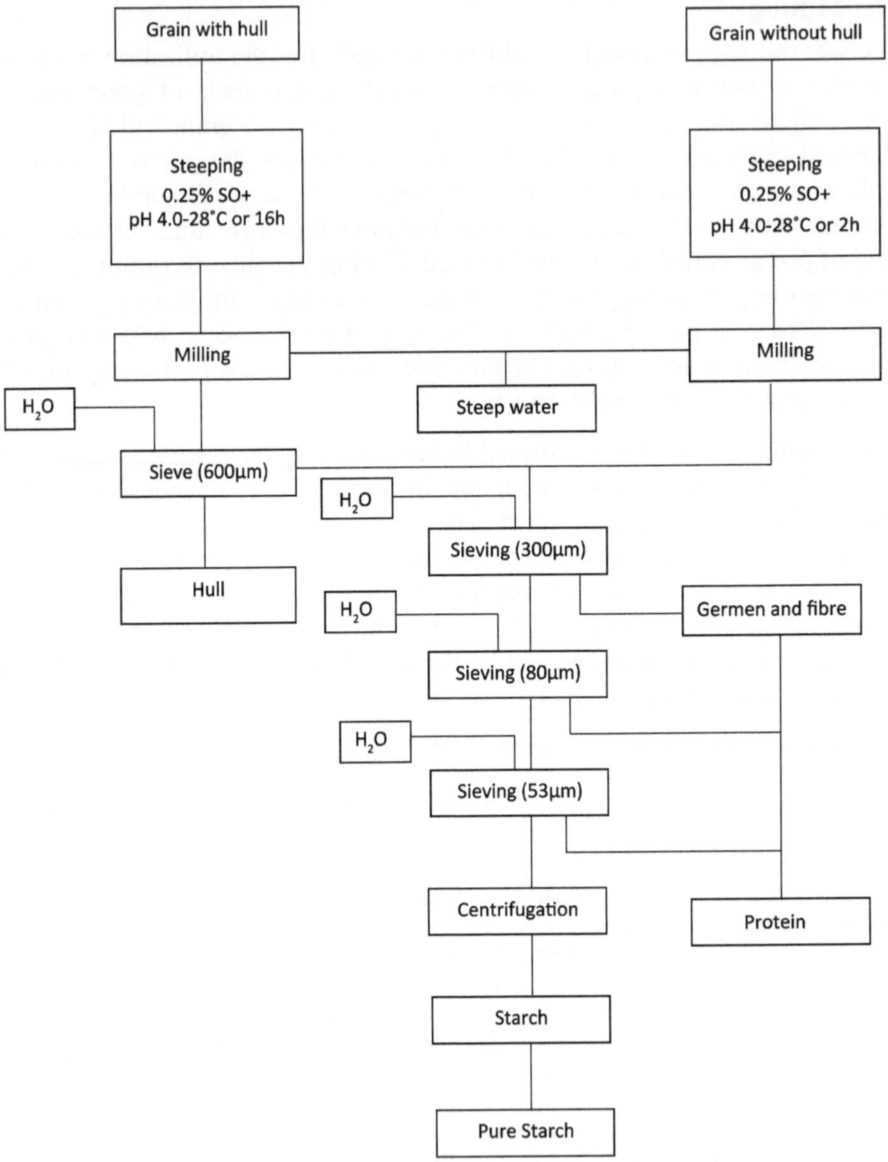

Fig. 4: Wet milling process flow chart of buckwheat (Wronkowaka and Haros, 2014)

Food Uses

Pseudocereals are the grains which possess beneficial nutritional content with high amount of macro and micro nutrients (Codex Standards, 2008). They are found to be superior than cereals in many ways, be it be the nutritional profile of the pseudocereals or their possibility for being used by the consumer having celiac disease. Significantly higher fiber content of these grain makes them quite appropriate for the development of dietary fiber rich product with beneficial health effects. The altered consumption pattern towards pseudocereals has resulted in increased investigation towards finding out the possible ways of utilizing pseudocereal in the manufacturingand expansion of food products.

Among all described pseudocereal, buckwheat can be processed in same manner as that of wheat. But, because of their morphological difference amaranth and quinoa requires some specific processing before being utilized. usedquinoa and buckwheat needs to be hulled to manipulate the saponins and antinutritional content of quinoa and buckwheat hull. Conversely, amaranth can effectively be used as a whole grain or milled to generate flour. After dehulling quinoa can be puffed, ground, sprouted or milled into flour. The quinoa flour cannot be consumed alone due to its unfavorable sensory characteristic but can be mixed with cereal flours for example corn and wheat. This will enhance the nutritional quantity and superiority of the flour.

Pseudocereals can also be used in bakery product like bread, biscuits, cookies, pancakes etc. Buckwheat flour in combination with refined wheat flour can be utilized for the development of white bread with lower GI. Amaranth and quinoa, individually or in combination can replace (percent basis) refined wheat flour for producing bakery products with improved nutritional content (Iglesias-Puig *et al.*, 2015). According to the previous researches quinoa can efficientlybe used at 30 percent replacement levels for the development of bread without effecting much of consumer acceptability. Further, amaranth can also be used at 40 percent replacement levels with mere drop of 10 percent in final loaf volume of bread (Tomoskozi *et al.*, 2011). The addition of amaranth in bakery products tend to produce reddish to brown colorand adversely affect sensory attributes. Hence, it is suggested to use the amaranth at levels upto 40 percent in bread and 30 percent in biscuits only (Sandak and El-Hofi, 2000; Tomoskozi *et al.*, 2011).

Buckwheat can significantly replace barley in beer making as the volatile flavoring component in beer made with fermented buckwheat werecomparable to that of barley. Quinoa and amaranth also finds their use to produce fermented traditional beverage called*chichi*(Taylor and Parker, 2002). In addition, these grains can be used to produce non-dairy milk and dairy imitating beverages with improved nutritional quality and targeted consumer of vegan society as well as those who are having lactose intolerance. (Dezelak *et al.*, 2014; Pineli *et al.*, 2015; Coman *et al.*, 2013).

Extrusion processing involves expansion of starch and denaturation of protein. Addition of these starch rich pseudocereals affected few of physical properties namely color, bulk density and product density of extruded food material. But, they did not affect the degree of expansion and also increased the oxidative stability of product during storage. Incorporation of these cereals also reduced the readily digestible carbohydrate content of extruded snack and resulted in slow release of glucose in contrast with snacks made from pure starch material (Kraus *et al.*, 1988; Chavez-Jauregui *et al.*, 2000; Brennan *et al.*, 2012; Ramos-Diaz *et al.*, 2013).

Starch from these grains can efficiently be used in the preparation of soups, stew, filler, stabilizer etc. Buckwheat being higher in resistant starch can effectively help in developing food with low GI. Buckwheat can partially or fully replace the starch from corn and potato in developing soups with better functional properties (Singer *et al.*, 1990; Whistler and Bemiller, 1997). Celiac disease is defined as the inability of individual in digesting the gluten protein. This results in individual suffering from irritation in gut resulting in abdominal pain and other digestion related problems. All the pseudocereals contain very low level of prolamin (major gluten protein) and therefore they can be effective in manufacturing the gluten free products like bread, biscuit, cookies, cakes, pastries, pancakes, snacks, noodles, pasta and breakfast cereals (Calderon de la Barca *et al.*, 2010; Ballabio *et al.*, 2011; Calasso *et al.*, 2012; Gallagher, 2013)

Conclusion

All three buckwheat, amaranth and quinoa are recognized as potential grains with significant nutritional qualities. They are observed as vital source of protein, minerals, vitamins, PUFA, dietary fiber and bioactive component namely rutine, fagopyrins, phenolics and betalains. Buckwheat starch also found to be rich in resistant starch. Therefore, these grains or their various fractions can be utilized for developing various food stuffs with improved functional and nutritional qualities. Because of the existence of significant resistant starch and dietary fiber, buckwheat starch mayfind their way in the production of nutraceutical products and development of possible probiotic food like yoghurt. These grains can also be used for preparation of gluten free product. These gluten free commodities are necessary for consumers suffering from celiac as they cannot consume wheat based products. Nonetheless for the expansion of future food market, the bioactive compounds of pseudocereals require further analysis and elaborated studies for their possible use as nutraceuticals.

References

Abalone, R., Cassinera, A., Gastón, A., and Lara, M. A. (2004). Some physical properties of amaranth seeds. Biosystems Engineering, 89(1), 109–117. doi:10.1016/j.biosystemseng.2004.06.012

Abreu, M., Hernández, M., Castillo, A., González, I., González, J., and Brito, O. (1994). Study on the complementary effect between the proteins of wheat and amaranth. Food / Nahrung, 38(1), 82–86. doi:10.1002/food.19940380114

Abugoch James, L. E. (2009). Quinoa (*Chenopodium quinoa* Wild.): Composition, chemistry, nutritional and functional properties. Advances in Food and Nutrition Research, 58, 1–31. doi:10.1016/S1043-4526(09)58001-1

Acquistucci, R., and Fornal, J. (1997). Italian buckwheat (*Fagopyrum esculentum*) starch: Physicochemical and functional characterization and *in vitro* digestibility. Nahrung, 41(5), 281–284. doi:10.1002/food.19970410506

Ahamed, N. T., Singhal, R. S., Kulkarni, P. R., and Pal, M. (1996). Physicochemical and functional properties of *Chenopodiumquinoa* starch. Carbohydrate Polymers, 31(1–2), 99–103. doi:10.1016/S0144-8617(96)00034-3

Ahmed, A., Khalid, N., Ahmad, A., Abbasi, N. A., Latif, M. S. Z., and Randhawa, M. A. (2014). Phytochemicals and biofunctional properties of buckwheat: A review. Journal of Agricultural Science, 152(3), 349–369. doi:10.1017/S0021859613000166

Alemayehu, F. R., Bendevis, M. A., and Jacobsen, S. -E. (2015). The potential for utilizing the seed crop amaranth (*Amaranthus* spp.) in East Africa as an alternative crop to support food security and climate change mitigation. Journal of Agronomy and Crop Science, 201(5), 321–329. doi:10.1111/jac.12108

Al-Hakkak, J., and Al-Hakkak, F. (2007). New non-destructive method using gluten to isolate starch from plant materials other than wheat. Starch - Stärke, 59(3–4), 117–124. doi:10.1002/star.200600564

Alvarez-Jubete, L., Arendt, E. K., and Gallagher, E. (2010). Nutritive value of pseudocereals and their increasing use as functional gluten-free ingredients. Trends in Food Science and Technology, 21(2), 106–113. doi:10.1016/j.tifs.2009.10.014

Alvarez-Jubete, L., Holse, M., Hansen, Å, Arendt, E. K., and Gallagher, E. (2009). Impact of baking on vitamin E content of pseudocereals amaranth, quinoa, and buckwheat. Cereal Chemistry Journal, *86*(5), 511–515. doi:10.1094/CCHEM-86-5-0511

Ando, H., Chen, Y.-C., Tang, H., Shimizu, M., Watanabe, K., and Mitsunaga, T. (2002). Food components in fractions of quinoa seed. Food Science and Technology Research, 8(1), 80–84. doi:10.3136/fstr.8.80

Arendt, E., and Zannini, E. (2013). Cereal grains for the food and beverage industries (pp. 369–466). Cambridge: Woodhead Publishing.

Atwell, W. A., Patrick, B. M., Johnson, L. A., and Glass, R. W. (1983). Characterization of quinoa starch. Cereal Chemistry, 60, 9–11.

Babić, L., Babić, M., Turan, J., Matić-Kekić, S., Radojčin, M., Mehandžić-Stanišić, S., . . . Zoranović, M. (2011). Physical and stress–strain properties of wheat (*Triticum aestivum*) kernel. Journal of the Science of Food and Agriculture, 91(7), 1236–1243. doi:10.1002/jsfa.4305

Badami, R. C., and Patil, K. B. (1976). Studies on vegetable oils of genus Amaranthus. Journal of the Oil Technologists Association of India, *8*, 131.

Ballabio, C., Uberti, F., Di Lorenzo, C., Brandolini, A., Penas, E., and Restani, P. (2011). Biochemical and immunochemical characterization of different varieties of amaranth (*Amaranthus ssp*. L.) as a safe ingredient for gluten-free products. Journal of Agricultural and Food Chemistry, 59(24), 12969–12974. doi:10.1021/jf2041824

Baltensperger, D. D. (2003). Cereal grains and pseudo-cereals. In S. H. Katz and W. W. Weaver (Eds.). Retrieved from http://www.encyclopedia.com/doc/1G2-3403400119.html, Encyclopedia of Food and Culture. New York, NY: Scribner.

Barba de la Rosa, A. P., Fomsgaard, I. S., Laursen, B., Mortensen, A. G., Olvera-Martínez, L., Silva Sánchez, C., and De León-Rodríguez, A. (2009). Amaranth (*Amaranthus hypochondriacus*) as an alternative crop for sustainable food production: Phenolic acids and flavonoids with potential impact on its nutraceuticals quality. Journal of Cereal Science, 49(1), 117–121. doi:10.1016/j.jcs.2008.07.012

Becker, R. (1989). Preparation, composition and nutritional implications of amaranth seed oil. *Cereal Foods World*, 34, 950–953.

Becker, R. (1994). Amaranth oil: Composition, processing, and nutritional qualities. In O. Paredes-Lopez (Ed.), *Amaranth—Biology, Chemistry, and Technology* (pp. 133–142). London: CRC Press Press.

Becker, R., Wheeler, E. L., Lorenz, K., Stafford, A. E., Grosjean, O. K., Betschart, A. A., and Saunders, R. M. (1981). A compositional study of amaranth grain. *Journal of Food Science*, 46(4), 1175–1180. doi:10.1111/j.1365-2621.1981.tb03018.x

Belton, P. S., and Taylor, J. R. N. (2002). Pseudocereals and less common cereals. Berlin: Springer.

Berghofer, E., andSchonelechner, R. (2002). Grain amaranth. In P. Belton and J. Taylor (Eds.), Pseudocereals and less common cereals. Grain properties and utilization potential (pp. 219–260). Berlin: Springer-Verlag.

Betschart, A. A., Irving, D. W., Shepherd, A. D., and Saunders, R. M. (1981). *Amaranthus cruentus*: Milling characteristics, distribution of nutrients within seed components, and the effects of temperature on nutritional quality. Journal of Food Science, 46(4), 1181–1187. doi:10.1111/j.1365-2621.1981.tb03019.x

Biacs, P., Aubrecht, E., Leder, I. *et al.* (2002). Buckwheat. In Cereals (Ed.) P. Belton and J. Taylor), *Pseudocereals and less common* (pp. 123–152). Berlin: Springer-Verlag.

Bodroža-solarov, M., Filipčev, B., Kevrešan, Ž, Mandić, A., and Šimurina, O. (2008). Quality of bread supplemented with popped Amaranthus Cruentus Grain. Journal of Food Process Engineering, *31*(5), 602–618. doi:10.1111/j.1745-4530.2007.00177.x

Bohn, L., Meyer, A. S., and Rasmussen, S. K. (2008). Phytate: Impact on environment and human nutrition. A challenge for molecular breeding. Journal of Zhejiang University. Science. *B*, *9*(3), 165–191. doi:10.1631/jzus.B0710640

Bonafaccia, G., and Kreft, I. (1994). Technological and qualitative characteristics of food products made with buckwheat. *Fagopyrum*, *14*, 35–42.

Bonafaccia, G., Marocchini, M., and Kreft, I. (2003). Composition and technological properties of the flour and bran from common and Tartary buckwheat. Food Chemistry, *80*(1), 9–15. doi:10.1016/S0308-8146(02)00228-5

Brady, K., Ho, C. T., Rosen, R. T., Sang *et al.* (2007). Effect of processing on nutraceuticals potential of quinoa. Food Chemistry, 52, 6503–6509.

Brennan, M. A., Menard, C., Roudaut, G., and Brennan, C. S. (2012). Amaranth, millet and buckwheat flours affect the physical properties of extruded breakfast cereals and modulates their potential glycaemic impact. Starch - Stärke, 64(5), 392–398. doi:10.1002/star.201100150

Bressani, R. (1994). Composition and nutritional properties of amaranth. In Paredes-Lopez (Ed.), Amaranth—Biology, Chemistry, and Technology (pp. 185–206). London: CRC Press Press.

Bressani, R., Elias, L. G., and Garcia-Soto, A. (1989). Limiting amino acids in raw and processed amaranth grain protein from biological tests. Plant Foods for Human Nutrition, *39*(3), 223–234. doi:10.1007/BF01091933

Bressani, R., Velasquez, L., and Acevedo, E. (1990). Dietary fibre content in various grain amaranth species and effect of processing. Amaranth Newsletter, 1, 5–8.

Brinegar, C. (1997). The seed storage proteins of quinoa. Advances in Experimental Medicine and Biology, 415, 109–115. doi:10.1007/978-1-4899-1792-8_8

Bruni, R., Medici, A., Guerrini, A., Scalia, S., Poli, F., Muzzoli, M., and Sacchetti, G. (2001). Wild Amaranthus caudatus seed oil, a nutraceutical resource from Ecuadorian flora. Journal of Agricultural and Food Chemistry, *49*(11), 5455–5460. doi:10.1021/jf010385k

Cai, Y., Sun, M., and Corke, H. (2003). Antioxidant activity of betalains from plants of the Amaranthaceae. Journal of Agricultural and Food Chemistry, 51(8), 2288–2294. doi:10.1021/jf030045u

Calasso, M., Vincentini, O., Valitutti, F., Felli, C., Gobbetti, M., and Di Cagno, R. (2012). The sourdough fermentation may enhance the recovery from intestinal inflammation of coeliac patients at the early stage of the gluten-free diet. European Journal of Nutrition, 51(4), 507–512. doi:10.1007/s00394-012-0303-y

Calderon de la Barca, A. M., Rojas-Martínez, M. E., Islas-Rubio, A. R., and Cabrera-Chávez, F. (2010). Gluten-free breads and cookies of raw and popped amaranth flours with attractive technological and nutritional qualities. Plant Foods for Human Nutrition, 65(3), 241–246. doi:10.1007/s11130-010-0187-z

Caselato-Sousa, V. M., and Amaya-Farfán, J. (2012). State of knowledge on amaranth grain: A comprehensive review. Journal of Food Science, 77(4), R93–104. doi:10.1111/j.1750-3841.2012.02645.x

Chaturvedi, A., Sarojini, G., Nirmala, G. I., Nirmalamma, N., and Satyanarayana, D. (1997). Glycemic index of grain amaranth, wheat and rice in NIDDM subjects. Plant Foods for Human Nutrition, 50(2), 171–178. doi:10.1007/BF02436036

Chauhan, G. S., Eskin, N. A. M., and Tkachuk, R. (1992). Nutrients and antinutrients in quinoa seed. *Cereal Chemistry*, 69, 85–88.

Chavez-Jauregui, R. N., Silva, M. E. M. P., and Arěas, J. A. G. (2000). Extrusion cooking process for amaranth (*Amaranthus caudatus* L.). Journal of Food Science, 65(6), 1009–1015. doi:10.1111/j.1365-2621.2000.tb09408.x

Christa, K., and Soral-Śmietana, M. (2008). Buckwheat grains and buckwheat products—Nutritional and prophylactic value of their components—A review. Czech Journal of Food Sciences, 26(3), 153–162. doi:10.17221/1602-CJFS

Codex standard for foods for special dietary use for persons intolerant to gluten. (2008). Retrieved from http://www.fao.org/fao-who-codexalimentarius/standards/listofstandards/en

Coman, M. M., Verdenelli, M. C., Cecchini, C., Silvi, S., Vasile, A., Bahrim, G. E., . . . Cresci, A. (2013). Effect of buckwheat flour and oat bran on growth and cell viability of the probiotic strains *Lactobacillus rhamnosus* IMC 501 (R), *Lactobacillus paracasei* IMC 502 (R) and their combination SYNBIO (R), in synbiotic fermented milk. International Journal of Food Microbiology, 167(2), 261–268. doi:10.1016/j.ijfoodmicro.2013.09.015

Coulter, L., and Lorenz, K. (1990). Quinoa-composition, nutritional value, food applications. *LWT- Food Technology*, 123, 203–207.

D'Amico, S., Jungkunz, S., Balasz, G. *et al.* (2015). Abrasive milling of quinoa and characterization of milling fractions. Journal of Cereal Science.

De la Cruz Torres, E., Mapes Sanchez, C., Laguna Cerda, A. *et al.* (2008). Ancient underutilised pseudocereals—Potential alternatives for nutrition and income generation. In J. Smartt and N. Haq (Eds.). RPM Print and Design, New crops and uses: Their role in a rapidly changing world (pp. 186–203). Chichester, UK.

Deželak, M., Zarnkow, M., Becker, T., and Košir, I. J. (2014). Processing of bottom fermented gluten-free beer-like beverages based on buckwheat and quinoa malt with chemical and sensory characterization. Journal of the Institute of Brewing, 120, n/a–n/a. doi:10.1002/jib.166

Dodok, L., Modhir, A. A., Buchtová, V., Halásová, G., and Poláček, I. (1997). Importance and utilization of amaranth in food industry. Part 2. Composition of amino acids and fatty acids. *Food / Nahrung*, 41(2), 108–110. doi:10.1002/food.19970410211

Drzewiecki, J., Delgado-Licon, E., Haruenkit, R., Pawelzik, E., Martin-Belloso, O., Park, Y. S., Gorinstein, S. (2003). Identification and differences of total proteins and their soluble fractions in some pseudocereals based on electrophoretic patterns. Journal of Agricultural and Food Chemistry, 51(26), 7798–7804. doi:10.1021/jf030322x

Escudero, N. L., De Arellano, M. L., Luco, J. M., Giménez, M. S., and Mucciarelli, S. I. (2004). Comparison of the chemical composition and nutritional value of Amaranthus cruentus Flour and Its Protein Concentrate. Plant Foods for Human Nutrition, 59(1), 15–21. doi:10.1007/s11130-004-0033-3

Farfan, J. A., Ciacco, C. F., Fernandes, J. L. A., and Ruiz, W. A. (1978). Removal of saponins from quinua (*Chenopodium quinoa* Wild) grain by milling. International Union of Food Science and Technology, Fifth International Congress of Food Science and Technology, pp. 101.

FDA. (2004). FDA allows qualified health claim to decrease risk of coronary heart disease. Retrieved from http://www.fda.gov/NewsEvents/Newsroom/PressAnnouncements/2004/up.cm108368 htm (accessed 8 July 2019).

Gallagher, E. (2013). Overview on the present state of gluten-free processing. Third International Symposium on Gluten-free Cereal Products and Beverages. June 12–14, Vienna p. 30.

Gamel, T. H., Linssen, J. P., Alink, G. M., Mosallem, A. S., and Shekib, L. A. (2004). Nutritional study of raw and popped seed proteins of *Amaranthus caudatus* L. and *Amaranthus cruentus* L. Journal of the Science of Food and Agriculture, 84(10), 1153–1158. doi:10.1002/jsfa.1781

Gamel, T. H., Mesallam, A. S., Damir, A. A., Shekib, L. A., and Linssen, J. P. (2007). Characterization of amaranth seed oils. Journal of Food Lipids, 14(3), 323–334. doi:10.1111/j.1745-4522.2007.00089.x

García-Mantrana, I., Monedero, V., and Haros, M. (2014). Application of phytases isolated from bifidobacteria in the development of cereal-based products with amaranth. European Food Research and Technology, 238(5), 853–862. doi:10.1007/s00217-014-2167-2

García-Mantrana, I., Monedero, V., and Haros, M. (2015). myo-inositol hexakisphosphate degradation by Bifidobacterium pseudocatenulatum ATCC 27919 improves mineral availability of high fibre rye-wheat sour bread. Food Chemistry, 178, 267–275. doi:10.1016/j.foodchem.2015.01.099

Glorio, P., Repo-Carrasco, R., Velezmoro, C. (2008). Dietary fiber in fruits, roots, tubers, cereals and pulses from Peruvian cultivars. Revista de la Sociedad Química del Perú, 74, 46–56.

Gallardo, C., Jiménez, L., and García-Conesa, M. T. (2006). Hydroxy-cinnamic acid composition and in-vitro antioxidant activity of selected grain fraction. Food Chemistry, 99(3), 455–463.

Gonzalez, J. A., Konishi, Y., Bruno, M., Valoy, M., and Prado, F. E. (2012). Interrelationships among seed yield, total protein and amino acid composition of ten quinoa (*Chenopodium quinoa*) cultivars from two different agroecological regions. Journal of the Science of Food and Agriculture, 92(6), 1222–1229. doi:10.1002/jsfa.4686

Gonzalez-Roberto, N., Fernåndez-Espinar, M. T., Gil, J. V., and Haros, M. (2015). Development and optimisation quinoa wet milling process. First International Congress of Students of Food Science and Technology, Valencia, Spain, March 5–6.

Gordillo-Bartidas, E., Rizzolo-Diaz, D. A., Roura, E., Massanes, T., and Gomis, R. (2016). Quinoa (*Chenopodium quinoa* wild), from nutritional value to potential health benefits: An integrative review. Journal of Nutrition and Food Sciences, 6. doi:10.4172/2155-9600.1000497, PubMed: 1000497

Gorinstein, S., Drzewiecki, J., DelgadoLicon, E., Pawelzik, E., Ayala, A. L. M., Medina, O. J., Trakhtenberg, S. (2005). Relationship between dicotyledonaeamaranth, quinoa, Fagopyrum, soybean and monocots sorghum and rice based on protein analyses and their use as substitution of each other. European Food Research and Technology, 221(1–2), 69–77. doi:10.1007/s00217-005-1208-2

Gorinstein, S., Jaramillo, N. O., Medina, O. J., Rogriques, W. A., Tosello, G. A., and Paredes-Lopez, O. (1999). Evaluation of some cereals, plants and tubers through protein composition. Journal of Protein Chemistry, 18(6), 687–693. doi:10.1023/a:1020610424625

Guo, X., and Yao, H. (2006). Fractionation and characterization of Tartary buckwheat flour proteins. Food Chemistry, 98(1), 90–94. doi:10.1016/j.foodchem.2005.05.055

Guo, X., Wu, C., Ma, Y. et al. (2007). Comparison of milling fraction of Tartary buckwheat for the phenolics and antioxidant properties. Food Research International, 49, 231–258.

Guo, X., Wu, C., Ma, Y., Parry, J., Xu, Y., Liu, H., and Wang, M. (2012). Comparison of milling fractions of Tartary buckwheat for their phenolics and antioxidant properties. Food Research International, 49(1), 53–59. doi:10.1016/j.foodres.2012.07.019

Guzman-Maldonado, S. H., and Paredes-Lopez, O. (1998). Functional products of plants indigenous to Latin America: Amaranth, quinoa, common beans, and botanicals. In G. Mazza (Ed.), Functional foods: Biochemical and processing aspects (pp. 293–328). Lancaster, PA: Technomic Publishing.

Hager, A. S., Wolter, A., Jacob, F., Zannini, E., and Arendt, E. K. (2012). Nutritional properties and ultra-structure of commercial gluten free flours from different botanical sources compared to wheat flours. Journal of Cereal Science, 56(2), 239–247. doi:10.1016/j.jcs.2012.06.005

Hirose, Y., Fujita, T., Ishii, T., and Ueno, N. (2010). Antioxidative properties and flavonoid composition of Chenopodium quinoa seeds cultivated in Japan. Food Chemistry, 119(4), 1300–1306. doi:10.1016/j.foodchem.2009.09.008

Horbowicz, M., and Obendorf, R. L. (1992). Changes in sterols and fatty acids of buckwheat endosperm and embryo during seed development. Journal of Agricultural and Food Chemistry, 40(5), 745–750. doi:10.1021/jf00017a007

Hu, F. B. (2003). The Mediterranean diet and mortality-olive oil and beyond. New England Journal of Medicine, 348(26), 2595–2596. doi:10.1056/NEJMp030069

Iglesias-Puig, E., Monedero, V., and Haros, M. (2015). Bread with whole quinoa flour and bifidobacterial phytases improve contribution to dietary mineral intake and their bioavailability without substantial loss of bread quality. LWT – Food Science and Technology, 60(1), 71–77. doi:10.1016/j.lwt.2014.09.045

Ikeda, K. (2002). Buckwheat: Composition, chemistry, and processing. Advances in Food and Nutrition Research, 44, 395–434. doi:10.1016/s1043-4526(02)44008-9

Ikeda, K., Shida, K., and Kishida, M. (1994). Alpha-amylase inhibitor in buckwheat seed. Fagopyrum, 14, 3–6.

Izydorczyk, M. S., McMillan, T., Bazin, S., Kletke, J., Dushnicky, L., and Dexter, J. (2014). Canadian buckwheat: A unique, useful and under-utilized crop. Canadian Journal of Plant Science, 94(3), 509–524. doi:10.4141/cjps2013-075

Jahaniaval, F., Kakuda, Y., and Marcone, M. F. (2000). Fatty acid and triacylglycerol compositions of seed oils of five Amaranthus accessions and their comparison to other oils. Journal of the American Oil Chemists' Society, 77(8), 847–852. doi:10.1007/s11746-000-0135 0

Kakabouki, I., Bilalis, D., Karkanis, A., Zervas, G., Tsiplakou, E., and Hela, D. (2014). Effects of fertilization and tillage system on growth and crude protein content of quinoa (Chenopodium quinoa Wild.): An alternative forage crop. Emirates Journal of Food and Agriculture, 26(1), 18–24. doi:10.9755/ejfa.v26i1.16831

Karpati, I., and Banyai, L. (1980). The buckwheat and the Tartary buckwheat—Fagopyrum esculentum, F. tataricum (L.). Cultur flora of Hungary. Akademia, Budapest (in Hungarian).

Kent, N. L., and Evers, A. D. (1994). Kent's technology of cereals (4th ed) (pp. 74–75, 178, 284–295). Kidlington, UK: Pergamon Press Science.

Kim, S. L. K., Kim, S. L. K., Lee, Y. H., and Park, C. H. (2001). Varietal difference of fatty acid composition and vitamin E content in buckwheat grains. Proceedings of the Eighth Symposium on Buckwheat, July 8–13 (pp. 523–531). Zurich: International Society of Biomechanics.

Kim, S. L., Kim, S. K., and Park, C. H. (2004). Introduction and nutritional evaluation of buckwheat sprouts as a new vegetable. Food Research International, 37(4), 319–327. doi:10.1016/j.foodres.2003.12.008

Kong, X., Kasapis, S., Bao, J., and Corke, H. (2012). Influence of acid hydrolysis on thermal and rheological properties of amaranth starches varying in amylose content. Journal of the Science of Food and Agriculture, 92(8), 1800–1807. doi:10.1002/jsfa.5549

Koziol, M. J. (1991). Afrosimetric estimation of threshold saponin concentration for bitterness in quinoa (Chenopodium quinoa Wild). Journal of the Science of Food and Agriculture, 54(2), 211–219. doi:10.1002/jsfa.2740540206

Koziol, M. J. (1992). Chemical composition and nutritional evaluation of quinoa (Chenopodium quinoa Wild.). Journal of Food Composition and Analysis, 5(1), 35–68. doi:10.1016/0889-1575(92)90006-6

Koziol, M. J. (1993). Quinoa: A potential new oil crop. In J. Janick, J. Simon and E. (Eds.), New crops (pp. 328–336). New York: Wiley.

Kraus, S. V., Linnichenko, V. T., Kriventseva, L. D., and Tupitsa, Y. Y. (1988). Extruded buckwheat and rice flours for baby foods. Izv Vyssh Uchebn Zavedenii Pishch Tekhnoll, 62–64.

Kreft, I., Fabjan, N., and Yasumoto, K. (2006). Rutin content in buckwheat (Fagopyrum esculentum Moench) food materials and products. Food Chemistry, 98(3), 508–512. doi:10.1016/j.foodchem.2005.05.081

Krkošková, B., and Mrázová, Z. (2005). Prophylactic components of buckwheat. Food Research International, 38(5), 561–568. doi:10.1016/j.foodres.2004.11.009

Kuhn, M., Wagner, S., Aufhammer, W., Lee, J. H., Kuebler, E., and Schreiber, H. (1996). Effects of cultivation on the mineral content of amaranth, buckwheat, quinoa and oats. Dtsch. Lebensm. Rundsh, 92, 147–152.

Lakshminarayana, G., Pantulu, A. J., and Rao, K. S. (1984). Lipid class and fatty acid compositions of young Amaranthus gangeticus L. leaves. Journal of Agricultural and Food Chemistry, 32(6), 1361–1363. doi:10.1021/jf00126a036

Lehman, J. (1988). Carbohydrates of amaranth. Legacy, 1 (pp. 4–8). Amer: Amraranth Institute Bricelyn, MN.

Li, W., Lin, R., and Corke, H. (1997). Physicochemical properties of common and Tartary buckwheat starch. Cereal Chemistry Journal, 74(1), 79–82. doi:10.1094/CCHEM.1997.74.1.79

Linsberger-Martin, G., Lukasch, B., and Berghofer, E. (2012). Effects of high hydrostatic pressure on the RS content of amaranth, quinoa and wheat starch. Starch - Stärke, 64(2), 157–165. doi:10.1002/star.201100065

Lintschinger, J., Fuchs, N., Moser, H., Jäger, R., Hlebeina, T., Markolin, G., and Gössler, W. (1997). Uptake of various trace elements during germination of wheat, buckwheat and quinoa. Plant Foods for Human Nutrition, 50(3), 223–237. doi:10.1007/BF02436059

Longvah, T., Ananthan, R., Bhaskaracharya, K., and Venkaiah, K. (2017). Indian food composition tables (pp. 3–6). Hyderabad: National Institute of Nutrition, Indian Council of Medical Research.

Mabry, T. J., Taylor, A., and Turner, B. L. (1963). The betacyanins and their distribution. Phytochemistry, 2(1), 61–64. doi:10.1016/S0031-9422(00)88016-4

Maksimovic, V. R., Varkonji-Gasic, S. R., and Savic, A. P. (1966). The biosynthesis of 13S buckwheat seed storage protein. Journal of Plant Physiology, 147, 759–761

Makus, D. J. (1990). Composition and nutritive value of vegetable amaranth as affected by the stage of growth, environment and method of preparation. Proceedings of the Fourth Amaranth Symposium Minnesota Ext. Serv. St. Paul, MN: Minnesota Agriculture, University of Minnesota.

Marcone, M. (2005) E. Abdel-Aal, P. Wood and P. (Eds.). Amaranth, in specialty grains for food and feed (pp. 351–374). St. Paul, MN: American Association of Cereal Chemists.

Marcone, M. F. (1999). Evidence confirming the existence of a 7S globulin-like storage protein in A. hypochondriacus seed. Food Chemistry, 65(4), 533–542. doi:10.1016/S0308-8146(98)00221-0

Marcone, M. F., Beniac, D. R., Harauz, G., and Yada, R. Y. (1994). Quaternary structure and model for the oligomeric seed globulin from Amaranthus hypochondriacus K343. Journal of Agricultural and Food Chemistry, 42(12), 2675–2678. doi:10.1021/jf00048a005

Marcone, M. F., and Kakuda, Y. (1999b). A comparative study of the functional properties of amaranth and soybean globulin isolates. Nahrung/Food, 43(6), 368–373. doi:10.1002/(SICI)1521-3803(19991201)43:6<368::AID-FOOD368>3.0.CO;2-R

Martirosyan, D. M., Miroshnichenko, L. A., Kulakova, S. N., Pogojeva, A. V., and Zoloedov, V. I. (2007). Amaranth oil application for coronary heart disease and hypertension. Lipids in Health and Disease, 6, 1. doi:10.1186/1476-511X-6-1

Mattila, P., Pihlava, J. M., and Hellström, J. (2005). Contents of phenolic acids, alkyl- and alkenylresorcinols, and avenanthramides in commercial grain products. Journal of Agricultural and Food Chemistry, 53(21), 8290–8295. doi:10.1021/jf051437z

Mazza, G. (1993). Storage, processing, and quality aspects of buckwheat seeds. In New Crops (Ed.) Janick, J. and simon, J.E.) (pp. 251–254). New York: Wiley.

Mazza, G., and Oomah, B. D. (2003) B. Caballero (Ed.). Buckwheat in encyclopedia of Food Science and Nutrition (pp. 692–699). Oxford: Academic Press.

McKevith, B. (2004). Nutritional aspects of cereals. British Nutrition Foundation Nutrition Bulletin. Nutrition Bulletin, 29(2), 111–142. doi:10.1111/j.1467-3010.2004.00418.x

Mickowska, B., Socha, P., Urminská, D., and Cieślik, E. (2013). Immunodetection, electrophoresis and amino acid composition of alcohol soluble proteins extracted from grains of selected varieties of pseudocereals, legumes, oat, maize and rice. Cereal Research Communications, 41(1), 160–169. doi:10.1556/CRC.2012.0029

Miranda, M., Delatorre-Herrera, J., Vega-Galvez, A., Jorquera, E. (2014). Antimicrobial potential and phytochemical content of six diverse sources of quinoa seeds (Chenopodium quinoa Wild). Agricultural Sciences, 5, 1015–1024

Miranda, M., Vega-Gálvez, A., Martínez, E. A., López, J., Marín, R., Aranda, M., and Fuentes, F. (2013). Influence of contrasting environments on seed composition of two quinoa genotypes: Nutritional and functional properties. Chilean Journal of Agricultural Research, 73(2), 06–07. doi:10.4067/S0718-58392013000200004

Miroshnichenko, L. A., Sobolev, S. N., and Vashchenko, Y. U. E. (2009). Russkaya Oliva Co Ltd (Russ-Soviet Institute). Amaranth seeds processing method involves purification, milling of seeds, separation of obtained mixture of particles according to fractions and extraction of oil from germs fraction. RU2363724-C2. Derwent Primary Accession Number. (2009–), P03950.

Montoya-Rodríguez, A., Gómez-Favela, M. A., Reyes-Moreno, C., Milán-Carrillo, J., and González de Mejía, E. (2015). Identification of bioactive peptide sequences from amaranth (*Amaranthus hypochondriacus*) seed proteins and their potential role in the prevention of chronic diseases. Comprehensive Reviews in Food Science and Food Safety, 14(2), 139–158. doi:10.1111/1541-4337.12125

Morita, N., Maeda, T., Sai, R., Miyake, K., Yoshioka, H., Urisu, A., and Adachi, T. (2006). Studies on distribution of protein and allergen in graded flours prepared from whole buckwheat grains. Food Research International, 39(7), 782–790. doi:10.1016/j.foodres.2006.02.005

Mosha, A. C., and Svanberg, U. (1990). The acceptance and intake of bulk-reduced weaning foods: The Luganga village study. Food and Nutrition Bulletin, 12(1), 1–6. doi:10.1177/156482659001200102

National Center for Health Statistics. (2014). Health, United States, 2013 with special feature on prescription drugs. Retrieved from http://www.cdc.gov/nchs/data/hus/hus13. Pdf.

Negi, N., Padmashree, A., Wadikar, D. D., Semwal, A. D., and Sharma, G. K. (2018). Nutrition aspects of pseudocereals-a review. Indian Food Industry, 37, 15–31.

Obara, T., and Miyata, N. (1969). Studies on the conjugated lipids of Fagopyrum sagittatum (buckwheat) flour sanako. Journal of the Agricultural Chemical Society of Japan, 43(2), 95–98. doi:10.1271/nogeikagaku1924.43.95

Palombini, S. V., Claus, T., Maruyama, S. A., Gohara, A. K., Souza, A. H. P., Souza, N. Ed, Matsushita, M. (2013). Evaluation of nutritional compounds in new amaranth and quinoa cultivars. Food Science and Technology, 33(2), 339–344. doi:10.1590/S0101-20612013005000051

Paredes-López, O., Bello-Pérez, L. A., and López, M. G. (1994). Amylopectin: Structural, gelatinization and retrogradation studies. Food Chemistry, 50(4), 411–417. doi:10.1016/0308-8146(94)90215-1

Paśko, P., Bartoń, H., Zagrodzki, P., Chłopicka, J., Iżewska, A., Gawlik, M., Gorinstein, S. (2011). Effect of amaranth seeds in diet of oxidative status in plasma and selected tissue of high fructose-fed rats. Food Chemistry, 126(1), 85–90. doi:10.1016/j.foodchem.2010.10.081

Pedersen, B., Kalinowski, L. S., and Eggum, B. O. (1987). The nutritive value of amaranth grain (Amaranthus caudatus). I. Protein and minerals of raw and processed grain. Qualitas Plantarum Plant Foods for Human Nutrition, 36(4), 309–324. doi:10.1007/BF01892352

Perez, E., Bahnassey, Y. A., and Breene, W. M. (1993). Some chemical, physical, and functional properties of native and modified starches of Amaranthus hypochondriacus and Amaranthus cruentus. Starch - Stärke, 45(6), 215–220. doi:10.1002/star.19930450606

Pineli, LdLdO., Botelho, R. B. A., Zandonadi, R. P., Solorzano, J. L., de Oliveira, G. T., Reis, C. E. G., and Teixeira, DdS. (2015). Low glycemic index and increased protein content in a novel quinoa milk. LWT – Food Science and Technology, 63(2), 1261–1267. doi:10.1016/j.lwt.2015.03.094

Pomeranz, Y., and Robbins, G. S. (1972). Amino acids composition of buckwheat protein. Journal of Agricultural and Food Chemistry, 20(2), 270–274. doi:10.1021/jf60180a029

Praznik, W., Mundigler, N., Kogler, A., Pelzl, B., Huber, A., and Wollendorfer, M. (1999). Molecular background of technological properties of selected starches. Starch - Stärke, 51(6), 197–211. doi:10.1002/(SICI)1521-379X(199906)51:6<197::AID-STAR197>3.0.CO;2-K

Prego, I., Maldonado, S., and Otegui, M. (1998). Seed structure and localization of reserves in Chenopodium quinoa. Annals of Botany, 82(4), 481–488. doi:10.1006/anbo.1998.0704

Przybylski, R., Chauhan, G. S., and Eskin, N. A. M. (1994). Characterization of quinoa (Chenopodium quinoa) lipids. Food Chemistry, 51(2), 187–192. doi:10.1016/0308-8146(94)90255-0

Qian, J., and Kuhn, M. (1999b). Physical properties of buckwheat starches from various origins. Starch - Stärke, 51(2–3), 81–85. doi:10.1002/(SICI)1521-379X(199903)51:2<81::AID-STAR81>3.0.CO;2-I

Qian, J. Y., and Kuhn, M. (1999). Characterization of Amaranthus cruentus and Chenopodium quinoa starch. Starch - Stärke, 51(4), 116–120. doi:10.1002/(SICI)1521-379X(199904)51:4<116::AID-STAR116>3.0.CO;2-R

Radosavljevic, M., Jane, J., and Johnson, L. A. (1998). Isolation of amaranth starch by diluted alkaline-protease treatment. Cereal Chemistry Journal, 75(2), 212–216. doi:10.1094/CCHEM.1998.75.2.212

Radovic, R. S., Maksimovic, R. V., Brkljacic, M. J., Varkonji Gasic, I. E., and Savic, P. A. (1999). 2S albumin from buckwheat (Fagopyrum esculentum Moench) seeds. Journal of Agricultural and Food Chemistry, 47(4), 1467–1470. doi:10.1021/jf980778s

Radović, S. R., Maksimović, V. R., and Varkonji-Gašić, E. I. (1996). Characterization of buckwheat seed storage proteins. Journal of Agricultural and Food Chemistry, 44(4), 972–974. doi:10.1021/jf950655x

Ramos Diaz, J. M., Kirjoranta, S., Tenitz, S., Penttilä, P. A., Serimaa, R., Lampi, A., and Jouppila, K. (2013). Use of amaranth, quinoa and kaniwa in extruded corn-based snacks. Journal of Cereal Science, 58(1), 59–67. doi:10.1016/j.jcs.2013.04.003

Ranhotra, G. S., Gelroth, J. A., Glaser, B. K. et al. (1993). Composition and protein nutritional quality of quinoa. Cereal Chemistry, 70, 303–305.

Rayas-Duarte, P., Majewska, K., and Doetkott, C. (1998). Effect of extrusion process parameters on the quality of buckwheat flour mixes. Cereal Chemistry Journal, 75(3), 338–345. doi:10.1094/CCHEM.1998.75.3.338

Reguero, M., and Haros, C. M. (2017). Structure and composition of kernels. In Haros and Schoencechner (Eds.), Pseudocereals chemistry and technology p. 36. New York, NY: Wiley.

Repo-Carrasco, R., Espinoza, C., and Jacobsen, S. -E. (2003). Nutritional value and use of the Andean crops quinoa (Chenopodium quinoa) and kañiwa (Chenopodium pallidicaule). Food Reviews International, 19(1–2), 179–189. doi:10.1081/FRI-120018884

Repo-Carrasco-Valencia, R., and Arana, J. V. (2017). Carbohydrates of kernels. In Haros and Schoencechner (Eds.), Pseudocereals chemistry and technology p. 53. New York, NY: Wiley.

Repo-Carrasco-Valencia, R. A., and Serna, L. A. (2011). Quinoa (*Chenopodium quinoa* Wild.) as a source of dietary fiber and other functional components. Ciencia e Tecnologia de Alimentos, 31(1), 225–230. doi:10.1590/S0101-20612011000100035

Repo-Carrasco-Valencia, R., Hellström, J. K., Pihlava, J. M., and Mattila, P. H. (2010). Flavonoids and other phenolic compounds in Andean indigenous grains: Quinoa (*Chenopodium quinoa*), kaniwa (*Chenopodium pallidicaule*) and kiwicha (*Amaranthus caudatus*). Food Chemistry, 120(1), 128–133. doi:10.1016/j.foodchem.2009.09.087

Rosero, O., Marounek, M., Brenova, N., and Lukesova, D. (2013). Phytase activity and comparison of chemical composition, phytic acid P content of four varieties of quinoa grain (*Chenopodium quinoa* Willed.). Acta Agronomica, 62, 13–20.

Ruales, J., and Nair, B. M. (1993b). Saponins, phytic acid, tannins and protease inhibitors in quinoa (*Chenopodium quimoa,* Willed) seeds. Food Chemistry, 48(2), 137–143. doi:10.1016/0308-8146(93)90048-K

Ruales, J., and Nair, B. M. (1993a). Content of fat, vitamins and minerals in quinoa (*Chenopodium quinoa,* Willd) seeds. Food Chemistry, 48(2), 131–136. doi:10.1016/0308-8146(93)90047-J

Sandak, R. N., and El-Hofi, A. A. A. (2000). Utilization of barley and Amaranthus seeds in biscuit processing. Egyptian Journal of Agricultural Research, 78, 1241–1252.

Sandberg, A. S., Hulthén, L. R., and Türk, M. (1996). Dietary Aspergillus niger phytase increases iron absorption in humans. Journal of Nutrition, 126(2), 476–480. doi:10.1093/jn/126.2.476

Sano, M., Nakagawa, M., Oishi, A., Yasui, Y., and Katsube-Tanaka, T. (2014). Diversification of 13S globulins, allergenic seed storage proteins, of common buckwheat. Food Chemistry, 155, 192–198. doi:10.1016/j.foodchem.2014.01.047

Sanz-Penella, J. M., Wronkowska, M., Soral-Śmietana, M., and Haros, M. (2013). Effect of whole amaranth flour on bread properties and nutritive value. LWT – Food Science and Technology, 50(2), 679–685. doi:10.1016/j.lwt.2012.07.031

Saturni, L., Ferretti, G., and Bacchetti, T. (2010). The gluten-free diet: Safety and nutritional quality. Nutrients, 2(1), 16–34. doi:10.3390/nu20100016

Saunders, R. M., and Becker, R. (1984). Amaranthus: A potential food and feed resource. In Y. Pomeranz (Ed.), Advances in cereal science and technology (VI) (pp. 357–396). MN: American Association of Cereal Chemists St. Paul.

Scanlin, L. A., Stone, M. B., and Burnett, C. (2010). Quinoa protein concentrate, production and functionality [Patent]. PubMed: 20100184963, July 22.

Schoenlechner, R., Siebenhandl, S., and Berghofer, E. (2008). Pseudocereals. In E. K. Arendt and F. Dal Bello (Eds.), Gluten-free cereal products and beverages (pp. 149–190). Oxford: Elsevier.

Segura-Nieto, M., Barba-de la Rosa, A. P., and Paredes-Lopez, O. (1994). Biochemistry of amaranth proteins. In O. Paredes-Lopez (Ed.), Amaranth—Biology, chemistry, and technology (pp. 75–101). London: CRC Press Press.

Shelton, R., and Jong Lee, W. (2000). Cereal carbohydrates. In Handbook of cereal science and technology (K. Kulp, J. Ponte, Jr.) (2nd ed) (pp. 385–416). New York, NY: Marcel Dekker, Inc.

Singer, N. S., Tang, P., Chang, H. H., and Dunn, J. M. (1990). Carbohydrate cream substitute. European Patent 0403696A1.

Skrabanja, V., and Kreft, I. (1998). Resistant starch formation following autoclaving of buckwheat (*Fagopyrum esculentum* Moench) groats. An in vitro study. Journal of Agricultural and Food Chemistry, 46(5), 2020–2023. doi:10.1021/jf970756q

Skrabanja, V., Liljeberg Elmståhl, H. G., Kreft, I., and Björck, I. M. (2001). Nutritional properties of starch in buckwheat products: Studies in vitro and *in vivo*. Journal of Agricultural and Food Chemistry, 49(1), 490–496. doi:10.1021/jf000779w

Soral-Smietana, M., Fornal, Ł., and Fornal, J. (1984). Characteristics of lipids in buckwheat grain and isolated starch and their changes after hydrothermal processing. Food/Nahrung, 28(5), 483–492. doi:10.1002/food.19840280504

Stalknecht, G. F., and Schulz-Schaffer, J. R. (1993). Amaranth rediscovered. In J. Janick and J. E. Simon (edn), New crops (pp. 211–218). New York, NY: Wiley.

Steadman, K. J., Burgoon, M. S., Lewis, B. A., Edwardson, S. E., and Obendorf, R. L. (2001b). Minerals, phytic acid, tannin and rutin in buckwheat seed milling fractions. Journal of the Science of Food and Agriculture, 81(11), 1094–1100. doi:10.1002/jsfa.914

Steadman, K. J., Burgoon, M. S., Schuster, R. L., Lewis, B. A., Edwardson, S. E., and Obendorf, R. L. (2000). Fagopyritols, D-chiro-inositol, and other soluble carbohydrates in buckwheat seed milling fractions. Journal of Agricultural and Food Chemistry, 48(7), 2843–2847. doi:10.1021/jf990709t

Stone, L. A., and Lorenz, K. (1984). The starch of Amaranthus—Physico-chemical properties and functional characteristics. Starch - Stärke, 36(7), 232–237. doi:10.1002/star.19840360704

Tang, Y., and Tsao, R. (2017). Phytochemicals in quinoa and amaranth grain and their antioxidant, anti-inflammatory, and potential health beneficial effects: A review. Molecular Nutrition and Food Research, 61(7), 1–16, 1600767, doi:10.1002/mnfr.201600767

Tang, Y., Li, X., Zhang, B., Chen, P. X., Liu, R., and Tsao, R. (2015). Characterization of phenolics, betanins and antioxidant activities in seeds of three Chenopodium quinoa wild. Genotype. Food Chemistry, 166, 380–388. doi:10.1016/j.foodchem.2014.06.018

Taylor, J. R. N., and Parker, M. L. (2002). Quinoa. In P. S. Belton and J. R. N. Taylor (Eds.), Pseudocereals and less common cereals. Grain properties and utilization potential (pp. 93–122). Berlin: Springer-Verlag.

Tiara, H. (1974). Buckwheat. In A. M. Jonson and M. J. Peterson (Eds.), Encyclopedia of food Technology p. 139. Westport, CT: Avi Publishing, Co.

Tohver, M., Shukla, S., Rajan, S., and Ohri, D. (2007). High Molecular Weight (HMW) Glutenin Subunit Composition of Some Nordic and Middle European Wheats. Genetic Resources and Crop Evolution, 54(1), 67–81. doi:10.1007/s10722-005-1885-5

Tömösközi, S., Gyenge, L., Pelcéder, A., Abonyi, T., and Lásztity, R. (2011). Effects of flour and protein preparations from amaranth and quinoa seeds on the rheological properties of wheat flour dough and bread crumb. Czech Journal of Food Sciences, 29(2), 109–116. doi:10.17221/45/2010-CJFS

Tosi, E. A., Re, E. D., Lucero, H., and Masciarelli, R. (2000). Amaranth (*Amaranthus* spp.) grain conditioning to obtain hyperproteic flour by differential milling. Food Science and Technology International, 5, 60–63.

Uriyapongson, J., and Rayas-Duarte, P. (1994). Comparison of yield and properties of amaranth starches using wet and dry-wet milling processes. Cereal Chemistry, 71, 571–577.

Valencia-Chamorro, S., Maldonado-Alvarado, P., and Sotomayor-Grijalva, C. (2017). Lipids of kernels. In Haros and Schoencechner (Eds.), Pseudocereals chemistry and technology p. 53. New York, NY: Wiley.

Vashishth, R., Semwal, A. D., Pal Murugan, M., and Sharma, G. K. (2017). Post-harvest processing, cooking and textural properties of horse gram (*Macrotyloma uniflorum*) varieties. Journal of Food Characterization and Measurement, 12, 257–261.

Vega-Gálvez, A., Miranda, M., Vergara, J., Uribe, E., Puente, L., and Martínez, E. A. (2010). Nutrition facts and functional potential of quinoa (*Chenopodium quinoa* willed.), and ancient Andean grain: A review. Journal of the Science of Food and Agriculture, 90(15), 2541–2547. doi:10.1002/jsfa.4158

Venskutonis, P. R., and Kraujalis, P. (2013). Nutritional components of amaranth seeds and vegetables: A review on composition, properties, and uses. Comprehensive Reviews in Food Science and Food Safety, 12(4), 381–412. doi:10.1111/1541-4337.12021

Vidueiros, S. M. M., Curti, R. N. N., Dyner, L. M. M., Binaghi, M. J., Peterson, G., Bertero, H. D., and Pallaro, A. N. (2015). Diversity and interrelationships in nutritional traits in cultivated quinoa (*Chenopodium quinoa* Willd.) from Northwest Argentina. Journal of Cereal Science, 62, 87–93. doi:10.1016/j.jcs.2015.01.001

Vilche, C., Gely, M., and Santalla, E. (2003). Physical properties of quinoa seeds. Biosystems Engineering, 86(1), 59–65. doi:10.1016/S1537-5110(03)00114-4

Vojtiskova, P., Kmentová, K., Kubán, V., and Krácmar, S. (2012). Chemical composition of buckwheat plant (*Fagopyrum esculentum*) and selected buckwheat products. Journal of Microbiology, Biotechnology and Food Sciences, 1, 1011–1019.

Wei, Y. M., Hu, X. Z., Zhang, G. Q., and Ouyang, S. H. (2003). Studies on the amino acid and mineral content of buckwheat protein fractions. Nahrung, 47(2), 114–116. doi:10.1002/food.200390020

Whistler, R., and BeMiller, J. (1997). Bulking agents, fat mimetics and carbohydrate nutrition. In R. Whistler and J. B. Miller (Eds.), Carbohydrate chemistry for food scientists (pp. 217–224). St. Paul, MN: Eagan Press.

Whistler, R. L., and Be-Miller, J. N. (1997b). Carbohydrate chemistry for food scientists (pp. 124–125, 220–221). St. Paul, MN: Eagan Press.

Widmer, R. J., Flammer, A. J., Lerman, L. O., and Lerman, A. (2015). The Mediterranean diet, its components, and cardiovascular disease. American Journal of Medicine, 128(3), 229–238. doi:10.1016/j.amjmed.2014.10.014

Wijngaard, H. H., and Arendt, E. K. (2006). Buckwheat. Cereal Chemistry Journal, 83(4), 391–401. doi:10.1094/CC-83-0391

Wilhelm, E., Themeier, H. W., and Lindhauer, M. G. (1998). Feinkornstärken und hydrophile Polymereals Komponenten für neue biologisch abbaubare Zweiphasenwerkstoffe für Spezielle Anwendungen, 1. Separations- und Aufbereitungstechniken für Feinkornstärken von quinoa und amaranth. Starke, FL: Starch, 50, 7–13.

Wolter, A., Hager, A. S., Zannini, E., and Arendt, E. K. (2013). In vitro starch digestibility and predicted glycaemic indexes of buckwheat, oat, quinoa, sorghum, teff and commercial gluten-free bread. Journal of Cereal Science, 58(3), 431–436. doi:10.1016/j.jcs.2013.09.003

Wronkowska, M., and Haros, M. (2014). Wet milling of buckwheat with hull and dehulled—The properties of the obtained starch fraction. Journal of Cereal Science, 60(3), 477–483. doi:10.1016/j.jcs.2014.09.004

Wronkowska, M., Haros, M., and Soral-Śmietana, M. (2013). Effect of starch substitution by buckwheat flour on gluten-free bread quality. Food and Bioprocess Technology, 6(7), 1820–1827. doi:10.1007/s11947-012-0839-0

Wu, H., and Corke, H. (1999). Genetic diversity in physical properties of starch from a world collection of *Amaranthus*. Cereal Chemistry Journal, 76(6), 877–883. doi:10.1094/CCHEM.1999.76.6.877

Yanez, G. A., Messinger, J. K., and Walker, C. E. (1986). Amaranthus hypochondriacus: Starch isolation and partial characterization. Cereal Chemistry, 63, 273–276.

Yoshimoto, Y., Egashira, T., Hanashiro, I., Ohinata, H., Takase, Y., and Takeda, Y. (2004). Molecular Structure and some physicochemical properties of buckwheat starches. Cereal Chemistry Journal, 81(4), 515–520. doi:10.1094/CCHEM.2004.81.4.515

Zhang, Z., Zhou, M., Tang, Y., Li, F., Tang, Y., Shao, J., Wu, Y. (2012). Bioactive compounds in functional buckwheat food. Food Research International, 49(1), 389–395. doi:10.1016/j.foodres.2012.07.035

Zielinski, H., Ciska, E., and Kozlowska, H. (2013). The cereal grains: Focus on vitamin E. Czech Journal of Food Sciences, 19(5), 182–188. doi:10.17221/6605-CJFS

15

Amaranth (*Amaranthus* sps.): Chemical Composition, Nutritional Attributes and Processing Innovations

Ritu Sindhu*[1] *and B. S. Khatkar*[2]

[1]Centre of Food Science and Technology, CCSHAU, Hisar, (Haryana), India
[2]Department of Food Technology, Guru Jambheshwar University of Science and Technology, Hisar -125001 (Haryana), India

Introduction

Amaranth crop stands in the group of pseudocereals and it is popular for its significant nutrient content as well as attractive inflorescence. The word Amaranthus came from a Greek word "Anthor" meaning eternal or unwilting. Amaranthus belongs to family of dicotyledonous plants Amaranthaceae and is one of the multipurpose rare plants whose leaves are consumed in the form of vegetable while the grains are consumed as cereals. Amaranthus being an ancient crop has been included in the human diet since the pre-colonized South American civilizations. Later, the farming of amaranth declined in some geographical regions of world due to unknown reasons and its records fell into forgetfulness. The new interest in amaranth crop started in early seventy's; and for that reason, it is classified nowadays as a new, forgotten, ignored and optional crop of huge dietary significance. It was consumed as main food by ancient Aztecs and Incas who assumed that amaranth grains had supernatural powers. They used amaranth as foodstuff as well as facet of religious rituals and for creation of religious sculpture (Kelly and Martin, 1983). Amaranth is called by different names worldwide such as Ramdana (grain of God) in India, Huautli by Aztecs and Xtes by Mayas. Despite the high nutritional value, strong agronomic features of amaranth crop produce it as substitute of major crops like cereals and vegetables in regions of dry soils, high altitudes, and high temperatures. Amaranth is cultivated in gardens and homes as ornamental plant due to its attention-grabbing colourful inflorescence. Color of inflorescence depends on varieties and/or cultivars, and that can be yellow, green, bluish-purple, brownish, orange, and pink and also double shaded inflorescence Fig 1. About 400 species of amaranth are spread worldwide and from these only 20 are found in India. Depending on the utilization amaranth species are categorized into two groups leafy or vegetable amaranth and grain amaranth. Few species of amaranth have gained more attention because of their extraordinary agricultural characters for instance fast growth, tolerance to water deficiency, high yield with better nutritional quality. Majorly consumed vegetable species are *A. tricolor* and *A. lividis* whereas *A. caudatus, A. cruentus* and *A. hypochondriacus* are mainly cultivated grain amaranth species (Topwal, 2019). Grain color of amaranth may vary from light or dark with yellow, cream and red shade, depending of the plant varieties. Generally seed color of *A. hypochondriacus* is cremish yellow and *A. caudatus* give seed of red color. Amaranth grains are lenticular in shape with diameter ranging from 0.8 to1.5mm and weigh 0.6 to1.3mg per seed (Sindhu and Khatkar, 2016). In amaranth seeds, the embryo is campylotropous that encases the starchy perisperm like a hoop and jointly with the seed coat signify the bran fraction, that is comparatively rich in fat and protein (Bressani, 2003; Taylor and Parker, 2002). It is appreciated in diet for good protein content (14 to 17%) with ideally balanced composition of amino acid and specially containing appreciable amount of lysine that is generally limiting in cereal proteins. Vegetable and grain amaranth contain satisfactory level of minerals including calcium, iron, potassium and phosphorus as well as various bioactive components required for good health. Seeds are suitable for different various kinds of processing such as popping, flaking, extrusion and flour forming similar to major cereals processing. Processed amaranth grains are consumed as snacks and cereals or in combination with cereals. Flour and starch of amaranth is utilized for the formulation of a variety of bakery and confectionary products like biscuits, cookies, soups and breads. Due to lacking of gluten in composition, amaranth grains are well-liked by consumers having wheat or gluten allergies. All parts of amaranth plant are traditionally used worldwide in remedies for curing of a number of diseases. The oil fraction of the amaranth seeds makes it industrial utilization due to its extraordinarily high level of squalene, an organic compound that is a very important component in synthesis of all plant and animal sterols.

Fig. 1: Image of amaranth grains and inflorescence on plants of different amaranth species

Composition

Carbohydrate

Carbohydrates are the important components of human diet and give upto 70 percent of dietary energy. Grain amaranth contains carbohydrates as starch, sugars and dietary fibres; and their concentration differs according to varieties and crop growing environment. Starch situated in perisperm part of amaranth grains is major constituent of carbohydrates. Amount of starch in amaranth varies from 49 to 69 percent of the dry grain matter varying according to the variety and crop growing atmospheric circumstances. Starch is the most important constituent of human diet contributing to the energy value and also supports healthy digestive system by interacting with the gut micro-flora (Wronkowska *et al.*, 2006). Amylopectin content of amaranth starch varying from 88.9 to 99.9 percent that is more than that of normal cereal starches normally contain 17 to 24 percent amylopectin content. Shape of granules can be oval, polygonal or spherical with diameter in the range of 0.8 to 2.5μm indicating their smaller size as compared with granules of starch from different sources like rice (3 to 8 μm), wheat (3 to 34 μm) and maize (5 to 25 μm). Applicability of amaranth starch depends on its unique characteristics like high crystallinity, viscosity and gelatinisation temperature range, great water absorption ability and swelling power. Dietary fibres are required in our diet for intestinal health benefits and generally total dietary fibres present in foods are categorised in two fractions- soluble dietary fibres and insoluble dietary fibres. Amaranth is a known to have good level of dietary fibres and accounted to have 14.2 percent total dietary fibre with 8.1 percent soluble and 6.1 percent insoluble dietary fibres in flour from grains of *A. cruentus* (Tosi *et al.*, 2001). Dietary fibre content of amaranth grain is between 9.8 to 14.5 percent which is comparable to quinoa but comparatively lesser than that of wheat. Monosaccharide and disaccharide are reported to be present in very small amount in grain amaranth. Sucrose is dominating form of sugar present in amaranth, but glucose, galactose, fructose, maltose, raffinose, stachyose and inositol are also present. *A. cruentus* and *A. caudatus* are reported to have concentration of sucrose in the range of 0.58 to 0.75 percent, glucose 0.34 to 0.42 percent, fructose 0.12 to 0.17 percent, maltose 0.24 to 0.28 percent, raffinose 0.39 to 0.48 percent, stachyose 0.15 to 0.13 percent, and inositol 0.02 to 0.04 percent of dry grain matter (Gamel *et al.*, 2006a).

Table 1: Chemical composition of amaranth grains and comparison with that of some cereals and pseudo cereals (per 100gm)

Crops	Protein (gm)	Lipid (gm)	Dietary fibre (gm)	Ash (gm)	Calcium (mg)	Iron (mg)	Phosphorus (mg)
Amaranth	11-18	5-10.2	3-20	2-3.6	160-212	7-17.4	455
Maize	10	4-5.5	9.3	1.2-1.8	5-20	1.8-3	256
Wheat	11.5-14	1-2.7	2.4-12.1	1.7-2.2	33-35	2.5-3.5	158
Buckwheat	10- 12.5	2.1-3.4	6.3-29.5	1.8-2.6	33-61	2.8-4.7	282
Quinoa	14.5-16.5	5.2-6.1	13-14.2	2.7-3.8	32-48	5.5	457
Rye	11.7	1.8	13.6	1.5	38	2.6	376

Personal elaboration using information from sources: Pedersen *et al* . (1987), Rastogi and Shukla (2013); Ahmed *et al.* (2014); Mburu *et al.* (2011); Alvarez-Jubete *et al.* (2009); Valcárcel-Yamani (2012)

Protein

Protein ranging 13 to 18 percent of dry matters constitutes the second major component of amaranth seed. Protein content is more in case of amaranth grains as compare to major cereals as well as buckwheat and quinoa (Table 1). Amaranth leaves contain comparatively lesser protein (average 1.2%) than grains though leaves serve as better source of protein than another green leafy vegetables such as spinach (Table 2). In grains of cereals protein is located mainly in endosperm whereas protein is dispersed as 65 percent in embryo portion and only 35 percent in perisperm of amaranth grains (Senft, 1979, Betschart *et al.*, 1981). This grain is also favoured in diet due to appreciable digestibility (about 90%) of its protein with suitable balancing of amino acids. Occurrence and ratio of amino acids in protein from amaranth stands close to the most favourable protein outline suggested for inclusion in human food as per FAO/WHO recommendations. Index value of essential amino acid is upto 90.4 percent that indicates its protein quality equal to egg protein and produces amaranth protein as a good alternative to meat (Písaříková *et al.*, 2005). Generally lysine is the present in very little quantity in protein of major cereals while amaranth grains contain 4.9 to 6.1 gm of lysine per 100 gm of total protein. Limiting level of some other amino acids like leucine, isoleucine and valine in amaranth protein does not create any complexity in consumption of amaranth as a mixture with cereals because these amino acids are found in appreciable amount in mostly cereals. Bakery products formulated using composite flour constituted by wheat and amaranth flours have enhanced nutritional value. Nutritional score of mixed flour raised upto 100 on adding flours from amaranth and maize in equal amount (Academy of Sciences, 1984; Bressani, 1989). Amaranth protein is primarily composed of albumins (40%), globulins (20%) and glutelins (25 to 30%) while content of prolamins is very less (2 to 3%) as compare to wheat and other cereals (Schoenlechner *et al.*, 2010). A very little content or absence of prolamins in amaranth produces it as foodstuff free from gluten required for patients of celiac disease.

Lipid

In different varieties of amaranth the content of lipid in grains varies from 6 to 9 percent which is higher than that of cereals as well as other pseudocereals. Lipid in amaranth is composed of fatty acids, triglycerides, sterols, phospholipids, glycolipids, tocopherols, and hydrocarbons. Amaranth oil is clear in appearance having delicate taste and aroma, pourable at low temperature with -27°C melting temperature and rich in omega fatty acids. Digestibility of amaranth oil is roughly similar to cottonseed oil except *A. cruentus* oil having lower digestibility (Rastogi and Shukla, 2013). Linoleic acid is the dominating one in amaranth lipid and constitutes greater than 50 percent of total fatty acid followed by oleic, palmitic and linoleic fatty acids. Amaranth oil is known for its high level of unsaturated fatty acids and the ratio of saturated to unsaturated fatty acid varies from 0.26 to 0.31 in oil of grain species and from 0.33 to 0.36 in oil from all other amaranth species. However, composition and unsaturation level of amaranth oil can vary depending on varieties, cultivars and cropping conditions. In lipids from different amaranth grain species the ratio of saturation to unsaturation ranged from 0.26 to 0.32 and fatty acids constituted by linoleic, oleic and palmitic acids in the range of 36.7 percent to 55.9 percent, 18.7 percent to 38.9 percent, and 19.1 to 23.4 percent, respectively (He *et al*., 2002). Despite the high degree of unsaturation, amaranth oil is less susptible to oxidation due to presence of tocopherols. Amaranth oil contains squalene in the range from 2 to 7 percent or higher in refined oil which is comparatively higher than other plant sources such as oils from wheat germ, rice bran and olives containing 0.1 to 1.7 percent squalene (Leon-Camacho *et al.,* 2001; Berghofer and Schoenlechner, 2002). Squalene, a triterpene, is highly unsaturated hydrocarbon and documented as the necessary biochemical predecessor of sterols. Commercial sources of squalene include industrial synthesis as well as liver of sharks, whales, and some other marine species containing 33 to 66 percent squalene in their oils. Squalene is used as lubricant in cosmetics, electronic instruments and pharmaceuticals. The anti-oxidative, anti-bacterial and anti-tumor effects of squalene are scientifically confirmed and accepted. Still consumption of squalene in food industry is conflicting as squalene rich diet increased LDL-cholesterol level of blood plasma in some investigations while other showed no change in lipid and cholesterol level of blood serum and noticeable increment recorded in bile acids and faecal cholesterol along with its non-polar derivatives (Kopicová and Vavreinová, 2007). Phospholipids mainly composed of lecithin; cefalin and phosphoinositol are present in very low concentration (about 5% of total lipid) in amaranth lipid (Becker, 1994). Amaranth seed contains very low concentration of sterols (0.27 to 0.32 mg/g) constituted by chondrila sterol and clerosterol as dominating phytosterols in oil (Leon-Camacho *et al*., 2001).

Table 2: Nutrients of amaranth leaves and comparison with similar leafy vegetables (per 100gm)

Leaves (boiled and trained	Protein (gm)	Lipid (gm)	Fiber (gm)	Calcium (mg)	Ascorbic acid (mg)	Iron (mg)
Amaranth	2.5-3.5	0.31-0.5	1.3	215-267	80	2.3-3.9
Spinach	3.2	0.3	0.6	93	51	3.1
Swiss Chard	2.4	0.3	0.8	88	32	-
Basella	1.8	0.3	0.7	109	102	1.2

Personal elaboration from sources: Uusikua *et al* . (2010), Topwal (2019)

Minerals and vitamins

Amaranth is fairly rich in minerals particularly iron, magnesium, calcium and zinc. The quantity of minerals in amaranth is more in comparison to major cereals but lesser than that of buckwheat grains. Amaranth is reported to contain 7.2 to 17.4 mg iron, 130 to 285 mg calcium, 16 to 48 mg sodium, 230 to 336 mg magnesium and 3.62 to 4 mg zinc per 100 gm of grains (Becker *et al.*, 1981). The ratio of calcium to phosphorus ranging from 1:1.9 to 2.7 is attuned to the suggested calcium to phosphorus ratio of 1:1.5 by nutritionists. The appreciable content of calcium and iron make it suitable, to be included in diet, for patients of iron-deficiency anaemia and osteoporosis. Leaves and grains of amaranth species are good source of vitamins including carotene, vitamin B_6, Vitamin B_2, ascorbic acid and vitamin B_9. However the overall content of vitamins in amaranth grains is lower than buckwheat. Amaranth contains 0.19 to 0.23 mg riboflavin, 4.5 mg ascorbic acid, 1.17 to 1.45 mg niacin, and 0.07 to 0.1 mg thiamine per 100 gm of flour (Becker *et al.*, 1981).

Anti-Nutrients

Along with the important nutrients amaranth grains contain some components such as phytates, saponins, nitrate and oxalic acids which are considered as inhibitors of nutrients. Chemically, saponins are glycosides having triterpenoidal or steroidal structure. Saponins are toxic and very bitter in taste. Saponins have surfactant like properties and they are able to form complexes with different food components like proteins, lipids, zinc and iron and in consequence decline their absorption. Surfactants can react with lipid layer of epithelial cell membranes in small intestine in this way alter its permeability for various components. As saponins content in amaranth grains is low varying from 0.09 to 0.1 percent on dry basis and having very low toxicity; their presence in amaranth is not hazardous to the consumers. Saponins can be removed from grains by washing before cooking or consumption. However, saponins have some positive health effects like blood cholesterol reducing and anti-carcinogenic properties.

Phytates are present uniformly in whole seeds of amaranth and characterized to have inhibitory effects on starch digestion as well as ability to bind with iron and other minerals, and reduce their absorption in intestine. Phytates are also known for their positive health effect on cholesterol lowering in blood system. The content of phytates is reported to be in the range from 4.8 and 34.0 μmol/gm, and in few studies it is measured upto 21.1 μmol/gm of amaranth grains (Sanz-Penella *et al.*, 2012). As phytates are distributed throughout the amaranth grains, these cannot be removed by water extraction or fractionation; however their level can be reduced by seed germination (Gamel *et al.*, 2006b).

Presence of oxalates and nitrates is considered to be a cause of kidney stones as these compounds reduce the absorption of zinc and calcium. Oxalates content ranges from 0.5 to 1.9 percent in vegetable species and 0.3 to 1.6 percent in grain species (Prakash and Pal, 1991, Gelinas and Seguin, 2007). Oxalates and nitrates in amaranth can be removed by heating the grains or leaves in boiling water for 5 to 6 minutes before consumption. In comparison to major cereals, amaranth seeds have lower level of protease inhibitor (trypsin and chymotrypsin). The content of trypsin and chymotrypsin in seeds of *A. caudatus* is reported to be 4.34 and 0.21 AIU/mg, respectively while these values are 3.05 and 0.26 AIU/mg respectively for *A. cruentus* (Gamel *et al.*, 2006b). These protease inhibitors are heat labile and can be reduced or in some cases removed, by cooking, popping, or germination. Tannins present in amaranth grains inhibit protein digestion by binding with proteins and digestive enzymes. Tannins have also ability to reduce digestion of starch. As tannins are concentrated in hull portion of amaranth grains these can be easily reduced during de-hulling process.

Processing Innovations

All parts of amaranth plant including leaves, grains, roots and flowers are processed and utilized in various ways globally depending on the knowledge, requirements and traditions of population (Table 3). After harvesting the grains of amaranth are dried and cleaned to prevent spoilage during storage or further processing. Amaranth grains are mostly sun dried in open space in farms and sometimes indoor drying using heat source is also practiced. Cleaning of grains is done by passing through screen or by winnowing using any blowing device or by combination of both screening and winnowing. Dehulling is not required as grain of amaranth has no hulls unlike common cereals and legumes. Amaranth grains are very small in size (0.81 mm diameter) with very thin and smooth seed coat (Sindhu and Khatkar, 2016). Storage tissues are present in the two layers and the embryo and accounts 25 percent of the kernel is present in peripheral part. Therefore milling operations and equipments used for common cereals are not suitable for milling of amaranth due to extremely small size and unique morphology of seeds. Betschart *et al.* (1981) worked on the modified Strong-Scott barley pearler and found it to be the most suitable for making amaranth flour fractions of varying chemical compositions. Round or sphere shaped whole starchy perisperm was produced after the complete removal of seed coat and germ parts during the five consecutive passes through the pearler. The fraction having germ portion and seed coat constituted one fourth of total flour production indicated approximately 75 percent flour extraction rate. Overall concentration of nutrients was 2 to 3 times higher in bran-germ fraction than whole seed flour of amaranth. Becker *et al.* (1986)

efficiently produced amaranth flour fractions using modified mill stone with 0.75 to 0.50 mm gap distance and amplified spindle speed. Like milling of common cereals moisture content of seed plays important role in production of amaranth flour fractions using stone mill.

However, fractionation of amaranth flour by milling is not favourable for the retention of maximum nutrients in flour due to unique seed morphology. Processing for refinement of amaranth flour causes reduction in content of protein and amino acids; when extraction level reaches upto 40 percent, the protein content decreases to 4 percent in the flour (Marcılio *et al.*, 2005). Flour of amaranth is used alone or in blended form for preparation of various products like biscuits, cookies and pancakes. Extruded flours of amaranth produced by precooking the flour suspension using extrusion technology are most suitable for instant food products during extrusion starch in flour gets modified and shows increased swelling capacity under conditions of high and low temperature. Extruded flours of amaranth form more viscous paste with lesser lumps as compare to untreated flour. Whole seeds of amaranth are consumed as expanded or puffed cereals and for production of pre-gelatinized flour from puffed grains. In Mexico, grains of *A. cruentus* cultivars are puffed and used to prepare a traditional food product "Alegria" by mixing with honey or molasses (Irving *et al.*, 1981). Toasted amaranth seeds – as whole seeds or in grounded form are used for preparation of "Atole" and "Pinole" type food products. In Asian countries, amaranth is used in culinary as a source of nutrients and as a substitute of wheat as well as a compatible ingredient in traditional food products (Dixit *et al.*, 2011). Puffed amaranth grains mixed with syrup or jaggery are used to prepare "laddoo" in India which are consumed mostly in fasting periods.

Amaranth grains being rich in good quality protein and starch are processed to produce protein concentrates and starch isolates. Amaranth seeds can be treated with glucoamylase and α-amylase to produce protein rich flour constituted by 26 percent to 28 percent protein, 40 percent to 52 percent starch and 10 percent to 16 percent fat (Paredes-L´opez *et al.*, 1990). It is reported in some studies that amaranth protein concentrates may be produced by air classification methods as well as isoelectric precipitation of protein-rich liquid formed during wet-milling of grains (González *et al.*, 2007 Bejosano and Corke, 1998). Extremely small size of granules, high gel strength, excellent freeze-thaw stability and clear paste of amaranth starch have presented it as a suitable ingredient for various food applications in the form of stabilizers, thickeners as well as non-food applications such as dusting powders, cosmetics and biodegradable plastics (Sindhu and Khatkar, 2018a,b).

Table 3: Processing and possible utilisation of amaranth

Process	Products	Utilisation or applications
Grains		
Cooking	Cooked seeds	As sweet or salty porridge, in multigrain porridge
Puffing /popping	Popped, expanded seeds	As snacks; combined with sugar syrup, jaggery or honey for laddoo preparation
Milling	Whole seed flour	Preparation of gluten free food products; in composite and blended flours for improving the nutritional and sensory quality of food products
Milling with fractionation of flour	Flour fractions, protein-rich or starch-rich flour fraction	As source of protein or starch; preparation of gluten free with high nutrients food items; in composite and blended flours for improving the nutritional and sensory quality of food products
Cooking and flaking	Flakes	As snacks; as light meal or breakfast when consumed with milk or fruits
Drum drying	Pre-gelatinised flour	Used as flour in combination or alone for development of easy cooked or instant food products with nutritional value of amaranth
Cooking extrusion and milling	Pre-gelatinised flour	Used as flour in combination or alone for development of easy cooked or instant food products with nutritional value of amaranth
Cooking extrusion and flaking	Flakes	As snacks; as light meal or breakfast when consumed with milk or fruits
Germination	Sprouts, malt	Sprouts consumed as healthy breakfast; malt used for fortification of beverages of food products; development of amaranth malt based beverages
Direct starch hydrolysis	Protein concentrates, glucose syrup	As direct source of high quality protein and glucose
Isolation of starch	Starch powder	In food industry; pharmaceuticals and cosmetics; in packaging industry like development of edible films
Extraction of lipids	Amaranth oil	As eatable oil, as source of squalene
Leaves		
Cooking /simmering	Cooked leaves	As green leafy vegetable; high nutrient diet
Decoction	Leaves extract	Traditionally used in medicines
Roots		
Drying	Root powder	Traditionally used in medicines
Cooking and grinding	Paste	Traditionally used in medicines
Whole plant		
Chopping of plants with soft stem	Chopped plant	As feed for animals
Drying	Dried plants	As fuel in rural area

Health Benefits of Amaranth

Leaves and seeds of amaranth from different varieties are used in various herbal remedies due to their nutraceutical properties. Presence of essential nutrients in considerable amount and different bioactive components in amaranth enhances its utility in curing and prevention of diet related health problems. Different parts of amaranth plants are utilized in preparation of various Ayurvedic and homoeopathic medicines due to following health benefits –

Gluten-free food: All species of amaranth contain very low content or no prolamins in protein which make it gluten free. As patients suffering from celiac disease or having allergy to gluten cannot eat cereal proteins constituting gluten. Therefore, amaranth attracts the interest of consumers as an alternative source of protein especially for celiac patients and those seeking low gluten diets.

Muscle health: High quality protein is required in diet for building of cells and tissues, maintenance of muscular health and immune system in human body. Amaranth grains contain appreciable level of protein with perfectly balanced amino acids. Owing to high content of lysine, amaranth flour is suitable for blending with cereal flours to make perfect ratio of amino acids in mixed flour for preparation of food products.

Bone health: Mineralization is important process for the maintenance of bone health and recovery from bone injuries. Calcium is the most important mineral for the formation of healthy bones because it plays a key role in mineralization. Higher calcium content of amaranth than other cereals make it a supportive food for development of healthy bones as well as prevention of bone diseases.

Hypocholesterolemic Activity: Amaranth lipids are highly unsaturated and have ability to affect cholesterol metabolism in animals. Unsaturated fatty acids in amaranth oil have cholesterol lowering properties and therefore support prevention of cardiovascular diseases. Phytosterols in plants are found to reduce the body's absorption of cholesterol. Manganese is present in amaranth in good quantity; and it is required to maintain normal cholesterol level in blood. Squalene in amaranth oils increases the faecal elimination of steroids by interfering with absorption of cholesterol and thus lowers cholesterol level. Amaranth protein isolate and oil fraction are reported to decrease the total and LDL-cholesterol level in blood without increasing the level of liver cholesterol (Pogojeva *et al.*, 2006, Mendonça *et al.*, 2009).

Fights against migraines: Consumption of amaranth is helpful in fighting against migraines and related symptoms. Magnesium, present in amaranth in considerable amount, is effective in relaxation of blood vessels and reduces constriction and bounce back dilation.

Anti-inflammatory activity: Consumption of amaranth is helpful in reduction or prevention of chronic diseases caused by inflammation. Peptides present in amaranth proteins are biologicallay active against inflammation. Several pro-inflammation indications are reduced by action of biologically active peptides of protein hydrolyzates formed during extrusion of amaranth. The ethanol and petroleum ether extracts of whole plant of *Amaranthus spinosus Linn.* showed anti-inflammatory properties (Bavarva and Narasimhacharya, 2013). It is considered that activity of amaranth plant extract is equivalent to standard drug Ibuprofen against inflammation. The bioactive compounds including steroids, alkaloids and flavonoids present in amaranth plant extracts are considered to be responsible for anti-inflammation action.

Cure duodenal peptic ulcer: Amaranth oil is useful in treatment of duodenal peptic ulcer and chronic gastritis caused by *Helicobacter pylori.*

Fights diabetes: Consumption of amaranth in daily diet is helpful in preventing diabetes. Manganese helps in regulating blood glucose and it is present in adequate level in amaranth grains. Therefore, if amaranth containing high content of manganese is consumed in sufficient amount it is possible to regulate blood glucose level or to avoid diabetes.

Food for pregnant women: Folic acid supports the generation and development of new cells. Therefore it is recommended in the diet of pregnant women to prevent defects in the body organs of foetus. Amaranth grains are good source of folic acids that makes amaranth a recommendable food for pregnant women.

Prevent constipation: Dietary fiber helps in smoothening the function of digestive system. Amaranth contains sufficient level of dietary fiber with higher content of insoluble fiber as compare to soluble fiber. So intake of amaranth adds in proper functioning of digestive system and avoids constipation. Amaranth starch with higher water binding capacity is also helpful in reducing the chances of constipation occurrence.

Anti-oxidative activity: Free radicals produced in body can combine with proteins, lipids or nucleic acids and cause oxidative stress. Antioxidants are effective in preventing or limiting the damage or negative effects of free radicals in body system. Compounds like polyphenols, β-carotene, phenolic acids, tocopherols and L-ascorbic acid having antioxidative properties are found in good amount in amaranth. Therefore intake of amaranth as food or food ingredient is helpful in fighting against free radicals or oxidative stress in body system of consumers. The ethanolic extracts of amaranth plant showed considerable level of anti-oxidative activity in β-carotene and linoleate model

(Klimczak *et al.*, 2002). Cooking of amaranth does not have negative effects on antioxidative activities. However, simmering and blanching of amaranth generally improve the anti-oxidative activity by increasing the content of β-carotene and lutein in amaranth.

Uses of Amaranth

Domestic uses

Amaranth, an excellent source of nutrients, is a multipurpose crop used in various ways such as preparation of soups, vegetables, pops, flakes, flour and as forage in different parts of world. In Asian and African countries, mainly amaranth leaves are used like green leafy vegetables while in North and South America amaranth grains are used most commonly. Amaranth grain known as 'Ramdaana' and 'King Grain' is supposed to have religious importance and generally consumed in fasting periods in India. Smooth and green leaves of amaranth are used in soups, vegetables and salads. As good source of protein and pigments, amaranth leaves are used in preparation of extracts rich in protein and natural colorants. Leafy vegetable of amaranth have appreciable content of fibre and used in diet for smoothening the digestive system. Soft stems or whole plants of amaranth are chopped and used as animal feed. Soft and smooth stem of amaranth are added in curry as drum stick. Popped grains of amaranth can be consumed like poncorn as a snack or mixed with milk or fruit as nutritious breakfast. In northern India, Laddoos or candy balls are prepared by mixing popped amaranth grains with sugar syrup or honey. Sweet porridge or 'Kheer' is prepared by cooking amaranth grains in milk and sugar. Amaranth grains mixed in water containing spices are cooked in pressure cooker or in open pan to form savoury 'polenta' or porridge. The fermented beverage named 'Chichi' is prepared from amaranth grains (Early, 1990). Amaranth grains are processed to produce flour of high nutritive value. Amaranth flour is generally blended with cereals flour for preparation of soups, biscuits, pancakes and cookies. The purpose of amaranth flour blending with other flours is to increase the nutritional value and modify the sensory properties of products prepared from blends. Whole amaranth flour is used to prepare gluten-free food products especially for gluten-allergic people or celiac patients. Amaranth flour can be used in preparation of gluten-free biscuits and cracker with higher content of protein and high energy value. These products are recommended as dietetic complements to enlarge and enrich the range of food items for celiac patients as well as for other consumers (Hozová *et al.*, 1997). White amaranth flour is blended with flour of maize and finger millet for making of 'Parathas'and 'Chapaties'. The amaranth flour from grains or flaked grains is mixed with wheat flour to prepare different bakery products like bread, cookies and similar food items. Generally it is suggested that amaranth grain flour should be add upto 20 percent of the total composite flour. However it has been found in some studies that amaranth flour added up to 75 percent in the composite flour maintained functional behaviour and as well as sensory properties of final flour. Colored bread with acceptable sensory properties were developed using composite flour containing amaranth flour upto 50 percent by Rosell *et al.* (2009). Bread prepared from mixed flour of *A. caudatus* grain and rice in ratio of 50:50 showed improved level of fat, dietary fiber and magnesium than wheat flour bread (Alvarez-Jubete *et al.*, 2009). Mineral content of amaranth leaves is on top of the green vegetables; lower content of oxalates and nitrates makes amaranth leaves highly recommendable for human consumption.

Traditional uses

All parts of amaranth plant have remarkable medicinal properties and this crop has been used worldwide in different ways for treatment of various diseases since very long in history. In Ayurveda, different names in Sanskrit language are given to various species of amaranth used in preparation of medicines. *A. spinosus* is called as Meghnath and Vishagna while *A. Blitum* is called as Marisa and Vaspaka these plants are explained as diuretic and recommended in treatment of cough, vata, throat and urinary infections, and in extreme bile secretion (Bhandari, 1938). *A. caudatus* is called as Rajadri and Rajagiri in Ayurveda and unitized as Pittanashka and Rucikara. It is also suggested for curing the goitre, urinary problems and as antiscorbutic. The whole plant of this species is used in diet for good health and purification of blood. Ramasitalika is the Sanskrit name of *A. tricolor* and it is recommended in case of Raktapradara as well as Raktatisara (Chopra *et al.*, 1955). It is used in preparation of lotion for skin problems, as pain reliever in pregnancy and as a cooling agent in urinary problems. Plant roots of this species are used with bark of Ashok tree, Daru Haldi and amla for the treatment of leucorrhoea. Unani method of medication uses the juice of amaranth plant root for as curative for gastric problems, leucorrhea and menorrhea. Root juice is also utilised in treatment of snake bite or scorpion poison. Plant leaves are helpful in fighting against urinary problems and especially in kidney stones. In India, the tea prepared from *A. caudatus* leaves is consumed as diuretic and blood purifier, for treatment of pulmonary problems as well as piles (Gupta *et al.*, 2004). Lal bhaji is used by traditional people for the treatment of obesity and management of body weight as well as curing of internal and external health problems. They also suggest amaranth as remedy for the anaemia, epistaxix and gynaecological problems. The fresh juice of green leaves is used in curing the earache and ear diseases. In Himachal Pradesh, the decoction of amaranth grains is used in treatment of foot and mouth diseases in animals. In east Africa, the amaranth plant leaves called as Mchicha are recommended to increase the count of red blood cells. In South Africa, leaves of *A. caudatus* are used as an abortifacient and in treatment of pulmonary diseases. The tea prepared from dried flowers of amaranth is helpful in contraception and too much blood flow during menses. The cooked leaves of amaranth are consumed in stomach disturbance and swelling (Michael, 2002). Chinese people use roots of amaranth plants as diuretic, wound healer and to cure cold. They consume amaranth in summer to fight internal heat. Amaranth plant is recommended in Chinese diet as curative for several diseases like inflammation, diarrhea, sore, throat and mouth associated problems and allergic reactions (Lin *et al.*, 2005). For the treatment of jaundice and dysentery, cooked leaves are consumed in combination with cuttle fish and honey, respectively. In Nepal, amaranth root juice is used in treatment of diseases including fever, diarrhea, menorrhagia, stomach disorders, urinary problems, gonorrhea, eczema, colic lactogogue and dysentery. Cooked amaranth leaves and roots are consumed by children as laxative and also used as bandage in erupted or burnt organs and snake bite. Other uses of amaranth include preparation of snuff from plant ash, as an astringent, as mouthwash in toothache, as face wash to cure pimples, as wormicide and in impotency. Grains of amaranth are used for their high nutritional value, anti-oxidative and cholesterol lowering properties to improve the immunity. Therefore, amaranth plants of different species are widely used in medications according to traditions or cultures of the region.

Agro-industrial uses

Amaranth grains have great nutritional significance which present is as most usable food or food ingredient for all age groups. Similar to other pseudocereal crops, amaranth is capable to grow in harsh environmental conditions; so it can be a potential food crop in regions where growth of other cereal crops is not possible. Interest of farmers, researchers, industrialists and consumer is growing towards amaranth. Amaranth Food Company has released amaranth based new food products named Amarlac and Amarameal having excellent dietary profile in Guatemala. It indicates the outcome of few years' research on processing and utilization of amaranth grain in development of nutrient rich foods for pregnant women and adolescents (Topwal, 2019). These kinds of products can be tried in other countries too. Amaranth contain good amount of oil which can be used as eating oil. Squalene in amaranth has anti-oxidative and cholesterol lowering properties. Commercially squalene is used in pharmaceutical and cosmetic products and its sources are liver oil of shark and some other ocean animals. Plant sources of squalene are wheat germ oil and rice bran oils but their content is comparatively lower than amaranth oil. So, amaranth can be utilized as commercial source of squalene. Starch is the important component that mainly contributes to the energy value of diet. Amaranth grains or flour can be used for production of energy rich meal because starch is the major component of amaranth seeds ranging from 48 to 69 percent in different varieties. Starch granules of amaranth are extremely small in size and starch is waxy type with high crystallinity and high water absorption capacity. These entire features make amaranth starch suitable for specific application in different food and non-food industries. Amaranth protein is rich in lysine and has perfect balance of amino acids. Amaranth grains being an excellent source of good quality proteins are utilizable for preparation of protein concentrates. Amaranth grains are processed for making the different starch-rich and protein-rich flour fractions. Green leaves and colourful inflorescence present amaranth as ornamental plant as well as good source of natural pigments or colorants.

Conclusion

Amaranth has been recognised for its nutritional and health benefits since ancient times. All plant parts of amaranth have great nutraceutical properties and have been used widely in homoeopathic and Ayurvedic medicines. Amaranth grains are processed and utilized in number of ways for food and non-food applications. It provides a possibility of fruitful farming as an alternative crop in areas where climatic conditions are not suitable for cultivation of major cereals and vegetables. Amaranth crop has great potential to boost the income of farmers as well as to provide nutritional and food security to the increasing population.

References

Ahmed, A., Khalid, N., Ahmad, A, Abbasi, N., Latif, M. and Randhawa, M. (2014). Phytochemicals and biofunctional properties of buckwheat: a review. The Journal of Agricultural Science, 152, 349-69.

Alvarez-Jubete, L., Arendt, E. and Gallagher, E. (2009). Nutritive value and chemical composition of pseudocereals as gluten-free ingredients. International Journal of Food Sciences and Nutrition, 60, 240-57.

Bavarva, J. and Narasimhacharya, A. (2013). Systematic study to evaluate anti-diabetic potential of Amaranthus spinosus on type-1 and type-2 diabetes. Cellular and Molecular Biology (Noisy-le-Grand, France), 59, OL1818-25.

Becker, R. (1994). Amaranth oil: Composition, Processing and Nutritional Qualities. In: Paredes-Lopez O., (Ed.), Amaranth: Biology, Chemistry and Technology (133–141). CRC Press, Boca Raton.

Becker, R., Irving, D. and Saunders, R. (1986). Production of debranned amaranth flour by stone milling. Lebensmittel-Wissenschaft and Technologie, 19, 372-75.

Becker, R., Wheeler, E. L., Lorenz, K., Stafford, A. E., Grosjean, O.K., Betschart, A. A. and Saunders, R. M. (1981). A compositional study of amaranth grain. Journal of Food Science, 46, 1175-80.

Bejosano, F. P. and Corke, H. (1998). Protein quality evaluation of Amaranthus wholemeal flours and protein concentrates. Journal of the Science of Food and Agriculture, 76, 100-06.

Berghofer, E. and Schoenlechner, R. (2002). Grain Amaranth In: P. S. Belton, (Ed.), Pseudocereals and Less Common Cereals, Grain Properties and Utilization Potential (pp. 219-60). JRN Taylor, Verlag: Springer.

Betschart, A. A., IGrving, D. W., Shephard, A. D. and Saunders, R. (1981). Amaranthus cruentus: milling characteristics, distribution of nutrients within seed components, and the effects of temperature on nutritional quality. Journal of Food Science, 46, 1181-87.

Bhandari, C. (1938). Vanaushadhi Chandrodaya: An Encyclopedia of Indian Botanics and Herbs. Varanasi, U.P., India

Bressani, R. (1989). The proteins of grain amaranth. Food Reviews International, 5, 13-38.

Bressani, R. (2003). Amaranth. Encyclopedia of Food Sciences and Nutrition, Ten-Volume Set, 166-73.

Chopra, R. and Chopra, I. C. (1955). A review of work on Indian medicinal plants. Pharmacopoeia of India. (Special Report Series No. 30). New Delhi: Indian Council of Medical Research.

Dixit, A. A., Azar, K. M., Gardner, C. D. and Palaniappan, L.P. (2011). Incorporation of whole, ancient grains into a modern Asian Indian diet to reduce the burden of chronic disease. Nutrition Reviews, 69, 479-88.

Early, D. K. (1990). Amaranth production in Mexico and Peru. Advances in New Crops (OR: 140-42). Portland: Timber Press.

Gamel, T. H., Linssen, J. P., Mesallam, A. S., Damir, A. A. and Shekib, L. A. (2006b). Seed treatments affect functional and antinutritional properties of amaranth flours. Journal of the Science of Food and Agriculture, 86, 1095-1102.

Gamel, T. H., Linssen, J. P., Mesallam, A. S., Damir, A. A., Shekib, L. A. (2006a). Effect of seed treatments on the chemical composition of two amaranth species: oil, sugars, fibres, minerals and vitamins. Journal of the Science of Food and Agriculture, 86, 82-89.

Gelinas, B. and Seguin, P. (2007). Oxalate in grain amaranth. Journal of Agricultural and Food Chemistry, 55, 4789-94.

González, R., Carrara, C., Tosi, E., Añón, M. and Pilosof, A. (2007). Amaranth starch-rich fraction properties modified by extrusion and fluidized bed heating. LWT-Food Science and Technology, 40, 136-43.

Gupta, A. and Neeraj, T. (2004). Reviews on Indian medicinal plants. Volume 1 (Abe-Alle). Reviews on Indian Medicinal Plants. Volume 1 (Abe-Alle).

He, H.P., Cai, Y., Sun, M. and Corke, H. (2002). Extraction and purification of squalene from Amaranthus grain. Journal of Agricultural and Food Chemistry, 50, 368-72.

Hozova, B., Buchtova, V., Dodok, L. and Zemanovič, J. (1997). Microbiological, nutritional and sensory aspects of stored amaranth biscuits and amaranth crackers. Food/Nahrung 41, 155-58.

Irving, D.W., Betschart, A. and Saunders, R. (1981). Morphological studies on Amaranthus cruentus. Journal of Food Science, 46, 1170-74.

Kelly, G. and Martin, L. (1983). Amaranth: Grain and Vegetable Types. ECHO Technical Note.

Klimczak, I., Małecka, M. and Pachołek, B. (2002). Antioxidant activity of ethanolic extracts of amaranth seeds. Food/Nahrung, 46, 184-86.

KopiCoVá, Z. and VaVreiNoVá, S. (2007). Occurrence of squalene and cholesterol in various species of Czech freshwater fish. Czech Journal of Food Sciences, 25, 195.

León-Camacho, M., García-González, D. L. and Aparicio, R. (2001). A detailed and comprehensive study of amaranth (*Amaranthus cruentus* L.) oil fatty profile. European Food Research and Technolog, 213, 349-55.

Lin, B. F., Chiang, B. L. and Lin, J. Y. (2005). Amaranthus spinosus water extract directly stimulates proliferation of B lymphocytes *in vitro*. International Immunopharmacology, 5, 711-22.

Marcılio, R., Amaya-Farfán, J. and Silva, M. (2005). Avaliaçao da farinha de amaranto na elaboraçao de biscoito tipo "cookie" para celıacos. Braz J Food Technol, 8, 175-81.

Mburu, M. W., Gikonyo, N. K., Kenji, G. M. and Mwasaru, A. M. (2011). Properties of a complementary food based on amaranth grain (Amaranthus cruentus) grown in Kenya. Journal of Agriculture and Food Technology, 1, 153-78.

Mendonça, S., Saldiva, P. H., Cruz, R. J. and Arêas, J. A. (2009). Amaranth protein presents cholesterol-lowering effect. Food chemistry, 116, 738-42.

Michael, H. (2002). South American Medicinal Plants: Botany, Remedial Properties and General Use. Berlin: Springer.

National Academy of Science. (1984). Amaranth: Modern Prospects for an Ancient Crop (92 pp). Washington, DC: The National Academies Press.

Paredes López, O., Barba De La Rosa, A. and Cárabez-Trejo, A. (1990). Enzymatic production of high protein amaranth flour and carbohydrate rich fraction. Journal of Food Science, 55, 1157-61.

Pedersen, B., Kalinowski, L. and Eggum, B. (1987). The nutritive value of amaranth grain (*Amaranthus caudatus*). Plant Foods for Human Nutrition (Formerly Qualitas Plantarum), 36, 309-24.

Písaříková, B., Kráčmar, S. and Herzig, I. (2005). Amino acid contents and biological value of protein in various amaranth species. Czech Journal of Animal Science, 50, 169-74.

Pogojeva, A., Gonor, K., Kulakova, S., Miroshnichenko, L. and Martirosyan, D. (2006). Effect of amaranth oil on lipid profile of patients with cardiovascular diseases. Book Functional Foods for Chronic Diseases, 35-45.

Prakash, D. and Pal, M. (1991). Nutritional and antinutritional composition of vegetable and grain amaranth leaves. Journal of the Science of Food and Agriculture, 57, 573-83.

Rastogi, A. andShukla, S. (2013). Amaranth: a new millennium crop of nutraceutical values. Critical Reviews in Food Science and Nutrition, 53, 109-25.

Rosell, C. M., Cortez, G. and Repo-Carrasco, R. (2009). Breadmaking use of andean crops quinoa, kañiwa, kiwicha, and tarwi. Cereal Chemistry, 86, 386-92.

Sanz-Penella, J. M., Laparra, J.M., Sanz, Y. and Haros, M. (2012). Bread supplemented with amaranth (*Amaranthus cruentus*): effect of phytates on in vitro iron absorption. Plant foods for human Nutrition, 67, 50-56.

Saunders, R., and Becker, R. (1984). Amaranthus: a potential food and feed resource. In: Pomeranz Y (Ed.), Advances in cereal science and technology (vol 6, 357). American Association of Cereal Chemists, St Paul, MN.

Schoenlechner, R., Wendner, M., Siebenhandl-Ehn, S. and Berghofer, E. (2010). Pseudocereals as alternative sources for high folate content in staple foods. Journal of Cereal Science, 52, 475-79.

Sindhu, R. and Khatkar, B. S. (2016). Physical properties of grains of some pseudocereals. Annals of Agri Bio Research, 21, 174-78.

Sindhu, R. and Khatkar, B. S. (2018a). Amaranth starch isolation, oxidation, heat-moisture treatment and application in edible film formation. International Journal of Advanced Engineering Research and Science, 5,136-141.

Sindhu, R. and Khatkar, B. S. (2018b). Development of edible films from heat-moisture treated and acetylated starches of amaranth. International Advanced Research Journal in Science, Engineering and Technology, 5, 80-85.

Taylor, J. R. N. and Parker, M. L. (2002). Pseudocereals and less common cereals: Grain properties and utilization. P. S. Belton (Ed.), (93-122). Berlin: Springer Verlag.

Topwal, M. (2019). A Review on Amaranth: Nutraceutical and Virtual Plant for Providing Food Security and Nutrients. Acta Scientific Agriculture, 3, 09-15.

Tosi, E., Ré, E., Lucero, H. and Masciarelli, R. (2001). Dietary fiber obtained from amaranth (*Amaranthus cruentus*) grain by differential milling. Food Chemistry, 73, 441-43.

Uusikua, N. P., Oelofse, A., Duodu, K. G. and Bester, M. J. (2010). Nutritional value of leafy vegetables of sub-Saharan Africa and their potential contribution to human health: A Review. J. Food Comp. Anal, 23, 499-509.

Valcárcel-Yamani, B. and Lannes, S. D. S. (2012). Applications of quinoa (*Chenopodium quinoa* Willd.) and amaranth (*Amaranthus* spp.) and their influence in the nutritional value of cereal based foods. Food and Public Health, 2, 265-75.

Wronkowska, M., Soral-Śmietana, M., Krupa, U. and Biedrzycka, E. (2006). In vitro fermentation of new modified starch preparations—changes of microstructure and bacterial end-products. Enzyme and Microbial Technology, 40, 93-99.

16

Buckwheat (*Fagopyrum esculentum*) Chemistry, Nutritional Attributes Processing Innovations, Traditional and Modern Foods and Beverages

***Charu Goel*[1], *Rahul Vashishth*[2], *Anil Dutt Semwal*[1] and *Gopal Kumar Sharma*[1]**

[1]*Defence Food Research Laboratory, Mysore-570011*
[2]*Vignan's Foundation for Science, Technology and Research, Guntur- 522213*

Introduction

Buckwheat (*Fagopyrum esculentum Möench*) is an annual pseudocereal crop belonging to the genus *Fagopyrum* of the family Polygonaceae. Its grains belong to cereals because of their similar use and chemical composition (Campbell, 1997). Among a number of buckwheat species nine species possed agricultural and nutritional trials common buckwheat (*F. esculentum*) and tartary buckwheat (*F. tartaricum*) are frequently cultivated around the world (Krkošková and Mrázová, 2005).

While buckwheat is available almost everywhere in the world but its cultivation is majorly confined to the Northern hemisphere. It originated from North or East Asia, being cultivated since 1000 BC in China and then introduced to North America by the colonists. The main producers of buckwheat are China, Russian Federation, Ukraine and Kazakhstan (Li and Zhang, 2001; Bonafaccia *et al.*, 2003). It is also produced in Slovenia, Poland, Hungary and Brazil (Kreft *et al.*, 1999).

Dehulled seeds (raw groats of buckwheat) are primarily utilized for human consumption as breakfast cereal products, processed flour (at 0.3-60% level) for manufacturing different bakery products (bread, biscuit, cookies, pancakes, snacks and noodles). The consumption of buckwheat and buckwheat-enriched products wide range of health benefits like hypocholesterolemic, hypoglycemic, anticarcinogenic and anti-inflammatory. Buckwheat proteins and phenolic compounds are presumed to be responsible partly for these health benefits. Buckwheat protein is of excellent quality and is high in the essential amino acid lysine, unlike the protein of the common cereals. It is also regarded as a multipurpose crop. The buckwheat plant's leaves and shoots are used as leafy vegetables and the flowers and green leaves are used for rutin extraction which is used as a medicinal ingredient. Buckwheat flower can also be utilized to produce good quality honey.

Origin and Growth Conditions

Buckwheat absorbs less water from soil and fewer nutrients than other major crops (Li and Zhang, 2001). It has very strong adaptability to adverse environments (low fertility soil, water shortage) and a very short growing span. Several varieties are grown all over the world, the majority of which are found in the northern hemisphere. *F. esculentum* Moench is grown in many countries including Russia, China, the United States, Canada, France, Germany, Italy, Poland, Japan, Korea and Nepal. *F. tartaricum* or Tartary buckwheat, a congener which is also used as a food crop, is also available in some mountainous regions of Asia. Tartary buckwheat is usually grown at higher altitudes due to its frost resistance whereas common buckwheat is cultivated at lower altitudes.

Short duration buckwheat cultivators complete the cycle from sowing to seed ripening within 80 to 90 days. Crops with longer cycles will mature within 100-110 days. The sowing season should be determined for the plant to reach physiological maturity, considering that buckwheat thrives in cool, humid climates and is not frost tolerant. Buckwheat can be grown with varying fertility rates in a wide variety of soil types. Better yields are obtained from fertile and well drained soils (Tkachuk *et al.*, 1996). Buckwheat tolerates acidic soli compared to cereals crops. It is effective in extracting phosphorous from low phosphorous soils.

Production

Buckwheat cultivation has seen a decline in past years but recently cultivation has shown an increase owing to its health promoting properties. According to data collected by FAO/UN, the production of buckwheat in the world reached 2,905,294 tonnes in 2018. Buckwheat-producing countries are also the largest consumers: China produces 55 percent of the world's total production; Russia, 20 percent; Ukraine, 15 percent; France, 6 percent and Poland, 3 percent. The United States of America, Canada and Brazil are exporters of buckwheat.

Kernel Structure

The main parts of buckwheat kernels (where perisperm is absent) include the hull (which may be glossy or dull, brown, black or grey), the testa, the protein-rich aleurone layer, the central part of the starchy endosperm that occupies the major portion of the seed and the embryo, embedded in the centre of the endosperm and includes the two cotyledons (Figure 1).

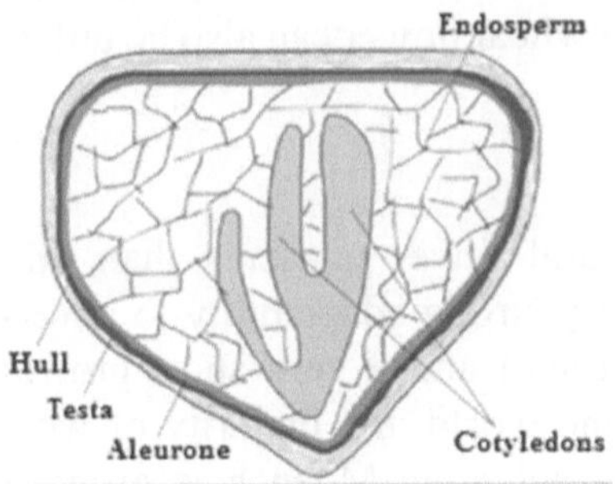

Fig. 1: Structure of buckwheat kernel

Chemical and Nutritional Components of Buckwheat Grains

Buckwheat grains contain major compounds comprising of proteins, polysaccharides, dietary fibre, lipids, minerals, vitamins and bioactive compounds (Kim *et al.*, 2004). Comparative chemical composition of buckwheat and wheat is shown in Table. 1. The nutritional composition depends on the variety or environmental factors (Bárta *et al.*, 2004).

Table 1: Chemical Composition of buckwheat kernels and wheat

Unit	Buckwheat	Wheat
Protein	10.9[c] 11.3-14.6 [e] 11.9-14.2 [g] 12.3-15.2 [f] 12.5[a]	11.6[b] 11.7, 14.3[d]
Lipids	1.3-2.3[e], 2.1[a] 2.1-2.6[g], 2.7[c], 3.0-3.4[e]	1.7[b], 2.0[c], 2.3 [d]
Starch	58.5/69.4 [f], 58.9[a], 67.2[c]	61.0[c], 78.4[d]
Dietary Fiber	6.7-9.9[e], 29.5[a]	2.8[d],
Ash	1.4-1.9[g], 2.1[a], 3.9/1.7[f]	1.4[b], 2.2[d]

a. Data from Alvarez-Jubete *et al* . (2009);
b. García-Mantrana *et al.*(2014) of *A. cruentus* and *T. aestivum* L.;
c. Souci *et al* .(2000) of *A. cruentus, abrased quinoa, F. escutentum* and *T. aestivum* L.;
d. Koziol (1992),
e. Izydorczyk *et al.* (2014) of whole groat buckwheat;
f. Wronkowska and Haros (2014) of *F. escutentum* (common buckwheat) with/without hulls;
g. Mazza (1993) of dehulled buckwheat cultivars.

Proteins

The protein content of buckwheat whole grains has been reported to be 8.5 percent to 18.8 percent depending on the variety, climate and fertilization (Table 2) (Christa and Soral Śmietana, 2008; Fornal, 1999).

Table 2: Protein content of buckwheat grains

N (% d.m.) × 6.5	Authors
12.0 -13.0	Steadman *et al.* (2001)
12.11	Li and Zhang (2001)
13.30-15.55	Wei *et al* . (2003)
8.51-18.87	Krkoškova and Mrazova (2005)
12.2	Hager *et al.* (2012)

The protein content of all fractions of buckwheat grain increases from the inner fraction to the outer fraction of the grain in sequence. The protein content of the inner fraction is only 1-2 percent while the outer fraction comprises about 40 percent of the proteins. In the embryo it reaches values up to 55.9 percent (Morita *et al.*, 2006) whereas the protein concentration in the hull is about 4 percent (Pomeranz and Robbins, 1972; Christa and Soral Śmietana, 2008).

All proteins can be classified into four classes based on their solubility; (i) albumins (soluble in water and dilute buffers at neutral pH); (ii) globulins (soluble in salt solutions but insoluble in water); (iii) glutelins (soluble in dilute acid or alkali solutions) and (iv) prolamins (soluble in aqueous alcohols of 70-90%) (Bewley and Black, 1985). Prolamins are the major storage proteins in cereals, while globulins are mainly found in buckwheat. The quantity of the individual protein fractions varies in a wide range depending on the species (Aubrecht and Biacs, 2001). Common buckwheat seeds comprise of 64.5 percent globulin, 12.5 percent albumin, 8.0 percent glutelin and a small percentage of prolamins, 2.9 percent (Ikeda *et al.*, 1991). Radovic *et al.* (1999) have also reported similar ranges. The proportions of albumin and globulin (38-44%), prolamin (2-5%), glutelin (21-29%) and residual protein (28-37%) in four buckwheat species were measured by Tahir and Farooq (1985). Wei *et al.* (2003) reported 16.8-30.3 percent albumin, 4.96-21.6 percent globulin, 3.08-7.01 percent prolamin and 11.5-16 percent glutelin. According to Guo and Yao (2006) albumin was the predominant protein fraction of around 43.8 percent, followed by 14.6 percent glutelin, 10.5 percent prolamin and 7.82 percent globulin.

Globulins consist of 12-13 subunits from 16 kDA to 66 kDA molecular weights (Krkošková and Mrázová, 2005). Buckwheat grain's main storage protein is 13S globulin, which accounts for about 43 percent of the total seed protein (Aubrecht and Biacs, 1999; Krkošková *et al.*, 2001).

The main storage proteins, globulins, consist of two major families that vary in molecular weight and sedimentation coefficients (S values) of approximately 7 (average 7-8) and 11 (11-13) during ultracentrifugation. Both are composed of regularly assembled subunits (Bewley and Black, 1985). In buckwheat, 13S globulin and 8S globulin have been identified. 13S globulin is a major storage protein in buckwheat and contributes 33 percent of total seed protein (Radovic *et al.*, 1996). The smaller 8S globulin is a trimer. It consists of sub-units with a molecular weight of 57,000-58,000 and is similar to vicilin like storage proteins. The 8S globulin contributes ~7 percent of total seed proteins (Milisavljevic *et al.*, 2004). The 13S protein was only found in the cotyledons while the 8S globulin was found in both cotyledons and in the endosperm. In buckwheat, 2S albumins have also been identified in addition to globulins. These water-soluble proteins are single-chain polypeptides and have molecular weight ranging from 8000 to 16000 (Elpidina *et al.*, 1991; Radovic *et al.*, 1999).

Variations between buckwheat and wheat contribute to differences in food functional properties and quality of processing. Due to low level of prolamin content in buckwheat and based on chemical and immunological research, buckwheat protein is a valuable source of dietary protein for gluten-sensitive individuals (celiac disease).

Buckwheat protein addition leads to weakened starch gel structure which may be due to preferential interactions between starch granules and proteins during gelatinization and retrogradation. Furthermore, buckwheat protein concentrate was examined for direct utilization in an emulsion-type meat product comprising lean beef, pork fat, salt and water which considerably affected emulsion and cooked meat gel properties. Buckwheat protein showed favorable effects to be suitable as a meat extender as compared to soy protein. Buckwheat protein concentrate has also been tested for direct application in other food systems such as wheat dough and noodles where it could improve the mixing efficacy of wheat flour and have a positive influence on noodle formulation and final product quality.

Amino Acids

Amino acid composition of buckwheat protein is nutritionally well balanced and is rich in essential types (lysine and arginine) as compared with most cereals. Table 3 shows that all amino acids are higher in buckwheat grain than in wheat grain except for glutamic acid and proline.

Table 3: Amino acid composition of buckwheat (

Amino acid	Buckwheat (g/100 g protein)	Wheat (g/100 g protein)	Recommended daily intake (mg/kg/day)
Asp	5.2-9.5	3.08	-
Thr*	1.9-4.04	1.83	15
Ser	2.4-4.9	2.87	-
Glu	9.7-19.38	18.6	-
Pro	2.6-7.93	6.21	-
Gly	4.2-6.23	2.45	-
Ala	3.0-4.82	2.26	-
Cys	2.06-3.27	1.59	-
Val*	3.4-4.97	2.76	26
Met*	0.99-2.3	0.94	10.4
Ileu*	2.6-3.41	2.04	20
Leu*	2.8-6.12	4.17	39
Tyr*	1.5-3.03	1.87	25
Phe*	2.0-4.42	2.82	25
Lys*	4.9-6.7	1.79	30
His	1.4-2.52	1.43	-
Arg	5.4-11.6	2.88	-

Source: Christa and Soral Śmietana, 2008) compared to wheat (FAO) and the WHO recommendation of essential amino acid daily intake for adults.* essential amino acids

Buckwheat protein consists of mainly globulins and albumins (almost 50% of total proteins). Buckwheat proteins contain a wide variety of amino acids. They show a high level of lysine (6.1% of 100g amino acid recovered), arginine (9.7%), aspartic acid (11.3%) and low levels of proline (3.9%) and glutamic acid (18.6%) compared to cereal grains (Pomeranz and Robbins, 1972; Javornik *et al.*, 1981). Relatively low levels of proline and glutamic acid can be explained by the fact that buckwheat contains very small quantities of prolamins, which are high in proline and glutamic acid (Ikeda *et al.*, 1991). In buckwheat species, the first and second limiting amino acids are threonine and methionine respectively (Pomeranz and Robbins, 1972). The lysine/arginine and methionine/glycine ratios in buckwheat proteins are lower than in other plant proteins. A balanced amino acid profile can be achieved by mixing buckwheat with other cereal grains (Krkošková *et al.*, 2001). The amino acid composition differs in the various sections of the buckwheat seed and a non uniform distribution of proteins within buckwheat seeds (Krkošková *et al.*, 2001).

Nutritional Quality of Proteins

Buckwheat grain not only contains high protein levels but also good protein quality and nutritional balance. The nutritional quality of a protein is a reflection of its essential amino acid content, digestibility and amino acid bioavailability (FAO/UN, 1990). Eggum *et al.* (1980) reported two different buckwheat varieties, both with high protein quality, namely biological values of 93.1 percent and 90.5 percent respectively compared to egg albumin (100), soybean protein (68) and wheat protein (63). High biological value can be due to high concentration of most essential amino acids especially lysine, threonine, tryptophan and sulphur-containing amino acids. However, when the buckwheat varieties were fed to rats, values for true protein digestibility obtained were 79.9 percent and 78.8 percent (Eggum *et al.*, 1980) which are low compared to cereals (FAO/UN, 1990). It is primarily due to the presence of exogenous antinutrients (e.g. dietary fibre, protease inhibitor and tannin-type compounds) and protein susceptibility to proteolytic action (Pomeranz and Robbins, 1972; Bonafaccia and Kreft, 1994).

Buckwheat protein also exhibits numerous bioactive effects such as cholesterol lowering effect, antihypertensive effect, antitumor activity and improvement in constipation and obesity conditions by functioning similarly to dietary fiber and interrupting in vivo metabolisms.

Thiamin-binding proteins (TBP) act as transporters of B^1 vitamins in the plant and stabilise it during technological processing. They can also increase thiamin stability and its bioavailability during storage. Mitsunaga *et al.* (1986) were the first to isolate the TBP from buckwheat grains. TBP may be used in the cases of thiamin deficiency and difficulties in its storage (Wanatabe *et al.*, 1999; Krkošková and Mrázová, 2005).

Buckwheat Allergy

Buckwheat proteins can cause a hypersensitive reaction (allergy). This immunoglobulin (IgE) mediated hypersensitive response can cause serious symptoms including anaphylactic shock.

Buckwheat allergies are not common, however, it is considered to be a highly potent allergen, particularly in children. An increasing incidence of allergy manifestations has been observed in people who frequently and in large quantities consume buckwheat-containing foods mainly in Europe, North America and Japan as well as some other Asian countries (Wang *et al.*, 2004; Hirao *et al.*, 2005; Morita *et al.*, 2006). During buckwheat grain milling and food processing, some workers have allergic reactions, such as nose itching, sneezing, rhinorrhea, dyspnea, asthma and even more serious symptoms. Several researchers identified and characterized the allergens in buckwheat (Table 4).

Table 4: Main buckwheat allergens

Name of the allergen	Family	Molecular weight	Biological function
Fag e 1	13S Globulin	24 kDa	Legumin-like protein
Fag e 2	2S Albumin	16 kDa	2S Albumin
Fag e 3	7S Vicin	19 kDa	7S Vicilin-like globulin
Fag e TI	Trypsin inhibitor	9 kDa	Trypsin inhibitor

Source: Heffler *et al.*, 2014

The allergens varied among the patients Sera and electrophoretic immunoblotting measurements confirmed that IgE antibody binding to the seed storage proteins of buckwheat showed varying patterns between the patients.

Polysaccharides

Total carbohydrate content in buckwheat is 68-73 percent of which 54.5 percent was found to be starch (Li and Zhang, 2001; Steadman *et al.*, 2001). The percentage of starch (57.4%) in tartary buckwheat is marginally higher than in common buckwheat.

Buckwheat starch has relatively high amylose content 18.3-47 percent (Qian *et al.*, 1998; Yoshimoto *et al.*, 2004; Wolter *et al.*, 2013) with no consistent variations between apparent amylose percentages of common and tartary buckwheat (Li *et al.*, 1997). Yoshimoto *et al.* (2004) stated a big difference between actual amylose content (15.6-17.9%) and apparent amylose content (25.5-26.5%) due to high iodine affinity of long chain buckwheat amylopectin (2.21-2.48%) affecting the gelatinization properties. The water-binding capacity of buckwheat starch (109.9%) is higher than the water binding capacity of wheat and corn starch (Qian *et al.*, 1998) due to the small size of buckwheat starch granules (2.9-9.3 µm) (Qian and Kuhn, 1999; Wolter *et al.*, 2013) and pores in starch granule making it more susceptible to fungal α-amylase than corn and wheat starch (Qian *et al.*, 1998). Buckwheat starch exhibits a higher peak viscosity than common cereal starches and its pasting behaviour resembles more closely to that of root and tuber starches (Qian *et al.*, 1998; Arendt and Zannini, 2013). These high viscosity values can be due to supermolecular glucan structures and high tendency of buckwheat starch to swell and gel (Yoshimoto *et al.*, 2004). The peak gelatinization temperature varies from 57-70°C among different cultivars (Li *et al.*, 1997; Noda *et al.*, 1998; Qian *et al.*, 1998).

Buckwheat starch granules are polygonal, spherical and oval in shape with noticeable flat areas due to compact packing in the endosperm, the granule size distribution ranges from 2.9 to 9.3 µm (Wolter *et al.*, 2013). Buckwheat starch has its own unique characteristics; some properties correspond to tuber starches (high viscosity values) and others to cereal starches (shape and composition) (Wijngaard and Arendt, 2006).

Buckwheat starch has a relatively high resistant starch (RS) content; according to Skrabanja and Kreft (1998) the RS content in buckwheat groats is about 35 percent of total starch. Resistant starch is not absorbed in the small intestine and is partly or completely available for fermentation by microflora in the large intestine. It could exhibit similarity to dietary fibre and thus can have positive nutritional effects. The factors influencing starch availability include its botanical origin and physical properties, the ratio of amylose to amylopectin and starch interactions with other constituents like protein and lipid (Kreft *et al.*, 1999). High amylose/amylopectin ratios result in increased levels of resistant starch upon heating.

Dietary Fibre

Dietary fibre is the edible part of a plant or analogous carbohydrates that is resistant to digestion and absorption in the human small intestine but is partly or completely fermented by microflora in the large intestine (AACC Report, 2001). They are concentrated in tissues with thicker cell walls, aleurone, seed coat and hulls. On the basis of affinity to water, total dietary fibre (TDF) is classified as either insoluble dietary fibre (IDF) or soluble dietary fibre (SDF). IDF includes cellulose, lignins and certain non-cellulosic polysaccharides while SDF includes pectins and some associated non-cellulosic polysaccharides (Brennan, 2005). The whole buckwheat grains contain 7 percent TDF, bran with hull fragments has 40 percent TDF (of which 25% is SDF) while bran without hull fragments has 16 percent TDF (of which 75% is SDF) as fibre is primarily present in the outer seed coverings such as the seed coat and hull (Steadman *et al.*, 2001). SDF may contribute positively to human health by reducing levels of blood cholesterol (Roberfroid, 1993). Total dietary fibre in buckwheat bran milled without hulls is comparable to that of oat bran (17%) (Lee *et al.*, 1992). Bonafaccia *et al.* (2003) reported that common and tartary buckwheat contains similar amounts of total dietary fibre. A higher proportion of soluble fibre was found in tartary buckwheat than in common buckwheat. Dziedzic *et al.* (2012) stated that buckwheat fibre has a low SDF/IDF ratio (0.5-0.28). Dietary fibre exhibits the most effective physiological action at a SDF/IDF ratio of 1:2 (Jaime *et al.*, 2002). The authors recommend the combination of buckwheat fibre with fruit and vegetable fibre, which is characterized by a much higher ratio of the SDF/IDF fractions.

Zhu *et al.* (2014) reported that SDF increased from 16 percent to 26.6 percent while IDF decreased from 70.27 percent to 56.63 percent after ultrafine grinding of buckwheat hulls, suggesting that ultrafine grinding causes a redistribution of fibre components in TDF (86.83%). This suggested that mini-type airflow pulverization instruments could effectively reduce the particle size of buckwheat hull and redistribute fibre components. Buckwheat hull can be micronized which results in higher SDF content and could be used as an alternative to produce functional foods and nutraceuticals.

DF may also play a negative role by binding proteins and minerals or inhibiting digestive enzymes, thereby, lowering digestibility or absorption of proteins and minerals (Ikeda and Kusano, 1983; Steadman *et al.*, 2001).

Lipids

Buckwheat contains 2.4-3.2 percent lipids (Pomeranz and Lorenz, 1983; Mazza, 1988; Ryan *et al.*, 2007; Tang, 2007). Buckwheat embryo is composed of 8.2 percent oil, followed by the testa with 2.0 percent, then the pericarp with 0.5 percent and finally the endosperm with 0.4 percent. The embryo contains around 66.66 percent of the total oil in the seed (Dorrell, 1971). Dorrell (1971) and Kim *et al.* (2002) have detected lower total lipid levels in common buckwheat varying from 1.9 to 2.4 percent.

In buckwheat, 18 fatty acids have been recognized, of which 14 appear in all seed tissues. The eight main acids (oleic, linoleic, palmitic, linolenic, lignoceric, stearic, behenic and arachidic) represent 93 percent of the total fatty acids detected (Dorrell, 1971). The fatty acid composition of common and tartary buckwheat is shown in Table 5.

Table 5: Fatty acid composition of common and tartary buckwheat

Fatty acid	Common buckwheat (%)	Tartary buckwheat (%)
Myristic (C14 : 0)	0.0	0.0
Palmitic (C16 : 0)	15.6	19.7
Palmitoleic (C16 : 1)	0.0	0.0
Stearic (C18 : 0)	2.0	3.0
Oleic (C18 : 1)	37.0	35.2
Linoleic (C18 : 2)	39.0	36.6
Linolenic (C18 : 3)	1.0	0.7
Arachidonic (C20 : 0)	1.8	1.8
Eicosaenoic (20 : 1)	2.3	2.0
Behenic (C22 : 0)	1.1	0.8
Saturated	20.5	25.3
Unsaturated	79.3	74.5
Unsaturated/saturated	3.87	2.94

The embryo contains most of the unsaturated fatty acids while hull contains high level of saturated fatty acids (Dorrell, 1971). Unsaturated fatty acids constitute ~75-80 percent of total fatty acids. Linoleic acid is the major fatty acid present in buckwheat and is particularly present in high amount in the seed coat (Taira *et al.*, 1986; Mazza, 1988). Total saturated fatty acids are present around 0.46 percent of the edible portion (Pomeranz and Lorenz, 1983). The pericarp has higher saturated fatty acid content which are usually present in common buckwheat than tartary buckwheat (Dorrell, 1971)

Fatty acids content differs among buckwheat samples which may be attributed to a number of factors such as the cultivars, growth location and seeding time (Tsuzuki *et al.*, 1991; Bonafaccia *et al.*, 2003). Germination may increase levels of certain fatty acids. The level of linoleic acid and linolenic acid increased from 36.8 percent to 51 percent and from 2.7 percent to 19 percent, respectively after germination of buckwheat seeds (Kim *et al.*, 2004).

Total tocopherol content in buckwheat is about 50 mg/kg. The presence of tocopherols in this grain is also related to a high antioxidant activity. There is a positive correlation between the tocopherol content and fatty acid content.

Minerals and Vitamins

Minerals are essential for various physiological functions in the human body. Buckwheat grain is rich in minerals including Mg, Se, Fe, K, P, Ca, Cu, Mn and Zn (Table 6). Studies revealed that buckwheat flour contains more Mg (11-13 times), Sr (5-36 times), Li (5-7 times), Fe (3-4 times) and K (2.9-3.6 times) than wheat flour. High Se content is also detected in tartary buckwheat flour (0.43 mg/ kg).

Table 6: Mineral composition of buckwheat groats

Cluster	Element	Mineral composition (mg/100g db)
Cluster I	K	565 ± 2.9
	P	490 ± 5.8
	Mg	267 ± 0.9
	Ca	19.7 ± 0.3
Cluster II	Fe	3.03 ± 0.11
	Zn	2.92 ± 0.01
	Mn	1.64 ± 0.02
	Cu	0.71 ± 0.03
	Mo	0.09 ± 0.02
	Co	0.01 ± 0.01
Cluster III	B	0.67 ± 0.02
	Al	0.36 ± 0.13
	Ni	0.24 ± 0.01

*The first cluster of minerals contains major minerals for humans, while the second cluster contains trace elements and the third remaining minerals. *Source*: Steadman *et al.* (2001).

Variations in mineral composition between cultivars and growth locations have been reported (Ikeda and Yamashita, 1994; Bonafaccia *et al.*, 2003). Buckwheat is generally a richer mineral source than many cereals such as rice, sorghum, millet and maize (except for Ca). Mg, Zn, K, P and Co are mainly stored as phytate in protein bodies and are therefore not accessible to humans (Elpidina *et al.*, 1991; Steadman, 2001). Phytin is hydrolysed and metal ions are released during germination which allows the accessibility of the metals for their physiological functions in the human diet (Bewley and Black, 1985). Mg, Zn, K, P and Co are generally present in embryo tissues and in the aleurone layer, minerals such as Fe, Zn, Mn, Cu, Mo, Ni and Al are primarily localized in both hull and seed coat, Ca and B are present in hull fractions (Steadman *et al.*, 2001a, b; Bonafaccia *et al.*, 2003; Skrabanja *et al .*, 2004).

Vitamins are a group of organic compounds which are essential for the normal functioning of the human body in very small amounts. Buckwheat grain contains vitamin B^1 (thiamine), B^2 (riboflavin), E (tocopherol) and PP (nicotinic acid/amide) in higher levels than most cereals. Generally, tartary buckwheat has more vitamin B^1, B^2 and PP, but less vitamin E than common buckwheat (Pomeranz and Lorenz, 1983; Bonafaccia *et al.*, 2003). The vitamin content of common buckwheat groats is depicted in Table 7 (Gabrovska *et al.*, 2002; Bonafaccia *et al.*, 2003).

Table 7: Vitamin content of buckwheat grains

Vitamins	Concentration	Reference
Thiamine (vitamin B_1)/seeds	2.2-3.3 µg/gDW	Bonafaccia *et al.*(2003) Wijngaard and Arendt (2006) Zielińska and Zieliński (2009)
Riboflavin(vitamin B_2)/seeds	10.6 µg/g DW	
Niacin (vitamin B_3)/seeds	18 µg/g	
Pantothenic acid (vitamin B_5)/seeds	11 µg/g	
Pyridoxine (vitamin B_6)/seeds	1.5 µg/g	
Vitamin C/seeds	50 µg/kg DW	
Vitamin C/sprouts	250 µg/kg DW	

After germination, vitamin C level in buckwheat sprouts can increase up to 25 mg/100g (Lintschinger *et al.*, 1997; Kim *et al.*, 2004). Buckwheat groats exhibited the maximum amount of tocopherols (5.46 mg/100g) when compared with wheat, barley, oats and rye.

Milling of Buckwheat

Primary processing of buckwheat includes cleaning (remove soil, stones, weed, chaff, dust, seeds of other crops, metallic particles, small and immature buckwheat seeds), dehulling and milling. Hammer mills, stone mills, pin mills, disk mills and roller mills have been used to mill buckwheat. Among these, stone mills and roller mills are most extensively used today.

The milling process may be of two types. In the first and commonly practised type, the whole seeds are first dehulled and then milled. In the second type, the seeds are milled and then screened to remove the hulls. When dehulling and milling are separate operations, the seeds are segregated according to size and may be steamed and dried prior to dehulling. Dehulling is carried out by impact or abrasion against emery stones or steel followed by air or screen separation of groats and hulls. The ease of dehulling and percentage recovery of undamaged groats depends on variety and moisture content. From the dehuller, the groats are separated over sieves of different mesh for sizing into whole groats and two or more sizes of broken groats. Flour is produced by passing the groats through stone and/or roller grinders.

When buckwheat seed is to be processed only into flour, the seeds are ground on break rolls or stone mills and then screened to separate the coarse flour from the hulls. The coarse flour is further reduced by a series of size reduction rolls, each grinding operation followed by sifting to fractionate the mixture of particles according to size. The flour yield ranges from 50 to 75 percent depending on the size, shape, condition of the seeds, efficiency of the dehulling and milling operations.

Processing Innovations of Buckwheat

New processing techniques like thermal (microwave heating, extrusion, autoclaving), high pressure, ionization and radiation induce minimum chemical changes in food product and cause minimum losses of flavor, nutritional and bioactive compounds and thus are utilised for buckwheat processing in food industries.

Thermal Processing

Several new thermal technologies are being availed in the food industry to enhance the quality of buckwheat functional food. Extrusion processing has become a significant process in the production of pasta, ready-to-eat cereals, snacks, pet food and textured vegetable protein. Microwave heating has gained popularity in food processing due to its ability to achieve high heating rates, substantial reduction in cooking time, safe handling and more uniform heating. This technique might change flavor and nutritional qualities of food to a less significant extent as opposed to conventional heating during the cooking process (Chandrasekaran *et al.*, 2013). Mostly research focused on determining the effect of heat treatment on total phenolic content, flavonoid content and antioxidant potential due to their role in human health and disease management.

Heat treatment of buckwheat was found to induce changes in chemical composition and functional properties of selected bioactive compounds. After thermal processing, common buckwheat groats, cakes and bread exhibited detrimental effect on flavonoid which might reduce the antioxidant capacity

(Kreft *et al.*, 2006; Vogrinćić *et al.*, 2010). Sensoy *et al.* (2006) reported that antioxidant activity of buckwheat flour was decreased by roasting but not after extrusion treatment. Contrarily, Randhir *et al.* (2008) reported an increase in total antioxidant activity in buckwheat sprouts and seeds after autoclaving. Also, total phenolic compounds observed 20 percent increase in sprouts and 7 percent decrease in seeds. Contradictorily, Zieliński *et al.* (2006) showed a reduction in the antioxidant capacity in extruded groats accompanied by a decrease in rutin and isovitexin, but at the same time an increase in free phenolic acids and those released from ester bonds.

The potential beneficial effects of buckwheat phytochemicals may be related to the inherent antioxidant capacity of these compounds. Kreft *et al.* (2006) described consequential association between rutin content and the antioxidant activity of buckwheat seeds and buckwheat-based food products. Zhang *et al.* (2010) reported that roasting, microwave heating and particularly pressurised steaming leads to decreased antioxidant capacity of tartary buckwheat flour accompanied by a reduction of total flavonoids and total phenolic content. These results showed that phenolics are the main antioxidant compounds in tartary and common buckwheat. Zieliński *et al.* (2009) reported that roasting buckwheat groats showed a decrease in the antioxidant capacity along with a decrease in the protein quality and content in groats. During thermal treatment, maillard compounds are generated due to a chemical reaction between free amino groups of lysine and carbonyl groups of reducing sugars which can contribute to an increase in the antioxidant activity masking the real decrease in the total phenolic compounds (Delgado-Andrade, 2014; Zicliński *et al.*, 2009; Zhang *et al.*, 2010). The antioxidant capacity can be increased as a result of the phenol disassociation and release of bound phenols due to thermal processing followed by some polymerization/oxidation of the phenolic constituents or the byproducts generated (Randhir *et al.*, 2008). Furthermore, more studies are necessary to validate the effect of thermal processing (time and temperature) on functional compounds, including protein and phenolic compounds in order to obtain buckwheat products with high quality and biologically active compounds.

High Pressure Processing

High pressure (HP) has been shown to offer a viable alternative to thermal processing without any off-flavor formation, vitamin and phytochemical properties loss and discoloration (Norton and Sun, 2008). Błaszczak *et al.* (2013) observed that the antioxidant capacity and rutin content differed in case of raw and roasted buckwheat groats after HP treatment. The pressurization of roasted groats increased the total reducing capacity due to hydrophilic antioxidants by 18 percent and decreased rutin concentration after treatment. Further, detailed studies investigating the effect of pressurization on bioactive compounds are required to determine which conditions are the most adequate to produce high-quality buckwheat products.

Ionization and Radiation

Food irradiation is a method of treating food to control surface microorganisms and to extend its shelf life without affecting the nutritional quality. Hayashi *et al.* (1998) reduced the microbial load to a lower level, exposing buckwheat grains to low-energy electrons without affecting their quality. Recently, there have been published studies where aqueous chlorine dioxide, fumaric acid, modified atmosphere packaging enriched in CO_2 and ultraviolet (UV)-C radiation were combined to treat buckwheat sprouts to reduce total aerobic bacteria, yeast and mould, and enterobacteria to low levels without affecting the sensory acceptability. An increase in the concentration of rutin was shown after treatment (Chun and Song, 2013).

A comparative study carried out by Orsak *et al.* (2001) revealed the effect of different UV doses, microwave and γ-irradiation on polyphenol and rutin content in three buckwheat samples. Hossen (2007) described that rutin and flavones C-glucosides are enhanced in sprouts after exposition to light emitting diodes. Therefore, irradiation could offer a method to increase the half-life of food, maintaining the sensory quality, reducing the microbial load and enhancing the nutritional value due to the increase in the bioactive compounds in buckwheat-derived products.

Bioactive Compounds in Buckwheat

The buckwheat grain shows prolonged shelf life without any chemical changes due to the presence of several natural antioxidants including tocopherols, phenolic acids and flavonoids (Biacs *et al.*, 2002). Whole buckwheat contains 2-5 times more phenolic compounds than oats or barley, whereas buckwheat bran and hulls have 2-7 times higher antioxidant activity than barley, triticale and oats (Holasova *et al.*, 2002; Zdunczyk *et al.*, 2006). Tartary buckwheat hull and bran with high amounts of phenolic compounds are considered to be excellent materials for cereal-based food processing with significant health benefits (Guo *et al.*, 2012).

The primary antioxidants present in buckwheat are rutin, quercetin, hyperin and catechins (Morishita *et al.*, 2007). Generally, flavonoids content of F. *tartaricum* (about 40 mg/g) is higher than that of F. *esculentum* (4-13 mg/g) (Oomah and Mazza, 1996; Li and Zhang, 2001, Chao *et al.*, 2002). Rutin, quercetin, orientin, vitexin, isovitexin and isoorientin were identified in buckwheat hulls (Biacs *et al.*, 2002). Kitabayashi *et al.* (1995) reported that rutin content in buckwheat grains ranges from 12-36 mg/100g dry basis. Grapes and buckwheat are the most important rutin-containing foods (Kreft *et al.*, 2006; Zhang *et al.*, 2012). Rutin has been shown to be able to antagonize the increase of capillary fragility associated with haemorrhagic disease of hypertension (Biacs *et al.*, 2002). Rutin also helps in lowering blood pressure levels, reducing lipid levels and enhancing antioxidant activity.

Biacs *et al.* (2002) reported therapeutic uses of buckwheat. Daily doses of rutin of between 180-350 mg have been reported to have clinically demonstrated positive effects. A daily intake of 100 g buckwheat flour or bran in food would cover 10 percent of the therapeutic dose of rutin. Watanabe (1998) detected four catechins with antioxidant activity in buckwheat groats. These catechins were epicatechin, catechin 7-O-β-D-glucopyranoside, epicatechin 3-O-p- hydroxybenzoate and epicatechin 3-O-(3,4-di-O-methyl) gallate. These catechins showed higher antioxidant activity than that in rutin.

Intragastric administration of buckwheat concentrate containing D-chiro-inositol (DCI), myo-inositol and fagopyritols lowered serum glucose concentrations in diabetic rats (Kawa *et al.*, 2003). D-chiro-inositol is an inositol isomer that is probably the main mediator of insulin metabolism. It enhances insulin activity and decrease blood pressure, plasma triglycerides and glucose concentration (Fonteles *et al.*, 2000; Ueda *et al.*, 2005; Zhang *et al.*, 2012). D-chiro-inositol has been found in buckwheat ranging from 20.7 to 41.7 mg 100 of dry wt. mostly in the form of fagopyritols (Steadman *et al* ., 2000). Fagopyritol A1 and Fagopyritol B1 are the most remarkable of all the fagopyritols. Fagopyritol A1 may be used in the treatment of diabetes and polycystic ovarian syndrome (PCOS) (Horbowicz *et al.*, 1998; Janet *et al.*, 2005). Buckwheat concentrate has been suggested as a natural product in treating diabetes (Arendt and Zannini, 2013).

Qiu *et al.* (2014) isolated and identified cytoprotective agents from nonpolar extracts of buckwheat flour. Three pure compounds were isolated, identified and evaluated for bioactivity named ferulic acid ethyl ester, furaneol, protocatechuic acid. Binary mixtures of these three isolated components acted borderline additively / antagonistically in the quinone reduction bioassay. The compounds identified in this study can be added to a growing list of bioactive agents in buckwheat, headed up by quercetin, rutin and additional quercetin glycosides.

Choi *et al.* (2013) investigated the protective effects of common and tartary buckwheat on amyloid-β-induced impairment of cognition and memory function in vivo in order to identify potential therapeutic agents against

Alzheimer's disease (AD) and its associated progressive memory deficits, cognitive impairment and personality changes. Results of behaviour tests in AD models showed that oral administration of methanol extracts of buckwheat improved cognition and memory function, inhibit lipid peroxidation and decreased nitric oxide levels in tissues.

When buckwheat is processed, flavonoid levels and antioxidant activity can be affected. Heat treatment at 150°C for 10 min significantly reduced the flavonoid concentration by about 20 percent (Arendt and Zannini, 2013). Kreft *et al.* (2006) discovered that there was much less rutin in noodles (78 mg/kg, dm) than in the dark buckwheat flour (218 mg/kg, dm) from which they were produced. In raw (uncooked) groats, there was 230 mg/ kg (dm) of rutin and in precooked groats, 88 mg/kg (dm). In buckwheat beer and vinegar, there was a negligible content of rutin. Buckwheat leaf flour contained about 2700 mg/kg (dm) rutin, and could thus be a suitable material for enriching different kinds of food products.

Buckwheat protein shows high biological value due to a well-balanced amino acid pattern and is rich in lysine and arginine (Krkoskova and Mrazova, 2004). In humans, consumption of buckwheat is associated with a lower prevalence of hyperglycemia and improved glucose tolerance in people with diabetes (Zhang *et al.*, 2007). Buckwheat protein extracts may help to almeirate some chronic diseases, such as diabetes, hypertension, hypercholesterolemia and many other cardiovascular diseases (Li and Zhang, 2001). The lysine/arginine and methionine/glycine ratios in buckwheat proteins are lower than those in most other plant proteins. The lower the lysine/arginine and methionine / glycine ratios are better in cholesterol-lowering effect. The cholesterol-lowering effect of buckwheat proteins is also attributed to the low digestibility of buckwheat proteins and other dietary fibre-like components in buckwheat (Kayashita et al., 1997).

Buckwheat protein extract has hypocholesterolemic, anticonstipation, antidiabetic and antiobesity properties (Li and Zhang, 2001; Zhang *et al.*, 2012). Buckwheat protein has a protective effect against induced colon carcinogenesis in rats by reducing cell proliferation (Tomotake *et al.*, 2006).

Phytosterols found in buckwheat seeds at low levels show positive effects in lowering the blood cholesterol level. Phytosterols show pharmaceutical effects for many chronic diseases. Plant sterols were reported to have antiviral effects improving the immunological status of the tested subjects (Krkoskova and Mrazova, 2004; Li and Zhang, 2001). Buckwheat contains several kinds of phytosterols, for example, β-sitosterol, campesterol, stigmasterol, and isofucosterol. β-Sitosterol is the major sterol component, accounting for about

70 percent of total sterol content, occuring in the endosperm and embryo of buckwheat grains. It cannot be absorbed in human body and displays a considerable competitive inhibitory effect on cholesterol absorption in vivo (Bryngelsson *et al.*, 1999, Normén, 2001; Krkošková and Mrázová, 2005).

Fagopyrins are naturally occurring substances in the buckwheat plant. Its chemical structure contains a naphthodianthrone skeleton similar that of to hypericin. Fagopyrins in buckwheat grain can be utilised in the treatment of type II diabetes (Li and Zhang, 2001; Bonafaccia *et al.*, 2003, Horbowicz and Obendorf, 2005).

C-Glycosylflavons present in buckwheat seedling cotyledons are vitexin, isovitexin, orientin and isoorientin. Condensed catechins, phenolic acids including hydrobenzoic acids, synigric, p-hydroxy-benzoic, vanillic and p-coumaric acids are present in the bran-aleurone layer of buckwheat grains (Przybylski *et al.*, 1998). Buckwheat grain contains soluble oligomeric condensed catechins which along with phenolic acids provide astringency and affect the colour and biological activity of buckwheat products.

Other Phenolic Compounds (free and bound phenolic acids, phenylpropanoid glycosides, anthocyanins and procyanidin) have been identified in buckwheat, particularly in tartary buckwheat. Some phenolic compounds have been recently isolated and identified for the first time in tartary buckwheat, for example, 1, 3, 6, 60-tetra-feruloyl sucrose, 3, 6-di-p-coumaroyl-1, 60-di-feruloyl sucrose, 1, 6, 60-tri-feruloyl-3-p-coumaroyl sucrose, N-trans-feruloyl tyramine, quercetin-3-O-(β-D-xyloxyl-(1-2)-a-L-rhamnoside) and proanthocyanidins (procyanidin A).

Antinutritional Factors

Saponins

Saponins form complexes with proteins, lipids, zinc, iron and possess a haemolytic effect. They are absorbed in small amounts and their main effect is restricted to the intestinal tract (Schoenlechner *et al.*, 2008). They can increase membrane permeability, thus enabling their use for increasing food intake at the intestinal level or even for drug assimilation (Vega-Gálvez *et al.*, 2010). Saponins also have health-promoting effects as anticarcinogenic, antibiotic/fungistatic, hypocholesterolemic, immune modulating and anti-inflammatory effects (Koziol, 1992, Schoenlechner *et al.*, 2008).

Phytic Acid

Buckwheat seed generally contain higher amounts of phytic acid than legumes and cereals grains and is mainly concentrated in the bran (Steadman *et al.*,

2001). Whole grains contain significant amounts of phytic acid (myo-inositol (1,2,3,4,5,6)-hexakisphosphate, Ins P6) or its salts (phytates). Phytic acid intake has been reported to have favourable effects, such as antioxidant function, prevention of heart diseases and anticarcinogen effect, which it performs through its hydrolysis products (Haros *et al.*, 2001, Kumar *et al.*, 2010). However, it acts as an inhibitor of mineral (such as Ca, Mg, Zn, Cu and Fe), protein and trace element bioavailability due to formation of insoluble complexes which are not absorbed by human gastrointestinal tract (Hurell *et al.*, 2003). However, phytic acid can be enzymatically degraded by endogenous and exogenous phytases (myo-inositol hexakisphosphate phosphohydrolase), improving the nutritional value of the cereal/pseudocereal products (García-Mantrana *et al.*, 2014; Iglesias-Puig *et al.*, 2015).

During pasta or bread processing, optimal conditions for the degradation of phytate are rarely reached (Hager *et al.*, 2012; Sanz-Penella *et al.*, 2013, García-Mantrana *et al.*, 2014; Iglesias-Puig *et al.*, 2015). However, sourdough fermentation provides optimum pH condition for phytase activity as alternative for phytic acid hydrolysis in bread (García-Mantrana *et al.*, 2015).

Inhibitors

Protease inhibitors are proteins that form stable complexes with proteolytic enzymes. The poor digestibility of buckwheat protein is due to different susceptibility of proteolytic action of buckwheat fractions and antinutritional components such as tannins and inhibitors (Wijngaard and Arendt, 2006). Several inhibitors were identified in buckwheat seeds such as protease inhibitors and α-amylase inhibitor (Wijngaard and Arendt, 2006).

Industrial Application and General Food Uses

Buckwheat has become popular owing to the gluten-free foods market, improved nutritional and technological quality of products (Wronkowska *et al* ., 2013) and can be processed very similarly to wheat.

Hull is used in the production of therapeutic mattresses / pillows and cushions, which adapt to the position of the body, quickly absorb moisture, do not heat up and are always cool (Pomeranz, 1983). Hull inhibits the development of harmful micro-organisms; mites, mould, bacteria and fungi due to the presence of tannins. These byproducts of buckwheat processing are characterized by a high carbon and hydrogen content and are used as a raw material for the production of granular biofuels (Borkowska and Robaszewska, 2012). However, buckwheat hull can also be used in the preparation of tea.

The buckwheat groats can be ground into several fractions with varying levels of the aleurone layer remaining (Zondag, 2003). Coarsely ground groats are called grits and can be used for porridges or in breads. Roasted groats (kasha) are used in Eastern European ethnic dishes (Vinning, 2001; Zondag, 2003). Buckwheat flour made from the aleurone layer of the groats is called Farinetta and can be used in breads, bakery products and pancakes (Zondag, 2003). Flour made from the entire buckwheat groats is called Supreme flour which can be used in preparation of bakery products, breads, extruded snacks, pancakes and pasta (Zondag, 2003). Buckwheat flour can thus be utilised for development of various food stuffs like pancakes, biscuits, bread, muffins, pastas and noodles, puffed buckwheat, flakes, drinks, tea and honey (Handoyo *et al.*, 2006). Nowadays, buckwheat in various proportions is also used in composite flour with wheat, rice or maize flour for the development of bakery products (Dutta, 2004; Petr *et al.*, 2004). Various buckwheat based bakery and non- bakery products are described in Fig 2 and 3.

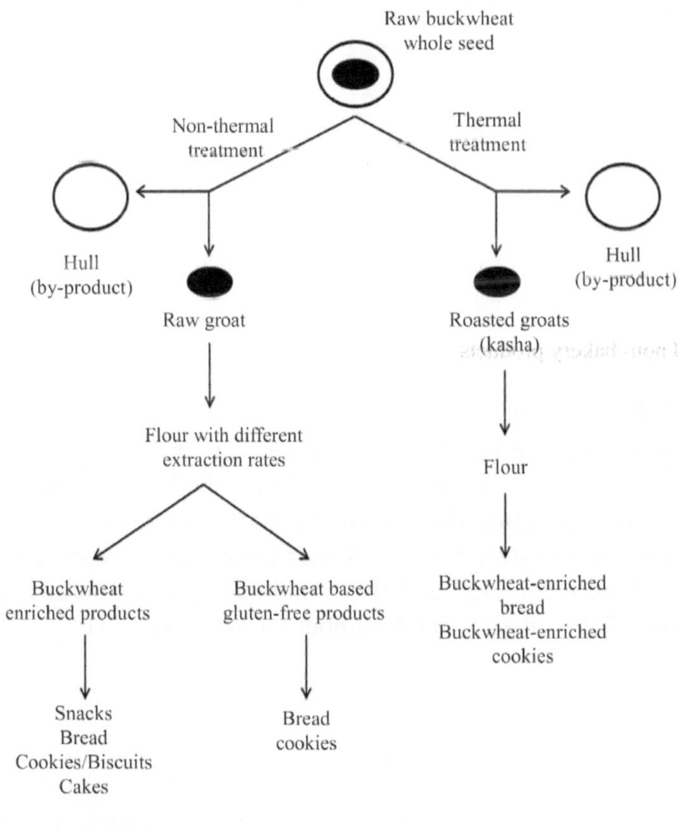

Fig. 2. Buckwheat based bakery products

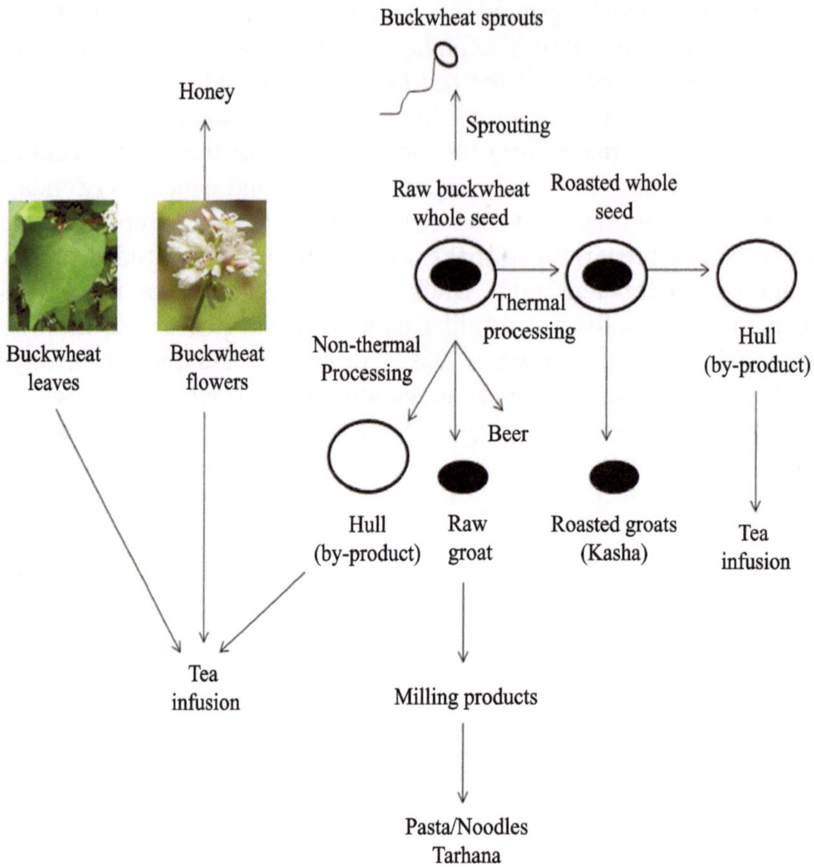

Fig 3. **Buckwheat based non- bakery products**

Most Popular Traditional Food

Buckwheat was domesticated and cultivated in southeast Asia around 6000 years ago and spread to Central Asia and Tibet and then to the Middle East and Europe. In Japan, Korea, and China, 100 percent buckwheat or a mixture with wheat flour is used for making soba noodles. Today buckwheat is commonly used as an important functional food and the traditional Asiatic buckwheat noodles is very popular and well known throughout the world (Fig 4).

Fig. 4. Asiatic buckwheat noodles

In Central and Eastern Europe, especially Poland, Russia and Ukraine, kasha is one of the oldest known dishes, made with roasted whole-grain buckwheat or buckwheat groats boiled in water or milk as a type of porridge (Marshall and Pomeranz, 1982). In Italy there are pizzoccheri, short, flat-ribbon noodles made with buckwheat and wheat flour and France is known for its traditional buckwheat pancakes (crêpes). In India, buckwheat pancakes called kuttu ki puri and kuttu pakoras (potato slices dipped in buckwheat flour and deep-fried in oil) are also famous.

Bakery Products

Bread

Buckwheat can be incorporated into bread to provide bread with increased sugar and total free amino acids content, a stronger umami taste and characteristic aroma than white bread (Lin *et al.*, 2009).

The polyphenols content of wheat bread was increased by incorporation of 10 percent buckwheat sourdough leading to higher specific volume and softer crumb (Moroni *et al.*, 2012). The nutritional value of the bread supplemented with buckwheat seeds was enhanced (even at 40% supplementation level) with regard to the amounts of minerals, proteins, fats and fibre and had excellent sensory characteristics (Demin *et al.*, 2013).

Whole buckwheat flour was used in leavened flat bread (bazlama) and unleavened flat bread (yufka) formulations to improve the nutritional status of flat breads (Yildiz and Bilgiçli, 2015). Taste and odour scores decreased above the 20 percent buckwheat level for bazlama and 30 percent level for yufka compared with control samples. Buckwheat is also used to prepare chillare (sweet puddings), an unleavened bread fried with butter or ghee and mixed with potatoes to prepare stuffed parathas (Dutta, 2004). Chapattis can be made with buckwheat flour mixed with wheat or barley flour (Dutta, 2004).

Buckwheat can be utilized for the development of lower glycemic index foods (Skrabanja *et al.*, 2001). The resistant starch level in bread products based on different proportions of buckwheat flour (30–70%) varied from 0.9 percent to 4.4 percent and the rate of in vitro amylolysis was significantly lower ($p < 0.05$) in comparison with the reference white wheat bread (Skrabanja *et al.*, 2001).

Angioloni and Collar (2011) reported that the quality profile of associated mixtures of oat, rye, buckwheat and common wheat flours (20:20:20:40 w/w/w/w) was suitable for making highly nutritious baked goods meeting functional and sensory standards with lower digestible starch, improved dietary fibre fractions, minerals and antioxidant activity.

Mariotti *et al.* (2013) studied the effect of 40 percent substitution with dehulled flour and puffed buckwheat flour in development of gluten-less bread. The inclusion of 40 percent dehulled buckwheat flour improved the baking performances of the commercial gluten-free flour mixtures and presence of small amount of puffed flour limit both the diffusion and the loss of water from the bread crumb and the interactions between starch and protein macromolecules resulting in a softer gluten free bread crumb and reduced staling kinetics during storage.

A blend of buckwheat, amaranth, chickpea and quinoa flours (weight ratio 1:1:5.3:1) was selected and fermented with baker's yeast or with *Lactobacillus plantarum* sourdough and compared with wheat flour bread started with baker's yeast (Coda *et al.,* 2010). Phenolic compounds and antioxidant activity was highest in the *Lactobacillus plantarum* sourdough bread, and the rate of in vitro starch hydrolysis was the lowest. In the study of Wolter *et al.* (2014) sourdoughs of buckwheat, quinoa and other gluten-free cereals were produced using *Lactobacillus plantarum* and added to a basic bread formulation of flour from the same grain type. Sourdough fermentation changed the protein profiles in all sourdoughs and its addition led to decreased dough strength, increased crumb porosity, reduced staling rate. Wronkowska *et al.* (2015a,b) demonstrated that the solid-state fermentation with *Rhizopus oligosporus* improved protein digestibility, increased protein content, minerals, water-soluble vitamins as well as tocopherols and acceptable sensory score making it a potential method to develop a new buckwheat product.

Long fermentation of buckwheat flour carried out with naturally occurring lactobacilli and yeasts to obtain buckwheat sourdough gained attention for wheat bread production (Moroni *et al.*, 2012). The application of the strains *Lactobacillus fermentum* AB 15, *L. plantarum* AB 16, *L. vaginalis* AB 17 and *L. crispatus* AB 19 improved the quality of the product. Acidification, aroma formation and leavening were the main functions of lactic acid bacteria and yeasts. The fermentation during buckwheat sourdough preparation decreased the content of phytic acid of buckwheat flour, without hindering the nutritional properties of buckwheat. The incorporation of buckwheat sourdough in wheat bread also improved the crumb structure and volume of the bread, prolonged shelf life and increased polyphenols content (Moroni *et al.*, 2012).

Biscuits, Cookies and Cakes

Schoenlechner *et al.* (2006) blended buckwheat with 25 percent, 50 percent and 75 percent common bean flour to develop biscuits (short dough biscuits). Biscuits of 100 percent buckwheat flour and 100 percent bean flour biscuits were developed for comparison. Results revealed that addition of bean flour increased the crispness and nutritional composition (dietary fiber and proteins) of biscuits. As an alternative to white bean, addition of chickpea (at 40:60 chickpea: buckwheat flour) could increase the acceptability of the amaranth and buckwheat based gluten-free cookies by the consumer test panel as determined for spread factor and hardness (Yamsaengsung *et al* . 2012).

Gambus *et al.* (2009) supplemented gluten-free cakes (sponge cake, carrot cake and coconut cake) and biscuits with amaranth and/or buckwheat flour which showed increased nutritional value (protein, dietary fiber, mineral) and higher consumer scores.

Wheat flour in ginger-nut biscuits (popular traditional biscuits with honey) was replaced with buckwheat flour which resulted in increased spread factor, hardness, fracturability and sensory acceptance of biscuits (Filipčev *et al.*, 2011). The opposite behaviour was found by Baljeet *et al.* (2010)and Chopra *et al.*, 2014, who reported that as the concentration of buckwheat flour increased, the spread ratio, fracture strength, sensory scores (for texture, appearance and flavour) of cookies decreased. Spread ratio and hardness values were also significantly affected by flour combinations (buckwheat flour with rice and corn flour at different levels) in gluten-free cookie formulations (Altõnda *et al.*, 2015). However, addition of transglutaminase resulted in increased spread ratio and fracturability but decreased hardness values (Altõnda *et al.*, 2015).

Substitution with buckwheat flour resulted in a significant increase in protein, Zn, Fe, Mn, Cu, total polyphenols and antioxidative and chelating activity (Filipčev *et al.*, 2011; Vranac *et al.*, 2013; Altõnda *et al.*, 2015; Bhavsar *et al* . 2015; Jan *et al.*, 2015). Most of the rutin present in buckwheat flour can be recovered in biscuits as very little amount of rutin is degraded to quercitin during dough preparation and baking (Brunori *et al.*, 2009).

Gluten free crackers made from buckwheat flours (refined and wholegrain) were significantly higher in total phenolic content and antioxidant activity than wheat crackers, with high concentrations of tocopherols without affecting sensory acceptability (Sedej *et al.*, 2011). According to Pandey *et al.* (2015), biscuits developed from husked buckwheat flour were found to be more acceptable than those made with dehusked flour. In general, the addition of low amount of pseudocereal to wheat flour enhances nutritional and functional value of products without impairing rheological properties of dough or overall quality of final products.

Others

There are many homemade and commercial bakery products that use buckwheat such as pastries, pretzels, muffins, croissants, pies, fritters, pancakes, waffles and crêpes.

Muffins are prepared with various amounts of buckwheat powder (0-50%) and the specific gravity is not affected by the addition of buckwheat powder up to 30 percent. No significant difference was observed in muffins containing buckwheat with regard to baking loss rate, specific volume, cohesiveness, springiness and gumminess, whereas hardness decreased. The sensory score for colour decreased with increasing buckwheat amounts whereas differences in grain, flavour, taste, texture and overall acceptances were insignificant between formulations (Bae and Jung, 2013).

Leng and Shi (2006) also developed pancakes using buckwheat and various grains (millet, soybean, wheat and mung bean) with several flavouring materials such as fennel fruit, star anise, xanthoxylum and coriander seed, with good taste and high nutritive value. Kwon (2003) developed a buckwheat pancake method that involves soaking buckwheat in water, maturing, pulverizing with water, extracting buckwheat syrup, filtering the extract, and mixing with kimchi (traditional Korean food made from vegetables) and Welsh onion in boiling water. The final composition of the buckwheat pancake comprises 80 percent of buckwheat suspension, 17 percent of kimchi, and 3 percent of Welsh onion.

Snacks and Breakfast Cereals

The use of buckwheat or amaranth flours in the preparation of extruded breakfast cereals altered some of the nutritional quality and physical characteristics such as product density and bulk density, without reducing the degree of expansion. In addition, the products resulted in a reduction in readily digestible carbohydrates in extruded breakfast cereals as compared to control samples (Brennan *et al.*, 2012). The use of buckwheat flour in extruded snacks and ready-to-eat breakfast cereal products (with wheat flour or corn meal and with non-fat milk) offers a desirable variation in flavour and can take advantage of the nutritional quality of buckwheat. Wójtowicz *et al.* (2013) showed a good acceptability of buckwheat enriched snacks, at less than 30 percent level proposing corn-buckwheat snack as an attractive type of appetizer with increased nutritional properties.

Beverages/ Drinks

Buckwheat has been used as a substitute for other grains in the manufacture of gluten-free beer. According to Dezelak *et al.* (2014), buckwheat beer appears quite similar to barley beer in terms of fermentable carbohydrate content and volatile compound content. Sensory analysis indicated that buckwheat beer was acceptable with regard to odour, purity of taste, mouthfeel, tingling and bitterness (Nic Phiarais *et al.*, 2010).

Buckwheat malt can also be used for preparing whiskey after distilling and aging in oak casks. There are other drinks made with buckwheat such as country liquor, which is prepared from buckwheat grains by tribal people residing in remote high-altitude areas in India (Dutta, 2004) or shōchū, which is a distilled beverage made from buckwheat (or barley, potatoes or rice) in Japan (stronger than wine or sake but weaker than vodka or whiskey).

Optimized malting and mashing conditions for 100 percent buckwheat malt have been described in a number of studies (Hager *et al.*, 2014), which demonstrated that production of wort from 100 percent buckwheat malt is feasible by using commercial enzymes. These authors showed that the utilization of commercial cellulase, amyloglycosidase and α-amylase can sufficiently increase extract levels, fermentability, total fermentable extract, total soluble nitrogen and free amino nitrogen.

De Meo *et al.* (2011) performed micro-malting experiments on buckwheat, quinoa and amaranth using different malting parameters to study their brewing behaviour. Alkaline steeping has been applied, which showed an increase in total soluble nitrogen and free amino nitrogen. From this study, the authors conclude that pseudocereals can successfully be employed for gluten-free beer production and the alkaline steeping seems to be a useful process, which is variable for the optimization of malt quality.

Mixes of buckwheat flour and oat bran were investigated as prebiotics in the production of probiotic fibre-enriched fermented milks (Coman *et al.*, 2013). Lee and Park (2013) varied the concentrations of buckwheat saccharified solution added to milk, followed by fermentation with commercially available mixed strains of lactic acid bacteria. They observed that undesirable compounds, such as acetic acid and 2-butanone decreased as the buckwheat solution concentration increased, thereby improving the flavour quality of plain yogurt (Lee and Park, 2013). Tartary buckwheat sprout was also mixed with mung beans, black rice and skimmed milk with sweeteners to develop a yogurt with health-promoting properties (Wang *et al.*, 2013).

Pasta Products

Schoenlechner *et al*. (2010) investigated the possibility of producing amaranth, quinoa and buckwheat pasta with good textural quality and concluded that combining amaranth, quinoa and buckwheat (20:20:60) is an advantageous formulation in which negative effects on texture firmness and cooking time were minimized and product was acceptable on sensory basis.

Alamprese *et al.* (2007) developed fresh gluten-free egg pasta based on buckwheat and rice flours (raw, precooked or pregelatinized). Samples containing the precooked rice flour gave the best results in terms of workability, break strain and weight increase during cooking. Verardo *et al.* (2011) produced gluten-free spaghetti form buckwheat. Common buckwheat is often used as an important functional food and is ground to make buckwheat noodles (Lin *et al.*, 2009a).

Buckwheat Honey

Buckwheat honey is a reddish brown product with a strong animal aroma, low viscosity (Juszczak and Fortuna, 2006), higher antioxidant capacity as well as flavonoid and total phenolic content than other honeys (Nagai *et al.*, 2006; Vit *et al* ., 2010; Wilczyńska, 2010; Socha *et al.*, 2011). Buckwheat honey has been used as a protective agent against lipid oxidation in ground turkey (McKibben and Engeseth, 2002). It has been also used as an inhibitor of heterocyclic aromatic amine formation in fried ground patties, fried beef steak and chicken breast (Shin *et al.*, 2003; Shin and Ustunol, 2004) and also as a potent inhibitor of the growth of foodborne pathogens and food spoilage organisms (Mundo *et al.*, 2004; Nagai *et al.*, 2006).

Commercial honey processing includes controlled heating to destroy yeast and to dissolve dextrose crystals, combined with fine straining or pressure filtration. Wang et al. (2004) indicated that processing dramatically reduced antioxidant capacity of buckwheat honey by 33.4 percent as compared to raw one and also during storage for a long period (6 months or more) which may be due to the reduction in phenolics of buckwheat honey (vanillic acid or p-hydroxibenzoic acid) by 37 percent. Furthermore, the use of filters, such as celite, during honey filtration may have nonspecifically bound phenolics which can perhaps explain the reduction of the antioxidant capacity. Other compounds generated during thermal treatment of buckwheat honey such as hydroxymethylfurfural

(HMF), melanoidins and other maillard reaction products may influence the antioxidant activity (Brudzynski and Miotto, 2011).

Buckwheat honey has proven to be an effective antioxidant which may be used as a source of natural antioxidants. The optimization of different steps during the processing is essential to minimize the loss of antioxidant capacity.

Buckwheat Tea

Buckwheat tea is a popular health product in Asia and Europe made from flowers, leaves and/or hulls (Qin *et al.*, 2013; Zielińska *et al.*, 2013). It is produced by soaking, steaming, dehulling and baking buckwheat seeds. The final buckwheat tea product is in granule form with an attractive yellowish colour and pleasant baked flavour. The content of rutin and the inhibitory activity against formation of advanced glycation end products (AGEs) was higher in tartary buckwheat tea than in common buckwheat tea (Zielińska *et al.*, 2013). Buckwheat tea shows lower antioxidant capacity and lower content of total phenolic compounds than the green tea. This product can be offered to the consumers as a new type of tea enriched in flavonoids, especially rutin and other compounds, such as quercetin and flavone C-glucosides. Therefore, the buckwheat groat or hull may be used for tea preparation as the main single tea ingredient or as a mixed component.

Buckwheat Tarhana

Tarhana is a traditional fermented cereal food prepared by mixing wheat products, yoghurt, some vegetables and spices. The addition of buckwheat improved the nutritional contents of this soup in terms of fat, ash, protein, mineral (K, Mg and P) and lysine content. Higher buckwheat flour levels negatively affected fermentation loss, color values, water and oil absorption capacity of tarhana. It was observed that 40 percent was the optimal concentration to yield a product with adequate physical, functional and sensory properties (Bilgiçli, 2009 a,b).

Buckwheat Sprouts

Sprouting enhances nutritional content, antioxidative capacity and phytochemical content (Kim *et al.*, 2004; Álvarez-Jubete *et al.*, 2010) which helps in the prevention of human diseases and/or in health maintenance (Brinskin, 2000; Espín *et al.*, 2007). Kim *et al.* (2004) used the mass production system to harvest buckwheat sprouts. After 7 days of seedling, rutin and quercetrin concentrations were notably increased and chlorogenic acid was moderately enhanced.

In 2010, Álvarez-Jubete *et al.* (2010) investigated polyphenol composition and antioxidant properties of methanolic extracts from buckwheat after sprouting. In this case, the germination time was 96 hours and each seeds was subjected to alternate wet and dry cycles of 3 hours for over 24 hours. Following sprouting the phenol content was two times higher (670.2 mg gallic acid equivalent/100 g) and the antioxidant capacity was increased. However, more studies are necessary to find out the optimal culture conditions to enhance the concentration of nutrients and to obtain healthier sprouts.

Conclusion

Buckwheat is a popular gluten free pseudocereal which can be exploited to make various food products with anti-diabetic, low insulinemic, hypocholesterolemic and therapeutic effects. Though, further deep research is needed on effect of processing on antioxidant and phytochemical content of buckwheat based products so as to develop a product with maximum beneficial health effects.

References

AACC. 2001. The definition of dietary fiber. Report of the Dietary Fiber Definition Committee to the Board of Directors of the American Association of Cereal Chemists. Cereal Foods World 46(3), 112-126.

Alamprese, C., Casiraghi, E., & Pagani, M. A. (2007). Development of gluten-free fresh egg pasta analogues containing buckwheat. European Food Research and Technology, 225(2), 205-213.

Altonda, G., Certel, M., Erem, F., & Konak, U. I. (2015). Quality characteristics of gluten free cookies made of buckwheat, corn, and rice flour with/without transglutaminase. Food Science and Technology International, 21, 213-220.

Alvarez-Jubete, L., Arendt, E. K., & Gallagher, E. (2009). Nutritive value and chemical composition of pseudocereals as gluten-free ingredients. International Journal of Food Science and Nutrition, 4, 240-257.

Álvarez-Jubete, L., Wijngaard, H., Arendt, E.K., & Gallagher, E. (2010). Polyphenol composition and in vitro antioxidant activity of amaranth, quinoa buckwheat and wheat as affected by sprouting and baking. Food Chemistry, 119, 770-778.

Angioloni A, & Collar, C. (2011). Nutritional and functional added value of oat, Kamut R, spelt, rye and buckwheat versus common wheat in breadmaking. Journal of the Science of Food and Agriculture, 91, 1283-1292.

Angioloni, A., & Collar, C. (2011). Nutritional and functional added value of oat, Kamut®, spelt, rye and buckwheat versus common wheat in breadmaking. Journal of the Science of Food and Agriculture, 91(7), 1283-1292.

Arendt E, & Zannini, E. (2013). Cereal Grains for the Food and Beverage Industries, Woodhead, Cambridge, 369-466.

Aubrecht, E., & Biacs, P. Á. (2001). Characterization of buckwheat grain proteins and its products. Acta Alimentaria, 30, 71-80.

Aubrecht, E., & Biacs, P.A. (1999). Immunochemical analysis of buckwheat proteins, prolamins and their allergenic character. Acta Alimentaria, 28, 261-268.

Bae, J. H., & Jung, I. (2013). Quality characteristics of muffin added with buckwheat powder. The East Asian Society of Dietary Life, 23, 430-436.

Bae, J. H., & Jung, I. C. (2013). Quality characteristics of muffin added with buckwheat powder. Journal of the East Asian Society of Dietary Life, 23(4), 430-436.

Baljeet, S.Y., Ritika, B.Y., & Roshan, L.Y. (2010). Studies on functional properties and incorporation of buckwheat flour for biscuit making. International Food Research Journal, 17,1067-1076.

Bárta, J., Kalinová, J., Moudrý, J., & Čurn, V. (2004). Effects of environmental factors on protein content and composition in buckwheat flour. Cereal Research Communications, 32, 541-548.

Bewley, J. D., & Black, M. (1985). Seeds. In Seeds (pp. 1-27). Springer, Boston, MA.

Bewley, J. D., & Black, M. (1985). Seeds: Physiology of Development and Germination. New York: Plenum Press.

Bhavsar, G. J., Sawate, A. R., & Kshirsagar, R. B. (2015). Effect of replacement of refined wheat flour with buckwheat flour on quality characteristics of cookies. The Indian Journal of Nutrition and Dietetics, 52, 88-96

Biacs, P., Aubrecht, E., Leder, I., & Lajos, J. (2002). Buckwheat, in Pseudocereals and Less Common Cereals (eds P. Belton and J. Taylor). Springer-Verlag, Berlin, 123-152.

Bilgiçli, N. (2009a). Effect of buckwheat flour on chemical and functional properties of tarhana. LWT-Food Science and Technology, 42, 514-518.

Bilgiçli, N. (2009b). Enrichment of gluten-free tarhana with buckwheat flour. International Journal of Food Science and Nutrition, 60, 1-8.

Błaszczak, W., Zielińska, D., Zieliński, H., Szawara-Nowak, D., & Fornal, J. (2013). Antioxidant properties and rutin content of high pressure-treated raw and roasted buckwheat groats. Food Bioprocess Technology, 6, 92-100.

Bonafaccia, G., & Kreft, I. (1994). Technological and qualitative characteristics of food products made with buckwheat. Fagopyrum, 14, 35-42.

Bonafaccia, G., Marocchini, M., & Kreft, I. (2003). Composition and technological properties of the flour and bran from common and tartary buckwheat. Food Chemistry, 80(1), 9-15.

Borkowska, B., & Robaszewska, A. (2012). Use of buckwheat grain in various industry branches. Scientific notebooks of the Maritime Academy in Gdynia, 73, 43-55.

Brennan, C. S. (2005). Dietary fibre, glycaemic response and diabetes. Molecular Nutrition & Food Research, 49, 560–570.

Brennan, M. A., Menard, C., Roudaut, G., & Brennan, C. S. (2012). Amaranth, millet and buckwheat flours affect the physical properties of extruded breakfast cereals and modulates their potential glycaemic impact. Starch / Stärke, 64, 392-398.

Briskin, D. P. (2000). Medicinal plants and phytomedicines. Linking plant biochemistry and physiology to human health. Plant Physiology, 124, 507-514.

Brudzynski, K., & Miotto, D. (2011). Honey melanoidins: Analysis of the compositions of the high molecular weight melanoidins exhibiting radical-scavenging activity. Food Chemistry, 126, 1155-1163.

Brunori, A., Baviello, G., Zannettino, C., Corsini, G., & Sandor, G. (2009). The use of Fagopyrum tataricum gaertn. Whole flour to confer preventive contents of rutin to some traditional Tuscany biscuits. Annals of the University Dunarea de Jos of Galati, 34, 38-41.

Bryngelsson, S., Jonhsson, M., Normén, L., Dutta, P., Andersson, H. (1999). Plant sterols in cereal products, In: Sanberg, A.S., Andersson, H., Amado, R., Schlemmer, H., Serra, F. (eds.): Bioactive Inositol Phosphates and Phytosterols in Food. Cost 916. Office for Official Publications of the European Communities, Luxemburg: 131-138.

Campbell, C. G. (1997). Buckwheat, Fagopyrum esculentum Moench. Rome: IPGRI

Chandrasekaran, S., Ramanathan, S., & Basak, T. (2013). Microwave food processing-A review. Food Research International, 52, 234-261.

Chao, P. L., Hsiu, S., & Hou, Y. (2002). Flavonoids in herbs: biological fates and potential interactions with xenobiotics. Journal of Food and Drug Analysis, 10, 219-228.

Choi, J., Cho, E., Song Lee, H., Lee, J., Yoon, Y., & Lee, S. (2013). Tartary buckwheat improves cognition and memory function in an in vivo amyloid-b-induced Alzheimer model. Food and Chemical Toxicology, 53, 105–111.

Chopra, N., Dhillon, B., & Puri, S. (2014). Formulation of buckwheat cookies and their nutritional, physical, sensory and microbiological analysis. International Journal of Advanced Biotechnology and Research, 5, 381-555.

Christa, K., & Soral-Šmietana, M. (2008). Buckwheat grains and buckwheat products - nutritional and prophylactic value of their components - a review. Czech Journal of Food Science, 3, 153-162.

Chun, H. H., & Song, K. B. (2013). Optimisation of the combined treatments of aqueous chlorine dioxide, fumaric acid and ultraviolet C for improving the microbial quality and maintaining sensory quality of common buckwheat sprout. International Journal of Food Science and Technology, 49, 121-127.

Coda, R., Rizzello, C. G., & Gobbetti, M. (2010). Use of sourdough fermentation and pseudo- cereals and leguminous flours for the making of a functional bread enriched of γ- aminobutyric acid (GABA). International Journal of Food Microbiology, 137, 236-245.

Coman, M.M., Verdenelli, M.C., Cecchini, C., Silvi, S., Vasile, A., Bahrim, G. E., Orpianesi, C., & Cresci, A. (2013). Effect of buckwheat flour and oat bran on growth and cell viability of the probiotic strains Lactobacillus rhamnosus IMC 501 (R), Lactobacillus paracasei IMC 502 (R) and their combination SYNBIO (R), in synbiotic fermented milk. International Journal of Food Microbiology, 167, 261-268.

De Meo, B., Freeman, G., Marconi, O., Booer, C., Perretti, G., & Fantozzi, P. (2011). Behaviour of malted cereals and pseudocereals for gluten-free beer production. Journal of the Institute of Brewing and Distilling, 117(4), 541-546.

Degradation of rutin and polyphenols during the preparation of Tartary buckwheat bread. Journal of the Science of Food and Agriculture, 58, 4883-4887.

Delgado-Andrade, C. (2014). Maillard reaction products: some considerations on their health effects. Clinical Chemistry and Laboratory Medicine, 52, 53-60.

Demin, M. A., Radovic, B. V., Banjac, N. R., Tipsina, N. N., & Milovanovic, M. M. (2013). Buckwheat and Quinoa seeds as supplements in wheat bread production. Hemijska Industrija, 67, 115-121.

Dezelak, M., Zarnkow, M., Becker, T., & Kosir, I. J. (2014). Processing of bottom fermented gluten-free beer-like beverages based on buckwheat and quinoa malt with chemical and sensory characterization. Journal of the Institute of Brewing, 120, 360-370.

Dorrell, D. G. (1971). Fatty acid composition of buckwheat seed. Journal of the American Oil Chemists Society, 48(11), 693-696.

Dutta, M. (2004). Buckwheat improvement in India: current status and future prospects. Proceedings of the 9th International Symposium on Buckwheat, Prague, 302-312.

Dziedzic, K., Górecka, D., Kucharska, M., & Przybylska, B. (2012). Influence of technological process during buckwheat groats production on dietary fibre content and sorption of bile acids. Food Research International, 47(2), 279-283.

Eggum, B. O., Kreft, I., & Javornik, B. (1980). Chemical composition and protein quality of buckwheat (Fagopyrum esculentum Moench). Qualitas Plantarum - Plant Foods for Human Nutrition, 30, 175-179.

Elpidina, E. N., Yoskoboynikova, N. E., Belozersky, M. A., & Dunaevsky, Y. E. (1991). Localization of a metalloproteinase and its inhibitor in the protein bodies of buckwheat seeds. Planta, 185, 46-52.

Espín, J. C., García-Conesa, M. T., & Tomás-Barberán, F. A. (2007). Nutraceuticals: Facts and fiction. Phytochemistry, 68, 2986-3008.

FAO/UN. (1990). Protein quality evaluation: report of a Joint FAO/WHO Expert Consultation. Rome: FAO.

Filipčev, B., Šimurina, O., Sakac, M., Sedej, I., Jovanov, P., Pestoric, M., & Bodroza-Solarov, M. (2011). Feasibility of use of buckwheat flour as an ingredient in ginger nut biscuit formulation. Food Chemistry, 125, 164-170.

Fonteles, M., Almeida, M., & Larner, J. (2000). Antihyperglycemic effects of 3-methyl-D-chiro-inositol and D-chiro-inositol associated with manganese in streptozotocin diabetic rats. Hormone and Metabolic Research, 32, 129-132.

Fornal, Ł. (1999). Cougrantic seed chemistry and directions of food use. Scientific Bulletin. University of Warmia and Mazury in Olsztyn, 4, 7-17.

Gabro Vská, D., Fi Edlerova, V., Holasova, M., Maskova, E., Smrcinov, H., Rysova, J., Winterova, R., Michalova, A., & Hutar, M. (2002). The nutritional evaluation of underutilized cereals and buckwheat. Food and Nutrition Bulletin, 23, 246–249.

Gambus, H., Gambuś, F., Pastuszka, D., Wrona, p., Ziobro, R., Sabat, R., Mickowska, B., Nowotna, A., & Sikora, M. (2009). Quality of gluten-free supplemented cakes and biscuits. International Journal of Food Sciences and Nutrition, 60, 31-50.

García-Mantrana, I., Monedero, V., & Haros, M. (2014). Application of phytases isolated from bifidobacteria in the development of cereal-based products with amaranth. European Food Research and Technology, 238, 853–862.

García-Mantrana, I., Monedero, V., & Haros, M. (2015). Myo-inositol hexakisphosphate degradation by Bifidobacterium pseudocatenulatum ATCC 27919 improves mineral availability of high fibre rye-wheat sour bread. Food Chemistry, 178, 267–275.

Guo, X., & Yao, H. (2006). Fractionation and characterization of tartary buckwheat flour proteins. Journal of Food Chemistry, 98, 90-94.

Guo, X., Wu, C., Ma, Y., Parry, J., Xu, Y., Liu, H., & Wang, M. (2012). Comparison of milling fractions of tartary buckwheat for their phenolics and antioxidant properties. Food Research International, 49, 53-59.

Hager, A., Wolter, A., Jacob, F., Zannini, E., & Arendt, E. K. (2012). Nutritional properties and ultrastructure of commercial gluten free flours from different botanical sources compared to wheat flours. Journal of Cereal Science, 56, 239-247.

Handoyo, T., Maeda, T., Urisu, A., Adachi, T., & Morita, M. (2006). Hypoallergenic buckwheat flour preparation by Rhizopus oligosporus and its application to soba noodle. Food Research International, 39, 598-605.

Haros, M., Rosell, C. M., & Benedito, C. (2001). Fungal phytase as a potential breadmaking additive. European Food Research and Technology, 213, 317–322.

Hayashi, T., Takahashi, Y., & Todoriki, S. (1998). Sterilization of foods with low-energy electrons ("soft-electrons"). Radiation Physics and Chemistry, 52, 1-6.

Heffler, E., Pizzimenti, S., Badiu, I., Guida, G., & Rolla, G. (2014). Buckwheat allergy: an emerging clinical problem in Europe. Journal of Allergy and Therapy, 5(2), 168-171.

Hirao, T., Imai, S., Sawada, H., Shiomi, N., Hachimura, S., & Kato, H. (2005). PCR method for detecting trace amounts of buckwheat (*Fagopyrum* spp.) in food. Bioscience, Biotechnology and Biochemistry, 69, 724-731.

Holasova, M., Fiedlerova, V., Smrcinova, H., Orsak, M., Lachman, J., & Vavreinova, S. (2002). Buckwheat – the source of antioxidant activity in functional foods. Food Research International, 35, 207–211.

Horbowicz, M., & Obendorf, R. L. (2005). Fagopirytol accumulation and germination of buckwheat seeds matured at 15, 22, 30°C. Crop Science, 45, 1–11.

Horbowicz, M., Brenac, P., & Obendorf, R. L. (1998). Fagopyritol B1, O-α-Dgalactopyranosyl-(1-->2)-D-chiro-inositol, a galactosyl cyclitol in maturing buckwheat seeds associated with desiccation tolerance. Planta, 205, 1-11.

Hossen, Z. (2007). Light emitting diodes increase phenolics of buckwheat (Fagopyrum esculentum) sprouts. Journal of Plant Interactions, 2, 71-78.

Hurrell, R. F., Reddy, M. B., Juillerat, M. A., & Cook, J. D. (2003). Degradation of phytic acid in cereal porridges improves iron absorption by human subjects. American Journal of Clinical Nutrition, 77(5), 1213–1219.

Iglesias-Puig, E., Monedero, V., & Haros, M. (2015). Bread with whole quinoa flour and bifidobacterial phytases improve contribution to dietary mineral intake and their bio availability without substantial loss of bread quality. LWT – Food Science and Technology, 60, 71-77.

Ikeda, K., & Kusano, T. (1983). In vitro inhibition of digestive enzymes by indigestible polysaccharides. Cereal Chemistry, 60, 260–263.

Ikeda, K., Sakaguchi, T., Kusano, T., & Yasumoto, K. (1991). Endogenous factors affecting protein digestibility in buckwheat. Cereal Chemistry, 68, 424-427.

Ikeda, S., & Yamashita, Y. (1994). Buckwheat as a dietary source of zinc, copper and manganese. Fagopyrum, 14, 29–34.

Izydorczyk, M. S., McMillan, T., Bazin, S., Kletke, J., Dushnicky, L., & Dexter, J. (2014). Canadian buckwheat: a unique, useful and under-utilized crop. Canadian Journal of Plant Science, 94, 509–524.

Jaime, L., Mollá, E., Fernández, A., Martin-Cabrejas, M. A., Lopez-Andreu, F. J., & Esteban, R. M. (2002). Structural carbohydrate differences and potential source of dietary fiber of onion (Allium cepa L.) tissues. Journal of Agricultural and Food Chemistry, 50, 122-128.

Jan, U., Gani, A., Ahmad, M., Shah, U., Baba, W. N., Masoodi, F. A., Maqsood, S., Gani, A., Wani, I. A., & Wani, S. M. (2015). Characterization of cookies made from wheat flour blended with buckwheat flour and effect on antioxidant properties. Journal of Food Science and Technology, 52(10), 6334-6344.

Janet, M., Horbowicz, M., & Obendorf, R. L. (2005). Cyclitol galactosides in embryos of buckwheat stem-leaf-seed explants fed D-chiro-inositol, myo-inositol or D-pinitol. Seed Science Research, 15, 329–338.

Javornik, B., Eggum, B. O., & Kreft, I. (1981). Studies on protein fractions and protein quality of buckwheat. Genetika, 13, 115-121.

Juszczak, L., & Fortuna, T. (2006). Rheology of selected Polish honeys. Journal of Food Engineering, 75, 43-49.

Kawa, J. M., Taylor, C. G., & Przybylski, R. (2003). Buckwheat concentrate reduces serum glucose in streptozotocin-diabetic rats. Journal of Agricultural and Food Chemistry, 51(25), 7287-7291

Kayashita, J., Shimaoka, I., Nakajoh, M., Yamazaki, M., & Kato, N. (1997). Consumption of buckwheat protein lowers plasma cholesterol and raise faecal neutral sterols in cholesterol-fed rats because of its low digestibility. Journal of Nutrition, 127, 1395-1400.

Kim, S. J., Kim, S. K., & Park, C. H. (2004). Introduction and nutritional evaluation of buckwheat sprouts as a new vegetable. Food Research International, 37, 319 327.

Kim, S. L., Kim, S. K., & Park, C. H. (2002). Comparisons of lipid, fatty acids and tocopherols of different buckwheat species. Food Science and Biotechnology, 11, 332- 336.

Kitabayashi, H., Ujihara, A., Hirose, T., & Minami, M. (1995). Varietal differences and heritability for rutin content in common buckwheat Fagopyrum esculentum Moench. Japanese Journal of Breeding, 45, 75-79.

Koziol, M. J. (1992). Chemical composition and nutritional evaluation of quinoa (Chenopodium quinoa Willd.). Journal of Food Composition and Analysis, 5, 35–68. Kreft, I., Srabanja, V., Ikeda, S., Ikeda, K., & Bonafaccia, G. (1996). Dietary value of buckwheat. Research Reports Biotechnical Faculty of the University of Lubljana, 67, 73- 78.

Kreft, I., Fabjan, N., & Yasumoto, K. (2006). Rutin content in buckwheat (*Fagopyrum esculentum* Moench) food materials and products. Food Chemistry, 98, 508-512.

Kreft, I., Plestenjak, A., Golob, T., Skrabanja, V., Rudolf, M., & Draslar, K. (1999). Functional value of buckwheat as affected by the content of inositol, phosphate, minerals, dietary fiber and protein. In: Sanberg, A. S., Andersson, H., Amado, R., Schlemmer, H., Serra, F. (eds): Bioactive Inositol Phosphates and Phytosterols in Food. Cost 916. Office for Official Publications of the European Communities, Luxemburg, 69-72.

Kreft, S., Knapp, M., & Kreft, I. (1999). Extraction of rutin from buckwheat (*Fagopyrum esculentum* Moench) seeds and determination by capillary electrophoresis. Journal of agricultural and food chemistry, 47(11), 4649-4652.

Krkošková, B., & Mrazova, Z. (2005). Prophylactic components of buckwheat. Food Research International, 38(5), 561-568.

Krkoškova, B., Li, S., & Zhang, Q. H. (2001). Advances in the development of functional foods from buckwheat. Critical Reviews in Food Science and Nutrition, 41, 451-464. Krkoškova, B., & Mrazova, Z. (2005). Prophylactic components of buckwheat. Food Research International, 38, 561-568.

Kumar, V., Sinha, A. K., Makkar, H. P. S., & Becker, K. (2010). Dietary roles of phytate and phytase in human nutrition: A review. Food Chemistry, 120, 945-959.

Kwon, O. N. (2003). Buckwheat-pancake manufacture method and buckwheat-pancake composition. Patent KR2003041892-A, 24 March.

Lee, B.S., & Park, S.K. (2013). Volatile aromatic compounds and fermentation properties of fermented milk with buckwheat. Korean Journal of Food Science and Technology, 45, 267-273.

Lee, S., Prosky, L., & DeVries, J. (1992). Determination of total, soluble and insoluble dietary fibre in foods - Enzymatic-gravimetric method, MES-TRIS buffer: Collaborative study. Journal of AOAC International, 75, 395-416.

Leng, G., & Shi, D. (2006). Thin pancake preparing method, involves graining millet, soybean, wheat, mung bean and buckwheat and several flavoring materials of fennel fruit, star anise, xanthoxylum and coriander seed. Patent CN1775075-A, 24 May.

Li, S. Q., & Zhang, Q. H. (2001). Advances in the development of functional foods from buckwheat. Critical Reviews in Food Science and Nutrition, 41, 451-464.

Li, W., Lin, R., & Corke, H. (1997). Physicochemical properties of common and Tartary buckwheat starch. Cereal Chemistry, 74, 79-82.

Lin, L. Y., Hsieh, Y. J., Liu, H. M., Lee, C. C., & Mau, J. L. (2009). Flavor components in buckwheat bread. Journal of Food Processing and Preservation, 33, 814-826.

Lintschinger, J., Fuchs, N., Moser, H., Jager, R., Hlebeina, T., Markolin, G., & Gossler, W. (1997). Uptake of various trace elements during germination of wheat, buckwheat and quinoa. Plant Foods for Human Nutrition, 50, 223-237.

Mariotti, M., Pagani, M. A., & Lucisano, M. (2013). The role of buckwheat and HPMC on the breadmaking properties of some commercial gluten-free bread mixtures. Food Hydrocolloids, 30, 393-400.

Marshall, H. G. (1982). Buckwheat: description, breeding, production, and utilization. Adv. Cereal Sci. Technol., 5, 157-210.

Mazza, G. (1988). Lipid content and fatty acid composition of buckwheat seed. Cereal Chemistry, 65(2), 122-126.

Mazza, G. (1993). Buckwheat. Encyclopaedia of Food Science and Nutrition (Macrae R, Robinson RK, Sadler MJ eds.). Academic Press, Toronto, New York, 516-521.

McKibben, J., & Engeseth, N. J. (2002). Honey as a protective agent against lipid oxidation in ground Turkey. Journal of Agricultural and Food Chemistry, 50, 592-595.

MilisavljEvic, M. D., Timotijevic, G. S., Radovic, S. R., Brkljacic, J. M., Konstantinovic, M. M., & Maksimovic, V. R. (2004). Vicilin-like storage globulin from buckwheat (Fagopyrum esculentum Moench) seeds. Journal of Agricultural and Food Chemistry, 52, 5258-5262.

Mitsunaga, T., Matsuda, M., Shimizu, M., & Iwashima, A. (1986). Isolation and properties of a thiamine-binding protein from buckwheat seed. Cereal Chemistry, 63, 332–335.

Molecular structure and some physicochemical properties of buckwheat starches. Cereal Chemistry, 81, 515-520.

Morishita, T., Yamaguchi, H., & Degi, K. (2007). The contribution of polyphenols to antioxidative activity in common buckwheat and tartary buckwheat grain. Plant Production Science, 10, 99–104.

Morita, N., Maeda, T., Sai, R., Miyake, K., Yoshioka, H., Urisu, A., & Adachi, T. (2006). Studies on distribution of protein and Allergen in graded flours prepared from whole buckwheat grains. Food Research International, 36, 782-790.

Moroni, A.V., Zannini, E., Sensidoni, G., & Arendt, E. K. (2012). Exploitation of buckwheat sourdough for the production of wheat bread. European Food Research and Technology, 235, 659-668.

Mundo, A., Padilla-Zakour, O. I., & Worobo, R. W. (2004). Growth inhibition of foodborne pathogens and food spoilage organisms by select raw honeys. International Journal of Food Microbiology, 97, 1-8.

Nagai, T., Inoue, R., Kanamori, N., Suzuki, N., & Nagashima, T. (2006). Characterization of honey from different floral sources. Its functional properties and effects of honey species on storage of meat. Food Chemistry, 97, 256-262.

Nic Phiarais, B. P., Mauch, A., Schehl, B. D., Zarnkow, M., Gastl, M., Herrmann, M., Zannini, E., & Arendt, E. K. (2010). Processing of a top fermented beer brewed from 100% buckwheat malt with sensory and analytical characterization. Journal of the Institute of Brewing, 116, 265-274.

Noda, T., Takahata, Y., Sato, T., Suda, I., Morishita, T., Ishiguro, K., & Yamakawa, O. (1998). Relationships between chain length distribution of amylopectin and gelatinization properties within the same botanical origin for sweet potato and buckwheat. Carbohydrate Polymers, 37, 153–158.

Normén, L. (2001). Phytosterols – intake and metabolic effect. In: Sanberg, A. S., Andersson, H., Amado, R., Schlemmer, H., Serra, F. (eds.): Bioactive Inositol Phosphates and Phytosterols in food. Cost 916. Office for Official Publications of the European Communities, Luxembur, 185-189

Norton, T., & Sun, D. W. (2008). Recent advances in the use of high pressure as an effective processing technique in the food industry. Food Bioprocess Technology, 1, 2-34.

Oomah, B., & Mazza, G. (1996). Flavonoids and antioxidant activities in buckwheat. Journal of Agricultural and Food Chemistry, 44, 1746–1750.

Orsák, M., Lachman, J., Vejdová, M., Pivec, V., & Hamouz, K. (2001). Changes of selected secondary metabolites in potatoes and buckwheat caused by UV, gamma - and microwave irradiation. Rostlinna Vyroba, 47, 493-500.

Pandey, S., Senthil, A., & Fatema, K. (2015). Effect of hydrothermal treatment on the nutritional and functional properties of husked and dehusked buckwheat. Journal of Food Processing and Technology, 6, 1-7.

Petr, J., Kalinova, J., Moudry, J., & Michalova, A. (2004). Historical and current status of buckwheat culture and use in the Czech Republic, in Proceedings of the 9th International Symposium on Buckwheat - Advances in Buckwheat Research (eds I. Faberova, V. Dvořaček, P. Čepkova et al .), Research Institute of Crop Production, Prague. Critical Reviews in Food Science and Nutrition, 19(3), 213-258.

Pomeranz, Y., & Robbins, G. S. (1972). Amino acids composition of buckwheat protein. Journal of Agricultural and Food Chemistry, 20, 270-274.

Przybylski, R., Lee, Y. C., & Eskin, N. A. M. (1998). Antioxidant and radical-scav enging activities of buckwheat seed components. Journal of the American Oil Chemists Society, 75, 1595– 1601.

Qian, J., & Kuhn, M. (1999). Physical properties of buckwheat starches from various origins. Starch/ Stärke, 51, 81-85.

Qian, J., Rayas-Duarte, P., & Grant, L. (1998). Partial characterization of buckwheat (Fagopyrum esculentum) starch. Cereal Chemistry, 75, 365-373.

Qin, P., Li, W., Yang, Y., & Guixing, R. (2013). Changes in phytochemical compositions, antioxidant and a-glucosidase inhibitory activities during the processing of tartary buckwheat tea. Food Research International, 50, 562-567.

Qiu, R., Wang, J., & Parkin, K. (2014). Isolation and identification of cytoprotective agents from nonpolar extracts of buckwheat flour. Food Research International, 66, 86-92.

Radovic, R. S., Maksimovic, M. J., Brakljacic, J. E., Varkonji, G., & Savic, P. A. (1999). 2S Albumin from buckwheat (*Fagopyrum esculentum* Moench) seeds. Journal of Agricultural and Food Chemistry, 47, 1467-1470.

Radovic, R. S., Maksimovich, R. V., & Varkonji-Gasic, J. E. (1996). Characterization of buckwheat seed storage proteins. Journal of Agricultural and Food Chemistry, 44, 972-974.

Randhir, R., Kwon, Y. I., & Shetty, K. (2008). Effect of thermal processing on phenolics, antioxidant activity and health-relevant functionality of selected grain sprouts and seedlings. Innovative Food Science and Emerging Technologies, 9, 355-364.

Roberfroid, M. (1993). Dietary fiber, inulin, and oligofructose – a review comparing their physiological effects. Critical Reviews in Food Science and Nutrition, 33, 103–148.

Ryan, E., Galvin, K., O'Connor, T.P., Maguire, A. R., & O'Brien, N. M. (2007). Phytosterol, squalene, tocopherol content and fatty acid profile of selected seeds, grains, and legumes. Plant Foods for Human Nutrition, 62(3), 85-91.

Sanz-Penella, J. M., Wronkowska, M., Soral-Śmietana, M., & Haros, M. (2013). Effect of whole amaranth flour on bread properties and nutritive value. LWT – Food Science and Technology, 50, 679–685.

Schoenlechner, R., Drausinger, J., Ottenschlaeger, V., Jurackova, K., & Berghofer, E. (2010). Functional properties of gluten-free pasta produced from amaranth, quinoa and buckwheat. Plant Foods for Human Nutrition, 65, 339-349.

Schoenlechner, R., Linsberger, G., Kaczyk, L., & Berghofer, E. (2006). Production of gluten-free short dough biscuits from the pseudocereals amaranth, quinoa and buckwheat with common bean. Ernaehrung / Nutrition, 30, 101-107.

Schoenlechner, R., Siebenhandl, S., & Berghofer, E. (2008). Pseudocereals, in Gluten-Free Cereal Products and Beverages (eds E. K. Arendt and F. Dal Bello). Elsevier, Oxford, 149-190.

Sedej, I., Sakac, M., Mandic, A., Misan, A., Pestoric, M., Simurina, O., & Canadanovic- Brunet, J. (2011). Quality assessment of gluten-free crackers based on buckwheat flour. LWT - Food Science and Technology, 44, 694-699.

Şensoy, Í., Rosen, R. T., Ho, C. T., & Karwe, M. V. (2006). Effect of processing on buckwheat phenolics and antioxidant activity. Food Chemistry, 99, 388-393.

Shin, H. S., & Ustunol, Z. (2004). Influence of honey-containing marinades on heterocyclic aromatic amine formation and overall mutagenicity in fried beef steak and chicken breast. Journal of Food Science, 69, 147-153.

Shin, H. S., Strasburg, G. M., & Ustunol, Z. (2003). Influence of different unifl oral honeys on heterocyclic aromatic amine formation and overall mutagenicity in fried ground-beef patties. Journal of Food Science, 68, 810-815.

Skrabanja, V., & Kreft, I. (1998). Resistant starch formation following autoclaving of buckwheat (Fagopyrum esculentum Moench) groats. An in vitro study. Journal of Agriculture and Food Chemistry, 46, 2020-2023.

Skrabanja, V., Kreft, I., Golob, T., Modic, M., Ikeda, S., Ikeda, K., Kreft, S., Bonafaccia, G., Knapp, M., & Kosmelj, K. (2004). Nutrient content in buckwheat milling fractions. Cereal Chemistry, 81, 172–176.

Skrabanja, V., Liljeberg Elmstahl, H. G. M., Kreft, I., & Bjorck, I. M. E. (2001). Nutritional properties of starch in buckwheat products: studies in vitro and in vivo. Journal of Agricultural and Food Chemistry, 49, 490-496.

Socha, R., Juszczak, L., Pietrzyk, S., Gałkowska, D., Fortuna, T., & Witczak, T. (2011). Phenolic profile and antioxidant properties of Polish honeys. International Journal of Food Science and Technology, 46, 528-53.

Souci, S. W., Fachmann, W., & Kraut, H. (2000). Food Composition and Nutrition Tables. Wissenschaft Verlags, Stuttgart.

Steadman, K. J., Burgoon, M. S., Lewis, B. A., Edwardson, S. E., & Obendorf, R. L. (2001b). Minerals, phytic acid, tannin and rutin in buckwheat seed milling fractions. Journal of the Science of Food and Agriculture, 81, 1094-1100.

Steadman, K. J., Burgoon, M. S., Lewis, B. A., Edwardson, S. E., & Obendorf, R.L. (2001a). Buckwheat seed milling fractions: description, macronutrient composition and dietary fibre. Journal of Cereal Science, 33, 271–278.

Steadman, K., Burgoon, M., Schuster, R., Lewis, B., Edwardson, S., & Obendorf, R. (2000). Fagopyritols, D-chiro-inositol, and other soluble carbohydrates in buckwheat seed milling fractions. Journal of Agricultural and Food Chemistry, 48, 2843-2847.

Tahir, I., & Farooq, S. (1985). Grain composition in some buckwheat cultivars (Fagopyrum spp.) with particular reference to protein fractions. Quality Plant Food Human Nutrition, 35, 153-158.

Taira, H., Akimoto, I., & Miyahara, T. (1986). Effects of seeding time on lipid content and fatty acid composition of buckwheat grains. Journal of Agricultural and Food Chemistry, 34(1), 14-17.

Tang, C. H. (2007). Thermal properties of buckwheat proteins as related to their lipid contents. Food Research International, 40(3), 381-387.

Tkachuk, J. J., & Gembarowsky, M. A. (1996). El alforfón. Cartilla N°11. Internal report, INTA. Estación Experimental Agropecuaria Cerro Azul, 2-3.

Tomotake, H., Yamamoto, N., Yanaka, N., Ohinata, H., Yamazaki, R., Kayashita, J., & Kato, N. (2006). High protein flour supresses hypercholesterolemia in rats and gallstone formation in mice by hypercholesterolemic diet and body fat in rats because of its low protein digestibility. Nutrition, 22, 166–173.

Tsuzuki, W., Ogata, Y., Akasaka, K., Shibata, S., & Suzuki, T. (1991). Fatty acid composition of selected buckwheat species by fluorometric high-performance liquid chromatography. Cereal Chemistry, 68(4), 365-369.

Ueda, T., Coseo, M., & Harrell, T. (2005). A multifunctional galactinol synthase catalyzes the synthesis of fagopyritol A1 and fagopyritol B1 in buckwheat seed. Plant Science, 168, 681- 690.

Vega-Gálvez, A., Miranda, M., Vergara, J., Uribe, E., Puente, L., & Martinez, E. A. (2010). Nutrition facts and functional potential of quinoa (*Chenopodium quinoa* willd.), and ancient Andean grain: a review. Journal of the Science of Food and Agriculture, 90, 2541-2547.

Verardo, V., Arráez-Román, D., Segura-Carretero, A., Marconi, E., Fernandez-Gutierrez, A., & Caboni, M. F. (2011). Determination of free and bound phenolic compounds in buckwheat spaghetti by RP-HPLC-ESI-TOF-MS: effect of thermal processing from farm to fork. Journal of Agricultural and Food Chemistry, 59, 7700-7707.

Vinning, G. (2001). Buckwheat: A Demand-Supply Analysis, Rural Industries Research and Development Corporation, Kingston, Australia.

Vit, P., Rodríguez-Malaver, A., Rondón, C., González, I., Di Bernardo, M. L., & García, M. Y. (2010). Bioactive indicators related to bioelements of eight unifloral honeys. Archivos Latinoamericanos de Nutrición, 60, 405-410.

Vogrinćić, M., Timoracka, M., Melichacova, S., Vollmannova, A., & Kreft, I. (2010).

Vogrincic, M., Timoracka, M., Melichacova, S., Vollmannova, A., & Kreft, I. (2010). Degradation of rutin and polyphenols during the preparation of tartary buckwheat bread. Journal of Agricultural and Food Chemistry, 58(8), 4883-4887.

Vranac, A., Muhovic, L., Savic, D., Orucevic, S., Begic-Akagic, A., & Selimovic, A. (2013). Effect of adding buckwheat whole grain flour on antioxidant and sensory properties of biscuits. Works of the Faculty of Agriculture University of Sarajevo, 58, 105-113.

Wanatabe, K., Shimizu, M., Adachi, T., & Mitsunaga, T. (1999). Characterization of thiamin-binding protein in buckwheat seeds. Journal of Nutritional Science and Vitaminology, 78, 323-328.

Wang, J. B., Zhao, L., Peng, L. X. et al . (2013). Development on tartary buckwheat compound yogurt. Food Science and Technology, 38, 167-171.

Wang, X. H., Gheldof, N., & Engeseth, N. J. (2004). Effect of processing and storage on antioxidant capacity of honey. Journal of Food Science, 69, 96-101.

Watanabe, M. (1998). Catechins as antioxidants from buckwheat (Fagopyrum esculentum Moench) groats. Journal of Agricultural and Food Chemistry, 46, 839–845.

Wei, Y., Hu, X., Zhang, G., & Ouyang, S. (2003). Studies on the amino acid and mineral content of buckwheat protein fractions. Nahrung / Food, 47, 114-116.

Wijngaard, H. H., & Arendt, E. K. (2006). Optimisation of a mashing program for 100% malted buckwheat. Journal of the Institute of Brewing, 112, 57–65.

Wijngaard, H., & Arendt, E. (2006). Buckwheat review. Cereal Chemistry, 83, 391-401.

Wilczyńska, A. (2010). Phenolic content and antioxidant activity of different types of Polish honey - A short report. Polish Journal of Food and Nutrition Sciences, 60, 309-313.

Wójtowicz, A., Kolasa, A., & Mościcki, L. (2013). Influence of buckwheat addition on physical properties, texture and sensory characteristics of extruded corn snacks. Polish Journal of Food and Nutrition Sciences, 63, 239-244.

Wolter, A., Hager, A. S., Zannini, E., & Arend, E. K. (2013). In vitro starch digestibility and predicted glycaemic indexes of buckwheat, oat, quinoa, sorghum, teff and commercial gluten-free bread. Journal of Cereal Science, 58, 431-436.

Wolter, A., Hager, A. S., Zannini, E., & Arendt, E. K. (2014). Impact of sourdough fermented with Lactobacillus plantarum FST 1.7 on baking and sensory properties of gluten-free breads. European Food Research and Technology, 239, 1-12.

Wronkowska, M., & Haros, M. (2014). Wet-milling of buckwheat with hull and dehulled - the properties of the obtained starch fraction. Journal of Cereal Science, 60, 477-483.

Wronkowska, M., Christa, K., Ciska, E., & Soral-Śmietana, M. (2015b). Chemical characteristics and sensory evaluation of raw and roasted buckwheat groats fermented by Rhizopus oligosporus. Journal of Food Quality, 38, 130–138.

Wronkowska, M., Haros, M., & Soral-Śmietana, M. (2013). Effect of starch substitution by buckwheat flour on gluten-free bread quality. Food and Bioprocess Technologies, 6, 1820-1827.

Wronkowska, M., Honke, J., & Piskuła, M. K. (2015a). Effect of solid-state fermentation with Rhizopus oligosporus on bioactive compounds and antioxidant capacity of raw and roasted buckwheat groats. Italian Journal of Food Sciences, 27, 424-431.

Yamsaengsung, R., Berghofer, E., & Schoenlechner, R. (2012). Physical properties and sensory acceptability of cookies made from chickpea addition to white wheat or whole wheat flour compared to gluten-free amaranth or buckwheat flour. International Journal of Food Science and Technology, 47, 2221-2227.

Yildiz, G., & Bilgicli, N. (2015). Utilisation of buckwheat flour in leavened and unleavened Turkish flat breads. Quality Assurance and Safety of Crops and Foods, 7, 207-215. Yoshimoto, Y., Egashira, T., Hanashiro, I., Ohinata, H., Takase, Y., & Takeda, Y. (2004).

Yoshimoto, Y., Egashira, T., Hanashiro, I., Ohinata, H., Takase, Y., & Takeda, Y. (2004). Molecular structure and some physicochemical properties of buckwheat starches. Cereal chemistry, 81(4), 515-520.

Zdunczyk, Z., Flis, M., Zielinski, H., Wroblewska, M., Antoszkiewicz, Z., & Juskiewicz, J. (2006). In vitro antioxidant activities of barley, husked oat, naked oat, triticale, and buckwheat wastes and their influence on the growth and biomarkers of antioxidant status in rats. Journal of Agricultural and Food Chemistry, 54, 4168–4175.

Zhang, H. W., Zhang, Y. H., Lu, M. J., Tong, W. J., & Cao, G. W. (2007). Comparison of hypertension, dyslipidaemia and hyperglycaemia between buckwheat seed-consuming and non-consuming Mongolian-Chinese populations in Inner Mongolia, China. Clinical and Experimental Pharmacology and Physiology, 34, 838-844.

Zhang, M., Chen, H., Li, J., Pei, Y., & Liang, Y. (2010). Antioxidant properties of tartary buckwheat extracts as affected by different thermal processing methods. LWT - Food Science and Technology, 43, 181-185.

Zhang, Z., Zhou, M., Tang, Y., Li, F., Tang, Y., Shao, J., Xue, W., & Wu, Y. (2012). Bioactive compounds in functional buckwheat food. Food Research International, 49, 389-395.

Zhu, F., Du, B., Li, R., & Li, L. (2014). Effect of micronization technology on physicochemical and antioxidant properties of dietary fiber from buckwheat hulls. Biocatalysis and Agricultural Biotechnology, 3, 30-34.

Zielińska, D., & Zieliński, H. (2009). Low molecular weight antioxidants and other biologically active components of buckwheat seeds. The European Journal of Plant Science and Biotechnology, 3, 29-38.

Zielińska, D., Szawara-Nowak, D., & Zieliński, H. (2013). Antioxidative and anti-glycation activity of buckwheat hull tea infusion. International Journal of Food Properties, 16, 228-239.

Zieliński, H., Michalska, A., Amigo-Benavente, M., Del Castillo, D., & Piskuła, M. K. (2009). Changes in protein quality and antioxidant properties of buckwheat seeds and groats induced by roasting. Journal of Agricultural and Food Chemistry, 57, 4771- 4776.

Zieliński, H., Michalska, A., Piskuła, M. K., & Kozłowska, H. (2006). Antioxidants in thermally treated buckwheat groats. Molecular Nutrition and Food Research, 50, 824-832.

Zondag, M. D. (2003). Effect of Microwave Heat-Moisture and Annealing Treatments on Buckwheat Starch Characteristics. Master thesis. Graduate School, University of Wisconsin– Stout, Menomie, WI.

17

CHIA (*Salvia hispanica*) A Super Food

Rahul Vashishth[1], Anil Dutt Semwal[2], Gopal Kumar Sharma[2]

[1]Vignan's Foundation for Science, Technology and Research, Guntur, Andhra Pradesh -522213, India
[2]Defence Food Research Laboratory, Siddarthanagar, Mysuru Karnataka - 570011, India

Introduction

Chia seed is an ancestral crop having fascinating history of 5500 years. It was the third most important grain of pre-columbian society. Chia seed is common name given to species of *Salvia. Salvia hispanica, Salvia columbariae* and *Salvia polystachya* are most important species among all *Salvia* seeds. *Salvia* belongs to family of *Lamiaceae*is, first native to Mexico and later introduced to Spain by Hernan Cortes (Ortiz de Montellano, 1978), but in next 200 years chia became virtually unknown to human and required research to establish the nutritional aspects of chia and reintegrate as food in modern diet.

In 1753, Swedish botanist Carl Von Linneo classified chia seeds and named it *Salvia* (save or cure) *hispanica (Spanish),* which in Latin means Spanish plant which can cure or save (Sosa *et al*, 2016). But, it was found that the name given was wrong and chia was native to Mexico not to Spain. Chia seeds were originally belongs to Aztec and people in Nahua language call chai as Chian, Chian word means oily, this was used by Aztec to refer all species of *Salvia* genus that possess high oil content (*Salvia hispanica, Salvia tiliifolia* V and *Salvia Columbariae* B). Chia was an important grain for Aztec, who during their conquest used to provide chia to their warriors as a fat rich food source for their survival.

At global stage, the integration and large acceptability of chia seeds in modern era is due to their adaptability to adverse agronomic and environmental conditions that other crop would hardly survive. In 1991, the Argentinean and EVA started a Regional Research project named as "Western Argentinean Regional Project" to study the nutritional aspect and benefits associated with the chia products which led to the integration of chia crops on global stage of modern agriculture. This impressive study resulted in increased cultivation land area of chia from 500 hectare in 1994 to 370,000 hectare in 2014 (established in 13 countries). Slowly and steadily chia found place in modern production and got spread in different agricultural region of Argentina and other countries such as Peru, Columbia, Bolivia and Ecuador. In current scenario chia seeds are produced in more than 14 countries around the world, in today's world chia has come up as seed with huge industrial potential. It offers great scope to food industry, animal feed industries, pharma-nutraceutical industry and cosmetic industries.

Botanical and Taxonomic Details

Chia is an annual herb commonly known as chia; Spanish sage; Mexican chia and black chia. It is a summer plant which is about 1 m tall, with opposite petiolate and, 4-8 cm long and 3-5 cm wide serrated leaves. Flower of chia plants are hermaphrodite with purple or white petals, which grows in cluster and protected by small bracts with pointed tips. Seeds are 1-2 mm in size generally oval with flattened centre, smooth and shiny surface colour of the seeds may vary from brown, grey, dark red and white. Stem is quadrangular with ribbed and hairy appearance. It can grow in various soil conditions such as light to medium clay and sandy soil and arid soil with good drainage and not excess wetness. Chia plants possess semi tolerance against acid soil and drought, and little tolerance to abiotic phenomena like freezing and sunless location due to which they are mainly grown in mountain regions. The plant can be grown without the use of pesticides and chemicals due to their tolerance against pest which is attributed to the essential oil present in leaves. Due to their resistance towards dry condition and other environment factors chia is grown in various developing countries such as Bolivia, Columbia and Argentina. The commercial chia plant yield around 500 to 600 kg of seeds per hectare whereas in control conditions such as optimum irrigation and nitrogen fertilizers the yield can be raised up to 2500 kg per hectare. Chia plant possibly be utilized as a rotational crop to maintain the nutrients level in soil which turns deplete during farming of certain crops like tobacco.

Nutritional Composition of Chia

Nutritional and functional potential of chia seeds are well known from ages. The domestication and human selection of the seeds resulted in diverse chia seeds morphology and genetic mapping among varieties of chia (Cahill and Provance, 2000; Cahill and Ehdaie, 2005). Seeds are mainly divided into three parts the testa or coat, embryo and endosperm. Seed coat provides the protection to the embryo and found rich in mucilage, a polysaccharide with potential to form a sticky gelatinous coating around seed when it comes in contact with water adhering to the seed. The principal function of this high amount of mucilage present in seed is still unknown however, it is assumed that it prevents the additional water loss from surface of seeds and render it capacity to grow in arid region as well (Valdivia-Lopez and Tecante, 2015).

Chia seeds were announced as functional food by European Parliament because of its excellent nutrition composition. It possesses 16-26 percent protein, 31-34 percent fat, 37-45 percent carbohydrate, out of which 32-40 percent is contributed by total dietary fibre, and 4-5 percent ash. Apart from the macro-nutrients, chia seeds also considered as a potential source of minerals (calcium, potassium, magnesium etc.), vitamin (B^1, B^2, B^3, B^6, vitamin C and vitamin A) and phytonutrients compounds such as phenolic acids. The wide range of nutritional components in these seeds can be attributed to their varietal difference, growing condition and agriculture practices. The major functional potential of chia seeds has been attributed to high fibre content, mucilage forming capacity, phytochemical concentration, high concentration of α-linolenic acid and high and balance concentration of amino acid and proteins.

Carbohydrates

The carbohydrate portion of chia seeds varies from 37-45 (percent dry) basis where 60-80 percent of it is contributed by the dietary fibre present in chia seeds. Dietary fibre of chia seeds is an imperative bio-component present in them which is specific to food prepared with whole grains, due to its potential positive effect on health. It is defined as a mixture of compounds consisting of plant carbohydrates such as oligosaccharides and polysaccharides e.g. cellulose, hemicelluloses, lignin, pectin components and various gums present as an adhesive associated with lignin and non-carbohydrate components such as phenols, tannins, waxes, phytates, minerals, proteins and other component (Elleuch *et al.*, 2011). Dietary fibre has been reported to be associated with minimizing the danger of coronary heart disease, diabetes mellitus type-II and various kind of cancer which is attributed to their capacity to bind these components and reduce the level of low density lipoproteins (LDC) present in body (Steinmetz and Potter, 1996; Lattimer and Haub, 2010; Kaczmarczyk *et al.*, 2012). It also possesses the capacity to increase the post-meal satiety and decrease subsequent hunger (Lee *et al.*, 2009; Vuksan *et al.*, 2010). American Dietetic Association (ADA) also mentioned in their report that fibre possesses health maintenance and disease prevention capacity. The optimum consumption of dietary fibre is established around 25-30 g/day and preferential insoluble to soluble dietary fibre ratio 3:1 by ADA. Chia seed contains fibre content of 30-40 percent out of which insoluble fibre concentration is around 85 percent and rest of 15 percent is contributed by soluble fibre. The fibrous portion of chia seeds is about twice the fibre content in bran, 4-5 times of almond and soya, quinoa or amaranth (Nutritional Nutrient Database for standard Reference, 2011). *Jalisco* and *Sinaloa* variety of Mexico chia seeds reported to contain 39.94 percent and 36.97 percent total dietary fibre (TDF) respectively. The soluble and insoluble dietary fibre was 6.84 and 34.90 g/100g for *Jalisco* and 6.16 and 32.87 g/100g for *Sinaloa* respectively and insoluble to soluble fibre ratio of 5:1 and 5:34 for *Jalisco* and *Sinaloa* respectively, which exhibit chia seeds as an excellent source of dietary fibre (DF). Dietary fibre is reported as the predominant fibre fraction with Klason lignin (KL) as its main component with 39-41 percent share of total dietary fibre (TDF). It is said to protect unsaturated fatty acid of chia by generating a steady and resistant structure in conjugation with the antioxidant compound present in grain. The concentrate of Neutral Sugar in IDF (NSIDF) ranges from 13.79-14.97 percent with lower uronic acid (UAIDF) with lower uric acid

The lignin present in chia seeds tends to absorb bile acids and reduces the occurrence of hypercholesterolemic condition. Chia seeds have a tendency to form a capsule of mucilage around seed during soaking which is the function of soluble fibre present in seeds. The soluble dietary fibre (SDF) fraction is mainly confirmed of neutrals sugars (NSSDF), that shows the occurrence of diverse carbohydrate with low concentration of uronic acid and no/or negligible pectic association with the mucilage. Lin, Daniel and Whistler, (1994) reported the primary tetrasaccharide repeating units of(1→4)-β-D-Xylp (1→4)α-D-Glcp(1→4)-β-D-Xylp (1→4), present in main backbone to which a substituent of 4-O-methyl-α-D- Glcp-A is linked to β-D-Xylp residue through a O_2 bond. The molecular weight of mucilage forming polymer varies from 0.8 to 2.0 μDa. On hydrolysis mucilage tends to produce β-D-Xylose, α-D-glucose and 4-O-methyl-α-D-glucuronic acid in ratio 2:1:1 respectively. The mucilage forming capacity of seed is reported to be affected by ionic strength, solution pH, temperature as well as seed to water ratio for example maximum hydration said to attained at 1:40 seed to water ratio, low mono and divalent salt concentration, pH ranges of 6.6 to 6.9 and 65-80°C with maximum hydration at 80°C. The isolated concentration of mucilage from seed ranges from 5 to 6.97 percent. The composition of isolated mucilage can vary from 9.37-11.08 percent moisture, 6.79-18.85 percent protein, 11.42-17.97 percent crude fibre, 0.88-3.22 percent oil, 9.75-10.27 percent ash and 56.24-64-61 percent nitrogen free extract. This wide variation in mucilage composition was accredited to the process of extraction and soaking condition of seeds. Capitani *et al.* (2012) reported rheological properties of these chia seed mucilage in terms of steady and oscillating flow under different conditions and concentration. They found out that at raw mucilage solution of 0.25-1.0 percent exhibit Non-Newtonian share thinning behaviour when kept at 25°C.

Table 1: Proximate composition of Chia Seeds.

S.No	Parameter	Reference 1	Reference 2	Reference 3	USDA
1	Moisture (%)	7.27	5.82	5-8	5.80
2	Carbohydrate (%)	42.52	34.57	37-45	42.12
3	Protein (%)	22.03	25.32	16-26	16.54
4	Fat (%)	23.66	30.22	31-34	30.74
5	Fibre (%)	33.47	37.50	32-40	34.40
6	Ash (%)	4.55	4.07	4-5	4.80

Source: Porras-Loaiza *et al.* 2014; Marineli *et al.* 2014; Marcinek and Krejpcio, 2017

The apparent viscosity for mucilage tends to increase and flow index decreases when polysaccharides is increased at a particular shear rate, which is similar to the typical hydrochloride solution behaviour. This expresses the overpowering non-Newtonian behaviour of mucilage solution under given condition. Another critical tendency of mucilage solution is their sensitivity towards mono and divalent cations i.e. sodium or calcium salt resulting in significant reduction in apparent viscosity with respect to salt-free solutions. This specific property of chia mucilage can be attributed to presence of glucuronic acid in primary structure. The thickening capacity of chia mucilage is not regarded at par with the thickening capacity of hydrocolloids due to the isothermal significant changes that occur in apparent viscosity of mucilage with share rate. Apart from the varying rheological properties, fibre content from chia was reported to possess high water holding capacity (WHC)15.41 g/g, water absorption capacity (WAC) 11.73g/g and organic constituent absorption capacity (OCAC) 1.09 g/g and low oil holding capacity (OHC) 2.02 g/g. With their high fibre content and beneficial function capability of fibre/mucilage, chia is considered as possible functional food with strong potential of controlling or preventing the occurrence of health diseases.

Protein

The protein, peptides and their monomeric unit such as amino acids are considered as necessary cell component of organisms and plant leading to their usual functioning. Protein content in chia seeds ranges from 16-26 percent contingent on environmental, geographical and harvesting conditions. The protein content of chia seed is greater than other common grain such as wheat, maize, barley and oats that normally have protein content of 14-16 percent. The protein from chia is categorized as gluten-free which provide them with a unique feature that they can be utilized by individual having celiac disease without creating any kind of irritation in their gut. Major protein portion in chia is contributed by prolamine with concentration of 538 g/kg of crude protein followed by glutelin (230 g/kg crude protein), globulins (70 g/kg crude protein) and albumin (39 g/kg of crude protein) (Guevara-Cruz *et al.*, 2012; Guiotto *et al.*, 2015; Arguelles-Lopez, *et al.*, 2018). The globulin portion of protein made of two major subunits 11S and 7S proteins and molecular size ranging from 15 to 50 kDa.

The amino acid content of chia protein was reported to be more complete than other common grains which generally be limited in two or more essential amino acids (Ayerza and Coates, 2001). Olivos-lugo, Valdivia-Lopez and Tecante, (2010), testified the amino acid profile for glutamic acid (123 g/kg of raw protein), arginine (80.6 g/kg of raw protein), and aspartic acid (61.3 g/kg of raw protein) along with its chemical score and in vitro protein digestibility. In spite of being superior to other cereals grains chia could not be suggested as a sole protein in case of lysine supplement for preschool children as the concentration of lysine in chia observed to be deficient in comparison with standard established by WHO/FAO/UNU, 1985 for preschool children. Hence it needs to be supplemented with lysine rich components. They also studied the gelling, foaming and water holding capacity of seed protein. The WHC and OHC of seed protein were 4.06 g/g and 4.04 g/g respectively, proving chia protein isolates a possible additive in bakery product and food industries for making food emulsion. Vazquez-Ovando *et al.* (2013), investigated the physico-chemical and functional potential of a protein rich fraction (PRF), isolated from chia (*Salvia hispanica*) using dry fractionation. They found emulsifying capacity/emulsifying activity and emulsion stability of 50-56 percent and 92 percent respectively, which was highest at pH 8 to 10. The foaming capacity of chia protein was 29 percent which render it as a poor foaming agent but conversely the foaming stability was 80 percent. These results express the fact that chia protein can efficiently be utilized as an effective ingredient for food emulsion and form stability. Protein from chia can also be combined with glycerol or sorbitol with a molecular weight between 4-50 kDa to form an elastic film with excellent mechanical properties such as permeability to oxygen 16.27 cm^{3}µm/m^2/dkPa) in comparison with other films prepared using synthetic material (LDPE 57.85 cm^{3}µm/m^2/dkPa) and natural macromolecules (whey protein 31.3 cm^{3}µm/m^2/dkPa) (Alvarado-Suarez, 2008). This superior oxygen barrier property of film made with chia seeds provide an advantage of being utilized for products prone to oxidation during storage and prevent the deterioration of product leading to change in colour, odour, flavour and nutrients as well as provide better quality product with extended shelf life.

Table 2: Amino acid composition of chia seeds and its protein isolate.

S.No.	Amino Acids	Reference 1 (g/16gN)	Reference 2 (g/100g protein)	USDA (2018) (g/100g seeds)	RDI (%) (As per USDA data)
Essential Amino Acids					
1	Histidine	2.65	1.37	0.531	22
2	Isoleucine	3.18	2.42	0.801	16
3	Leucine	5.54	4.15	1.371	14
4	Lysine	4.54	2.99	0.970	13
5	Methionine+Cysteine	2.10	2.78	0.995	63
6	Phenylalanine+Tyrosine	7.40	3.88	1.579	51
7	Threonine	3.93	1.80	0.709	19
8	Tryptophan	ND	ND	0.436	44
9	Valine	5.21	2.55	0.950	15
Non- Essential Amino Acid					
1	Alanine	4.36	2.68	1.044	
2	Arginine	7.91	4.23	2.143	
3	Aspartic Acid	7.24	4.73	1.689	
4	Glutamic acid	12.91	7.08	3.500	
5	Glycine	4.35	2.28	0.943	
6	Proline	4.45	1.99	0.776	
7	Serine	4.75	2.62	1.049	

Source: Melo-Ruiz *et al*., 2016; Grancieri *et al*., 2019; USDA database, release 28

Lipids

Lipids are group of bioactive substances which form structure elements in cell membrane like protein molecules and regulate the physiological functions. In addition, fat molecule provides maximum amount of energy during metabolism therefore, human organism requires lipid molecules to accumulate energy. The lipid content and their fraction of chia seeds are extensively reported in literature but there is some inconsistency regarding content and fraction of fatty acid present in chia seeds. Chia, was stated to possess 2:1 ratio of palmetic (16:0) to stearic acid (18:0), 5:1 ratio of oleic (18:1) to palmitoleic acid (16:1), (mono-unsaturated fatty acid) and 1:1 ratio of linolenic (18:2) to linolenic acid (18:3) (polyunsaturated fatty acid) with the concentration of 2.309 percent for mono and 23.665 percent for polyunsaturated fatty acid. Chia seed is considered as an excellent source of essential fatty acid because it contains concentration of α-linolenic acid (ALA ω-3) greater than any other seed. The essential fatty acids such as Omega-3 linolenic (n-3) and Omega-6-linoleic acid are well known for not being synthesized by human and needed through food. Chia contains 25-40 percent of fat portion in which majority of it is in the form of polyunsaturated fatty acid such that α-linolenic (ω-3) and α-linoleic acid (ω-6). α-linolenic is the major fatty acid in chia seeds representing 75 percent of its total weight. Other natural oil source with similar concentration are safflower (75%) and sunflower oil (67%) however, the α-linolenic concentration of these sources is significantly lesser as compare to chia seed, which can be because of higher concentration of oleic leading to significant oil content but lower α-linolenic acid. Fatty acid profile of chia can be elaborated as linoleic (17-26%) and linolenic (52-57%) with rest of consisting of oleic and other unsaturated fatty acid. The degree of saturation and its oil content is reported to be affected by the growing location, agriculture condition and variety of seeds. Ayerza (2010), mentioned the impact of seed colour on fatty acid fraction and concentration of two chia seed (*Salvia hispanica*) genotypes from Ecuador. He reported that chia seed coat colour had no influence on oil content whereas the concentration of oil was significantly affected by the location of oil. He also reported that fatty acid fraction of seed is not affected either by location or colour of the seed coat but significant change in fatty acid fractions among seeds grown at various region was observed. On the grounds of such statement and investigation, it can be concluded that environment and agronomic conditions significantly affect the oil concentration and fatty acid fraction of seed. Chia has also been claimed to contain bioactive compound; phytosterols, squalene and β-sitosterol. *Jalisco* and *Sinaloa* variety of chia seeds were reported to consist of around 5 g/kg of squalene, 7-17 g/kg of phytosterol with 74 percent of which is β-sitosterol. The oil extraction rate and potential for chia depends on extraction method used, as a high yield of 92.8 percent can also be obtained using supercritical extraction method using carbon dioxide in comparison with other physical methods. The recent researches unanimously endorsed chia as a potential source of oil with beneficial triglycerides (TGA), α-linolenic and α-linoleic fatty acid. Fatty acid profile in chia seeds on the basis of their ranking can be arranged as ALA>linoleic> palmitic ~ Oleic> stearic acid.

Table 3: Fatty acid composition of chia seed and its Oil.

S.No	Fatty Acid	Reference 1 (g/100g Oil)	Reference 2 (g/100 seeds)	Reference 3 (g/100 Oil)	USDS (2018) (g/100g seeds)
1	Myristic C14:0	0.07	ND	0.03	0.030
2	Palmitic C16:0	7.07	1.85	6.69	2.170
3	Palmitoleic C16:1	0.08	ND	0.09	0.029
4	Heptadecanoic C17:0	0.07	ND	ND	0.063
5	Stearic C18:0	3.36	1.03	2.67	0.912
6	Arachidic C20:0	0.29	ND	ND	0.093
7	Oleic C18:1	7.04	1.67	10.55	2.203
8	Linoleic C18:2	18.49	5.69	17.36	5.835
9	Linolenic C18:3	63.20	20.37	62.02	17.830

Source: Marineli *et al.* 2014; Silva *et al.* 2016; Coelho *et al.* 2014; USDA database, release 28

Minerals and Vitamins

Vitamin and minerals are essential micronutrients necessary for normal functioning, growth and development of organism. The optimum concentration of these components is necessary for proper growth and metabolic functions. Presence of micronutrients is indispensable for complete digestion and assimilation of macromolecules through the release of hormones, growth regulations and formation of cell and tissues. They also plays major role in reducing the oxidative stress in human body. Chia is regarded as potential source of B-Vitamins with concentration of thiamine (62 mg/100g), riboflavin (0.17 mg/100g), niacin (883 mg/100g) and folic acid (49 mg/100g). They also contain fat soluble vitamins such as vitamin A and Vitamin E. These both vitamins prevent the oxidative stress and vitamin E arrest free radical formation in oil, giving it higher shelf stability. The concentration of niacin present in chia seeds has been reported to be significantly higher in comparison with rice and maize. Thiamine and riboflavin in chia seeds were present in similar concentration to that of present in amaranth, quinoa and other oil seeds. Chia is considered as a fairly rich source of minerals with calcium (631 mg/100g), potassium (407 mg/100g), magnesium (335 mg/100g), phosphorus (860 mg/100g), selenium (55.2 μg/100g), copper (924 μg/100g), iron (7.72 μg/100g), manganese (2.72 μg/100g) and zinc (4.58 μg/100g) of seed. The concentration of calcium present in chia is almost 6 times the concentration of calcium present in 100g of milk. Similarly phosphorus concentration is 11 times and potassium is 4 times higher than 100 g of milk. Chia contains 7-fold higher concentrations of calcium, 2-12-fold higher phosphorus and 1.6-fold higher concentration of potassium than common cereals such as wheat, rice, oat and corn. The concentration of iron in chia was also reported to be significantly higher than other seeds, it was stated to be around 6 times higher than spinach, 1.8 times higher than lentils and 2.4 times higher than grains (Marcinek, *et al.*, 2017).

Table 4: Mineral profile of chia seeds

S.No.	Mineral	Reference 1 (mg/100g seeds)	Reference2 (mg/100g seeds)	Reference 3 (mg/100g seeds)	USDA 2018 (mg/100g seeds)
1	Iron	24.4	10.90	9.39	7.72
2	Zinc	6.9	6.0	3.65	4.58
3	Copper	ND	ND	ND	0.924
4	Manganese	ND	ND	4.05	2.723
5	Magnesium	369	403	ND	335
6	Sodium	ND	ND	15	16
7	Potassium	666	870	ND	407
8	Calcium	624	580	ND	631
9	Phosphorus	799	696	ND	860

Source: Bolanos *et al.*, 2016; Liorent- Martinez *et al.*, 2013; Silva *et al.*, 2016

Bioactive Components

Oxidation is considered as an important metabolic process that said to occurs during the production of energy from food in human organism. During this process the molecular oxygen is reduced to water. This results in formation of reactive oxygen molecules such as hydrogen peroxide, hydroxyl and peroxide radicals along with certain free radicals. The free radicals are held responsible for neurological diseases, cardiovascular diseases, mental imbalance (Alzheimer's and Parkinson's disease) and cancer. The increased oxidation stress in body can also lead to diabetes, thrombosis, inflammation etc. But these all life-threatening problems can be overcome by the ingestion of antioxidant in required amount and hence the increased knowledge regarding the functions of antioxidant has led to a significant change in consumption of antioxidant through food (Jeong *et al.*, (2004). These antioxidant molecules can prevent the ill effects of these radicals by neutralizing or chelating them. Chia is considered as decent source of these antioxidant which mainly consists of phenolic acid such as chlorogenic, caffeic, coumaric, ferulic acid, catechin, epicatechin, rosmarinic, procatechin and gallic acid, and isoflavanoids such as kaempeferol, diadzine and myricetin. The phenolic concentration present in some varieties of chia is reported as 0.9211 GAE mg/g of chia extract, for *Jalisco* and 0.8800 for *Sinaloa*. The presence of polyphenols in chia seeds does not vary significantly among varieties but region of crop production and harvesting tends to have impact on total phenolic concentration. As per the literature chia seeds are good source of flavones and its content varies from 0.379-0.590 mg/ml for crude extract. The major phenolic compound present in crude extract was the mixture of glycoside of quercetin and kaempeferol whereas, the aglycon of quercetin and kaempeferol were the predominant phenolic in hydrolysed extract. Even though chlorogenic and caffeic are predominant phenolic acid present in chia, but their concentration was quite lower in hydrolysed extract which could be accredited to sensitivity of phenolic acids toward hydrolysis. The antioxidant potential of chia seed in terms of ABTS* radical (2,2'-azino-bis- ethylbenzothiazoline-6-sulphonic acid radical cation), β-carotene and linoleic acid model were found comparable to commercial trolox (Rayes-Caudillo *et al*, 2008). The total vitamin E (tocopherol) content of chia (238-427 mg/kg) is also quite comparable with the vitamin E concentration of peanut oil 398.6 mg/kg but lower than soyabean (1797.6 mg/kg), sunflower (634.4 mg/kg) and linseed oil (588.5 mg/kg). Therefore, chia can be reckoned as potential source of bioactive component i.e. polyphenol, sterols, squalene etc.

Table 5: Bioactive component in chia seeds.

S.No	Component	Reference 1	Reference 2	Reference 3
1	Phenolic Compounds (mg GAE/g)	0.97-0.99	0.511-0.881	0.757
2	Tannins (mg GAE/g)	14.93-19.08	ND	ND
3	Phytates (g/100g)	0.96-1.16	ND	ND
4	Carotenoids (μg/100g)	57.01	ND	ND
5	Flavones (μg/100g)	6.07-16.03	ND	ND
6	Flavanones (μg/100g)	4.39-9.34	ND	ND
7	Vitamin E (mg/100g)	8.17-8.24	ND	ND

Source: Silva *et al.* 2017; Segura-Campos *et al.*, 2014; Reyes-Caudillo *et al.*, 2008

Table 6: Phenolic acids in chia seed

S.No.		Concentration (mg/g)	Analytical Method	Reference
1	Phenolic Acid Chlorogenic Acid 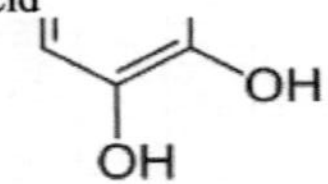	0.214-0.235 0.0459-0.102 0.007 X 10^{-3} mol/kg 0.226-0.218	HPLC HPLC HPLC HPLC	Ayerza, 2014 Reyes-Caudillo *et al.*, 2008 Ayerza and Coates, 2001 Ayerza, 2013b
2	Caffeic Acid 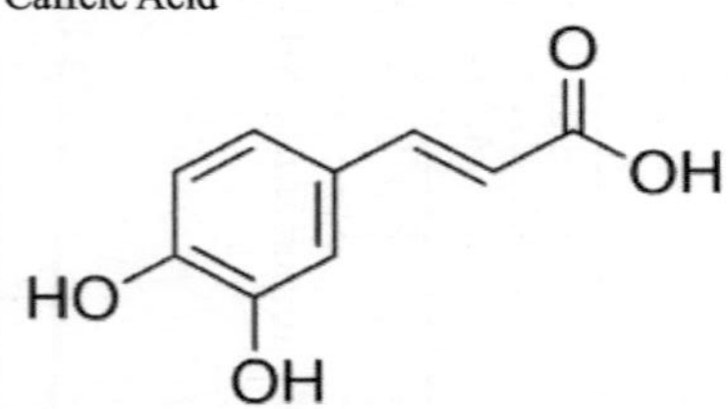	0.141-0.156 0.712-7.196 0.139-0.149 0.03089	HPLC HPLC HPLC UPLC	Ayerza, 2014 Pajak *et al.*, 2018 Ayerza, 2013b Coelho *et al.*, 2014
3	Rosmarinic Acid 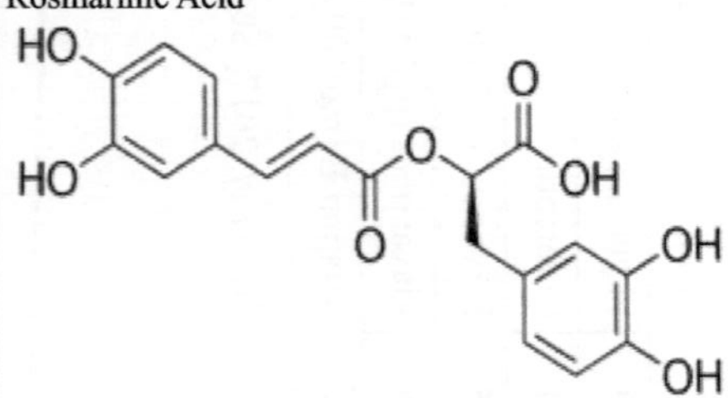	0.9267	UHPLC	Martinez-Cruz and Paredes-Lopez *et al.*, 2012

Flavanoids				
4	Quercetin 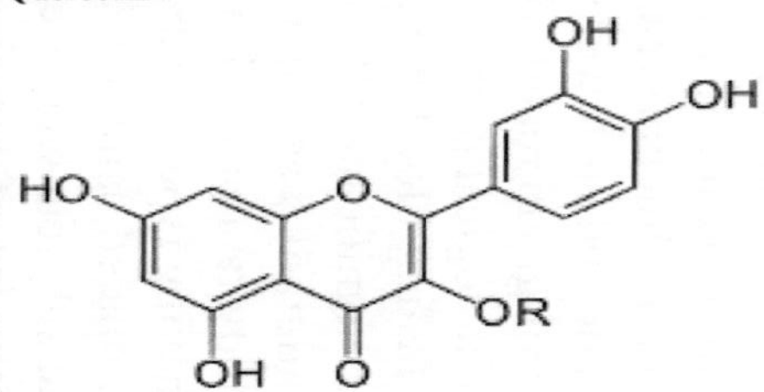	0.15-0.268 0.007–0.006 0.0018 0.2 X 10^{-3} mol/kg	HPLC HPLC HPLC HPLC	Reyes-Caudillo ***et al.***, 2008 Ayerza, 2013b Pajak ***et al.***, 2018 Ayerza and Coates ,2001
5	Myricetin 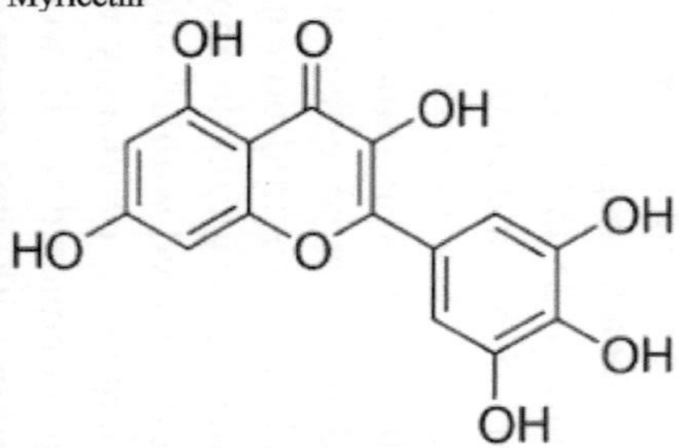	3.1 X 10^{-3} mol/kg 0.115–0.121 mg/g	HPLC HPLC	Ayerza and Coates, 2001 Ayerza ,2013b
6	Kaempferol 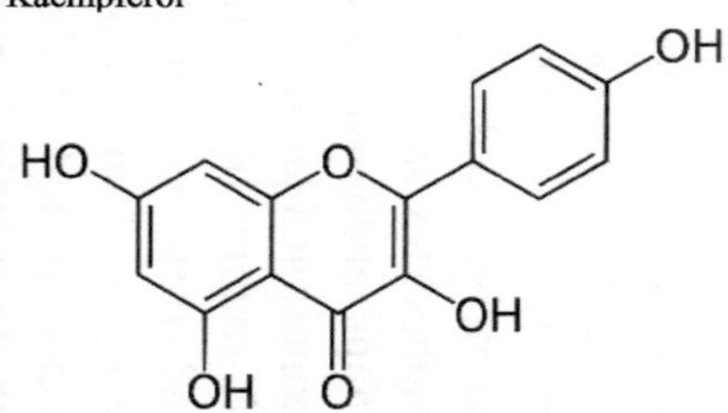	1.1 X 10^{-3} mol/kg 0.360-0.509 0.024-0.025	HPLC HPLC HPLC	Munoz ***et al.***, 2013 Reyes-Caudillo ***et al.***, 2008 Ayerza ,2014

Antinutritional Content and Toxicology

Chia is famously known for its nutritional and medicinal values due to the presence of phytochemicals hence, provided with designation of noble food and superfood. Certain researchers have reported concentration of tannins in range of 14.93-19.08 percent and phytate content in range of 0.96 to 1.16 percent. The presence of tannins and phytates in chia seed were also found region dependent as chia grown in Brazil found to possess some tannins and phytates content whereas these were absent (not detected) in Mexican chia seeds (Grancieri *et al.*, 2019). The antinutritional factors in chia seeds are either absent or present in extremely low concentration hence regarded as safe even at higher consumption. It is claimed to be safe on the basis of its "Safe use History" as there was no such evidence of adverse effect i.e. allergenicity, antinutritional factors and toxicity in major chia consuming countries.

Toxicological studies including animal studies on egg laying hens and broilers were performed by Ayerze *et al.* (2002). They observed no adverse effect on health at the treated concentration. However, reduced body weight was witnessed in broiler fed with 10 percent and 20 percent level dose of chia seeds in their diet for continuous 49 days. In another study performed with white and brown hens kept on 90 days diet of 28 percent chia showed reduction in egg production as well as egg weight in case of white hens (Ayerza and Coates, 2002). Ayerza and Coates (1999) in their study have shown that keeping laying hens on 30 percent chia seeds diet caused reduction in egg production but no allergenicity was induced in either of these studies. Along with animal studies there were some human studies carried out to establish safety of chia during long-term consumption. In 4 weeks, placebo-controlled study conducted with 100 male and female subjects having age between 21 to 65 years. Impact of chia was investigated on selected coagulation and immune function marker. During the study four groups of 25 individuals were made and each were fed with seed concentration of 2.5, 5.0 and 10.0 g/day, control-group received 4.0 g sunflower seeds/day. Individuals were tested for haematological parameters, lymphocytes and other physiological parameters with fasting blood samples. Groups analysed showed no adverse effect whereas significant effect on tiredness and fatigue was observed for mid-dose group and study was concluded with stating chia as safe for consumption. In another study randomised single blind crossover study was conducted with 20 subjects having type-II diabetes. Response of any supplement taken previously was eliminated using 4 weeks washout period. Study was conducted for 12 weeks and subjects were treated with 25 g/100g kcal portion with chia seeds. Test samples of blood were collected at week 0 and 12, and analysed for blood pressure regulation factor and inflammation. The results obtained were found in line to results of previous researches and no unfavourable impact was found in treated subjects. However, there was significant decrease observed in coagulation factors and C-reactive protein (CRP), a marker of inflammation for the blood lipids (HDL, LDL and triglycerides), glycaemic lipid enzyme, bleeding time and kidney function (EFSA, 2009.) The allergic likelihood of chia has been the area of investigation as none such conclusive report is available to prove the probability of chia seed to cause allergic response in human. In 2005, Members States considered possible chances of chia seeds to cause possible allergic responses hence they directed to conduct further investigation of chia seed for their possible allergic properties before being introduced into the market. EFSA, (2005) reported the possible allergic response of chia seeds on the basis of allergic reaction triggered by the crop; oregano and thyme as they share the same family with chia seeds (Benito *et al.*, 1996). A study conducted to find out IgE-binding responses in subjects reported to be allergic to peanut and tree nut. The binding of chia was variable, though specific. Further the common allergen detection test; skin prick was performed for chia seed protein with subjects sensitive to sesame seeds, also responded positively to chia (EFSA, 2005). Hence, possible chances of cross reactivity of chia and food allergenicity cannot be ruled out indicating the need to label the food containing chia seeds.

Chia Processing

Chia seeds have attained the popularity worldwide in recent years because of its excellent nutritive value. Chia is well known for its plentiful fibre, protein, mineral and phytochemical along with excellent fat content ranging from 25-40 percent. The composition of lipid in seeds are dominated by triglycerides and poly-unsaturated fatty acids. Major portion of fatty acid is contributed by α-linolenic acid and linoleic acid present in high concentration (Ixatina *et al.*, 2011; Palma and Dondum, 1947). Chia is considered among vegetable plant sources rich in ω-3 fatty acids. It plays vital role in physiological development during foetal and infant growth stages. It was also reported to produce beneficial effect against cardiovascular diseases, thrombosis, inflammation and arrhythmic disease since it contains phytochemical constituents such as tocopherol, phenols, phytosterol etc.

Similar to that of majority of oilseed, chia seed processing is also apportioned in two major parts. Complete cleaning, decontamination and collection of seeds and standardization of a suitable method. Method which can result in efficient oil extraction without altering overall quality and quantity of oil. Though, this sounds akin to other oilseeds processing but, it is little tricky as chia has tendency to produce mucilage when it comes in contact with water. Hence, the conventional processing with water is not appropriate for chia seeds as it makes it difficult to develop a stable and efficient processing technology for chia seed and deliver safe chia seeds oil. The principal processing step of chia seeds deals with the use of limited quantity of water to adjust the moisture content. Initial processing of chia involves cleaning the grains followed by drying it to desired moisture level of below 14 percent. Chia seeds then, passed through magnetic tunnel to remove any metal foreign material. Asymmetric cleaning unit are generally used to separate seeds on their shape and size basis. The graded seeds were packed and stored under control humidity and temperature. The processed grains of chia seeds were stored under control conditions with prior fumigation which majorly depends on the need and end use.

Some proprietary and patented information is also available for processing chia seed prior to oil extraction. The extraction process of chia involves conventional methods i.e. screw pressing, solvent extraction and supercritical fluid extraction in addition with newly practiced ultra-sonication which is reported to yield oil with better quality in terms of oil composition and longer shelf stability. It is preferred to employ an extraction process which can be made available to larger producers and involve fairly low investment and less tedious processing steps. In recent years, the resurgence of oil producing businesses has been towards using continuous mechanical screw pressing method for the manufacturing of oil because of low chances of solvent retention in oil. This also comes across as a method which is simple and reliable for processing small batches of seeds.

Commercial screw pressing method of oil is purely a mechanical and performed at low temperature, also recognized as "cold pressing". Thus, protecting the fatty acid composition, vitamin-E, phenolic profile and other phyto-nutrients which do not require additives to enhance the shelf stability of oil during storage. Chia seeds are first, partially milled and cleaned to get rid of impurities. The clean seeds are put inside screw press on the hydraulic press, extra care is given to not allow the temperature to rise above 45°C which ensures the molecular stability and composition of oil. The screw pressed oil generally devoid of waxes and other substances usually present in solvent extraction process. Further, the extracted oil is stored in stainless steel tanks. The stored oil is then pumped through the disposal cotton filters and collected in separate storage containers. The collected oil is analysed and packed in desired quantities (Rosas-Mendoza *et al.*, 2017).

Solvent extraction is the most utilised method for oil extraction in industries because of higher recovery percentage (98%) that can be achieved in comparison with other methods. This method can directly be applied to the grain having low oil content as well. Solvent extraction requires diffusion of solvent throughout the material due to which the particle size of seeds is reduced prior to solvent extraction by using method like grinding to boost oil extraction rate. The ground powder of seed is subjected to the chemical extraction using soxhlet equipment or a continuous shaker. The extraction rate and efficiency of extraction can be maximized using various organic solvent depending on diffusion capacity of solvent. In case of chia seed acetone is established as the superior solvent resulting in better oil quality than the other organic solvent. The organic solvent from the oil-solvent mixture is recovered using simple distillation method. Recovered oil is further cleaned and stored in suitable containers. Current researches in relation with oil extraction from chia seed have been designed to improve rate of oil extraction and standardizing the parameters for superior supercritical fluid extraction (SCFE). SCFE of oil using carbon dioxide (CO_2) is divided in three main modes: Dynamic, Static and Static-dynamic. Extraction of oil from chia seeds has been reported to utilise Static-dynamic mode (Uribea *et al.*, 2011). The extraction profile is divided in three distinct phases. Initial phase, this involves the highest collection rate depending upon the solubility of bulk material in supercritical fluid (Phase 1). Phase 2 is the intermediate region, where extraction rate of oil is lower than phase 1 because of lower diffusion rate of fluid through material. Phase 3 is extraction wise represent the portion where the process is truly diffusion control. The production cost for the extraction is estimated from first linear phase also known as the constant acceleration phase this involves high mass transfer domination which corresponds to high extraction rate of accessible solute. Extraction of chia seed using SCFE can be maximize at temperature range from 40-80°C and pressure range of 50-600 bar (Mangold *et al.*, 1983; list *et al.*, 1984; Reverchon *et al.*, 1999). Ixtania *et al*, 2010 reported three different pressure (250, 350 and 450 bars), temperature (40°C, 60°C and 80°C) and three retention time (60 min, 150 min and 240 min).

In another important study Uribea *et al.* (2011) explained the process of chia oil extraction at variable temperature (40, 60 and 80°C) and pressure (136, 272 and 408 bars). The operation involved investigation for finding out the parameters required to obtain best yield of oil. They used 99.99 percent pure carbon dioxide (CO_2) and 99.99 percent pure extra dry air for the process. Process involved filling the extraction chamber with milled and powdered chia seeds forming a bed like structure with void throughout the grain bed. Phase change was achieved for CO_2 by first cooling under pressurized conditions and later by increasing temperature. Extraction of oil is achieved using static-dynamic extraction, static phase involves soaking of ground chia by allowing CO_2 at its chosen supercritical stage to enter the extraction chamber for pre-determined time interval. Further, trailed by dynamic phase, this permits the flow of CO_2 through fixed-bed formed by ground seeds. The CO_2 containing extracted oil enters to the first separation chamber where it is allowed to expose and change phase at atmospheric pressure, here it gets condensed and collected. The decomposed CO_2 ejects from system through a rotameter (k).

Along with the extraction methods such as pressing, solvent extraction and SCFE, ultrasonication in combination with solvent extraction is also studied by Debrowski *et al.* (2018). The extraction process involved breaking the oil pocket in seed by using ultra-sonication method which will result in enhanced oil extraction rate and yield by improving rate of diffusion and extraction during chemical solvent extraction. Comprehensive researches in relation to chia are targeted towards developing a process of oil extraction which can provide high yield of oil with best retention of native oil present in seeds.

Chia Health Benefits

Chia seeds were used by Aztec for manufacturing various traditional food and beverages but in upcoming era somehow chia seeds became unknown and now, there is renewed interest of people in using chia seeds. This is subjected to its proven facts of being an exceptional source of protein, fat with higher unsaturated fatty acid specially Omega 3 fatty acids (α-linolenic), and dietary fibre content associated with various therapeutic health effects. These components tend to express functional properties and help in maintaining health by various means i.e. antioxidant, anticancer, anti-inflammatory, anti-allergic, anti-fatigue. It is reported to prevent chronic diseases; cardiovascular disease, heart attacks, cancer, hormonal/endocrine disorders, hyperlipidaemia, hypertension, stroke and vasodilatory problems. Some reports also suggested the probable role of these seeds as anticoagulant, antioxidant and antiviral.

Worldwide increasing of consumer awareness in regard with health concerns has increased burden on food industries to come up with food formulation which exhibits functional properties and adds on to boost metabolic and immune system in human with proven clinical trials. Development of medicinal food from conventional ingredients such as spices and herbs to arrest or reduced disease like obesity, cancer, autism and diabetes is now gaining momentum among industries and common people. Because of the presence of phytonutrients such as phenols, flavonoids and favourable omega 3and omega 6 essential fatty acids content in chia oil, enhanced utilization of these seeds has been seen for its various health benefits. The presence of bioactive components in chia is well-known but the quantity of chia seeds to be ingested for having prior mentioned health benefits yet to be established and confirmed by medical and scientific researches. Since clinical study-based researches on chia seed health benefits are still limited and prior reports did not come up with any kind of conclusive evidence.

Table 7: Health benefits of chia seeds

S.No	Study Period	Population Size and category	Dose	Conclusion	Reference
1	2 hours	Randomized, controlled, crossover study 15 healthy adults	25 g ground chia seeds with 50 g glucose or 25 g flax with 50 g glucose or alone 50 g glucose	Postprandial glycaemia significantly reduced in comparison with the control group. Chia significantly reduced the mean ratings of desire to eat, prospective consumption and overall appetite score, when compared with flax.	Vuskan *et al*, 2017
2	2 hours	11 healthy men and women	0, 7, 15, 24 g chia seeds added to bread /day	Postprandial glycaemia significantly reduced in comparison with the control group.	Vuskan *et al*., 2010
3	2 hours	Randomized double-blind trial, 13 healthy people	50 g bread with 0, 7, 15, 24 g chia seeds added	The blood test showed reduced postprandial glycaemia.	Ho *et al*., 2013
4	35-42 days	Hypercholesterolaemic rabbits	10 g chia oil/1kg diet [CD] or 10 g chia oil + 1g cholesterol/ 1 kg diet [HD-Cd]	Reduced concentration of triacylglycerols and increased content of α-linolenic acid in the serum in HD-cd group. Chia seed oil may have protective effect on blood vessels.	Sierra *et al*., 2015
5	49 days	10 women after menopause	25 g ground chia seeds / day	Increased concentration of α-linolenic and eicosapentaenoic acids in the serum of women supplemented with ground chia seeds.	Jin *et al*., 2012
6	42-84 days	36 young obese rats	133 g chia seeds/ 1 kg diet Or 40 g chia oil/ 1 kg diet	Chia seeds and oil reduced oxidative stress in vivo by improving the antioxidant status and reducing lipid peroxidation in diet-induced obese rats.	Marineli, *et al*., 2015

7	60 days	Randomized double-blind trial, 67 men and women aged 20-60 years	4 g of chia seeds mixed with palm, oats and soy powder diluted in 250 mL of water/2 per day + reduction diet	Reduced concentration of triacylglycerols, CRP and insulin resistance in the supplemented group.	Guevara-Cruz, *et al.*, 2012
8	70 days	Randomized double-blind trial, 62 overweight women aged 49- 75 years	25 g whole chia seeds/day or 25 g ground chia seeds /day	No influence of [whole/ground] seeds on inflammatory markers, blood pressure, body composition. Increased concentration of α-linolenic and eicosa-pentaenoic acids in the blood serum of obese women consuming ground seeds vs the control group and the group consuming whole chia seeds.	Nieman, *et al.*, 2012
9	84days	Randomized single-blind trial; 20 men and women aged 18-75 years with type 2 diabetes	37g ground chia seeds added to bread /day	Reduced systolic blood pressure and CRP concentration, a double increase in the concentration of α-linolenic and Eicosa-pentaenoic acids in the serum of patients supplemented with ground chia seeds in comparison with the control group.	Vuskan *et al.*, 2007
10	84 days	Single-blind trial, 76 obese people (placebo 37; chia seeds 39)	25 g chia seeds in 250 mL water twice a day	Increased concentration of α-linolenic acid in the serum of the group under study vs placebo. No influence of seeds on inflammatory markers, blood pressure, body composition	Nieman *et al.*, 2009
11	84 days	26 men and women aged 45-55 years (placebo 7; chia flour 19)	35 g chia flour/day	Decreased body weight in the group consuming chia flour, a greater decrease in obese people, no difference from the placebo group. Reduced total cholesterol and increased LDL cholesterol in the supplemented group.	Toscano, *et al.*, 2015

The limited studies conducted on chia dictate that the mode of action and functional benefits potential of chia seeds hinge on its mode of consumption. Health benefits of chia mucilage and soluble fibre can be efficiently utilized when they were ingested as whole whereas, more prominent benefits due to lipid and polyphenols were noticed during ingestion of powdered chia seeds. Nieman *et al.* (2012) studied the effect of both pounded and whole chia seeds on factors linked with risk of disease such as overweight and postmenopausal women. They detected the impact of chia in enhancing beneficial concentration of α-linolenic acid ALA) and eicosapentanoic acid (EPA) in plasma but, no effect was observed on factors which influence the disease. Espada *et al.* (2007) analysed that anti-proliferative and apoptotic impact of chia seed is due to its high α-linolenic acid content. They administered 25 g of chia powder diluted with 0.25 l of water into wistar rats and taken periodic observation twice daily, the observation was positive as tumour weight was expressively reduced in chia oil group in contrast with controlled rats.

Thus chia oil diet reduced tumour weight, supported apoptosis and restricted mitosis of cancerous cell. They attributed the anti-proliferating potential of chia to beneficial content of α-linolenic acid, leading to accumulation of ecosapentanoic acid (EPA) which might have a direct apoptotic effect by increasing caspase activity resulting in enhanced rate of cancerous cell death. Similar to the above study, Rosas-Ramirez *et al.* (2017), also investigated the influence of chia seeds oligosaccharide and mucilage on decreasing the resistance and proliferation of cancerous cell namely HeLa, MCF7 and HcpG2 cells. The results obtained were positive but, no claim was made regarding the specificity of seeds mucilage towards cancerous cells death and stated that degeneration of non-cancerous cells by mucilage yet to be studied.

Chia seeds potential to reduce cardiac and obesity problems are certain extensive studies performed in recent years. Several human and animal related studies were performed to investigate the potential of chia as whole seed and its oil against these modern era illnesses. Both ingesting chia as whole seed or its oil were found effective against cardiac diseases due to their anti-oxidation and blood thinning potential. The aforesaid potential of seeds was accredited to existence of α-linolenic acid, which aid in formation of important biochemical components such as prostaglandins, leukotrienes, and thromboxane which were analysed in previous researches to control several physiological functions (Pawlosky *et al.*, 2003). It also has capability of controlling the calcium and sodium channel functioning which helps in reducing hypertension, heart rate variability and improves parasympathetic, and protects ventricular arrhythmia (Ayerza and Coates, 2001). A research conducted at St. Michael Hospital in

Toronto recognized chia seeds among rich sources of omega-3 fatty acid, iron, calcium and magnesium (better than milk). Study also suggested that the consumption of 37g of seeds per day can alleviate the blood glucose in patients having diabetes It was stated to possess potential to prevent platelet aggregation resulting in reduce chances of myocardial Infarction, heart stroke and helps in lowering systolic blood pressure (SBP) up to 6 mmHg (Vuskan *et al.*, 2010; Vuskan *et al.*, 2017). In some others researches chia seeds were established as effective in reducing level of bad cholesterol such as LDL VLDL and increasing good cholesterol levels like HDL decreasing visceral adiposity, triglyceride and diastolic blood pressure (DSBP) from 66.1-61.5 mmHg (Vertomen *et al.*, 2005). Researches conducted in relation to obesity and body weight control were motivated due to the fact that chia seeds are excellent source of dietary fibres i.e. soluble and insoluble dietary fibre. Chia seeds also been reported to reduce sensitivity of insulin and control post-postprandial glucose levels. But the findings were inconclusive as equal number of reports are available reporting the adequacy and inadequacy of chia seeds or oil in controlling the body weight (Valenzuela *et al.*, 2014; Marinelli *et al.*, 2015; Chicco *et al.*, 2008; Fortno *et al.*, 2017). Creus *et al.* (2016) studied capacity of chia seeds on body weight and visceral adiposity. He observed increased activity of pyruvate dehydrogenase (45.1%), resulting in more responsive metabolism but no marked changes were observed in plasma metabolite, glucose infusion rate and insulin levels. In a study Lee *et al.* (2006) reported *S. hispanica* potential in controlling appetite and postprandial glucose in blood. In similar analysis Vuskan *et al.* (2010) investigated the reducing postprandial glycaemia. But no study reported the reduction in bodyweight complimenting the reduced appetite and postprandial glucose levels. Guevara-Cruz *et al.* (2012) reported the reduction in body weight of patients kept on chia seeds diet but the loss in body weight was found genetic makeup dependent as the subject who expressed ABC AC1 R230 gene experience significant weight loss. Hence, the potential of chia seeds in regulated blood glucose can be established effectively whereas, similar confirmation can't be given with weight loss or reduction in obesity. Although the proposed principles believed the clinical potential of chia seeds were simplistic in prior discovery phase, the hypothesis regarding health benefit generated interest and prompted extended investigation towards establishing physiological potential and health related benefits of chia on human.

Chia seeds were testified to have some unfavourable impact on human health such as excessive thinning of blood when consumer is already on blood thinning medication such as warfarin, aspirin or coumadin. This is attributed to anti-coagulating factors present in chia seeds. Although unlikely to occur but chia seeds allergic possibility were reported and assumed that if someone is allergic to sesame or mustard seeds then that person is likely to be allergic to chia seeds as well. Other symptoms for the allergy include hive, rashes and watery eyes. During severe exposure vomiting, diarrhoea, improper breathing and swelling of tongue may occur. High concentration of mucilage and dietary fibre in some cases also reported to cause the gastrointestinal problems such as constipation and stool, diarrhoea, excessive gas and bloating, these responses were accredited to excellent fluid absorption potential of chia. Chia seeds were stated as significant source of B group vitamins specifically B^{17}. Hence, if consumer is on B vitamin doses then ingestion of chia can lead to the overdose of vitamin B^{17} which has toxic effect and can cause sudden drop in blood pressure. The possible ill effects of chia seed are associated with the excessive consumption of chia seeds which is rare. The possibilities of side effects are fairly less and does not hold much concern. The researches have not been conducted in this regard and it demands basic research before establishing such problem possibilities, conversely studies on beneficial health effect of chia seeds are extensively conducted and well established.

Traditional Food and Beverages using Chia Seeds

Systematic scientific studies stated chia as a nutritionally dense seed with proven functional properties. It is widely accepted scenario that world population will reach 9 billion by 2050. Considering the present trend of modern era diseases such as diabetes, obesity, anxiety, anaemia and cardiovascular disorder, there is immense pressure on food technologist and nutritionist to design healthy and functional food using latest scientifically proclaimed knowledge. Chia can potentially be used as whole or as ingredient in development of processed food of choice with improved health benefits.

Recently, chia came up as subject of renewed interest. It has gathered importance of a novel functional food with widely reported nutritional properties. Indeed, there is remarkable interest of many countries from different parts of world in utilising chia seeds for manufacturing healthy food with commercial application and specific dietary needs. Nutritional potential of chia has been reported to be superior than most of commercial cereal grains owing to high protein (characterized with balanced essential amino acid), lipid (characterized by favourable fatty acid ratio), mineral content (iron, magnesium, calcium, potassium and zinc) and renowned dietary fibre (soluble and insoluble). Chia were used by Aztec people in various forms such as dried seed powder, whole seed, whole seed flour, soaked grain etc. in the raw cut vegetables, soup preparations and juices and other beverages. Increased awareness regarding chia resulted in current researches to develop food product using chia seed with improved quality and nutritional properties. Foremost, chia seeds were used in development of beverages technologies because of chia's potential to form mucilage, providing desirable consistency to the beverage contingent on quantity of seed used. Minatelli *et al.* (2009) was awarded with the patent, describing the novel method for developing chia seeds beverage. They developed Ready-to-Reconstituted (RTR) and Ready-to-Serve beverage. The prior was consisting of juice concentrate, flavouring components, sweetener, fruit smoothies, and chia seed powder. Whereas, later Ready-to-Serve consist of liquid fruit juice, water or naturally or artificially flavoured water and whole seed of *Salvia hispanica.* The beverage was reported to be effective in ensuring the gastrointestinal and heart conditions owing to exceptional fibre content of chia seeds. In earliest times, chia was used in manufacturing beverages commonly recognized as "Chia Fresca" usually made with different fruit juices for example lime juice in ratio ranging from 1:12 to 1:30 seed to juice w/w, with or deprived of additional sweetener. The product is still being prepared in Centre and South America. The ancient beverage was further explored by the researchers and stabilized to increase shelf stability. Minatelli *et al.* (2009) also specified in their patent about possible use of chia mucilage in developing gel enhanced beverages. Method involved hydration of fibre in chia seeds to form jelly mucilage. The gel mucilage is further isolated from seeds and utilised in juice to create gel enhanced beverage. Remaining seed were used either alone or mixed with other cereals or legume flour, meat-based seasoning, vegetable paste, dairy and non-dairy products to improve nutritional content.

Arguelles-Lopez *et al.* (2018) developed functional beverages from pseudocereal amaranth and chia flours. The grains were processed using germination, germination temperature of 29°C and 30°C and germination time of 197h and 78h were used for chia and amaranth seeds respectively. The optimized extrusion conditions for both the grain were same (extrusion temperature: 141°C and screw speed: 81rpm). The process was standardized to bring down the nutrition loss during processing and achieve better nutritional quality grain. Grains were investigated for their nutritional, antioxidant and antihypertensive capacity prior and after processing. A 200 mL of functional beverages, made with 25 g functional ingredient consisting 70 percent EAF+ 30 percent GCF or 70 percent GAF + 30 percent GCF mixtures. The nutritional composition of mixture was 3.90-4.53 g protein, 5.04-6.81 g dietary fiber, 95-96 kcal of energy, theoretical protein efficiency ratio = 2.52-2.69, antioxidant potential of 4,009-6,495 μmol Tocopherol equivalent, antihypertensive potential (IC_{50}) 0.43-0.47 μg extract/mL, and sensorial acceptability of above 7 on 9 point hedonic scale. Beverages were reported to possess immense capacity as functional beverage and exploited for their commercial use.

Felisberto *et al.* (2015) described process for the development of lyophilized chia mucilage (CMG) which was further utilised in the preparation of cake. Aim of investigation was to establish possible substitute of vegetable fat with chia mucilage at different level (25, 50, 75 and 100 g/g of fat). They reported successful substitution of ¼ vegetable fat by CMG without significantly altering the specific volume, symmetry and water activity. Higher level of substitution of CMG significantly affected the colour, crumb firmness and loaf volume of cake. Inglet *et al.* (2014) developed chia-oat composite cookies. They utilised three diverse formulations of chia with other grain, with substitution of wheat flour at 20 percent level: (1) Chia-nutrium composite 1:4, (2) chia-oat bran concentrate (OBC) 1:4 and (3) chia-whole wheat flour composite 1:4 (WOF). They successfully reported the 20 percent replacement of chia-OBC and chia-WOF without affecting the geometrical and textural properties of cookies whereas, significant decrease in cookies diameter along with enhanced height, and dough hardness was perceived in 20 percent chia-nutrium composite replacement. Villalobos *et al.* (2012), performed standardization of corn replaced chia tortillas. In their investigation they replaced corn flour at various level with chia flour (5%, 10%, 15%, and 20%). They reported effective substitution of chia flour at all levels without adversely influencing physicochemical and texture properties of tortillas. The nutritional profile of tortillas considerably improved as protein, lipid and fibre content for chia replaced tortillas were significantly higher ($p<0.001$) in comparison with control (corn) tortillas. The enzymatic starch hydrolysis rate for tortillas was reduced, resulting in reduced glycaemic index owing to complex carbohydrate and significant fibre content of chia seed. In conclusion they reported tortillas as a nutraceutical because of their high dietary fibre, lower assimilation rate and predicted glycaemic index. Coellus *et al.* (2014) studied the addition of chia seeds as whole or in powder form in two different bread production. Once chia flour was replaced at 7.8 g/100g fat content reduced by 0.9 g/100 g of bread, on replacement of 11.0 g/100g the fat was reduced 1.0 g/100g of bread, resulting in 27 and 24 percent reduction in saturated fat levels respectively. The poly-unsaturation to saturation ratio was also improved from 1.01 to 3.1 and 3.9 for chia flour and chia seed replaced bread respectively. The bread prepared using both these preparations also had significant dietary fibre and ω-fatty acids in it. The report represented that formulation can effectively be used in manufacturing product with reduced saturated fatty acid and increase the omega-3 fatty acid consumption. Segura-Compos *et al.* (2013) studied physiological favourable potential of *S. hispanica* protein hydrolysate and standardization of methods for functional foods like white bread and carrot cream. The investigation showed positive results with increased ACE-inhibitory activity (141.29-297.68; 0.24-1.71 μg protein/ml) and no significant impact on antioxidant properties of both preparations. The quality characteristic of product was not significantly influenced after incorporation of chia seed hydrolysate.

Though primarily replacement of chia seeds was performed in vegetarian product because of their similarity in composition. There are few studies which also reported substitution of chia seeds in meat-based products as well. Ding *et al.* 2018 developed the restructured ham-like product. The research included substitution of meat fraction with chia seed flour to decrease fat content without affecting product quality. Addition of chia 1 percent and 5 percent carrageen resulted in improved quality of ham restructured product with increased nutritional properties and unaffected sensorial properties. Incorporation of chia resulted in reduced centrifugation liquid loss, improved texture, lower lipid oxidation and other physico-chemical properties. Author reported that chia at 1 percent level and 5 percent carrageen can effectively be incorporated for preparation of restructured ham-like product. He also stated that chia possesses prodigious potential as an ingredient which can be used as fat substituent for manufacturing low fat healthy and good quality meat products. Fitzpatrick (2004), in his US patent 6,827,965 B1 reported probable usage of whole chia seeds or gluten-free agglutinant produced using chia seed for production of various foods like snacks, soups and breakfast product. In his claim, he reported method for producing dehydrated food slice comprising of fruit, sugar, agglutinant, vegetable, unsprouted grain, sprouted grain, honey and vegetable powder with whole chia seed or as an agglutinant. Leslie and Kenton (1984) studied the impact of particle size on final product quality. He also reported that the food preparation having chia seeds should be processed to minimize water activity of product to improve shelf stability of product as chia seeds has tendency to absorb and lock water which tends to improve water activity of product and result in reduction of shelf stability because of upsurge susceptibility of product against microbial attack. They also mentioned that chia seeds in fine particle form can also be used in the preparation of snacks like crackers, biscuits, cookies etc. without affecting the quality of food product whereas increased particle size tends to effect product quality significantly.

Market Potential and Industrial Application

The current market scenario vibrantly shifting towards functional foods as these foods have gained tremendous recognition worldwide due to increased awareness among consumers about beneficial health benefits. One of prime reason for shift in interest is because of around the world increased cases of individuals having chronic diseases like cardiovascular diseases, high blood pressure, obesity and diabetes, which are also known as modern day diseases. The common cause of these diseases is attributed to the inactive lifestyle, ignorant eating style, excessive ingestion of SFA and lower ingestion of PUFA which are stated as the reason for cardiovascular diseases. There has been significant shift in research from commercial to traditional grains with health benefits. At present, majority of the studies are concentrated on increasing nutraceutical quality of human diet. The functional potential of chia is primarily attributed to the availability of active ingredient and, bioactivity of compounds available in plant being stable and available even after processing for human consumption.

In recent years, chia has gained its status because of its high oil content, high concentration of α-linolenic acid and excellent nutritional composition of seeds. Claim of chia seed to be a functional food by European Parliament Committee has played important role in its enhanced popularity. Because of its beneficial nutritional qualities like high protein and dietary fibre, chia seeds currently used by various food industries like bakery, beverage, confectionery and nutraceutical industries. Chia seeds have been effectively supplemented for developing composite formulation of multipurpose flour with better nutritional composition and functional properties such as water holding capacity, gelling capacity etc. Chia in powder form is also added in numerous dough and batter formulation for preparing bread, biscuits, cookies, cakes, pastries, pasta and noodles. 5 percent addition of chia seeds in bread preparation is approved by European Commission. U.S. dietary guidelines recommended the use of chia seed as primary food not exceeding 48 g/day. Chia seeds owing to its various properties provide various opportunities for possible use in packaging industry as well. The film formed using chia seed were observed to have better water vapour transmission rate (WVTR) and gas transmission rate (GTR). The future usage of chia seeds expected to take commercial form as breakfast stuff, snacks and other products with desired functional and physiological potential with specified and targeted use. Chia seed can efficiently be utilized in development of both fermented and non-fermented beverages. Standardization and development of these functional foods will help in providing the variety in human diet with established beneficial health effects which will prevent modern lifestyle diseases and maintain various body function.

Conclusion

Chia seeds are well known for its immense nutritional quality with each of its component can be exploited for medicinal or functional uses. The exceptional nutritional potential of chia provides these grains as fascinating substitute to ensure and enhance diversity of food in diet. To realize this aim, however, contribution from agriculture side with respect to intense breeding efforts is necessary. Scientist and technologist need to put their efforts for developing better processing technologies which can make processing of chia seeds more economical, as conventional milling cannot be utilised for chia because it has tendency to form mucilage when it comes in contact with water utilized during traditional processing. The upsurge in research towards investigating the safety and possible allergic nature of chia will promote the incorporation of chia seeds in food preparations with improve functional and nutritional potential, and meet sensorial requirement of consumer, providing them with variety of food to choose from. In addition extended marketing efforts are also required to promote chia seeds, increase awareness and make chia more popular among consumers because at present scenario only a small percentage of world's population is much aware with this consumer beauty. Chia seeds with their high concentration of omega-3 & 6 fatty acid, is explored in food as well as pharma industries but, to realise this at large extent the aforementioned factors require consideration primarily to achieve commercially profitable venture of chia seeds. The right outline for evolution of chia seeds can provide the possible way to utilize the nutritional benefits of chia and provide it to majority of population.

References

Ali, N.M., Yeap, S.K., Ho, W.Y., Beh, B.K., Tan, S.W. and Tan, S.G. (2012). The promising future of chia, (*Salvia hispanica* L.). Journal of biomedicine and Biotechnology, doi.10.1155/2012/171956.

Alvarado-Sua´rez, L. A. (2008). Estudiosobrelaspropiedadesmeca´nicas y funcionales de biopelı´culasformadaș con proteı´na de semilla de limo´nmexicano (*Citrus aurantifolia*). Me´xico: Facultad de Quı´mica, UNAM.

Archer, B.J., Johnson, S.K., Devereux, H.M. and Baxter, A.L. (2004). Effect of fat replacement by inulin or lupin-kernel fibre on sausage patty acceptability, post-meal perceptions of satiety and food intake in men. British Journal of Nutrition, 91, 591-599.

Arguelles-Lopez, O.D., Reyes-Moreno, C., Gutiérrez-Dorado, R., Sanchez-Osuna, M.F., , Jaime Lopez-Cervantes, J., Cuevas-Rodríguez, E.O., Milan-Carrillo, J., Perales-Sanchez, J.X.K. (2018). Functional beverages elaborated from amaranth and chia flours processed by germination and extrusion. Biotecnia, 20, 135-145.

Ayerza, R. (1995). Oil content and fatty acid composition of chia (*Salvia hispanica* L.) from five northwestern locations in Argentina. Journal of the American Oil Chemist's Society,, 72, 1079-1081.

Ayerza, R. (2010). Effects of seed color and growing locations on fatty acid content and composition of two chia (*Salvia hispanica* L.) genotypes. Journal of the American Oil Chemist's Society,, 87, 1161-1165.

Ayerza, R.(2002). Chia as a new source of ω-3 fatty acids: advantage over other raw materials to produce ω-3 enriched eggs, in proceedings of the symposium on omega-3 fatty acids, evolution and human health,Washington, DC, USA.

Ayerza, R., and Coates, W. (2000).Dietary levels of chia: influence on yolk cholesterol, lipid content and fatty acid composition for two strains of hens,Poultry Science, 79, 724-739.

Ayerza, R., and Coates, W. (2001) Chia seeds: new source of omega-3 fatty acids, natural antioxidants, and dietetic fiber. Southwest Center for Natural Products Research & Commercialization. Office of Arid Lands Studies, Tucson

Ayerza, R., and Coates, W. (2001). Chia seeds:New source of omega-3 fatty acids, natural antioxidants and dietetic fiber. AZ, USA: Southwest Center for Natural Products Research & Commercialization Office of Arid Lands Studies, The University of Arizona.

Ayerza, R., and Coates, W. (2002) Dietary levels of chia: influence on hen weight, egg production and sensory quality, for two strains of hens. British Poultry Sciences, 43, 283-290.

Ayerza, R., and Coates, W. (2004) Composition of chia (*Salvia hispanica*) grown in six tropical and subtropical ecosystems of South America. Tropical Sciences, 44, 131-135.

Ayerza, R., and Coates, W. (2005) Chia: Rediscovering an ancient crop of the Aztecs. University of Arizona, Tucson

Ayerza, R., and Coates, W.(1999). "An ω-3 fatty acid enriched chia diet: influence on egg fatty acid composition, cholesterol and oil content," Canadian Journal of Animal Science, 79, 53–58.

Ayerza, R., Coates, W. and Lauria, M. (2002). Chia seed (*Salvia hispanica* L.) as an omega-3 fatty acid source for broilers: influence on fatty acid composition, cholesterol and fat content of white and dark meats, growth performance, and sensory characteristics,Poultry Science, 81, 826-37.

Ayerza, R.J. and Coates, W. (2007) Seed yield oil content and fatty acid composition of three botanical sources of & ohgr: 3 fatty acid planted in the Yugas ecosystem of tropical Argentina. Tropical Sciences, 47, 183-187.

Bolanos D., Marchevsky E.J. and Camina J.M. (2016). Elemental analysis of amaranth, chia, sesame, linen, and quinoa seeds by ICP-OES: Assessment of classification by chemometrics. Food Analytical Methods, 9, 477-484.

Cahill, J. P., & Ehdaie, B. (2005). Variation and heritability of seed mass in chia (Salvia hispanica L.). Genetic resources and crop evolution, 52(2), 201-207.

Capitani, M.I., Spotorno, V., Nolasco, S.M. and Tomas, M.C. (2012). Physicochemical and functional characterization of by-products from chia (*Salvia hispanica* L.) seeds of Argentina. LWT-Food Science and Technology. 45, 94-102.

Chicco, A.G., D'Alessandro, M.E., Hein, G.J., Oliva, M.E. and Lombardo, Y.B.(2009). Dietary chia seed (*Salvia hispanica* L.) rich in alpha-linolenic acid improves adiposity and normalises hyper-triacylglycerolaemia and insulin resistance in dyslipaemic rats. British Journal of Nutrition, 101, 41-50.

Ciftci1, O.N., Przybylski, R. and Rudzinska, M. (2012). Lipid components of flax, perilla and chia seeds. European Journal of Lipid Science and Technology. doi: 10.1002/ejlt.201100207.

Coelho, M.S., Salas-Mellado, M.D.L.M. (2015). Effects of substituting chia (*Salvia hispanica* L.) flour or seeds for wheat flour on the quality of the bread. LWT-Food Science and Technology, 60, 729-736.

Coelho, M.S., Salas-Mellado, M.D.L.M. (2015).Chemical characterization of chia (Salvia hispanica l.) For use in food products.Journal of Food and Nutrition Research, 2, 263-269.

Creus, A., Ferreira, M.R., Oliva, M.E. and Lombardo, Y.B. (2016). Mechanisms involved in the improvement of lipotoxicity and impaired lipid metabolism by dietary α-linolenic acid rich Salvia hispanica L. seed in the heart of dyslipemic insulin-resistant rats. Journal of Clinical Medicine, 5, E18.

Dabrowski, G., Konopka, I. and Czaplicki, S. (2018). Supercritical CO2 extraction in chia oils production: impact of process duration and co-solvent addition. Food Science and Biotechnology, 27, 677-686.

Deka, R. and Das, A. (2015). Advances in chia seed research. Advance of Biotechnology and Microbiology, doi: 10.19080/AIBM.2017.05.555662.

Ding, Y, Lin, H.W., Lin, Y.L., Yang, D.J., Yu, Y.S., Chen, J.W., Wang, S.Y. and Chen, Y.C. (2018). Nutritional composition in the chia seed and its processing properties on restructured ham-like products. Journal of Food and Drug Analysis, 26, 124-134.

Egras, M. A., Hamilton,W.R., Lenz,T. L. and Monaghan, M. S. (2011). An evidence-based review of fat modifying supplemental weight loss products. Journal of Obesity. doi:10.1155/2011/297315.

Elleuch, M., Bedigian, D., Roiseux, O., Besbes, S., Blecker, C., & Attia, H. (2011). Dietary fibre and fibre-rich by-products of food processing: Characterisation, technological functionality and commercial applications: A review. Journal of Food Chemistry, 124, 411-421. https://doi.org/10.1016/j.foodchem.2010.06.077

Espada, C.E., Berra, M.A., Martinez, M.J., Eynard, A.R.andPasqualini, M.E. (2007). Effect of chia oil (*Salvia hispanica*) rich in omega-3 fatty acids on the eicosanoid release, apoptosis and T-lymphocyte tumor infiltration in a murine mammary gland adenocarcinoma. Prostaglandins Leukot Essential Fatty Acids, 77, 21-8.

Estilai, A., Hashemi, A. and Truman, K. (1990). Chromosome number and meiotic behaviour of cultivated chia, (*Salvia hispanica* L). Horticulture Science.25, 1646-1647.

European Food and Safety Authority (2005). Opinion of the scientific panel on dietetic products, nutrition and allergies on a request from the commission related to the safety of chia (*Salvia hispanica* L.) seed and ground whole chia seed as a novel food ingredient intended for use in bread. European Food and Safety Authority Journal, 278, 1-12.

European Food Safety Authority, (2009). Opinion on the safety of 'chia seeds (*Salvia hispanica* l.) and ground whole chia seeds' as a food ingredient: scientific opinion of the panel on dietetic products, nutrition and allergies. The European Food Safety Authority Journal, 996, 1-26.

Falco, B.D., Amato, M. and Lanzotti, V. (2017). Chia seeds products: An overview. Phytochemical Review. 16, 745-760.

FAO/WHO/UNU (1985). Energy and protein requirements. Report of a joint FAO/WHO/UNU expert consultation: Technical report series, 724. Geneva: WHO.

Felisberto, M.H.F., Wahanik, A.L. Gomes-Ruffi, C.R, Clerici, M. T. P, Chang, Y. K, Steel, J.S. (2015). Use of chia (*Salvia hispanica* L.) mucilage gel to reduce fat in pound cakes. LWT-Food Science and Technology. 63, 1049-1055.

Fitzpatrick (2004). Food products containing whole chia seed or a gluten-free agglutinant derved there from and methods of making same. US Patent, US 6,827,965 B1.

Guevara-Cruz, M., Tovar, A.R., Aguilar-Salinas, C.A., Medina-Vera, I., Gil- Zenteno, L., Hernandez-Viveros, I., Lopez-Romero, P., Ordaz-Nava, G., Canizales-Quinteros, S., Pineda, L.E.G. and Torres, N. (2012). A dietary pattern includingnopal, chia seed, soy protein, and oat reduces serum triglycerides and glucose intolerance in patients with metabolic syndrome. Journal of Nutrition, 142, 64-69.

Guiotto, E.N., Capitani, M. I., Nolasco, S.M. and Tomas, M.C. (2015). Stability of oilinwater emulsions with sunflower (Helianthus annuus l.) and chia (Salvia hispanica l.) byproducts. Journal of the American Oil Chemist's Society, doi.10.1007/s11746-015-2746-9

Ho, H., Lee, A.S., Jovanovski, E., Jenkins, A.L., Desouza, R. and Vuksan, V. (2013). Effect of whole and ground salba seeds (*Salvia hispanica* L.) on postprandial glycemia in healthy volunteers: a randomized controlled, dose response trial. Europian Journal of Clinical Nutrition, 67, 105-110.

Imran, M., Nadeem, M., Manzoor, M.F., Javed, A., Ali, A. Akhtar, M. N., Ali, M. and Hussain, Y. (2016). Fatty acids characterization, oxidative perspectives and consumer acceptability of oil extracted from pre-treated chia (*Salvia hispanica* L.) seeds. Lipids in Health and Disease. doi. 10.1186/s12944-016-0329-x.

Inglett, G.I., Chen, D. and Sean, L. (2014). Physical properties of sugar cookies containing chia-oat composites. Journal of the Science of Food and Agriculture. 94, 3226-3233.

Ixtaina, V.Y., Martinez, M. L., Spotorna, V., Mateo, C. M., Maestri, D.M., Diehl, B.W.K., Nolasco, S.M. and Toamas, M.C. (2011). Characterization of chia seeds oils obtained by pressing and solvent extraction method. Journal of Food Composition, 24, 166-174.

Ixtaina, V.Y., Vega, A., Nolasco, S.M., Tomas, M.C., Gimeno, M., Barzana, E. and Tecante, A. (2010). Supercritical carbon dioxide extraction of oil from Mexican chia seed (*Salvia hispanica* L.): characterization and process optimization, Journal of Supercritical Fluids, 55, 192-199.

Jin, F., Nieman, D.C., Sha, W., Xie, G., Qiu, Y. and Jia, W. (2012). Supplementation of milled chia seeds increases plasma ALA and EPA in postmenopausal women. Plant Foods for Human Nutrition, 67, 105-110.

Julio, L.M., Ixtaina, V.Y., Fernandez, M., Sanchez, R.M.T., Nolasco, S.M. and Tomas, M.C. (2016). Development and characterization of functional O/W emulsions with chia seed (*Salvia hispanica* L.) by-products. Journal Food Science and Technology, 53, 3206-3214.

Kaczmarczyk, M.M., Miller, M.J. and Freund, G.G. (2012). The health benefits of dietary fiber: Beyond the usual suspects of type 2 diabetes mellitus, cardiovascular disease and colon cancer. Metabolism, 61, 1058-1066.

Lattimer, J.M. and Haub, M.D.(2010). Effects of dietary fiber and its components on metabolic health. Nutrients, 2, 1266-1289.

Lee, A.S., Jovanovski, E., Jenkins, A.L., Desouza, R. and Vuksan, V. (2009). Effect of whole and ground Salba Seeds (*Salvia hispanica* L.) on postprandial Glycemia in healthy Volunteers: a randomized controlled, dose-response Trial [Thesis]. Toronto, Canada, University of Toronto, 2009.

Leslie. and Kenton, S. (1984). "Raw Energy". Warner Books, New York, 264-266.

Lin, K.Y., Daniel, J. R. and Whistler, R. L. (1994). Structure of chia seed polysaccharide exudate. Carbohydrate Polymers, 23, 13-18.

Llorent-Martinez E.J., Cordova M.L.F, Ortega-Barrales P. and Ruiz-Medina A. (2013). Characterization and comparison of the chemical composition of exotic superfoods. Microchemistry Journal, 110, 444–451.

Mangold, H.K. (1983). Liquefied gases and supercritical fluids in oilseed extractionJournal of the American Oil Chemist's Society, 60,178-179.

Marcinek, K. and Krejpcio, Z. (2017). Chia seeds (salvia hispanica): health promoting properties and therapeutic applications – A review. Rocz Panstw Zakl Hig, 68, 123-129.

Marineli, R.D.S, Moraes, E.A., Lenquiste, S.A,Godoy, S.T., Eberlin, M.N. and Junior, M.R.M. (2014). Chemical characterization and antioxidant potential of Chilean chia seeds and oil (Salvia hispanica L.). LWT - Food Science and Technology. 59, 1304-1310.

Marineli, R.S., Lenquiste, S.A., Moraes, E.A. and Marostica, M.R. (2015). Antioxidant potential of dietary chia seed and oil (*Salvia hispanica* L.) in diet-induced obese rats. Food Research International, 76, 666-674.

Martinez, M.L., Marin, M.A., Faller, C.M.S., Revol, J., Penci, M. C. and Ribotta, P.D. (2012). Chia (*Salvia hispanica* L.) oil extraction: Study of processing parameters. LWT-Food Science and Technology,47, 78-82.

Masoero, G., Sala, G., Meineri, G., Cornale, P., Tassone, S. and Peiretti,P. G(2008). Nir spectroscopy and electronic nose evaluation on live rabbits and on the meat of rabbits fed increasing levels of chia (*Salvia hispanica* L.) seeds, Journal of Animal and Veterinary Advances, 7, 1394-1399.

Melo-Ruiz, V., Schettino-Bermudez, B., Rodriguez-Diego, J., Diaz-Garcia, R., Calvo-Carrillo, C. and Gazga-Urioste, C. (2016). Chia seeds (*Salvia hispanica* L.) wild plant rich in nutrients. Journal of Life Science, 10, 221-227.

Minatelli, J.A., Hill, W.S., Moerck, R. and Nguyen, U. (2009). Chia seed beverage and related method. US Patent US 2009/0181114 A1.

Munoz, L.A., Cobos, A., Diaz, O. and Aguilera, J. M. Chia seed (*Salvia hispanica*): an ancient grain and a new functional food. Food Review International. https://doi.org/10.1080/87559129.2013.818014.

Nieman, D. C., Gillitt, N., Jin, F., Henson, D. A., Kennerly, K., Shanely, R. A., *et al.* (2012). Chia seed supplementation and disease risk factors in overweight women: Ametabolomics investigation. Journal of Alternative and Complementary Medicine, 18, 700-708.

Nieman, D.C., Cayea, E.J., Austin, M.D., Henson, D.A., McAnulty, S.R. and Jin F. (2009). Chia seed does not promote weight loss or alter disease risk factors in overweight adults. Nutrition Research, 29, 414-418.

Nitrayova, S., Brestensky, M., Heger, J., Patras, P., Rafay, J. and Sirotkin, A. (2014). Amino acids and fatty acids profile of chia (*Salvia hispanica* L.) and flax (Linum usitatissimum L.) seed. Potravinarstvo Scientific Journal of Food Industry, 8, 72-76.

Nutritional Nutrient Database for Standard Reference, Release 24. Nutrient data laboratory home page. Us; 2011. Department of agriculture, agricultural research service.

Oliveira-Alves, S.C., Vendramini-Costa, D.B., Cazarin, C.B.B., Junior, M.R.M., Ferreira, J.P.B., Silva, A.B., Prado, M.A., and Bronze, M.R. (2017). Characterization of phenolic compounds in chia (*Salvia hispanica* L.) seeds, fiber flour and oil. Food Chemistry, 232, 295-305.

Olivos-Lugo, B. L., Valdivia-Lopez, M. A., and Tecante, A. (2010). Thermal and physicochemical properties and nutritional value of the protein fraction of Mexican chia seed (Salvia hispanica L.). Food Science and Technology International, 16, 89-96.

Ortiz de Montellano, B. R. (1978). Aztec Cannibalism: An Ecological Necessity? The Aztec diet was adequate in protein and cannibalism would not have contributed greatly. Science, 200(4342), 611-617.

Pająk, P., Socha, R., Broniek, J., Krolikowska, K. and Fortuna, T. (2018). Antioxidant properties, phenolic and mineral composition of germinated chia, golden flax, evening primrose, phacelia and fenugreek. Food Chemistry, https://doi.org/10.1016/j.foodchem.2018.09.081.

Palma, F. and Dondem,W. R. (1947). Fixed oils of Mexico I. Oil of chia-Salvia hispanica. Journal of the American Oil Chemist's Society, 24, 27-28.

Parker, J., Schellenberger, A. N., Roe, A. L., Oketch-Rabah, H. and Calderón, A. I. (2018). Therapeutic perspectives on chia seed and its oil: a review. Plant Medicine. 84, 606-612.

Pawlosky, R., Hibbeln, J., Lin, Y. and Salem, N. (2003). N-3 fatty acid metabolism in women. British Journal of Nutrition, 90, 993-994.

Porras-Loaiza, P., Jimenez-Munguia, M. T., Sosa-Morales, M.E., Palou, E. and Lopez-Malo, A. (2014). Physical properties, chemical characterization and fatty acid composition of Mexican chia (*Salvia hispanica* L.) seeds. International Journal of Food Science and Technology. 49, 571–577.

Reverchon, E.G., Porta, D. and Lamberti, G. (1999). Modelling of orange flower concrete fractionation by supercritical CO2, Journal of Supercritical Fluids, 14, 115-121.

Reyes-Caudillo, E., Tecante, A. and Valdivia-Lopez, M.A. (2008). Dietary fibre content and antioxidant activity of phenolic compounds present in Mexican chia (*Salvia hispanica* L.) seeds. Food Chemistry, 107, 656-663.

Rodea-Gonzaleza, D.A., Cruz-Olivaresa, J., Roman- Guerrerob, A., Rodriguez-Huezoc, M.E., Vernon-Carterd, E.J., Perez-Alonso, C. (2012). Spray-dried encapsulation of chia essential oil (*Salvia hispanica* L.) in whey protein concentrate-polysaccharide matrices. Journal of Food Engineering, 111, 102–109.

Rosas-Mendoza, M.E., Coria-Hernandez, J., Melendez-Perez, R. and Arjona-Roman, J.L. (2017). Characteristics of chia (*Salvia hispanica*) Seed oil extracted by ultrasound assistance. Journal of Mexican Chemist Society, 61, 326-335.

Rosas-Ramirez, D.G., Fragoso-Serrano, M., Escandon-Rivera, S., Vargas- Ramirez, A.L., Reyes-Grajeda, J.P. and Soriano-Garcia, M. (2017). Resistance-modifying activity in vinblastine-resistant human breast cancer cells by oligosaccharides obtained from mucilage of chia seeds. Phytotherapeutic Research, 31, 906-914.

Sandoval-Oliveros, M. R. and Paredes-Lopez, O. (2012). Isolation and characterization of proteins from chia seeds (*Salvia hispanica* L.). Journal of Agriculture and Food Chemistry. doi: 10.1021/jf3034978.

Schroth, A.B. (1996). An Ethnographic Review of Grinding, Pounding, Pulverizing, and Smoothing with Stones. Pacific Coast Archaeological Society. 32, 56-78.

Segura-Campos, M.R., Ciau-Solis, N., Rosado-Rubio, G., Chel-Guerrero, L., Betancur-Ancona, D. (2014). Chemical and functional properties of chia seed (*Salvia hispanica* L.) gum. International Journal of Food Science. doi:10.1155/2014/241053.

Segura-Campos, M.R., Salazer-Vega, I.M., Chel-Gurrerro, L. A., Betancur-Ancona, D.A. (2013). Biological potential of chia (*Salvia hispanica*) protein hydrolysates and their incorporation in functional foods. LWT-Food Science and Technology, 50, 723-731.

Sierra, L., Roco, J., Alarcon, G., Medina, M., Nieuwenhove, C.V., Peral de Bruno, M. and Jerez S. (2015). Dietary intervention with *Salvia hispanica* (Chia) oil improves vascular function in rabbits under hypercholesterolaemic conditions. Journal of Functional Foods, 14, 641-649.

Silva, B.P.D., Anunciacao, P.C., Matyelka, J.C.D.S., Lucia, C.M.D., Martino, H.S.D. and Pinheiro-Santana, H.M. (2016). Chemical composition of brazilian chia seeds grown in different places. Food Chemistry, http://dx.doi.org/10.1016/j.foodchem.2016.10.115.

Silva, C., Garcia, V.A.S. and Zanette, C.M. (2016). Chia (*Salvia hispanica* L.) oil extraction using different organic solvents: oil yield, fatty acids profile and technological analysis of defatted meal. International Food Research Journal, 23, 98-1004.

Sosa, A., Ruiz, G., Rana, J., Gordillo, G., Heather, W., Sharma, M., Liu, X., and Torre, R. R. R. D. L. (2016). Chia Crop (*Salvia hispanica* L.): Its history and importance as a source of polyunsaturated fatty acids omega-3 around the world: A review. Crop Research and Fertilizers. doi: 10.17303/jcrf.2016.104.

Steinmetz, K.A. and Potter, J.D. (1996). Vegetables, fruit, and cancer prevention: A review. Journal of American Dietetic.Association. 1996, 96, 1027–1039.

Toscano, L.T, Tavares, R.L, Silva, C.S.O and Silva A.S. (2015). Chia induces clinically discrete weight loss and improves lipid profile only in altered previous values. Nutrition Hospitalaria, 31, 1176-1182.

Ulbricht, C., Chao, W., Nummy, K., Rusie, E, Tanguay-Colucci, S., Iannuzzi, C.M., Plammoottil., Varghese and Weissner, W. (2009). Chia (*Salvia hispanica*): A systematic review by the natural standard research collaboration. Reviews on Recent Clinical Trials, 4, 168-174.

Ullah, R., Nadeem, M., Khalique, A., Imran, M., Mehmood, S., Javid, A. and Hussain, J. (2015). Nutritional and therapeutic perspectives of Chia (*Salvia hispanica* L.): A review. Journal of Food Science and Technology. doi. 10.1007/s13197-015-1967-0.

Uribea, J.A.R., Pereza, J.I.N., Kauila, H.C., Rubioa, J.R. and Alcocerb, C.G. (2011). Extraction of oil from chia seeds with supercritical CO2. The Journal of Supercritical Fluid. 56, 174-178.

Valdivia-López1, M. A. and Tecante, A. (2015). Chia (*Salvia hispanica*): A review of native Mexican seed and its nutritional and functional properties. http://dx.doi.org/10.1016/bs.afnr.2015.06.002.

Valenzuela, R., Barrera, C., Ayala, J.M., Sanhueza, J. and Valenzuela, A.B (2014). Vegetable oils rich in alpha linolenic acid allow a higher accretion of n-3 LCPUFA in the plasma, liver and adipose tissue of the rat. Grasas Aceites, 65, 26-28.

Valenzuela, R.B., Barrera, C.R., Gonzalez-Astorgo, M., Sanhueza, J.C. and Valenzuela, A.B. (2014). Alpha linolenic acid (ALA) from Rosa canina, Sacha inchi and Salvia hispanica oilsmay increase ALA accretion and its conversion into n-3 LCPUFA in diversetissues of the rat. Food Functions, 5, 1564.

Vazquez-Ovando, A., Betancur-Ancona, D., and Chel-Guerrero, L. (2013). Physicochemical and functional properties of a protein-rich fraction produced by dry fractionation of chia seeds (*Salvia hispanica* L.). CyTA-Journal of Food, 11, 75-80.

Vazquez-Ovando, A., Rosado-Rubio, G., Chel-Guerrero, L. and Betancur-Ancona, D. (2009). Physicochemical properties of a fibrous fraction from chia (*Salvia hispanica* L.). LWT-Food Science and Technology, 42, 168-173.

Vertommen, J., Van de Sompel, A.M., Loenders, M., Van der Velpen, C. and De Leeuw, I. (2005) efficacy and safety of 1month supplementation of salba (*Salvia hispanica* Alba) grain to diet of normal adults on body parameters, blood pressure, serum lipids, minerals status and haematological parameters. Results of a Pilot Study, the 23rd International Symposium on Diabetes and Nutiriton of the European Association for the Study of Diabetes.

Villalobos, R.R., Sánchez O.A, Javier Solorza-Feria, J. and Hernandez, C.A.T. (2012). Formulation, physicochemical, nutritional and Sensorial Evaluation of Corn Tortillas Supplemented with Chia Seed (*Salvia hispanica* L.). Czech Journal of Food Science. 30, 118-125.

Vuksan, V., Choleva, L., Jovanovski, E., Jenkins, A.L., Au-Yeung, F., Dias, A.G., Ho, H.V.T., Zurbau, A. and Duvnjak, L. (2017). Comparison of flax (*Linum usitatissimum*) and Salba-chia (*Salvia hispanica* L.) seeds on postprandial glycemia andsatiety in healthy individuals: a randomized, controlled, crossover study. European Journal Clinical Nutrition, 71, 234–238.

Vuksan, V., Jenkins, A.L., Dias, A.G., Lee, A.S., Jovanovski, E., Rogovik, A.L. and Hanna, A. (2010). Reduction in postprandial glucose excursion and prolongation of satiety:possible explanation of the long-term effects of whole grain Salba (*Salvia hispanica* L.). European Journal of Clinical Nutrition, 64, 436-438.

Vuksan, V., Whitham, D., Sievenpiper, J.L., Jenkins, A.L., Rogovik, A.L., Bazinet, R.P., Vidgen, E. and Hanna, A. (2007). Supplementation of conventional therapy with the novel grain Salba (*Salvia hispanica* L.) improves major and emerging cardiovascular risk factors in type 2 diabetes: results of a randomized controlled trial. Diabetes Care, 30, 2804-2810.

Wang, K., Lu, F., Li, Z., Zhao, L, Han, C. (2017). Recent developments in gluten-free bread baking approaches: A review. Food Science and Technology. doi: http://dx.doi.org/10.1590/1678-457X.01417.

Zotte, A.D. and Szendro, Z. (2011). The role of rabbit meat as functional food. Meat Science, 88, 319-331.

18

Quinoa (*Chenopodium quinoa* Willd) Chemistry, Nutritional Attributes, Antinutritional Factors and Health Benefits, Traditional and Modern Foods and Beverages, Processing Innovations Industrial Applications

Padmashree Ananthan, Gopal Kumar Sharma and Anil Dutt Semwal

Defence Food Research Laboratory, Siddarthanagar, Mysuru, Karnataka India -570011.

Introduction

Quinoa, a pseudocereal belongs to Amaranthaceae family, an Andean staple food, known to have been cultivated before 6000 BC in whole Andean region, Columbia, Peru, Equador, Bolivia before the conquest of Spanish. Incas have called this crop as a mother grain and was considered sacred. But in course of time, Catholics have suppressed the cultivation of crop after discovering it as a sacred drink during various religious ceremonies and hence the crop remained in the places where European could not reach (Vega Galvez, 2010). Archaeological findings of Argentina and Peru clearly mentions about the presence of seeds in the early stages of Christian era (Heisser and Nelson,

1974) and also been cultivated and used by the inhabitants of pre-Columbian culture as their diet (Heisser and Nelson, 1974; Mujica, 1992). The crop has its history of cultivation in Altiplano region 7000 years ago (Jacobson, 2003), less affected by environmental impact contributing to biodiversity and protecting the eco system.

Quinoa belongs to the class of dicotyledonaea, subfamily chenopodium, genus chenopodium, species quinoa and botanical name *Chenopodium quinoa* willd (Valencia Chamorro, 2003).

Due to the extreme adaptability of quinoa to divergent environmental conditions, it can be suitable for growing in wide range of temperature, with a relative humidity ranging from 40-88 percent and can withstand extreme cold temperature range of -4^0 C to 38^0 C. Quinoa can be grown at sea level and also at an altitude of 4000 m above the sea level (Sharma *et al.*, 2015) and can tolerate a wide range of soil pH from 6.0 to 8.5. The cultivation usually starts in the month of August and extends up to December in Andean highlands and in some areas it starts from January and extends till March. Seeds are planted in rows with a spacing range between 40-80 cm using mechanized agricultural practices and plant grows 1-3 m high. Quinoa matures after 7 to 8 months of sowing which mainly depends on the variety of the seeds and once flowering begins, it takes 70 to 90 days to reach physiological maturity. The yield is about 45-500 g/m^2 which further depend on variety and conditions of growing (Valencia Chamorro, 2003).Quinoa seeds comes in different colours from white, yellow, light brown to red. Seeds are round and flattened measures about 1.5mm in diameter weighing about 350 seeds/gram (Ruales and Nair, 1993).

Quinoa's exceptional adaptability to extreme growing conditions has made it an important crop for the diversification of agriculture as in high altitude regions of Himalayas and North eastern plains (Bhargava *et al.*, 2005). Quinoa is one among the few crops, which grows in the salt level of Northern Chile and Southern Bolivia (Jacobson *et al.*, 2000; Tagle and Planella, 2002), as the salt level has a greater influence on the growth of the plant, yield and quality of seeds, by increasing the content of protein and decreasing the content of total carbohydrates (Koyro and Aisa, 2007).

Though quinoa received a status of mother grain by Incas, the crop remained unrecognized among the urban population for several years due to various socio economic reasons, till US and Europeans opened their market for the food product (Sharma *et al.*, 2015) and now quinoa is cultivated in many countries including USA, Europe, India, China, Canada etc (Abugoch James. 2009). Currently, the leading countries which are growing quinoa are Peru (78657 tonnes), Bolivia (66792 tonnes) and Equador (1286 tonnes) (FAOSTAT, 2018). Quinoa contains quality protein with an excellent amino acid balance, particularly rich in lysine, makes it a more complete protein than many vegetables. Since quinoa is gluten free, it is highly beneficial to people of celiac disease and also to those who are allergic to wheat (Abugoch James, 2009). Despite of having all these attributes, little been known about the benefits of this wonder grain, may be due to the inadequate knowledge about its benefits to the common consumer or the high cost of the grain.

Removal of saponins, once considered as antinutritional, was a challenging task to the food processing industries, though several traditional methods have been followed by Incas population for use at house hold purpose. But with the invention of new processing technologies, the processing of quinoa has become much easier and was scaled up by several folds and the food processing companies are now capable of meeting the international demand for quinoa successfully.

This chapter presents comprehensive information about chemistry, nutritional attributes, processing innovation, traditional and modern foods and beverages, anti- nutritional and health benefits of quinoa.

Fig. 1: Quinoa plant

Fig. 2: Quinoa after harvesting

Fig. 3: Quinoa grain in varying colours

Chemistry and Nutritional Attributes

Protein

Quinoa has superior protein quality and quantity. Its protein digestibility is high with a gluten free property. It has a protein content in the range of 12.9 to 16.5 percent much higher than barley (10.8 to 11.0%), oat (11.6%), rice (7.5 to 9%) and maize (10.2 to 13.4%) and is comparable to that of wheat (14.3 to 15.4%). (Wright *et al.*, 2002; Repo Carrasco *et al.*, 2003; Comai *et al.*, 2007; Abugoch-James, 2009; Jancurova *et al.*, 2009; Pieretti *et al.*, 2013). Compared to other cereal grains, quinoa proteins are rich in lysine, a limiting amino acid in most of the cereals. Quinoa has an excellent amino acid balance than other cereals and legumes (Ruales and Nair, 1993), with higher amounts of lysine (5.1-6.4%) and methionine (0.4 to 1.0%) in them (Bhargava *et al.*, 2003; National Academy of Sciences, 1975; Prakash and Pal, 1998). Quinoa protein contains higher levels of histidine than barley, soy or wheat, while methionine and cystine found similar to that of barley and soy and lower than the amounts present in wheat (Abugoch, 2009).

The major storage proteins in quinoa are albumin and globulin containing little or no prolamin (Graf *et al.*, 2015). Chenopodin, a major storage protein, constitutes about 37 percent of the total protein, is an 11S type protein (Repo Carrsco *et al.*, 2003). This 11S protein is a hexameric protein with six pairs of acidic and basic subunits, connecting each subunit with disulfide bond. The similarities in the sequence of six binding regions suggest a similarity of 11S hexamer with glycinin (Barett, 2006). The acidic and basic polypeptides of quinoa globulin show a molecular mass of 20.5 KDa and 30-40 KDa respectively, linked by disulfide bonds (Abugoch *et al.*, 2008; Brinegar and Goundan, 1993). Chenopodin contains glutamine-glutamic acid, aspargine-aspartic acid, arginine, serine, leucine and glycine in higher amounts and meets the requirements of leucine, isoleucine and phenyl alanine + tyrosine according to FAO reference protein (Abugoch, 2009).

Quinoa represents other major protein constituting 35 percent of total protein is a 2S type protein albumin with a molecular mass of 8-9 KDa. Under reduced conditions, the purified form of quinoa 2S protein fraction is an electrophoretically heterogeneous collection of polypeptides with a molecular mass of 8-9 KDa showing higher levels of cystine, arginine and histidine (Brinegar *et al.*, 1996).

Quinoa's biological value is 73 percent, matches with that of animal protein derived from beef (74%) and higher than the protein of vegetarian sources attained from white rice (56%), wheat (49%) and corn (365) (Gordillo *et al.*, 2016). FAO (2011) reports that, essential amino acid composition of quinoa is near to that of casein and whole dried milk. The protein quality and the protein efficiency ratio (PER) of quinoa has been reported to be similar to that of casein (Ranhotra *et al.*, 1993) and processing found to affect its PER significantly showing 30 percent more in cooked quinoa in comparison to the uncooked one. Raw washed quinoa showed a protein digestibility of 83 percent, (casein 91%) as described by Ruales and Nair (1993). As per FAO/WHO recommendations, 180 percent requirements of all essential amino acids are met by quinoa by supplying them in adequate amounts.

Table 1: Essential amino acid profile of quinoa (g/100g)

	Koziol, 1992	**Repo-carrasco *et al.*, 2003**	**USDA, 2005**	**Abdellatif, 2018**
His	3.2	2.7	2.8	2.8
Ile	4.4	3.4	3.5	3.5
Leu	6.6	6.1	5.9	5.9
Met+ cys	4.8	4.8	3.6	3.6
Phe+ Tyr	7.3	6.2	6.0	6.0
Thre	3.8	3.4	2.9	2.9
Val	4.5	4.2	4.2	4.2
Lys	6.1	5.6	5.4	5.4
Trp	1.2	1.1	1.1	--

Carbohydrates

Carbohydrates are the important micronutrient in our diet and provide various physiological health benefits such as supply of energy, control of blood sugar, insulin metabolism, effect of satiety gastric emptying, cholesterol and glyceride metabolism etc (Abugoch, 2009). The major component of quinoa is a carbohydrate, having low glycemic index and its content varies from 58.1 to 64.2 percent of dry seed weight (Vega- Galvez *et al.*, 2010). Starch is the main carbohydrate in quinoa accounting 52-60 percent (Valencia Chamorro, 2003) of carbohydrate. Quinoa starch is highly branched having a granular size of less than 2 μm in diameter, much smaller than the one found in common cereals (Vega –galvez *et al.*, 2010). Starch consists mainly of D-xylose (120 mg/100g) and maltose (101mg/100g) with less amount of glucose (119 mg/100g) and fructose (19.6 mg/100g) (Bhargava *et al.*, 2006), which can be used in the formulation of malted drinks (Ogungbenle, 2003). Starch is mainly present in the perisperm of the seeds as simple units or spherical aggregates with an amylose content of about 11 percent, much lesser than the amylose content of rice (17%), wheat (22%) and barley (26%) (Valencia Chamorro, 2003). Size of the starch is very small and smaller size of the starch granules exhibits higher gelatization temperature (Valencia Chamorro, 2003). Rich amylopectin content of quinoa shows excellent freeze thaw stability (Berghofer and Schoenlechner,

2002) and hence can be used as a thickener in the preparation of frozen foods and other food applications where resistance to retrogradation is required (Ahamed *et al.*, 1998). The other carbohydrates such as monosaccharides (2%), disaccharides (2.35%), crude fibre (2.5-3.9%) and pentosans (2.9-3.65) are present in small amounts in quinoa (Valencia Chamorro, 2003).

The average molecular mass of quinoa starch is 11.3×10^6 g/mole, lower than that showed by waxy corn starch (917.4×10^6 g/ mole) or rice starch ($0.52\text{-}1.96 \times 10^8$g/mole) (Praznik *et al.*, 1999; Park *et al.*, 2007). The highly branched quinoa starch has a minimum degree of polymerization of 4600 glucan units and a maximum of 161,000 glucan units with an average weight of 70000 (Praznik *et al.*, 1999). Tang *et al.* (2002) reported that the average degree of polymerization of quinoa amylose is less (900) as compared to barley (1700) with an average of 116 chains/ molecule.

Quinoa has about 77.5 percent of amylopectin (Tari *et al.*, 2003). Quinoa's amylopectin fraction is comparable to some varieties of rice (Tukomane and Varavinit, 2008). Inouchi *et al.* (1999) reported that, amylopectin has a large number of short chains (8-12 units) and a small number of larger chains (13-20 units) in comparison to the starches of other cereals. X-ray diffraction studies indicatd that, quinoa has typical A type X-ray diffraction pattern reflection at 15.3^0, 17^0, 18^0, 20^0, and 23.4^0 angles, which is the characteristic of many cereal starches (Qian and Kuhn 1999; Lopez Rubio *et al.*, 2004; Watanabe *et al.*, 2007), with a degree of crystallinity between 35 and 45 percent (Qian and Kuhn, 1999; Tang *et al.*, 2002; Watanabe *et al.*, 2007), more than that of normal barley starch and near to waxy barley starch (Qian and Kuhn 1999; Tang *et al.*, 2002).

Quinoa has about 10 percent dietary fibre (Lamothe *et al.*, 2015). Dietary fibres have a major role to play in human body, highly essential for optimal digestive health and imparts various functional health benefits (Brownawell *et al.*, 2012). The insoluble dietary fibre content of quinoa accounting for 78 percent of total fibre is comprised mainly of galactouronic acid, arabinose, xylose, galactose and subunits of glucose, while soluble fibre content represents 22 percent of total fibre is comprised of glucose, galactouronic acid and subunits of arabinose. Quinoa's total fibre content is close to that of cereals, but the composition of manosaccharide subunit is matching closely with fruits, vegetables and legumes (Graf *et al.*, 2015).

The fermentable short chain carbohydrates referred to as 'Food maps' actuate the symptoms of irritable bowel syndrome, comprises of foods high in fructans and fructose content present is higher than the glucose, as present in wheat and rye. Quinoa comes under low food map diet that has been reported to impart marked beneficial effects on the irritable bowel syndrome (Biesiekierski *et al*., 2013), as it does not contain fructans and has a very low fructose content. Therefore the benefits of quinoa can be utilized to prevent certain diseases.

Lipids

The content of oil in quinoa is found in the range of 2 to 10 percent, on an average of 5-7 percent, higher than the lipid content (3-4%) of maize (Vega-Galvez, 2010). Studies on quinoa reported that, quinoa seed oil contains 89.4 percent unsaturated fatty acids and 54.2 to 58.3 percent of poly unsaturated fatty acid content (Tang *et al*., 2015). Quinoa lipid contains higher amount of neutral lipds in all the seed fractions, of which triglycerides are the major ones accounting for 50 percent of neutralipids and diglycerides comprises of 20 percent of neutral lipid fraction. Quinoa contains sufficient amount of phospholipids which includes lysophosphatidyl ethanolamine and phosphatidyl choline, accounting for 57 percent of total polar lipids (Przybylski *et al*., 1994). Quinoa lipid has 9 to 12.3 percent total saturated fatty acids mainly palmitic acid, 25-28.7 percent of total manounsaturated fatty acid mainly oleic acid and about 58.3 percent of poly unsaturated fatty acid containing linoleic acid (about 90%) as the predominanat one (Masson and Mella, 1985; Wood *et al*., 1993; Ranhotra *et al*., 1993; Oshodi *et al*., 1999; Ryan *et al*., 2007). Quinoa seed oil's fatty acid composition is very near to that of corn and soybean oil (Koziol, 1992; Oshodi *et al*., 1999; Youdim *et al*., 2000). Quinoa seed oil has a polyunsaturation index of 3.9-4.7, due to the presence of high amount of polyunsaturated fatty acids and hence regarded as high quality nutritious oil. (Abugoch, 2009). Also, quinoa has good amount of vitamin E (0.59 to 2.6 mg/100g) (Coultier and Lorenz, 1990; Ryan *et al*., 2007), a natural antioxidant that protects quinoa oil against lipid oxidation (Ng *et al*., 2007). Its whole flour also contains 3.1 to 5.5 mg/100g of $\beta+\gamma$ tocopherol (Ruales and Nair, 1993; Ryan *et al*., 2007).

The lipid fraction of quinoa contains 33.9 to 58.4 mg/g of squalene, an intermediatary in cholesterol biosynthesis (Jahaniaval *et al*., 2000; Ryan *et al*., 2007) and acts as a biochemical precursor for steroids apart from having antioxidant activity (Nesaretnam, 2008).

Quinoa also contains phytosterols, which are natural components of plant cell memebranes having anti inflammatory, antioxidant and anticarcinogenic activity. Their level in quinoa exceeds than that of barley, maize and pumpkin seeds. The levels of phytosterol in quinoa as reported by Ryan *et al*. (2007) showed 63.7 mg/100g of β- sitosterol, 15.6 mg/100g campesterol and 3.2 mg/100g stigmasterols. These sterols act as important dietary components and play a major role in maintenance of healthy heart by lowering LDL cholesterol (Berger *et al*., 2004).

Table 2: Proximate analysis of quinoa (g/100g dm) as compared to other cereals and legumes

	Quinoa[a]	Rice[a]	Wheat[a]	Oat[b]	Soy[c]	Maize[c]
Protein	16.3	8.8	14.8	11.6	36.1	10.22
Fat	7.0	3.2	2.8	5.2	18.9	4.7
Ash	2.7	1.7	1.8	2.9	5.3	11.7
Fibre	8.0	3.5	10.7	10.4	5.6	2.3
Carbohydrate	74	86.3	80.6	69.8	34.1	81.1
Energy (kcal/100g)	424.2	409.2	406.8	372	451	408

a: USDA (2013), b: Kozial (1992), c: Valencia-chamorro (2003).

Table 3: Fatty acid content of quinoa (g/100g)

	Saturated	Monounsaturated	Polyunsaturated
Masson and Mella, 1985	9.9-14.6	24.0-30.7	52.7-60.3
Palombini et al., 2013	12.28	25.53	54.98
Abdellatif, 2018	15.21	30.09	60.96

Vitamins

Humans require vitamins in their diet and they are the essential compounds needed for the growth and development of cells, act as co-factor of enzymes, protect cells against oxidative damage and play a vital role in various physiological activities (Fitzpatrick *et al.*, 2012). Quinoa has ideal amount of carotene, a vitamin A precursor (0.39 mg/100g), thiamine (0.4 mg/100g), foilc acid (78.1 mg/100g), riboflavin (0.39 mg/100g), niacin (1.06 mg/100g), pantothenic acid (0.61 mg/100g), pyridoxine (0.20 mg/100g), ascorbic acid (4.0-16.4 mg/100g) and vitamin E (3.7 to 6.0 mg/100g) (Ruales and Nair, 1993; Bhargava *et al.*, 2006; Vega-Galvez *et al.*, 2010; Tang *et al.*, 2015). The amount of γ- tocopherol (5.3 mg/100g) found in quinoa seeds is reported to be twice the amount of α-tocopherol (2.6 mg/100g) (Ruales and Nair, 1993). Quinoa also contains lutein and zeaxanthine in the concentrations ranging from 1.2 to 1.8 mg/100g)(Tang *et al.*, 2015).

Quinoa seeds showed higher concentrations of many vitamins than most of the cereal grains (Bhargava *et al.*, 2006; Vega Galvez *et al.*, 2010; Tang *et al.*, 2015), but thiamine was found lower than that of barley or oat, while niacin, riboflavin, pyridoxine and tocopherol were found in higher concentrations (Ranhotra *et al.*, 1993; USDA, 2005).

Table 4: Vitamin contents in quinoa (mg/100g dm)

Vitamins	Ranhotra *et al.*, 1993	Kozial, 1992	Ruales and Nair, 1992	USDA 2018
Ascorbic acid	NA	4.0	16.4	12.39
Alpha tocopherol	NA	5.37	2.6	2.44
Thiamine (B_1)	0.29	0.38	0.4	0.36
Riboflavin (B_2)	0.30	0.39	ND	0.32
Niacin (B_3)	1.24	1.06	ND	1.52

NA: Not available, ND: Not Detected

Minerals

Quinoa is rich in minerals, with a total mineral content of 3.4 percent higher than the amount present in rice (0.5%), wheat (1.8%) and other cereals (Bhargava *et al.*, 2006). Polishing and washing of quinoa seeds reduced the mineral content with 12-15 percent loss in iron, zinc and potassium, 27 percent in copper and 31 percent in magnesium (Valencia-chamorro, 2003). Calcium content was found in the range of 275 to 1487 mg/kg, iron 14 to 168 mg/kg, phosphorous 1400-5300 mg/kg, magnesium 260 to 5020 mg/kg, potassium 75 to 12000 mg/kg, zinc 28 to 48 mg/kg, copper 2 to 51 mg/kg (Repo-carrasco *et al.*, 2003; Bhargava *et al.*, 2006; Vega- Galvez *et al.*, 2010). Quinoa has higher contents of potassium, magnesium, calcium in comparison to oat and barley (Abugoch, 2009). Abrasion treatment used for the removal of saponin from quinoa has been found to reduce its calcium content. The distribution of phosphorous and magnesium were localized in embryonic tissue, while potassium and calcium were present in the pericarp with pectic compounds of the cell wall (Konishi *et al.*, 2004).

The content of iron in quinoa is found more than any cereal, but its bioavailability may be affected due to the interference of saponin and phytic acid. Iron bioavailability can be improved by the removal of saponin contents in quinoa. Based on animal experiments, the bioavailability of iron from quinoa based food was found in approximation with the bioavailability of iron from ferrous sulphate (Koziol, 1992; Ruales and Nair, 1993). Since quinoa contains less phytic acid of 1.18 g/100g, than many cereal crops and processing like cooking, soaking and fermentation further reduces the phytic acid content, iron bioavailability can be improved (Valencia chamorro, 1993).

Table 5: Mineral content in quinoa (mg/100g)

	Ca	Fe	Mg	P	K	Na	Zn	Cu	Mn	Se
Kozial, 1992	148.7	13.2	249.6	383.7	926.7	NA	4.4	5.1	ND	NA
Ranhotra *et al.*, 1993	70	6.3	161	462	855	2.7	3.2	0.7	3.5	ND
Konishi *et al.*, 2004	86.3	15	502	416	732	NA	4	ND	NA	NA
Bhargava *et al* 2006	127.4	2.0	ND	386.9	696.7	NA	4.8	ND	NA	NA
Sanders ,2009	56.5	1.4	176	468.9	1193	NA	2.8	0.2	NA	NA
Miranda *et al.*, 2013	110	7.80	175.34	367.38	2710.2	11.33	7.67	1.74	4.47	NA
USDA, 2018	47	4.57	197	457	563	5	3.10	0.59	2.03	8.5
Abdellatif, 2018	86.3	10.5	502	411	732	2.44	4.1	0.75	4.44	NA

NA: Not available

ND: Not Detected

Antinutritional Factors

Antinutritional factor, a secondary metabolite present in plant based foods, is seen as one of the major problems associated with the absorption of nutrients. Antinutritional factors are referred to as those compounds, originated from the plant based foods, which when consumed reduces the nutritive value of foods by interfering with the digestion, absorption and utilization of nutrients, thus causing harmful or negative effect on human health (Lopez *et al.*, 2009).

The major antinutritional factors present in quinoa are saponins, phytic acid, trypsin inhibitors, oxalates and tannins (Santos 2006; Lopes *et al.*, 2009; Borges *et al.*, 2010).

Saponins

Saponins are a group of glycosides, found mainly in plants in the outer layer of the grain (episperm). Its chemical structure is composed of fat soluble nucleus (Aglycone) which is either a triterpenoid (C-30) or neutral or alkaloid steroid (C-27) bound to one or more water soluble sugar (glycone) through ester linkages to the aglycone nucleus at different carbon sites (Hassan *et al.*, 2010). The name saponin is emanated from the plant genus*, saponaria*, whose root was used as soap (Sparg *et al.*, 2004). The presence of saponins in quinoa is attributed to its astringent or bitter taste (Tarade *et al.*, 2006). Saponins at higher concentrations are considered toxic to different organisms (Abugoch, 2009). All parts of quinoa plant contains saponins, especially seeds and leaves contains higher amounts, and also occurs to lesser extent in flowers and fruit part (Gupta and Wagle, 1988; Mastebroek *et al.*, 2000 ; Kuljana Bhagavad *et al.*, 2008) and their primary function is to protect the plant from external environment like insects, birds, pathogens, herbivores etc (Abugoch, 2009; Graf *et al.*, 2015). The four major saponin structures that have been identified in quinoa are oleonolic acid, hederagenin, phytolaccagenic acid and 30-*o*-methylsperulagenat (Zhu *et al.*, 2002). Glucose, galactose and arabinose are the main carbohydrates present. In addition to this, 20 triterpene saponins were also isolated from flowers, fruits, seed coats and seeds (Zhu *et al.*, 2002; Kuljana Bhagavad *et al.*, 2008).

Quinoa is classified based on the saponin content and the content of saponins present depends on the variety of quinoa. Sweet variety contains <0.11 percent of free saponins and bitter variety >0.11 percent (Koziol, 1992; Martinez *et al.*, 2009). The overall concentration of saponin varies between 1 and 5 percent (Valencia Chamorro, 2003).

Quinoa saponins possess strong surface active property, which is mainly due to the carbohydrate portion which is water soluble, while sapogenin is fat soluble

(Savage, 2003). The detergent properties of quinoa form foam which is strong and stable in water and reduce the superficial tension of aqueous solutions.

Various methods have been reported for saponin removal from quinoa to make it more palatable.Wet methods like hand scrubbing with alkaline water was one of the methods used traditionally to debitter quinoa seeds (Brady *et al.*, 2007). Slight alkalinity in water was recommended by Zhu *et al.* (2002) to debitter saponins in quinoa instead of neutral water. Dry methods like heat treatemt, roasting, extrusion or mechanical abrasion or a combination of both wet and dry methods were used to reduce the content of saponin in quinoa (Koziol, 1990; Dini *et al.*, 2001; Brady *et al.*, 2007; Comai *et al.*, 2007; Spehar, 2007, Farro, 2008; Jancurova *et al.*, 2009).

Quinoa possesses strong haemolytic activity and haemolysis may be due to the saponin interaction with the membranes, producing pores, thereby causing cell rupture (Seeman *et al.*, 1973). Valencia chamorro (2003) reported the formation of insoluble complex of saponins with minerals like zinc and iron, making it unavailable for the absorption in the gut.

Though saponins are considered as antinutrients, they do possess various biological properties such as antifungal activity (Armah *et al.*, 1999), anticholesterol activity (Oakenfull and sidhu, 1990), anti-inflammatory, analgesic and antiviral properties (Guclu- Utsundang and Mazza, 2007), insecticidal, antibiotic properties (Abugoch, 2009) etc.

Phytic Acid

The key storage form of phosphorous found in most of the plant tissues is a phytic acid, a saturated cyclic acid. The presence of high contents of phytic acid in foods binds minerals, rendering them unavailable for metabolism (Fardet, 2010; Gupta *et al.*, 2015). Low phytic acid content in foods known to exhibit various beneficial effects such as, acts as an antioxidant, suppresses damage of intestinal epithelium caused by oxidants, chelate various metals, interfere with the formation of ADP-Fe-O_2 complexes which initiate peroxidation of lipids, inhibit the formation of carcinogens, also blocks the interaction of cells with carcinogens, controls cell division, glucose, lipids and hepatic lipid levels, inhibits calcification of cardiovascular system and so on (Fardet, 2010). Phytic acid is mostly concentrated on the peels of most of the cereals and legumes at a level between 1 and 3 percent of dry matter. Also, found in some fruits and vegetables (Valencia-chamorro, 2003). In case of quinoa, phytic acid is located both on the pericarp (outerlayer) and endosperm. In five varieties of quinoa, the mean concentration of phytic acid has been reported to be 1.18 g/100g (Valencia Chamorro, 2003). Koziol (1992) also reports the value of phytic acid from 10.5 to 13.1 mg/g in 5 different varieties of quinoa, found close to the phytic acid content reported for barley (9.7 to 11.6 mg/g), wheat (96.2 to 13.5 mg/g), corn (8.9 to 9.9 mg/g).

No adverse effect has been noticed in the calcium and iron absorption, though quinoa showed higher content of phytic acid than other cereal grains as reported by Koziol, 1992. The phytate content in quinoa was found more in bran than its flour or peel, which accounted for 38-41 percent of total phytate from whole seeds (Chauhan *et al.*, 1992).

Various procedures have been adopted to reduce phytic acid from quinoa. The process like brushing and rinsing involved for the removal of saponins have reduced the phytic acid content approximately by 30 percent showing 10.4 mg/g in unprocessed one and 7.8 mg/g in processed ones. These values can be compared well with the values obtained for rye (7.7 mg/g), wheat (8.7 mg/g), faba bean (8 mg/g) (Ruales and Nair, 1993). Steeping, germination and fermentation showed significant reduction in phytic acid and the efficacy of its depletion was found more in case of germination and fermentation which promotes the activation of phytase (Ruales and Nair 1993a; Oliviera *et al.*, 2003; Khattab and Arntfield, 2009).

Trypsin Inhibitors

Trypsin or Protease inhibitors are distributed widely in the plant kingdom, form very stable complex with the protease enzymes. The presence of trypsin inhibitors reduces the action of trypsin in the intestinal tract, thus affecting the digestion of proteins (Khattab and Arntfield, 2009; Lopes *et al.*, 2009).

Significant reduction in trypsin inhibitors can be achieved in most of the legumes through soaking and washing procedures, if carried out for 18-22 followed by heat treatments such as boiling, autoclaving, roasting and microwave treatments (Khattab and Arntfield, 2009).

Trypsin inhibitors are concentrated more in the embryo of quinoa seeds, contributing to 89 percent of trypsin inhibition activity as compared to the whole grain (Ando *et al.*, 2002). The presence of trypsin inhibitor in quinoa does not pose any serious issue as the amount present is less than 5 ppm, which is much less than the amount found in commonly consumed cereals (Vega Galvez *et al.*, 2010).

The trypsin inhibition activity reported by Ruales and Nair (1993a) and Jancurova *et al.* (2009) in eight varieties of quinoa found in the range of 1.36 to 5.04 UTI/mg much lower than soybean which showed 24.5 to 41.5 UTI/mg.

Tannins

Another type of polyphenols which are widely distributed in the plant kingdom are tannins. Tannins form undesirable complexes with protein and starch, thus affecting the nutritional value of foods. Tannins also imparts undesirable colour to foods due to enzymic browning reactions. The astringent effect of tannins decreases the food palatability, also interfere with absorption, vitamin B_{12} and glucose (Santos, 2006).

The tannin content reported in quinoa found considerably less (0.53%) as compared to the rice grain (1.3%) (Chauhan *et al.*, 1992; Jancurova *et al.*, 2009). Washing, adequate rinsing and cooking reduces the tannin content of quinoa and improves digestibility (Borges *et al.*, 2010).

Oxalates

Oxalates are toxic in nature and found in vegetables such as spinach, beets, swiss chard, tomatoes, nuts, rhubarb, etc. Most of the oxalates are excreted in the urine as it is not metabolized in the body. Intake of dietary oxalates considerably in higher amount is reported to interfere with the mineral absorption and play a major role in hyperoxaluria, a risk factor for calcium oxalate stone formation in the kidneys, since oxalates form insoluble complexes with the divalent cation in GI tract (Jancurova *et al.*, 2009; Lopes *et al.*, 2009).

The plant belonging to Amaranthaceae, polygonaceae and chenopodiaceae known to possess high oxalate concentrations. In these plant families, oxalates are accumulated in leaves, stems, hypocotyls roots and nuts, showing highest oxalate contents in stems and roots. The amount of soluble oxalate content in leaves and stems are found in the range of 258 to 1029 mg/100g and in roots and seeds, it ranged from 59 to 131 mg/100g. The total oxalate content was present in the range of 874 to 1954 mg/100g in stems and leaves and 143 to 232 mg/100g in roots and seeds (Siener *et al.*, 2006; Jancurova *et al.*, 2009). It has been reported by Lopes *et al.* (2009) that normal diet should contain 50-200 mg/ day of oxalates. Filho *et al.* (2017) stated that the level of oxalic acid found in some species of quinoa is scarce in the literature.

Health Benefits

Quinoa is a wonder grain with excellent functional and remarkable health benefits. The functional properties of quinoa is mainly related to the presence of antioxidants, minerals, fibres, vitamins, phytonutrients and fatty acids which helps in the protection of cell membranes, improvement in neuronal functions and stands best over the plant foods by providing better nutrition to humans for the maintainance of health (Vega galvez *et al.*, 2010 ; Repo carrasco *et al.*, 2010).

The rapid urbanization and globalization has greatly influenced the dietary patterns and life style behaviour of large segments of population, which replaced the traditional food patterns consisting of complex carbohydrates, fibre and phytochemical foods to the diets rich in animal fats and refined carbohydrates that ultimately led to the development of certain chronic diseases (Schaffer *et al.*, 2017). Therefore great devotion and attention was rendered by food researchers to identify the food and food components which provide maximum health benefits and therefore quinoa was listed as one of the healthy grains due to its nutritional and therapeutic effect.

Antioxidant Activity

Polyphenols are the bioactive secondary metabolites of plant kingdom consisting mainly of phenolic acids, flavonoids and tannins, which act as potent antioxidants (Repo carrasco *et al.*, 2010) and are receiving greater focus and attention due to its various health benefits which includes antioxidant, anti-inflammatory, anticarcinogenic and antiallergic properties (Nishibe *et al.*, 1996; Dini *et al.*, 2004; Aalinkeel *et al.*, 2008, Pasko *et al.*, 2008; Khan *et al.*, 2010).

Majority of the phenolic compounds of quinoa exhibited antioxidant activity (Gorinstein *et al.*, 2007; Pasko *et al.*, 2009; Repo de carrasco and Zelada, 2010) with the bitter variety showing strong antioxidant activity than the sweet variety and their activity mainly based on phenols, flavonoids and carotenoids present in them (Dini *et al.*, 2010). The health promoting isoflavones like daidzein and genistein considered as phyto estrogens are also mentioned in some local Chilean varieties as 0.7- 1.15 and 0.05 to 0.25 mg/100g respectively. The darker coloured seeds received from high Andean (Altiplano) regions showed higher content of these isoflavones demonstrating potential health promoting bioactive compounds (Lutz *et al.*, 2013) as strong antioxidants (Gopalakrishnan *et al.*, 2006; Lo *et al.*, 2007; Jiang *et al.*, 2010). Based on the *invitro, invivo* and several clinical studies, these isoflavones have found its use in the formulation of several functional foods to lessen symptoms related to cardiovascular diseases, neurodegenerative diseases, osteophoresis and cancer (Song *et al.*, 2007; Jiang *et al.*, 2010; Lutz *et al.*, 2011). Studies also described the use of food formulations containing isoflavones to improve maternal milk production during lactation (Zhang *et al.*, 1995; Liu *et al.*, 1999; Groot *et al.*, 2004). Studies demonstrated that, there was an improvement in lactation when quinoa used traditionally by Aymara population of Altiplano in Northern Chile (Lutz *et al.*, 2013).

Hypocholesterimic Activity

Though cholesterol is needed essentially for the normal metabolic process, high cholesterol and high low density lipoprotein levels are linked with the risk of developing coronary heart diseases (Quillez *et al.*, 2003). Several studies have indicated that plant sterols inhibit the absorption of cholesterol from the body, thereby help in lowering plasma, LDL and total cholesterol levels (Moghadasian and Frohilch, 1999; Moreau *et al.*, 2002). Quinoa contains phytosterols such as β-sitosterol (63.7 mg), campesterol (15.6 mg/100g) and stigmasterol (3.2 mg/100g) and the content of these sterols were found more than that of barley, maize, squash, but less than that of pea, lentils and sesame seeds (Ryan *et al.*, 2007).

The *invitro* studies conducted on protein isolates of quinoa showed bile acid binding activity and modulation of expression in the liver of 3- hydroxy 3-methyl glutaryl coenzyme A (HMG-co A) an enzyme crucial for cholesterol biosynthesis. Also, significant reduction in the plasma and liver total cholesterol levels were observed in mice fed with fat enriched diets after the administration of quinoa protein isolates (Takao *et al.*, 2005). This may be attributed to the control of synthesis of cholesterol and its catabolism and inhibition of bile acid reabsorption in the small intestine (Takao *et al.*, 2005; Fuentes and Gonzalez, 2015). Significant reduction in blood pressure levels were also observed in rats and mice fed with quinoa flour or hydrolyzed quinoa protein (Aluko and Monu 2003; Ogawa *et al.*, 2001). *Invivo* studies have suggested that, incorporation of quinoa grain (3%) in the dietary supplement of mice also reduced plasma and liver cholesterol levels significantly (Konishi *et al.*, 2000). The administration of fructose enriched food products induced an adverse alteration in metabolism and oxidative status, leading to the conditions of hypertriglyceridemia, impaired glucose tolerance, increased blood pressure and insulin resistance (Auckerman *et al.*, 2005; Tappy *et al.*, 2010; Pasko *et al.*, 2010a). Studies conducted on the biochemical parameters of tissues and plasma of rats fed with quinoa supplemented diet, after feeding with high fructose diet indicated that, the diet containing quinoa seeds significantly affected the oxidative status by the decrease in plasma malondialdehyde (MDA) and antioxidant enzyme acivity viz glutathione peroxidase (GPX), Superoxide-dismutase (SOD) and catalase (CAT) in plasma, heart, kidney, testicles, lungs and pancreas. Moreover, quinoa supplementation also reduced total cholesterol, triglycerides, low density lipoprotein, glucose levels and lowered total plasma protein level without affecting HDL levels, suggesting protective effect of quinoa on the changes induced by high fructose diet by reducing lipid peroxidation and improved antioxidant activity in plasma (Pasko *et al.*, 2010a,b).

Saponin Activity on Health

From the pharmacological point of view, quinoa's saponin is reported to have many beneficial effects though it was considered as an antinutrient once. Studies have revealed that saponin possess antifungal activity, anti viral activity, cholesterol reducing activity, increased absorption of drug through mucous membranes, increased intestinal permeability to enable increased food intake at intestinal level (Meyer *et al.*, 1990; Bomford *et al.*, 1992; Mahato and Kundu, 1994; Estrada *et al.*, 1998; Armah *et al.*, 1999; Woldemichael and Wink, 2001; Stuardo and Sanmartin, 2008; Kuljinabhagavad and Wink, 2009). It has been demonstrated that saponins of quinoa can act as immunological adjuvant to boost the response of gene specific antibodies (Estrada *et al.*, 1998; Verza *et al.*, 2012) and also shown to bring about cytotoxicity and apoptosis in cancer cell lines (Kuljinabhagavad *et al.*, 2008) thus demonstrating greater therapeutic applications.

Anti-inflammatory Activity

Traditionally, quinoa grain especially black variety, has been used by Andean people for treating various anti-inflammatory diseases such as muscle sprains and strains by placing poultices made from quinoa grain, mixing with alcohol and covering on the affected area (FAO, 2011). Saponins present in quinoa may be responsible for its anti-inflammatory effect (Mujica, 1994), especially the monodesmosidic saponin (one carbohydrate chain) which could be related to monodesmosidic 3-o-β- D glucopyranosyl hederagenin (Kuljina Bhagavad and Wink, 2009).

Effect on Celiac Pateints

Celiac disease, a chronic auto immune enterophathy, is triggered by the ingestion of gluten containing grains such as wheat, rye, barley and also oats (Abugoch James, 2009). The intake of gluten containing diet causes self perpetuating damage to intestinal mucosa, thereby causing impairment in its functionality (Catassi and Fasano, 2008), affecting 1 percent of the general population worldwide (Catassi and Fasano, 2008; Mc Gough and Cummings, 2005). Life long elimination of gluten from the diet and strict adherence to the gluten free diet may be the only treatment for the celiac disease. Hence quinoa grain with high biological value and low prolamin concentration of ≤7 percent can act as a worthy alternative for the patients with celiac disease. As per the studies conducted by De Vincenzi *et al.* (1999); Berti *et al.* (2004); Bergamo *et al.*, (2011), it was inferred that, quinoa can be used as a very safe gluten free diet. Immunological effect of 15 quinoa cultivars evaluated by Zavallos *et al.* (2012) by testing the proliferation of T/ interferon-γ (IFN-γ) cells drawn from celiac patient's small intestine and the production of IFN-γ/IL-15 by cell culturing of duodenum biopsy samples revealed that, quinoa is safe for consumption by celiac patients.

Antiallergic Activity

With regard to the studies on the allergenicity of quinoa, so far only one case has been reported in the literature by Astier *et al.*, 2009 in a 52 year old man who has developed symptoms including dysphagia, dysphonia, urticaria and angioedema, following the ingestion of quinoa along with fish and bread. However, symptoms got resolved after the administration of antihistamins and intravenous corticosteroids. The cutaneous allergy test using ground quinoa seeds was observed positive with an irritation diameter of 15 mm, in comparison to just 5 mm in control and results for bread and fish were found negative. The immunoblotting test conducted, by taking the protein extracts of both raw and cooked quinoa and serum from allergic and non allergic patients revealed the reactivity of immunoglobulin E in the serum of an allergic patient through the detection of 35 KDa band for both raw and cooked protein extracts and no band was found in control samples of non allergic patients (Astier *et al.*, 2009; Fuentis and Gonzalez, 2015).

Anticarcinogenic Activity

Phytochemicals present in various foods have presented chemo preventive effect against cancer both in preclinical studies conducted on animal model system and epidemiological studies conducted on humans (Wang *et al.*, 2012; Lee *et al.*, 2013). Several studies have reported potential preventive mechanisms of many phytochemicals on certain types of cancers by blocking or suppressing multiple biological mechanisms related to carcinogenesis, which include metabolism of cancer, repair of DNA, apoptosis, cellular protection, angiogenesis and metastatic processes (Lee *et al.*, 2013).

The great diversity of compounds present in quinoa has resulted in the enhanced interest in exploring the effectiveness of these compounds especially as chemotherapeutic and phytotherapeutic agents. The use of MTT colorimetric assay to study the cytotoxicty of quinoa saponin compound in HeLa cervical cancer cells revealed a positive cytotoxic effect on the cervical cancer cells. The inhibitory effect of bidesmosidic saponins and their aglycones on the apoptosis and cellular growth of colorectal adeno carcinoma (Caco-2) was also reported (Kuljina Bhagavad *et al.*, 2008).

Antidiabetic Activity

Quinoa's low glycemic index (35-53) has been proven beneficial for its use by diabetic patients (Oshodi *et al.*, 1999). The *in vivo* hypoglycemic effect of quinoa has been suggested for its use for the manufacture of various gluten free products with low GI, which can act as a good alternative for the celiac diseased patients, as they have high probability of developing type 1 diabetes (Berti *et al.*, 2004).

Mineral and Vitamin Activity on Health

The antioxidant vitamin of quinoa ie vitamin E, omega 3 fatty acids, neuronal activity of tryptophan and vitamin B complex act as powerful aid in the function of brain. Quinoa consumption from experimental rats exhibited a powerful effect in protecting stressed neurons, with an evidential effect on neuronal gene expression under stressful conditions. Also reported, promotion of low anxiety and improvement in the spatial memory in the same animals (Munoz *et al.*, 2008).

Quinoa's rich contents of Ca, Fe, Mn, Zn and Mg can be used by different target populations such as adults and children to have the beneficial effect of calcium for bone formation and iron for blood function (Koziol, 1993; Repo- carasco *et al.*, 2003). These minerals are involved in carrying out various functions in the body as they act as enzyme cofactors, mineralization of bones, protecting cells and lipids in biological membranes. Low intake of minerals or decreased bioavailability may cause various deficiencies leading to the impairment in body functions (Schlenker and Williams, 2003).

Traditional and Modern Foods and Beverages

Quinoa's exceptional nutritional value has been slowly attracting the attention of different segments of population, though restricted only to Andean regions for several decades. Apart from providing various nutrients required, also offers health benefits to the body. It can act as a great food, especially among vegetarian population by providing a complete food with high quality protein with balanced amino acid composition, and complete benfits of micro and macronutrients.

Traditionally quinoa was used by Quechua and Aymara population of South America, Chicha and Mapuche (Vega Galvez *et al.*, 2010; Bhargava and Srivastava, 2013). The leaves, seeds and flour were used in various food preparations. The grain was cooked and consumed similar to rice, seeds were popped out to make breakfast cereals, soups were prepared and used, and the flour was used for making baked and toasted products like cookies, biscuits, bread, tortillas, flakes, noodles etc. (Popenoe *et al.*, 1989; Bhargava *et al.*, 2006). The leaves of quinoa were also consumed like spinach (Oelke *et al*, 1992). Both fresh and tender leaves and its tender shoots were cooked and eaten as vegetable (Schlick and Bubenheim, 1993; Jancurova *et al.*, 2009; Tang *et al.*, 2015b). Germinated quinoa was used for making salads (Schlick and Bubenheim, 1993). Beer, prepared from the fermented quinoa seeds was used traditionally as a ceremonial alcoholic beverage called Chicha in South America (Healy, 2001; FAO, 2011).

Various traditional foods and beverages consumed by Altiplano communities (Pacosillo and Chura, 2002, Vidaurre *et al.*, 2005; Apaza *et al.*, 2006; Flores *et al.*, 2008) are briefly described below

1. **Quinoa soup**: This is a thin quinoa soup made by using meat or dried meat, tubers and vegetables.
2. **Lawa**: A semi thick “Mazamorra” prepared close to porridge, using raw quinoa flour, animal fat and water with lime.
3. **Pésque:** It is made by cooking quinoa without adding salt and served with either milk or grated cheese as per their availability.
4. **Kispina**: It is a steamed quinoa bun having different size and shapes
5. **Tacti o tactacho**: It is a fried bun prepared from quinoa flour and Ilama fat similar to doughnut.
6. **Mucuna**: It is a steam cooked product of quinoa flour similar to humitas or tamales having seasoning at the centre.
7. **Phiri**: It is a slightly rough and dampened roasted quinoa flour.
8. **Phisara**:It is a slightly roasted and cooked quinoa grain
9. **Qúsa:** It is a cold drink prepared by macerated quinoa, named as Quinoa chichi,
10. **El Ullphu, Ullphi**: It is a cold drink prepared from roasted quinoa flour, can be diluted with the addition of water, and sugar as per taste can be added.
11. **Kaswira de quinua**: It is an oil fried flattened bread prepared by using white quinoa and katahui (lime).
12. **Kaswira de ajara**: It is also an oil fried flattened bread prepared from black quinoa or ajara and katahui (lime).
13. **Kápi kispina**: It is a steamed quinoa bun wherein quinoa grinding is done in K’ona and clay pot is used for cooking, offered commonly to saints.

14. **Turucha quispina o Polonca**: It is a steamed bread, large in size prepared by using slightly ground quinoa (Chama) in a K'ona and katahui, wherein cooking is done in clay pot.
15. **Mululsito quispina:** It is a smaller version of kispinas prepared from quinoa flour and katahui by cooking in a clay pot.
16. **Quichi quispina:** It is steamed bread prepared by using katahui and quinoa flour and fried in a pan.
17. **Juchacha:** It is an Andean soup made by using quinoa flour, roasted barley flour and katahui.
18. **Chiwa:** Tender leaves of quinoa are coimmonly called as Chiwa in Aymara and Lliccha in Quechua are used for preparing salads and soups as they are rich in phosphorous, calcium, iron and vitamins.

Quinoa products preparation in and around the Andean region clearly indicated the use of quinoa flour in various foods such as bread, where in quinoa flour was used from 10 to 40 percent, in cakes up to 60 percent, in pasta up to 40 percent and in biscuits up to 70 percent (Nieto and Wood, 1982; Ballon *et al.*, 1982; Nieto and Soria, 1991; Ruales and Nair, 1992; Jacobson, 1993; Sharma *et al.*, 2015). Robalino and Penaloza (1998) described the use of quinoa for making of tempeh (vegetable meat) by fermenting with Rhizopus oligosporus, having pleasant flavour and aroma with better sensory acceptance. Soria *et al.*(1990) also prepared quinoa based tempeh, with a pleasant aroma, consisting of quinoa grains joined by mycelium of fungus. Coulter *et al.* (1991) reported the utilization of quinoa and corn for extrusion. The use of quinoa at 10, 20 and 30 percent produced extruded products high in protein, fibre, ash and amino acids than the control extruded product prepared using 100 percent corn. Biscuits prepared using quinoa: flax seed: brown rice at 40:40:20 was found exceptional in terms of nutritive value (Sukeerthi and Singh, 2017). The use of quinoa and whey in bread preparation has improved its texture, sensory attributes and shelf life in comparison to those made from wheat flour alone (Salazar *et al.*, 2017). Ozer and Secen (2018) described the use of quinoa flour at 3, 5, 7 and 10 percent for making raw and cooked burgers to improve its quality characteristics with a shelf life of 3 months at -18^{0}c without affecting protein oxidation level. Bhathal and Kaur (2018) developed different food products like chapatti, laddoo, paratha, matri, pop-ups and cookies using quinoa flour. Developed quinoa products received high acceptability as compared to the control samples with improved nutritional value.

The health promoting factors and the functionality of quinoa has also attracted the researchers to formulate and develop various functional beverages singly and in combination with other cereals.

Various combinations of quinoa and soy (100Q:0S, 70Q:30S, 50Q:50S, 30Q:70S, 0Q:100S) extracts were used for the development of fermented symbiotic beverage using Fructo oligosaccharide and *Lactobacillus casie* (Bianchi *et al*., 2014). The viability of microorganisms was maintained throughout the storage period of 28 days at 5^0 C. The beverage prepared with the formulation of 70:30 ratio of quinoa and soy was found best in terms of physico- chemical and sensory attributes. Raw, soaked, germinated, and malted quinoa seeds were used for the preparation of gluten free healthy beverage (Kaur and Tanwar, 2016). Among the beverages developed, germinated quinoa beverage showed high acceptability, found close to the commercially available soy milk, while malted quinoa beverage showed higher antidiabetic and antihypertensive properties. A highly nutritious and a healthy fermented quinoa based beverage was successfully developed by Urquizo *et al*. (2017) using two varieties of quinoa (*Rosade de Huancayo* and *Pasankalla*) .The product was found free from phase separation, having low safe pH, with the retention of these properties for 28 days. Both the varieties were found suitable as a base for the development of fermented beverage by maintaining viability of bacteria during storage. The product can be highly useful for lactose intolerant and celiac population and can be a best alternative to general consumer. A fermented novel yogurt type beverage was developed by Lorusso *et al*. (2018) using quinoa and fermenting with different strains of lactic acid bacteria which includes, a probiotic *Lactobacillus rhamnosus* SP1, EPS producing *Weissella confuse* DSM 20194 and the one isolated from quinoa ie (*lactobacillus plantarum* T6B10). All the strains showed improved concentrations of free amino acids, availability of polyphenols, antioxidant activity up to 54 percent, γ- amino butyric acid (GABA), improved protein digestibility with good textural and organoleptic properties.

Though several products have been emanated from quinoa and now a days, many are available in the super market and health stores (Singh *et al*., 2016), the more research is under way for the development of different products based on quinoa which makes its consumption very attractive, useful and healthy.

Processing Innovations

The exceptional nutritional properties, combined with multiple benefits have made this Andean grain quinoa to attract enormous attention worldwide. In order to meet the increasing global demand, various countries in Asia, North America, Europe and Africa started cultivating this super crop (Jacobson, 2003).

Though several traditional methods are still in practice for processing of quinoa in Andean community, various innovations in industrial scale have also been carried out both in harvest and post harvest phases to replace the original traditional practices with particular emphasis on processing, specially linked to the area of saponin removal.

Traditionally, quinoa was harvested both manually and mechanically. Manual harvesting was carried out using stationery threshers involving steps like reaping or cutting, placing them in arcs or sheaves, threshing, winnowing and cleaning followed by drying, sorting and keeping them finally in bags for storage and transportation. While, in case of mechanical harvesting, reaping, threshing and winnowing are carried out simultaneously followed by sorting, bagging and storage (Quiroga *et al.*, 2015).

As per the survey conducted in Southern Altiplano region in the year 2008, 57 percent of quinoa was reaped by uprooting the plant, but gradually sickles, hoes, or mechanical mowers have been adopted for reaping, which reduced the contamination to a greater extent (Quiroga *et al.*, 2015).

Traditionally, Southern Altiplano population practiced sheaving of quinoa (piling together reaped plants) by forming an arc with the plants joined in the form of 'X' and pointing the panicle upwards, which made the drying process faster than the other methods used (Quiroga *et al.*, 2015).

In the slopy regions, grain were seperated from panicles by using huajtana, a stout pound paton this process is known as glomerulus.(Call and cartez, 2011) But, in plane areas, either a tractor or some other vehicle or stationery threshers were used to separate grains by passing the tyres of the tractors on the grains covered with tarpaulins to avoid soil / or sand contamination . Subsequently, different types of threshing machines were designed and developed which includes TR-C thresher, the MASEMA FAUTAPO thresher, modified vencedora thresher, tubular thresher etc. All these threshers have contributed successfully in separating grains from the chaffs with better yielding efficiency (Quiroga *et al.*, 2015).

Sifting and sieving was done manually (Quiroga *et al.*, 2015) which separates large impurities like bits of leaves, small stones, pedicels etc (Apaza *et al.*, 2006) by shaking the sieve back and forth to separate grains and husks from the chaff.

Traditionally, winnowing was achievd using wind energy to remove small and light impurities from the grain by keeping the grain on tray and pouring out in a stream transverse to the direction of the wind. Later mechanical winnowers, which can be operated either manually or by engine power were introduced. These winnowers do not require wind power, capable of generating regular air current and can be used for winnowing through out the year.The motorized winnower which can separate grain husk and chaffs was also developed by *Consultara y Taller Mecanico Aroni* in Uyuni, Bolivia having a yield capacity of 1600kg/hr (Quiroga *et al.*, 2015).

Sorting of grains is another important method followed to seperate the seeds of different colours which otherwise reduces price and quality of the grain. Grains of different colour of less than 1 percent was allowed as per the Bolivian standard NB NA0038 (IBNORCA, 2007).

In 2012-13 crop year, two types of engine driven combined harvesters were designed for medium and small sized plants, which can do shearing, threshing, sifting and cleaning simultaneously by avoiding the contamination with impurities (Quiroga *et al.*, 2015).

Grain has to be processed further before stacking them in polyethylene bags or other materials. The impurities like broken, damaged, coloured, germinated, covered and unripe seeds have to be removed and therefore grain has to undergo series of processing which includes preliminary sorting and removal of light and heavy impurities and also saponin removal which is normally carried out using combined methods of hulling (Dry method) and washing (Wet method) followed by drying, sorting by size, separation of different coloured grains and finally removal of residual impurities (Quiroga *et al.*, 2015).

Saponin removal is one of the important stages in the processing of quinoa grain, which otherwise makes the grain highly unpalatable. Various suitable technologies for saponin removal to the acceptable limit have been developed without affecting the nutritional quality of the grain (Quiroga *et al.*, 2015).

Many ecotypes or varieties of quinoa are designated as sweet or bitter based on the content of saponin which normally ranges from 1-3 percent for bitter variety and 0.0 - 0.1 percent for the sweet variety (Guclu-Ustundag and Mazza, 2007).

Both wet and dry methods have been followed in the Andean region for the removal of saponins. In dry methods, quinoa grains were roasted for 30-40 min in a metal container and the hot quinoa was mixed with an abrasive clay material known as 'Pojkera' and trodden for 30-60 min over the rough stone surface known as 'Saruna' for the removal of large quantity of saponins. Subsequently winnowing was followed for 20-40 min for the removal of remaining episperm and abrasives. Finally washing was done for 25 to 35 min for the final removal of the saponin and impurities consisting of small stones, seeds etc collected during harvest (Quiroga *et al.*, 2015).

In the wet process, quinoa was washed repeatedly in several stages for 25-30 min. The visual inspection of the washed water with the amount of foam formation indicates the quantity of saponin removal. The clear water without foam indicates the complete removal of saponin. Finally grain is dried for 2-4 h, sometimes winnowed and roasted again to obtain required moisture content (Quiroga *et al.*, 2015).

In north eastern Argentina, traditionally saponins are removed by placing 5-10 kg of grains in 50 kg bags made of clothes or some synthetic material, submerging them in a river and / or a brook by holding them at both the ends. The bags are moved up and down so that grains rub each other. The process is repeated several times until water becomes completely free from foam and finally grains are dried on zinc sheets (Quiroga *et al.*, 2015).

In Peru and Equador regions also, removal of saponins is performed by manual washing of the grain on the abrasive surface using excess of water until the outer layer of the grain is removed (Nieto and Valdividia, 2001).

Various modern methods have been introduced for the removal of saponin as the traditional methods explained above are highly tiresome and can be applicable for small batches for household consumption.

The Torres and Minaya huller in the year 1980, reduced the saponin content to 0.04-0.25 percent with 95 percent efficiency. A small scale abrasive dehulling machine developed in 1980 to stimulate abrasive action of industrial hullers (Reichhert *et al.*, 1986) was found successful in removing 85-95 percent of saponin from the grain. In the year 1983, Reggiàdro and Rodriguez developed pilot scale wet washing system which includes soaking, centrifuging, roasting and finally drying in a tunnel of warm air produced high quality quinoa. The removal of saponin coupled with hulling, washing and drying stages has been developed by Depric in 1988, which efficiently removes 65 percent hull layer, with less moisture absorption of 17-30 percent during washing which aides in faster and easier drying with low saponin concentration in the effluent. The method employs less water for washing, uses minimum amount of energy with good quality processed grain (Quiroga *et al.*, 2015). In 1992, Valdivieso and Rivadeneira developed a dry method of saponin removal, wherein saponin content was reduced to 0.026 percent.

Austudillo, (2007) explains the development of a small capacity saponin removal machine (12 kg/7 min) in Southern Altiplano, as a part of the project work in Bolivia which aimed at facilitating the consumption of quinoa for improving the nutritional status of quinoa producing rural communities. The machine uses traditional method of roasting, hulling, winnowing, washing and drying process which was well accepted by rural women. In 2010, a research group at Universidad Privada Boliviana (UPB) has developed a laboratory model spouted bed for dry saponin removal from the bitter quinoa. The reduction in saponin content was achieved to less than 0.01 percent in three commercial ecotypes of 'quinoa real' and their blends in less than 30 min with the application of the dry process, found in confirmity with the commercial export standards and less than 0.12 percent as prescribed by the Bolivian NB 063 standard. Complete recovery of powdered saponin was also achieved (Escalera *et al.*, 2010; Quiroga *et al.*, 2011) with reduced specific energy consumption to 0.23kWh/ kg (Obando *et al.*, 2011).

Though many types of equipment were used efficiently for the processing of quinoa, especially to saponin removal, but mainly, quinoa processing companies were using the existing technology which was suitable for processing wheat, soy and rice as the custom technology for quinoa processing was not developed by equipment manufacturers. Quinoa processing companies witnessed a sea change during the period between 2001 and 2004, in terms of processing when Centre for the Promotion of Sustainable Technologies (CPTS) has put in its dedicated effort to identify the problems existing in quinoa processing sectors with regard to the processing capacity and efficiency, and achieved major goals for improving the existing technology or replacement of the equipment with the best productive alternative existing in the market (Birbuet and Machicado, 2009).

After the cleaner production (CP) assessment in different quinoa processing companies, the main causes of the problems were identified with respect to loss of raw material with increased residual quantity, quality of grains used, consumption of water, electricity, processing cost, high content of saponins in waste water, pollution of water bodies, poor saponin recovery etc. (Birbuet and Machicado, 2009).

After assessing the features of the technology used, the most efficient, economical technological solutions at various stages of quinoa processing was proposed and implemented as follows (Birbuet and Machicado, 2009).

An efficient system of dry cleaning of quinoa was designed, constructed and implemented to minimize the loss in quality and quantity of the grain. Quick washing process for saponin removal has been initiated with the development of a washer which accomplished the simulation of laminar trajectory of the grain by using turbulent flow with the reduction of average residence time in the washer from 22 to 4.7 min. The older drying system with inefficient airflow was successfully replaced with the design of a highly efficient drying system. Also, in order to lessen the number of unit operations used for grain cleaning, drying and centrifugation process, a new technology has been designed (Birbuet and Machicado, 2009).

Funds granted in the year 2005 by United States Agency for International Development (USAID) for the execution of demonstration project which was mainly aimed at designing and constructing an entire quinoa processing plant based on the principles of cleaner production (CP) adopting to the needs of existing quinoa processing companies. CPTS has designed the technology, Industrial Metalicas Andinas (IMA) has constructed the technology and installed at Andean Valley of South America (AVSA). Eventually, the success of new tchnology wrt the quinoa processing plant has resulted in the establishment of the quinoa alliance between CPTS, IMA and AVSA.

The most significant contributions in quinoa processing industry is the technology developed by CPTS for the removal of saponin content, wherein the episperm of the seed is extracted using abrasive properties of the grain, thereby causing minimal damage to the grain when brushed or abrased. The machine consists of preliminary sorter for the removal of impurities followed by removal of saponin in a huller having dual compartments consisting of huller system and particle extraction and collection system. The huller efficiently removes 90-95 percent of saponins. Final removal of saponins was completed by wet washing of grains followed by centrifuging, which reduces saponin in the washed grain to 0.01 percent. The grain is dried in a drier comprising of natural gas operated warm air generator or LPG and dry grain is re-sorted to obtain most homogeneous grain. Different coloured grains are separated using optical pneumatic sorter and finally handpicked to check impurities if any, before being finally packed for the export (Quiroga *et al.*, 2015).

Using the above technology, quinoa's processing capacity was found to enhance by eight folds and nearly 75-80 percent of 'organic quinoa real' exported from Bolivia by using this technology. The implementation of cleaner production principles for designing and fabrication of equipment has greatly reduced the impact on the environment, especially in place of with respect to water and energy consumption. Also, enhanced residue reduction and recovery. The hulling and washing systems introduced have greatly reduced the material loss by maintaining nutritional quality. The combined technology has been used by most of the quinoa processing companies for saponin removal and complies with the established market standards (Quiroga *et al.*, 2015).

The technological alliance that began in 2006, has led the quinoa processing companies to satisfy the growing demand for quinoa and its export was raised by 747 percent during the period from 1998-2007. Since 2006, significant augmentation was noticed in quinoa export with the implantation of new processing technology and greater response capability was shown by national companies to meet international quinoa demand (Birbuet and Machicado, 2009).

The world wide production of quinoa showed a significant increase from 2010 to 2017. The production of quinoa was 80069 MT in 2010, increased significantly up to 193822 MT in the year 2015. Afterwards, years 2016 and 2017 showed decrease in the worldwide production showing 148720 MT in 2016 and 146735 MT in 2017 (FAOSTAT, 2018).

Industrial Applications

Quinoa has been receiving much popularity and attention during recent years especially among urban population due to its nutritional and health benefits. Many food industries have come forward in manufacturing various products and presently numerous recipes are available online featuring several quinoa preparations (Gardillo Bastidas *et al.*, 2016).

There are many ways to process quinoa to obtain products of our choice. Quinoa can be processed to produce pearled quinoa, flaked quinoa, quinoa granules, extruded and expanded products (Mujica *et al.*, 2006). It can be processed to get flakes and its size depends on the variety of quinoa used and the end use of the product (Mujica *et al.*, 2006). Food applications of quinoa flakes are many and it can go well with fruits, tarts, cakes and pies. Soups and juices can be prepared which requires less cooking and easy to consume (Quiroga *et al.*, 2015). A low volume flavoured and sweetened product can be developed from expanded quinoa, which can be used as a snack. It can also be used as a base for energy bars and the preparation of instant cereals. An extruded product of quinoa was found highly acceptable by consumers (Mujica *et al.*, 2006). Saponin free flour of quinoa was used in the preparation of different products by quinoa industry, and 40 percent was used for making bread and pasta, 60 percent for sponge cake and 70 percent for biscuits (Mujica *et al.*, 2006). The use of 19:81 ratio of quinoa and wheat flour was suggested for bread making (Reynaga *et al.*, 2013). Quinoa composite flour can be prepared with legume flours like broad beans and kidney beans and used in the preparation of various foods to improve nutritional quality especially in the food formulations of school going children and infants (Sharma *et al.*, 2015). Also, some varieties of quinoa flour can be a better option for making infant food formulations as quinoa starch takes less time for gelatinization (Bonafacio *et al.*, 2013). Soups and creams can be prepared from a white quinoa starch, called 'Pisankala' (Pumacahua *et al.*, 2013). Flour of quinoa is a best alternative in the pasta and noodles industry. The optimum quantity of 79 percent quinoa flour (Industrial quality Bolivian quinoa flour) and 21 percent rice flour was found to be best for making noodles. (Reynaga *et al.*, 2013). Also, highly acceptable gluten free pasta was prepared with the blend of 50 percent local Pisankala'flour and 50 percent rice flour. Addition of quinoa flour up to 75 percent also provided satisfactory results (Reynaga *et al.*, 2013). High lipid content and unsaturated fatty acids (89%) containing 50-56 percent linoleic acid, 21-26 percent oleic acid, 4.8-8.1 percent of linolenic acid (PROINPA 2011) and cholesterol reducing property (Takao *et al.*, 2005) can make industry to produce quinoa oil as a potential by-product.

In recent years, food researchers and industries are showing utmost interest in developing a quality food with high protein content. Since quinoa has about 12-18 percent protein (Schlick and Bubenheim, 1993) containing all essential amino acids, the concentrates and isolates of quinoa protein can be used as main ingredients in various food formulations especially for the undernourished children for the quick recovery from malnutrition

The excellent freeze thaw stability of quinoa starch finds its usefulness for the preparation of frozen foods and the opacity of the gelatinized starch is also useful in the manufacture of emulsified food products like salad dressings (Ahamed *et al.*, 1996).

The gluten free property of quinoa aided various food industries to use it as a main ingredient in the manufacture of gluten free food products to meet the growing international demand.

Conclusion

The cultivation and consumption of mother grain quinoa was restricted to the Andean population for several decades. As the awareness about the nutritional and health benefits of this superfood started increasing, quinoa cultivation started receiving greater attention and popularity from other parts of the world, mainly for its high protein quality containing all essential amino acids in a balanced amount equivalent to the biological value of milk protein. Grain, also represents higher amounts of vitamin B_2, B_3, B_6 and vitamin 'E' as compared to other cereals. Minerals like iron, zinc and calcium also present in appropriate amounts. Ideally suitable for consumption by the people with celiac disease due to its gluten free property. Also, suitable for those who are allergic to wheat. Quinoa's antinutritional factors like phytic acid, trypsin inhibitors, tannins etc can be removed easily by following easier processing steps like washing, soaking, germination etc. The main bitter principle, saponin concentrated in the episperm of the grain was removed traditionally by Incas by following the methods of roasting, washing, etc while the innovations brought about in processing technology has made its removal much easier and faster. Various bioactive compouinds like phenoilc antioxidants, phytosterols, phytoestrogens etc present in quinoa provides promising health benefits by reducing the risk of various diseases such as cardiovascular, anti-inflammatory, osteophorosis, cancer, diabetes etc., and hence has greater potential to use it as a complementary and alternative medicine. The benefits and popularity of this underutilized grain has drawn it from the Andean community, and now the grain and many of its recipes are available in many supermarket aisles and health food stores. Though modern technological innovations have been in use for processing of the grain in large scale to meet the growing international demand, but more attention is needed at the farmer level by motivating, educating and encouraging them to cultivate the environmental friendly crop especially in developing and underdeveloped countries, so as to reach the consumer at much affordable price, as the price of the grain is not up to the reach of the common consumer.

Also, grain promotion campaigns should be conducted to make aware of the nutritonal and functional benefits of quinoa at consumer level to promote greater consumption of a wonder grain, a functional super food for future generation.

References

Aalinkeel, Bindukumar, B., Reynolds, J., Stuyle, D., Mahajan, S., Chada, K., and Schwartz, S. (2008). The dietary bioflavonoids, quercetin, selectively induces apoptosis of prostate cancer cells by down regulating the expression of heat shock protein 90. Prostate, 68, 1773-1789.

Abdellatif, A.S.A., (2018). Chemical and technological evaluation of quinoa (*Chenopodium quinoa* willd) cultivated in Egypt. Acta Scientific Nutritional Health, 2, 42-53.

Abugoch James, L.E. (2009). Quinoa (*Chenopodium quinoa* Willd), composition, chemistry, nutritional and functional properties. Advances in Food and Nutrition Research. 58, 1-30.

Abugoch, J., Romero, N., Tapia, C., Silva, J., and Reviera, M. (2008). Study of some physico-chemical and functional properties of quinoa (*Chenopodium quinoa* Willd) protein isolates. Journal of Agricultural Food Chemistry. 56, 4745-4750.

Ackerman, Z., Oron-Herman, M., Grozovski, M., Rosenthal, T., Pappo, O., Link, G., and Sela, B.A. (2005). Fructose-induced fatty liver disease. Hepatic effects of blood pressure and plasma triglyceride reduction .Hypertension, 45, 1012-1018.

Ahamed, N.T., Singhal, R.S., Kulakarni, P.R., and Mohinder, P. (1998). A lesser known grain, chenopodium quinoa: review of the chemical composition of its edible parts. Food and Nutrition Bulletin, 19, 61-70.

Ahamed, N.T., Singhal, R.S., Kulkarni, P.R., and Pal, M. (1996). Physico- chemical and functional properties of Chenopodium quinoa starch. Carbohydrate polymers, 31, 99-103.

Aluko, R.E., and Monu, E. (2003). Functional and bioactive properties of quinoa seed protein hydrolysates. Journal of Food Science, 68, 1254-1258.

Ando, H., Chen, Y., Tang, H., Shimizu, M., Watanabe, K., and Mitsunaga, T. (2002). Food components in fractions of quinoa seed. Food Science and Technology Research, 8, 80-84.

Apaza., V, Rodriguez., D, Mujica., A, Canahua., A, and Jacobsen., S.(2006). Production de quinua de calidad, Estacion Experimental, Punoo, Peru, IIIpa. Pp 16.

Armah., C.N., Mackie, A.R., Roy, C.R.K., Osbourne, A.E, Bowyer., P and Ladha, S. (1999). The membrane permiabilizing effect of avanacin A-I involves the reorganization of bilayer cholesterol. Biphysics Journal, 76, 281-290.

Astier., C, Moneret-Vautrin., D.A, Puillandre., E, and Bihain., B.E. (2009). First case report of anaphylaxis to quinoa, a novel food in France. Allergy, 64, 819-820.

Astudillo., D.(2007). The potential of mall holder technology in quinoa producing communities of the Southern Bolivian Altiplano. Feature May 2007. Global facilitation unit for underutilized species, Mickey Leland International Hunger Fellowships. (http://www.underutilized.species.org/features/quinoa/quinoa _article. pdf).

Ballon., E.E, Paredes., E,and Coca., A.(1982). Comportamiento de la harina de quinua, variedades dulces y amargas (En mezlas de harinas compuestas para panificacion). En: Tercer Congreso Internacional de cultivas Andinos (Memorias) La PAZ, Bolivia. Pp 21-28.

Barett., M. (2006). Identification, sequencing, expression and evolutionary relationships of the 11S seed storage protein gene in Chenopodium quinoa Willd. Thesis of Master of Science. Department of Plant and Animal Sciences. Brigham Young University.

Benevides., C.M.J, Souza., M.V, Souza., R.D.E., and Lopes., M.V.(2011). Fatores antinutricionais em alimentos: revisao.Seguranca. Alimentr e Nutricional.18, 67-79.

Bergamo., P, Maurano., F, Mazzarella., G, Iaquinto., G, Vocca., L, Rivelli., A.R, De-Falco., E, Gianfrani., C, and Rossi., M.(2011). Immunological evaluation of the alcohol soluble protein fraction from gluten free grains in relation to coeliac disease. Molecular Nutrition and Food Research, 55, 1266-1270.

Berger., A, Jones., P, and Abumweis., S.(2004). Plant sterols: factors affecting their efficacy and safety as functional food ingredienys. Lipids Health and Disease, 3, 1-19.

Berghofer., E, and Schoenlechner., R.(2002). Grain Amaranth, In, Peudocereals and less common cereals, grain properties and utilization potential. Belton., P.S and Taylor., J.R.N (eds). Springer, Germany.

Berti., C, Ballabio., C, Restani., P, Porrini., M, Bonomoi., F and Lametti. S. (2004). Immunochemical and molecular properties of proteins in Chenopodium quinoa (*Chenopodium quinoa* Willd). Field Crop Research, 101, 101-116.

Bhargava., A and Srivastava., S. (2013). Quinoa: Botany, Production and Uses, Amity institute of technology, Amity University, Lucknow Campus, Lucknow, Inida, DOI:10.1079/9781780642260.0000,

Bhargava., A, Rana., T, Shukla., S, and Ohri., D. (2005). Seed protein electrophoresis of some cultivated and wild species of Chenopodium. Biologia Plantarum, 49,505-511.

Bhargava., A, Shukla., S, and Ohri., D. (2006). Chenopodium quinoa- an Indian perspective. Industrial Crops and Products, 23, 73-87.

Bhargava., A, Shukla., S, and Ohri., D.(2003). Genetic variability and heritability of selected traits during different cuttings of vegetable Chenopodium. Indian Journal of Genetics and Plant Breeding, 63, 359-360.

Bhathal., S.K, and Kaur., N. (2018). Nutritional analysis of gluten free products from quinoa (*Chenopodium quinoa* Flour). International Journal of Pure and Applied Sciences, 6, 826-836.

Bianchi., F, Rossi., EA, Gomes., R.G, and Siviera., K. (2014). Potentially symbiotic fermented beverage with aqueous extracts of quinoa (*Chenopodium quinoa* willd) and soy. Food Science and Techology International, 21, 403-415.

Biesiekierski., J.R, Muir., J.G, and Gibson., P.R. (2013). Is gluten a cause of gastrointestinal symptoms in people without celiac disease? Current Allergy and Asthma Report, 13,631-638.

Birbuet., J.C, Machicado., C.G.(2009). Technological progress and productivity in quinoa sector. Development Research Working Paper Series 07/ 2009. Institute for Advanced Developemnt Studies. Pp1-44.

Bomford., R, Stapleton., M, Winsor., S, Beesley., J.E, Jessup., E.A, Price., K.R, and Fenwink., G.R. (1992). Adjuvanticity and ISCOM formation by structurally diverse saponins. Vaccine, 10, 571-577.

Bonafacio., A, Vargas., A, Rojas., W. (2013). Mejaramiento papa la calidad Industrial de la quinua. In Memoria IV Congreso Mundial de la quinua, Ibarra, Equador, July 8-12. Calla., J and Cartez G., (2011). Guia Tecnica Currao- Taller “Post Cosecha y Transfermacion de Quinua Organica”. Puno, Peru. Pp 30.

Borges., J.T.S, Bonomo., R.C, Paula., C.D, Oliviera., L.C. and Cesario., M.C. (2010). Physico chemical and nutritional characteristics and uses of quinoa (*Chenopodium quinoa* Willd). Temas Agrarios, 15, 9-23.

Brady., K, Ho.,C.T, Rosen., R.T, Sang., S, and Karwe., M.V (2007). Effect of processing on the nutraceutical profile of quinoa. Food chemistry, 100, 1209-1216.

Brinegar., C, and Goundan., S. (1993). Isolation and characterization of chenopodin, the 11 S seed storage protein of quinoa (*Chenopodium quinoa*). Journal of Agricultural Food Chemistry, 41,182-185.

Brinegar., C, Sine., B, and Nwokocha., L. (1996). High cysteine 2S seed storage proteins from quinoa (*Chenopodium quinoa*). Journal of Agricultural Food Chemistry, 44: 1621-1623.

Brownawell., A.M, Caers., W, Gibson., G.R, Kendal., C.W.C, Lewis., K.D, Ringel., Y, and Slavin., J.L. (2012). Prebiotics and health benefits of fibre: current regulatory status, future research and goals. Journal of Nutrition, 142, 962-974.

Calla., J, and Cortez., G. (2011). Guia Tecnica Curso- Taller 'Post cosecha y Transfermacion de Quinua Organica'. Puno. Peru. Pp 30

Catassi., C, and Fasano., A. (2008). Celiac disease, In: Arendt., EK, Dal Bello., F (Eds), Gluten free cereal products and beverages, London: Academic Press.

Chauhan., G.S, Eskin., N.A.M, and Tkachur., R .(1992). Nutrients and antinutrients in quinoa seed. Cereal Chemistry, 69, 85-88.

Comai., S, Bertazzo., A, Bailoni., I, Zancato., M, Costa., C.V. I, and Allegri., G.(2007). The content of proteic, non proteic (free and protein bound) tryptophan in quinoa and cereal flours. Food Chemistry, 100, 1350-1355.

Coulter., L. A, and Lorenz., K. (1990). Quinoa composition, nutritional value, food applications. Lebensm –Wiss. U. Technol. 23, 203-207.

Coulter., L.A, and Lorenz., K. (1991). Extrudes corn grits- quinoa blends: II. Physical characteristics of extruded products. Journal of Food Processing and Preservation, 15, 231-259.

de Jesus Benevides, C. M., Souza, M. V., Souza, R. D. B., and Lopes, M. V. (2011). Fatores antinutricionais em alimentos: revisão. Segurança Alimentar e nutricional, 18(2), 67-79.

De Vincenzi., M, Silano., M, Luchetti., R, Carratú., B, Boniglia., C, and Pogna., N.E. (1999). Agglutinating activity of alcohol-soluble proteins from quinoa seed flour in celiac disease. Plant Foods for Human Nutrition (Formerly Qualitas Plantarum). 54, 93-100

Dini., A, Tenore., G.C, and Dini., A . (2004). Phenolic constituents of Kancolla seeds. Food Chemistry, 84, 163-168.

Dini., I, Schettino., O, Simioli., T, and Dini. A. (2001). Studies on the constituents of Chenopodium quinoa seeds: isolation and characterization of new triterpene saponins. Journal of Agricultural Food chemistry, 49, 741-746.

Dini., I, Tenore., G.C, and Dini., A. (2010). Antioxidant compound contents and antioxidant activity before and after cooking in sweet and bitter chenopodium quinoa seeds. LWT-Food Science and Technology, 43, 447-451.

Escalera., R, Quiroga., C, and Arteaga., L. (2010). Desarrollo y desempeno de un proceso de bonafaciado en seco de variedades amargas de quinoa basado en la aplicion de un Lecho Fluidizaso de Tipo Surtidor (LFTS). Universidad Privada Boliviana. Revista Investigacion and Desarrollo, 10, 5-22.

Estrada., A, Li., B, Laarveld., B.(1998). Adjuvant action of Chenopodium quinoa saponins on the introduction of antibody response to intragastric and intranasal administered antigensin mice. Comparative Immunoogy Microbiology and Infectious Diseases, 21, 225-236.

FAO. (2011). Quinoa: an ancient crop to contribute to world food security. http://www.fao.org/docrep/017/aq287e/aq287e.pdf.

Fardet., A. (2010). New hypothesis for the health protective mechanism of whole grain cereals: what is beyond fibre? Nutrition Research Reviews. 23, 65-134.

Farro., P.A.C. (2008). Desenvolvimento de filmes biodegradeveis a partir de derivados do grao de quinoa (*Chenopodium quinoa* Willdnow) da varsidade "Real" Thesis, Faculade de Engenharia de Alimentos, Universidade Estadual de Campinas, Brazil, Pp 303.

Filho., A.M.M, Pirozi., M.R, Da Silva Borge.,s J.T, Sant'ana., H.M.P, Chaves., J.B.P, and Coimbra., J.S.D.R. (2017). Quinoa: nutritional, functional and antinutritional aspects. Critical Reviews in Food Science and Nutrition, 57, 1618-1630.

Fitzpatrck., T.B, Basser., G.J.C, Borel., P, Carrari., F, Dellapenna., D, Fraser., P.D, Hellmann., H, Osorio., S, Rothan., C, Valpuesta., V, Caris-Veyrat., C, and Fernie., A.R. (2012). Vitamin deficiencies in humans: Can plant science help? Plant Cell, 24: 395-414.

Flores., J, Mamani., E, Alarcon., V, Paco., V, and Rojas., W.(2008). Caracterizacion de los conocimientos tradicionales asociodos a la agrobiodiversidad en coromata Media Y Santiago de okola. Informe. Octubre 2007- Junio 2008. NOS-I-FAD II. Fundacion PROINPA. La-Paz- Bolivia. Pp 129-143.

Fuentes., F.F, and Gonzalez., X.P. (2015). Nutraceutical perspectives of quinoa: biological properties and functional applications. In book .The state of the world's quinoa. ed: Bazile., D, Bertero., D, Nieto., C. Regional Office for Latin America and Caribbean at Food and Agriculture Organization (FAO). Pp 286-299.

Gonzalez., J.A, Roldan., A, Gallardo., M, Escudero., T and Prado., F.E. (1989). Quantitative determination of chemical compounds with nutritional value from Inca crops Chenopodium quinoa. Plant Foods for Human Nutrition, 39, 331-332.

Gopalakrishnan., A, Xu., C.J, Nair., S.S, Chen., C, Hebbar., V, and Kong., A.N.(2006). Modulation of activator protein-1 (AP-1) and MAPK pathway by flavonoids in human prostate cancer PC3 cells. Archives of Pharmacal Research, 29, 633-644.

Gordillo Bastidas., E, Diaza- Rizzolo., D.A, Roura., E, Massanes., T, and Gomis R. (2016). Quinoa (*Chenopodium quinoa* willd) from nutritional value to potential health benefits: An integrated review.6: 3. http:// dx.doi.org/10.4172/2155-9600.1000497.

Gorinstein., S, Medina Vargas., O, Jaramillo., N, Arnoa salas., I, Martinez Ayala., A, Arancibia Avila., P, Toledo., F, Katrich., E and Trakhtenberg., S. (2007). The total phenols and the antioxidant potentials of some selected cereals and peudocereals. European Food Research and Technology, 225, 321-328.

Graf., B.L, Silva., P.R, Rojo., L.E, Herrera., J.D, Baldeon., M.E, and Raskin., I. (2015). Innovations in health value and functional food development of quinoa (*Chenopodium quinoa* Willd). Comprehensive Reviews in Food Science and Food Safety, 14, 431-445.

Groot., M.J. (2004). Ohyto estrogenic activity of protein rich feeds for pigs. Project No 8017194701. Wageningen, The Netherlands, RIKILT-Institute of Food Safety.

Guclu- Ustundag., O, and Mazza., G. (2007). Saponins: properties, applications and processing. Critical Reviews in Food Science and Nutrition, 47, 231-258.

Gupta., K, and Wagle., D.S. (1988). Nutritional and antinutritional factors of green leafy vegetables. Journal of Agricultural Food Chemistry, 36, 472–474.

Gupta., R.K, Gangoliya., and S.S, Singh., N.K.(2015). Reduction of phytic acid and enhancement of bioavailable micronutrients in food grains. Journal of Food Science and Technology, 52, 676-684.

Hassan., S.M, Haq., A.U, Byrd., J.A, Berhow., M.A, Cartwright., A.I, Bailey., C.A. (2010). Haemolytic and antimicrobial activities of saponin contents from guar meal. Food Chemistry, 119, 600-605.

Healy., K. (2001). The quinoa trail: from South America salt flats to western health food stores. Llamas, weavings and organic chocolate: multicultural grassroots development in the Andes and Amazon of Bolivia, Notra Dame, In: University of Notra Dame Press.

Heisser., and Nelson.(1974). On the origin of cultivated Chenopods (Chenopodium), Genetic, 78, 503-505.

Inouchi., N, Nishi., K, Tanaka., S, Asai., M, Kawase., Y, Hata., Y, Konishi Yue., S, and Fuwa., H. (1999). Characterization of amaranth and quinoa starches. Journal of Applied Glycoscience, 469, 233-240.

Instituto Boliviano de Normalizacion y Calidad- IBNORCA 2007. Norma Boliviana Granos andinos- Pseudo cereales- Quinua en grano- Clasificacion y requisites, NB NA 0038. Comendino Normas Tecnicas y Gulas de Implementacion de Normas del Sector Quinua.1-7.

Jacobson., E.(1993). Quinoa, Chenopodium quinoa Willd. A novel crop of European Agriculture. Department of Agricultureal Science. The Royal Veterinary and Agricultural University. Denmark. Pp-145.

Jacobson., S, Quispe., H, and Mujica., A. (2000). Quinoa : An alternative crop for saline soils in the Andes. CIP Programme Report, 1999-2000, Pp 403-408.

Jacobson., S.E, Mujica., A and Ortiz., R. (2003). La importanacia de los cultivos andinos. Fermentum, 13, 14-24.

Jahaniaval., F, Kakuda., Y, and Marcone M. (2000). Fatty acid and triacylglycerol composition of seed oils of five amamranthus accessions and their comparison to other oils. Journal of American Oil Chemical. Society, 77, 847-852.

Jancurova., M, Minarovicova., I, and Dandar., A. (2009). Quinoa- A review. Czech Journal of Food Sciences, 27, 71-79.

Jiang., Q, Payton-Stewart., F, Elliott., S, Driver., J, Rhodes., L.V, Zhang., Q, Zheng., S, Bhatnagar., D, Boue., S.M, Collins- Burow., B.M, Sridhar., J, Stevens., C, McLachian., J.A, Weise., T.E, Burow., M.E and Wang., G. (2010). Effects of 7-O substitutions on estrogenic and anti estrogenic activities of daidzein analogues in MCF-and breast cancer cells. Journal of Medicinal Chemistry, 53, 6153-6163.

Kaur., I, and Tanwar., B. (2016). Quinoa beverages, formulation, processing and potential health benefits. Romanian Journal of Diabetes Nutrition and. Metabolic Diseases, 23, 215-225.

Khan., N, Mustafa., V, and Mukhtar., H. (2010). Apoptosis by dietary agents for prevention and treatment of prostate cancer. Endocrine Related Cancer. 17, R39-R52.

Khattab., R.Y, and Arntfield., S.D. (2009). Nutritional quality of legume seeds as as affected by some physical treatments. 2. Antinutrirional factors. LWT- Food Science and Technology, 42, 1113-1118.

Konishi., Y, Arai., N, Umeda., J, Gunji., N, Saeki., S, Takao., T, Minoguchi., R, and Kensho., G. (2000). Cholesterol lowering effect of the methanol insoluble materials from the quinoa seed pericarp. In K. Nishinari, ed, Hydrocolloids. Part 2. Pp. 417-422. Elsevier Science, B.V.

Konishi., Y, Hirano., S, Tsuboi., H, and Wada., M. (2004). Distribution of minerals in quinoa (*Chenopodium quinoa* willd) seeds. Bioscience Biotechnology and Biochemistry. 68, 231-234.

Koyro., H and Aisa., S. (2007). Effect of salinity on composition, viability and germination of seeds of (*Chenopodium quinoa* Willd). Plant and Soil, 302, 79-90.

Koziol MT (1993). Quinoa, Apotential new oil crop. Pp 328-336, In: JanickJ and Simon JE (eds), New crops, Wiley, New York.

Koziol., M. (1992). Chemical composition and nutritional composition of quinoa (Chenopodium quinoa willd). Journal of Food Composition and Anaalysis, 5, 35-68.

Koziol., M.J. (1990). Composicion quimica. In: quinua Hacia su Cultivo Comercial, Chap.VIII. Pp. 137-159. Ahli C. ed., Quito. Latinreco SA. Brazil.

Kuljanabhagavad., T, Thongphasuk., P, Chamulitrat., M. and Wink. (2008). Triterpene saponins from *Chenopodium quinoa* Willd. Phytochemistry, 69, 1919–1926.

Kuljinabhagavad., T and Wink., M.(2009). Biological activities and chemistry of saponins from *Chenopodium quinoa* willd. Phytochemistry Reviews. 8, 473-490.

Lamothe., L.M, Srichuwong., S, Reuhs., B.L, and Hamakar., B.R. (2015). Quinoa (*Chenopodium quinoa* Willd) and amaranth (*Amaranthus Caudatus* L) provide dietary fibres high in pectic substances and xyloglucans. Food Chemistry. 167, 490-496.

Lee., J.H, Khor., T.O, Shu., L, Su., Z.Y, Fuentes., F, and Kong., A.N. (2013). Dietary phytochemicals and cancer prevention: Nrf2 signaling, epigenetics and cell death mechanisms in blocking cancer initiation and progression. Pharmacology and Therapeutics. 137: 153-171.

Liu., G.T, Zheng., Y.L, Chen., W.H, Chen., J, and Han., Z.K. (1999). Effect of daidzein fed to pregnant cows on milk production and the levels of harmones in colostrums. Journal of Nanjing Agricultural University, 22, 69-72.

Lo., F.H, Mak., N.K, and Leung., K.N. (2007). Studies on the anti tumor activities of the soy isoflavone daidzein on murine neuroblastomacells. Biomedicine and Pharmacotherapy, 61, 591-595.

Lopes., C.O, Dessimoni., G.V, Dasilva., M.C, Vieira., G, and Pinto, N.A.V.D. (2009). Aproveitamento composicao nutricional e antinutricional da farinha de quinoa (*Chenopodium quinoa*)- Alimentos e Nutricao 20, 669-675

Lopez Rubio., A, Flanagon., B, Gilbert., E, and Gidley., M. (2004). A novel approach for calculating starch crystallinity and its correlation with double helix content. A combined XRD and NMR Study. Bio polymers, 89, 761-768.

Lorusso., A, Coda., R, Montemurro., M, and Rizello., C.G. (2018). Use of selected lactic acid bacteria and quinoa flour for manufacturing of novel yogurt like beverages. Foods, 7, 51-71.

Lutz., M, Martinez., A, and Martinez., E.A. (2013). Daidzein and genistein contents in seeds of quinoa (*Chenopodium quinoa* Willd) from local ecotypes grown in arid Chile. Industrial Crops and Products, 49,117-121.

Lutz., M. (2011). Soy isoflavonesas bioactive ingredients of functional foods. Centro de Investigación y Desarrollo de Alimentos Funcionales (CIDAF). Universidad de Valparaíso and Centro Regional de Estudios de Alimentos Saludables (CREAS).Valparaíso. Chile. Pp 329-360.

Mahato., S, and Kundu., P. (1994). 13C NMR spectra of pentacyclic triperpenoids-a compilation and some salient features. Phytochemistry. 37, 1517-1575.

Martinez., E.A, Veas., E, Jorquera., C, San Martin., R, and Jara P. (2009). Reintroduction of Chenopodium quinoa Willd into arid Chile: cultivation of two low land races under extremely low irrigation. Journal of Agronomy and Crop Science, 195, 1-10.

Masson., L, and Mella., M.(1985). Materias grasas de consumes habitual y potencial en Chile (Ed. Universitaria). Pp 23. Santiago.

Mastebroek., H.D, Limburg., H, Gilles., T, and Marvin., H. (2000). Occurrence of sapogenins in leaves and seeds of quinoa (*Chenopodium quinoa* Willd.). Journal of Science of Food and Agriculture. 80, 152–156.

Mc Cough., N, and Cummings., J.H. (2005). Coeliac disease: a diverse clinical syndrome caused by intolerance of wheat, barley and rye. Proceedings of the Nutrition Society, 64, 434-450.

Menuguetti., Q.A, Brenzan., M.A, Bastida., M.R, Bazotte., R.B, Silva., D.R, and Garcia Cortez., D.A. (2011). Biological effects of hydrolysed quinoa extracts from seeds of *Chenopodium quinoa* willd. Journal of Medicinal Foods, 14, 653-657.

Meyer., B.N, Heinstein., P.F, Burnouf-Radosevich., M, Delfel., N.E, Mc Laughlin., J.L. (1990). Bioactivity-directed isolation and characterization of quinoside A: one of the toxic/ bitter principles of quinoa seeds (*Chenopodium quinoa* Willd). Journal of Natural Products, 38, 205-208.

Miranda., A, Vega- Galvez., A, Martinez., E.A, Lopez., J, Marin., R, Aranda., M, and Fuentes., F. (2013). Influence of contrasting environments on seed composition of two quinoa genotypes: nutritional and functional properties. Chilean Journal of Agricultural Research, 73, 108-116.

Moghadasian MH and Frohlich JJ (1999). Effects of dietary phytosterols on cholesterol metabolism and atherosclerosis; clinical and experimental evidence, The American Journal of Medicine, 107, 588-594

Moreau., R, Whitaker., B, and Hick., K.(2002). Phytosterols, phytostanols and their conjugates in foods: structural diversity, quantitative analysis, and health promoting uses. Progress in Lipid Research, 41, 457-500.

Mujica., A, Ortiz., R, Bonafacio., A, Saravia., R, Corredor., G, Romero., A, and Jacobson., S.E. (2006). Agroindustria de la quinua (*Chenopodium quinoa* Willd) en los paises andinos. Proyecto Quinua: Cultiva multiproposito para los paises andinos Peru- Bolivia- Columbia, PNUD_CONCYTEC-UNA-PROINPA- U. Columbia. Puno, Peru. Pp 113.

Mujica., A. (1992). Granos y leguminosas andinas. In: J. Hernandez, J. Bermijoy, J leon (eds). Cultivos marginados: otra perspective de 1492. Organizacion de la Naciones Unidas para la agricultura yla alimentacion FAO, Roma. Pp 129-146.

Mujica., A. (1994). Andean grains and legumes. In Bermejo., J.E.H and Leon., J, eds. Neglected Crops: 1492 from a different perspective. Pp 131-148. Plant production and protection. Rome. FAO.

Munoz., P.L, Martinez., E.A, Wyneken., U, and Dagnino-Subiabre., A. (2008). Pseudocereal quinoa improves memory and decreases anxiety of stressed rats: behavioural and molecular approaches, in IIBRO/Larc Congress of Neurosciences of Latin America, Caribbean and Iberian Penensula, Buzios, Brazil, 1-4 September.

National Academy of Sciences. (1975). Underexploited tropical plants with promising economic value. Tress, Washington DC. Pp. 145.

Neito., C, and Soria M. (1991). Procesamiento de quinua en Ecuador. Proyecto 3p-85-0213. Informe final de labors. INIAP-DTA- CIID. Quito, Ecuador. Pp-94.

Neito., C, and Valdividia., R. (2001). Postocosecha Transformacion y Agroindustria. In Mujica., A, S-E. Jacobsoen, S.E, Zquierdo., J.I and Marathee., J.P, eds. Quinua (Chenopodium Quinoa Willd), Ancestral Cultivo andino, alimento del Presente y Futuro. (http:// www.ric. fao.org/es/ agricultura/produ/cdrom/contenido/libro03/home03.htm)

Nesaretnam., K. (2008). Multi targeted therapy of cancer by tocotrienols. Cancer Letters, 269, 388-395.

Ng., S, Anderson., A, Cokera., J, and Ondrua., M. (2007). Characterization of lipid oxidation products in quinoa (*Chenopodium quinoa*). Food Chemistry, 101, 185-192.

Nishibe., S, Takenaka., T, Fujikawa.,T, Yasukawa., K, Takido., M, Morimtsu., Y, Hirota., A, Kawamura., T, and Noro., Y. (1996) . Bioactive phenolic compounds from Catharanthus roseus and Vinca minor. Natural Medicine, 50, 378-383.

Oakenfull., D and Sidhu., G.S.(1990). Could saponins be a useful treatment for hypercholesterolemia? European Journal of Clinical Nutrition, 44, 74-88.

Obando., M, Escalera., R Quiroga., C, and Arteaga., L. (2011). Optimizacion del consume especifico de energia en el procesos de beneficiado en seco de variedades amargas de quinua, basado en la aplicacion de un Lecho Fluidizado de Tipo Surtidor (LFTS). Universidad Privada Boliviana. Investigacion and Desarollo, 11:19.

Oelke., E.A, Putnam., D.H, Teynor., TM, and Oplinger., E.S.(1992). Quinoa, alternative field crops manual. University of Wisconsin- Extension. http:// www. Hort. pirdue.edu/newcrop/ afcm/ quinoa.html.

Ogawa., H, Watanabe., K, Mitsunga., T, and Meguro., T. (2001). Effect of quinoa on blood pressure and lipid metabolism in diet induced hyperlipidemics in spantaneously hypertensive rats (SHR). Journal of Japan Society of Nutrition and Food Sciences, 54, 221-227.

Ogungbenle., H. (2003). Nutritional evaluation and functional properties of quinoa (Quinopodium quinoa) flour. International Journal of Food Science and Nutrition, 54, 153-158.

Oliviera., A.C, Reis., S.M.P.M, Carvalho., E.M, Pimenta., F.M.V, Rios., K.R, Paiva., K.C, Sousa., L.M, Almeida., M and Arruda., S.F (2003). Adicoes crescents de acido fitoco a dieta nao interferiram na digestibilidade da casein e no ganho de peso em ratos. Revista de Nutricao, 16, 69-77.

Oshodi., A.A, Oguungbenle., H.N, Oladimeji., M.O. (1999). Chemical composition, nutritionally valuable minerals and functional properties of beni seed (*Sesamum radiatum*) pearl millet (*Pennisetum typhoides*) and quinoa (*Chenopodium quinoa*) flours. International Journal of Food Science and Nutrition, 50: 325-331.

Ozer., C.O, and Secen., S.M. (2018). Effects of quinoa flour on lipid and protein oxidation in raw and cooked beef burger during long term frozen storage. 38 (suppl-1), 221-227.

Pacosillo., V, and Chura. 2002. Identifacion dee practices de procesamiento y obtencion de derivados de canahua y quinua a nivel familiar. En: Informe Tecnico Annual 2001-2002. Ano-1, Proyecto "Elevarr la contribucion que hacenlas species olvidadas y subutilizadas d la seguridad alimentaria y a los ingrresos de la poblacion rural de escasos recursos". IPGRI- IFAD. Fundacion PROINPA. La Paz. Bolivia. Pp 115-117.

Park., I, Ibanez., A, and Shoemaker., C. (2007). Rice starch molecular size and its relationship with amylose content. Starch/ Starke. 59, 69-77.

Pasko., P, Barton., H, Zagrodzk., I.P, Chlopicka J., and Gorinstein., S (2010b) . Effect of quinoa seeds (*Chenopodium quinoa*) in diet on some biochemical parameters and essential elements in blood of high fructose fed rats. Plant Foods for Human Nutrition, 65, 333-338.

Pasko., P, Barton., H, Zagrodzki., P, Izewska., A, and Folta., M. (2009). Anthocyanins, total polyphenols, and antioxidative activity in amaranth and quinoa seeds and sprouts during their growth. Food Chemistry, 115, 994-998.

Pasko., P, Barton., H, Zgrodzki., P, Izewska., A, Krosniak., M, Gawlik., M, and Gorinstein., S. (2010a) . Effect of diet supplemented with quinoa seeds on oxidative status in plasma and selected tissues of high fructose fed rats. Plant Foods for Human Nutrition, 65,146-151.

Pasko., P, Sajewicz., M, Gorinstein., S, and Zachwieja., Z.(2008). Analysis of selected phenolic acids and flavonoids in Amaranthus cruentus and chenopodium quinoa seeds and sprouts by HPLC. Acta Chromatographica, 20, 661-672.

Pieretti., P.G, Gai., F, and Tassone., S. (2013). Fatty acid profile and nutritive value of quinoa (*Chenopodium quinoa* Willd) seeds and plants at different growth stages. Animal Feed Science and Technology, 183, 56-61.

Popenoe., H, King., S.R, Leon., J, and Kalinowski., L S. (1989). Lost crops of the Incas. In: Vietmeyer., ND, ed. Little known plants of the Andes with promise for worldwide cultivation. Washington DC: National Academy Press.

Prakash., D, and Pal., M. (1998). Chenopodium seed protein, fractionation and amino acid composition. International Journal of Food Science and Nutrition, 49, 271-275.

Praznik., W, Mundigler, N, Kogler., A, Pelzi., B, Huber., A. and Wollendorfer M (1999). Molecular background of technological properties of selected starches. Starch/Starke, 51, 197-211.

Przybylski., R, Chauhan., GS, and Eskin., N. 1994. Characteriztion of quinoa (Chenopodium quinoa) lipid. Food Chemistry, 51,187-192.

Pumacahua., A, Lopez Filho., J.F, Telis Romero., J, and Demiate., M.I. (2013). Estudio del comportamiento de pasta de almidon de quinua (*Chenopodium quinoa* Willde) extraido por molienda humeda. In: Memoria IV Congreso Mundial de la Quinua, Ibarra, Ecuador, 8-12 July 2013.

Qian., J, and Kuhn., M. (1999). Characterization of Amaranth cruentus and *Chenopodium quinoa* starch. Starch/starke. 51, 116-120.

Quillez., J, Garcia-Lorda., P, and Salas-Salvadol., J. (2003). Potential uses and benefits of phytosterols in diet: present situation and future directions. Clinical Nutrition, 22, 343-351.

Quiroga., C, Escalera., R, Aroni., G, Bonifacio., A, Gonzalez., JA, Villica., M, Saravia R., and Ruiz A. (2015). State of the art report on quinoa around the world in 2013. Chapter 3.1, Traditional processes and technological innovations in quinoa harvesting, processing and industrialization ed : Bazile., D, Bertero., D and Nieto., C. Food and Agriculture Organization of the United Nations (FAO) and Centre de Coopération Internationale en RechercheAgronomique pour le Développement (CIRAD). Pp 218-249.

Quiroga., C, Escalera., R, Arteaga., L, Montano., F, and Nogales., R. (2011). Beneficiado en seco de la quinua-proyecto de prefactibilidad para el beneficiado en seco de quinua con un lecho tipo surtidor. In Fundacion PIEB. Investigacion Ambiental, Pp 26, La Paz.

Ranhotra., G.S, Gelroth., J.A, Glaser., B.K, Lorenz., K.J, and Johnson D. (1993). Composition and protein nutritional quality of quinoa. Cereal Chemistry, 7, 303-305.

Reichhert., R.D, Tatarynovich., J.T, and Tyler., RT. (1986). Abrasive dehulling of quinoa (*Chenopodium quinoa*): effect on saponin content as determined by an adapted hemolytic assay. Cereal Chemistry, 63, 471-475.

Repo- de Carrasco., R, Hellstrom., J, Pihlava., J, and Mattila., P. (2010). Flavonoids and other phenolic compounds in Andean indigenous grains: qunoa (*Chenopodium quinoa*), kaniwa (*Chenopodium pallicaule*) and Kiwicha (*Amaranthus cautadus*). Food Chemistry, 120, 128-133.

Repo-Carrasco., R, Espinoza., C, and Jacobsen., S.E. (2003). Nutritional value and use of the Andean Crops quinoa (*Chenopodium quinoa*) and kaniwa (*Chenopodium pallidicaule*). Food Review International, 19 (1and 2), 179-189.

Reynaga., A, Quispe., M, Huarachi., A, Calderon., I, Soto Quispe., J.L, and Torrez., M. (2013). Evaluacion de las cualidades agroindustriales de los granos de quinua real. In: Memorias del Congtreso Cientifico de la quinua, La Paz, Bolivia, 14-15 June 2013, Pp 525-534.

Robalino., D, and Penaloza., W. (1988). El Oso de la quinua en la elaboracionde tempeh. En: VI. Congress Internacional Sobre Cultivos andinos. Memorias. INIAP. Quito, Ecuador. Pp 20-24.

Ruales, J, and Nair., B. M. (1993a). Saponins, phytic acid, tannins and protease inhibitors in quinoa (*Chenopodium quinoa* Willd) seeds. Food Chemistry. 48,137-143.

Ruales., J, and Nair., B.M. (1993). Content of fat, vitamins and minerals in quinoa (*Chenopodium quinoa* Willd) seeds. Food Chemistry, 48, 131-136.

Ruales., J, and Nair., B. (1994). Properties of starch and dietary fibre in raw and processed quinoa (*Chenopodium quinoa* Willd) seeds. Plant Foods for Human Nutrition, 45, 223-246.

Ruales., J, and Nair., B.M. (1992). Nutritional quality of protein in quinoa (*Chenopodium quinoa* Willd) seeds. Plant Foods for Human Nutrition, 42, 1-12.

Ryan., E, Galvin., K, O' Connar., T, Maguire., A, and O'Brien., N.(2007). Phytosterol, squalene, tocopherol content and fatty acid profile of selected seeds, grains, and legumes. Plant Foods for Human Nutrition, 62, 85-91.

Salazar., D.M, Naranjo., M, Perez., L.V, Valencia., A.F, Acurio., L.P, Gallagas., L.M, Alvarez., F.C, Amancha., P.I, Valencia., M.P, Rodriguez., C.A, and Arancibia., M.Y. (2017). Development of newly enriched bread with quinoa flour and whey. 3rd international conference on Argiculture and Biological Sciences (AVS 2017). Earth and Environmental Science, 77 (2017)-012018.

Santos., M.A.T. (2006). Efeito do cozimento sobre alguns atores antinutricionais em folhas de brocou, couve-flor e couve. Ciencia e Zagroteccnologia, 30, 294-301.

Savage., G.I. (2003). Saponins, the Encyclopaedia of Food Science, Food Technology and Nutrition. 5095-5098.

Schaffer., LC, Lehmann., U, Ross., A.B, Roger., O, Eldridge., A.l, Ananta., E, Bietry., M.F, King., C.R, Moroni., A.V, Srichuwong., S, Wavreille., A.S, Redgwell., R, Labat., E, and Robin., F. (2017). Whole grain in manufactured foods: Current use, challenges and the way forward. Critical Reviews in Food Science and Nutrition, 24; 57, 1562-1568.

Schilck., G, and Bebenheim., D.L.(1996). Quinoa, candidate crop for NASA's controlled ecological life support systems. In: Janick J, editor, Progress in new crops, Arlington, VA: ASHS Press. Pp 630-640.

Schlenker., E, and Williams., S.R. (2003). Essentials of nutrition and diet therapy. Elsevier, Amsterdam

Schlick., G, and Bubenheim., D.I.(1993). (Ames Research Center, Moef fet Field, California). Quinoa: an emerging "New" Crop with potential for CLESS, NASA (US). Report No: 3422, Pp: 1-6.

Seeman., P, Cheng., D and Iles., G. (1973). Structure of memebrane holes in osmotic and saponin haemolysis. Journal of Cell Biology, 56, 519-527.

Sharma., V, Chandra., S, Dwivedi., P, and Patrukar., M. (2015). Quinoa (*Chenopodium quinoa* Willd): A nutritional healthy grain. International Journal of Advanced Research, 3, 725-736.

Siener., R, Honow., R, Seidler., A, Voss., H, and Hesse., A. (2006). Oxalate contents of species polygonaceae, amaranthaceae and chenopodiaceae families. Food Chemistry, 98, 220-224.

Song, W. O., Chun, O. K., Hwang, I., Shin, H. S., Kim, B. G., Kim, K. S., and Lee, S. G. (2007). Soy isoflavones as safe functional ingredients. Journal of medicinal food, 10(4), 571-580.

Soria., M, Marcial., M, and Penalosa., M. (1990). Lavado de quinua procesos y prototipos, seminario taller sorbe investigacion en posproducion de quinua en Ecuador.INIAP.UTA. CIID. Quito. Ecuador. Pp: 23-34.

Sparg., S, Light., M, and Van Staden., J. (2004). Biological activities and distribution of plant saponins. Journal of Ethnopharmacology. 94: 219-243.

Spehar., C.R. (2007). Quinoa: Alternativa para a Diversificacao Agricola Ee Alimentar. Planaltina. DF. ed., Tecnico. Embrapa Cerrados. Brasil. Pp 103.

Stuardo., M and San Martin., R. (2008). Antifungal properties of quinoa (*Chenopodium quinoa* Willd), alkali treated saponins against Botrytis cinerea. Industrial Crops and Products, 27, 296-302.

Sukeerthi., S.S, and Singh., T.K. (2017). Formulation and nutritional analysis of biscuits made from quinoa, flaxseed and brown rice. International Journal of Food Science and Nutrition, 2, 27-30.

Tagle., M.B. and Planella., M.T. (2002). La quinoa en la zona central de chile. Supervivencia de unatradicionprejispana (ed., Iku), Santiago, Pp 25-29.

Takao., T, Watanabe., N, Yuhara., K, Itoh., S, Suda., S, Tsuroaka., Y, Nakatsugawa., K and Konishi., Y. (2005). Hypocholesterolemic effect of protein isolated from quinoa (Cheno podium quinoa willd) seeds. Food Science and Technlogy Research, 11:161-167.

Tang., H, Watanabe., K, and Mitsunaga., T. (2002). Characterization of storage starches from quinoa, barley and adzuki seeds. Carbohydrate Polymer, 49, 13-22.

Tang., Y, Li., X, Chen., P.X, Zhang., B, Hernandez., M, Zhang., H, Marcone., M.F, Liu., R, and Tsao., R (2015) b. Characterization of fatty acid, carotenoid, tocopherol/ tocotrienol compositions and antioxidant activities in seeds of three *Chenopodium quinoa* willd, genotypes. Food Chemistry, 174, 502-508.

Tang.,Y, Li., X, Zhang., B, Chen., P.X, Liu., R, and Tsao., R.(2015). Characterisation of phenolics, betanins and antioxidant activities in seeds of three *Chenopodium quinoa* Willd, genotypes. Food Chemistry, 166,380–388.

Tappy., L, Le., K, Tran., C, and Paquot., N. (2010). Fructose and metabolic diseases: new findings, new questions. Nutrition, 26, 1044-1049.

Tarade., K.M, Singhal., R.S, Jayram., R.V, and Pandit., A.B. (2006). Kinetics of degradation of saponins in soybean flour (*Glycine max*) during food processing. Journal of Food Engineering, 76, 440-445.

Tari., T, Annapure., U, Singhal., R, and Kulkarni., P. (2003). Starch based spherical aggregates: screening of small granule sized starches for entrapment of a model flavoring compound, vanillin. Carbohydrate Polymer, 53, 45-51.

Tukomane., T, and Varavinit., S. (2008). Classification of rice starch amylose content from rheological changes of starch paste after cold recrystallization. Starch/starke, 60, 292-297.

Urquizo., F.M.L, Torres., S.M.G, Tolonen., T, Jaakkaola., M, Grazzia., M, Niebuhr., P, Wright., A.V, Valencia., R.E.C, Korhonen., H, and Ferrer., C.P. (2017). Development of a fermented quinoa based beverage. Food Science and Nutrition, 5, 602-608.

USDA, US Department of Agriculture, Agricultural Research Service (2005).USDA National Nutrient Database for Standard Reference, Release 18. Nutrient Data Laboratory Home Page. http:// www.nal.usda.gov/fnic/foodcomp.

USDA, US Department of Agriculture. Agricultural Research Service. 2013. USDA national nutrient database for standard reference, Release 3.1. Nutrient Data Laboratory Home Page. http:// www.ars.usda.gov/nutrientdata/flay.

USDA, US Department of Agriculture. Agricultural Research Service. 2018. USDA national nutrient database for standard reference, Release 1, Nutrient Data Laboratory Home Page. http:// fdc.nal.usda.gov/fdc-app.html#/food-details/168874/nutrients

Valencia-Chamorro., S.A.(2003). Quinoa In: CABALERO B: Encyclopedia of Food Science and Nutrition. Vol 8. Academic Press. Amsterdam. Pp 4895-4902.

Vega- Galvez., A.V, Miranda., M, Vergara., J, Uribe., E, Diaz., L.P, and Martinez., E.A.(2010). Nutritional facts and functional potential of quinoa (*Chenopodium quinoa* willd), an ancient Indian grain. Journal of Science of Food and Agriulture, 90, 2541-2547.

Verza., S.G, Silveira., F, Cibulski., S, Kaiser., S, Feriera., F, Gosmann Roehe., P.M, and Ortega., C.G. (2012). Immuno adjuvant activity, toxicity assays and determination by UPLC/ Q-TOF-MS of triterpenic saponins from *Chenopodium quinoa* seeds. Journal of Agricultural Food Chemistry, 60, 3113-3118.

Vidaurre., P.J, Alacoca., C, Flores., J, Soto., J.L, Pinto., M, and Rojas, Y.W. (2005). Participacion y apoyo en ferias de biodiversidad. En: Rojas, w. (Ed). Proyecto Manejo, conservacion y Uso Sostenible de los Recurso Genrticos de Granos Altoandinos, en el marco del SINARGEAA, Informe Final 2004-05. Fundacion PROINPA. LaPaz, Bolivia. 99, 140-147.

Wang., H, Khor., T.O, Shu., L, Su., Z.Y, Fuentes., F, Lee., J.H, and Kong., AN. (2012). Plants Vs Cancer: A review on natural phytochemicals in preventing and treating cancers and their druggability. Anticancer Agents in Medical Chemistry, 12, 1281-1305.

Watanabe., K, Peng., L, Tang., H, and Mitsunaga., T.(2007). Molecular structural characteristics of quinoa starch. Food Science and Technology Research, 13, 73-76.

Woldimichael., G.M, and Wink., M. (2001). Identification and biological activities of triterpenoid saponins from *Chenopodium quinoa.* Journal of Agricultural Food Chemistry, 49, 2327-2332.

Wood., S, Lawson., L, Fairbanks., D, Robinson., L and Anderson., W.(1993). Seed lipid content and fatty acid composition of three quinoa cultivars. Journal of Food Composition and. Analysis, 6, 41-44.

Wright., K, Pike., O, Fairbanks., D, and Huber., C. (2002). Composition of artiplex hortensis, sweet and bitter chenopodium quinoa seeds. Journal of Food Science, 67, 1380-1383.

Youdim., K, Martin., A, and Joseph., J. (2000). Essential fatty acids and the brain: Possible health implications. International Journal of Developmental Neuroscience, 18, 383–399.

Zevallos., V.F, Ellis., H.J, Suligoj., T, Herencia., L.I, and Ciclitira., P.J (2012). Variable activation of immune response by quinoa (*Chenopodium quinoa* Willd) prolamins in celiac disease. American Journal of Clinical Nutrition, 96, 337-344.

Zhang., R.Q, Han., Z.K, Chen., J, and Zhang., C.L. (1995). Daidzein diet promotes mammary gland development and lactation in rat. Journal of Zoology, 4, 332-338.

Zhu., N, Sheng., S, Sang., S, Jhoo., J.W, Bai., N, Karwe., M.V, Rosen., R.T and Ho C.T. (2002). Triterpene saponins from debittered quinoa (*Chenopodium quinoa*) seeds. Journal of Agricultural Food Chemistry, 50, 865-867.

19

Cockscomb (*Celosia* sps): Origin Nutritional Attributes, Processing Innovations and Health Benefits

Pal Murugan Muthaiah, Athulya Dev, Santosh Pal, Janifer Raj Xavier and Anil Dutt Semwal

Defence Food Research Laboratory, Siddarthanagar, Mysore-570011 Karnataka, India

---

Introduction

To constantly move towards advancement, points towards the addition to the tradition that converts to modernization. Today with the drawing waves risking health and diet of an individual, there is a constant demand of rolling back the traditional grains which lost its significance with the turning over of era. Family Amaranthaceae for a long time with the turn of events existed among the staple diets of people in different regions of Asia as China, Africa, America and some parts of India due to its complementary effects with different cuisines as well as the nutrient fulfillment (Olawuyi *et al.*, 2016).

The family Amaranthacea has almost 165 genera and 2040 species and is considered to be abundantly rich species lineage among the flowering plant order of caryophyllales (Muller K and Borsch, 2005). Most of the plants amongst the family Amaranthacae consist of annual or perennial plants, herbs or shrubs, small trees and vines. This family is widely available from the tropical region to cool temperate regions (Simpson, 2010). Amaranthacea family is native of tropical and sub-tropical areas of Central America, Africa and Australia. The agricultural habitats these families are temperate and sub-tropical regions. Native people in tropical and subtropical countries as well as temperate climate used Amaranthacea species as a source of traditional medicinal plants & phytofoods. (Sun *et al.*, 2011). Cockcomb; *Celosia* is one of the important species of Amaranthaceae family used traditionally in food and medicines

The Celosia species is a small room among the edible and ornamental plants belonging to Amaranthacea. The generic name "Celosia" derived from Greek word "kelos" meaning "burned", and refers to the flame-like flower heads (Müller and Borsch, 2005). *Celosia cristata* is commonly known as cockscomb, since the flower resembles rooster (cock) head (Ramesh *et al.*, 2013). It belongs to the class Magnoliopsida, order Caryophyllales and family Amaranthaceae. The most common regions known and recognized for cockscomb are Africa, South America, India and some parts of Asia (Rubini *et al.*, 2012). The most common names that prevails worldwide for *Celosia sps* are cockscomb, crested celosia, yellow toreador, red cockscomb, foxtail amaranth, fire-flame bush, shinaji tea and woodfordia. As that of amaranth crop, *Celosia sps* also stands in the group of pseudo cereals and it is popular for its significant nutrient content as well as attractive inflorescence.

Generally, Celosia species are an annual herb, hairless entirely. The stem is erect, thick, little branched, green or tinged with red, ridged and flat near the upper part. The simple leaves are alternate, petiolate; the blade is long-elliptical to ovally lanceolate, 5-13 cm long, 2-6 cm wide, acuminate or attenuate at the apex, gradually narrow and decurrent at the base, and entire marginally (Fact sheet, 1999). Occurring on top of the stem, the spikes are flat, succulent, crest-like or feather like; down their middle are numerous flowers; the perianthial segments are light red to purplish red, yellowish white or yellow, elliptically ovate, pointed at the tip, 5 in number; the bract, bractlet and perianthial segment are scarious; each flower has 5 stamens whose filaments are joined together to form a cup at the base (www.theflowerexpert). The fruit is an egg-shaped utricle, which is wrapped in the persistent perianth and becomes circumscissile when ripe. The seed is kidney-shaped, black, and lustrous (www.gsxti01.vicp). The most important Celosia species with economic importance are *Celosia argentea, Celosia Cristata* and *Celosia isertii.*

Celosia argentea

C.argentea is an annual herb with an erect, coarse, simple or branched stem and grows upto 0.5 to 1.5 m high. Leaves are alternate, linear to lanceolate, entire and are about 4 to 14 cm in long. The flowers are small in size and they are dense erect spikes, 8 to 12 mm in length, borne in solitary, erect, stout, dense, white, purple, or pink, glistening spikes. It contains membranaceous fruits. Seeds are shining and black and they are 1.5 mm in diameter. The plant of *Celosia argentea* is shown in Figure 1.

Celosia argentea commonly found in West Africa, Ethiopia, Somalia, and Kenya, other parts of East Africa, Mexico and Central Africa. It is an important cultivated vegetable in Africa. It grows as a weed during rainy season throughout India and other tropical regions of the world mainly Sri Lanka, Yeman, Indonesia, America and West Indies. (Aladesanwa *et al.*, 2001)

Fig. 1: *Celosia argentea*

Celosia cristata

It is herbaceous plant extensively seen in Africa, South America, India and some parts of Asia (Wang *et al.*, 2010) with height varies from 5 to 2 feet. Leaves are greenish purple or red in colour and arranged in alternate, simple and saggitate or arrow shaped. Cyanidin, a kind of anthocyanin was found in leaves. The flower of *Celosia cristata* is shown in Fig 2. Africa and India as well as dry stony regions of both North and South America are believed to be the geographic origins of Celosia.

Celosia isertii

It is a straggling herb, grows up to 3 m height and widely found in stream-banks, damp sites, clearings and rarely in Savanna. The leaves are boiled and applied hot for rheumatismin Sierra Leone (Anthony *et al*, 2013) . Traces of flavones have been found to be present in the entire plant from the Congo area. Recorded from Senegal to South Nigeria and Fernando and in Cameroun across central Africa to Tanganyika, Zambia and Angola. In West Africa the plant is often eaten as a vegetable or prepared in soups and sauces. The flower of Celosia isertii is shown in Figure 3.

Fig. 2: *Celosia cristata*

Fig. 3: *Celosia cristata*

As the constant rise in acceptability of *Celosia sps* has shown that different parts of cockscomb are medicinal impacts when consumed. Its popularity in the pharmaceutical industry is widely accepted along with the rise in the demand of an alternative source of pigment betalain which till date is extracted only from red beet, whereas cockscomb is now widely growing as a potential source of betalin and carotenoids (Cai and Corke, 2005). Since ages the seeds of *C. cristata* were used as demulcent; for painful micturition and for dysentery (Khare, 2008). The seeds of cockscomb are hypotensive and ophthalmic. It was used in the treatment of bloodshot eyes, blurring of vision, cataracts and hypertension (www.motherherbs). The seeds can also combat "liver-heat" improving eyesight, clearing wind and as an anti-inflammatory (Ramesh *et al.*, 2013). As the flowers were edible in India, they were also medicinally used in menorrhagia and as an astringent which is used to treat bloody stool, hemorrhoid bleeding and diarrhea. The flowers were used as astringent, styptic, depurative, uterine sedative, constipating, antibacterial and corrective of urinary pigments, febrifuge and alexiteric (National Research Council, 2006). Generally used in the cases of throat infection it can also be used as the cure of leprosy, burning sensation, skin diseases, diarrhea, dysentery, fever, headache, hemorrhoids, menorrhagia, ulcers and wounds. The flowers and seeds were astringent, hemostatic, parasiticide and poultice. To specify the medicinal benefits of the juice of leaves of cockscomb it's used as a stimulant for pregnancy (Sun *et al.*, 2011). The plant was also used in the treatment of fatigue, atherosclerosis, leucorrhoea and osteoporosis. As stated that the parts of cockscomb are edible it has a wide application in various industries as food, pharmaceutical, additives, etc. due its abundance of benefits provided to the body when consumed.

Origin

The exact origins of cockscomb in the wild are unknown, though some assign the geographical origins to the dry slopes of Africa and India as well as the dry rocky regions of both North and South America (Shanmugam *et al.*, 2011). *Celosia cristata* are widely believed to have developed from *Celosia argentea*, which is listed as native to India but is common in the wilds in China. In China it is a known ornamental flower used widely. In West Africa, Central Africa, and Southeast Asia, *Celosia argentea* is grown as a leaf vegetable and cereal crop. In Southern Nigeria, it is the most important leaf vegetable and is known as "soko yokoto" which means "make husbands happy and fat" (Hamzar *et al.*, 2018).

Cockscomb blooms from late summer to late fall and can be found in bright yellows, orange, red, pink, and purple. They are scent-free and strikingly beautiful. They symbolize silliness and affection, making them a perfect addition to flower arrangements and bouquets.

Use of Plant Parts

The different parts of *Celosia sps* that can be consumed in different ways includes leaves, flower and seeds. Amongst the 60 varieties of Celosia genus the most widely consumed ones are *C. argentea, C. cristata* as well as *C. trigyna.* As being a member of Amaranthus family its highly considered as the healthy option within leafy vegetables, as broad leaves with high protein content and brightly colored flower heads that looks like soft, fluffy plumes and remind the observer of crimson, scarlet or gold feathers. Because of its flavor, food value and familiarity the crop is widely consumed in several parts of Africa and plays a crucial role in health in parts of Nigeria and nearby countries (Kolade, 2017).

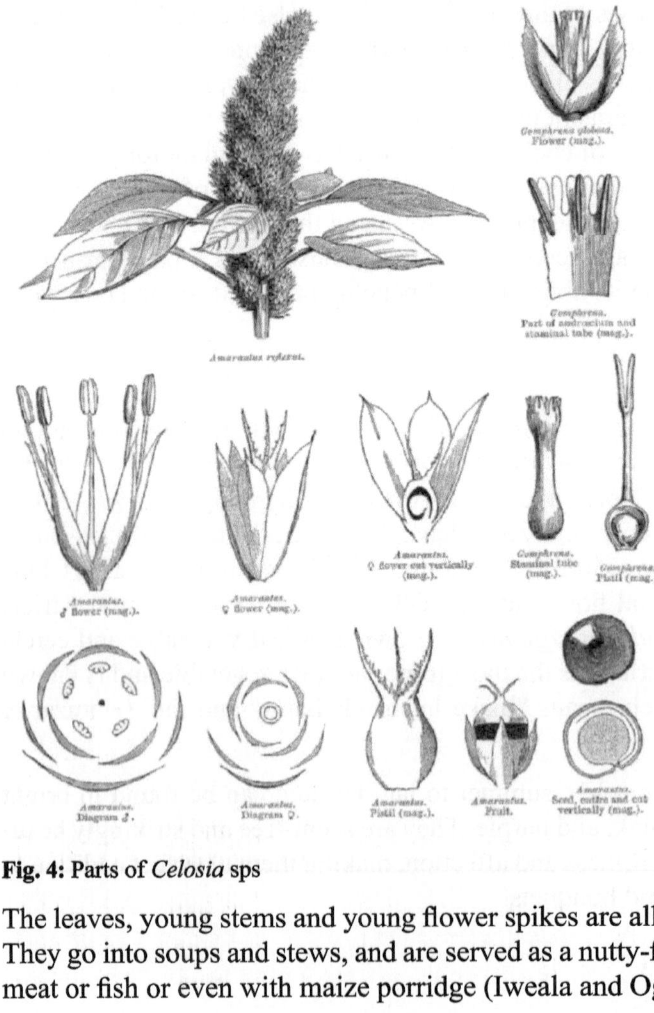

Fig. 4: Parts of *Celosia* sps

The leaves, young stems and young flower spikes are all handled like spinach. They go into soups and stews, and are served as a nutty-flavored side dish with meat or fish or even with maize porridge (Iweala and Ogidigo, 2015).

Leaves

Leaves along with the young stems and young inflorescences are eaten as pot-herbs (Spórna-Kucab *et al.*, 2018). They soften up readily and cook in minutes. The texture is soft with mild flavor. These boiled green leaves are often added in stews (Abubakar *et al.*, 2015). They are also pepped up with such things as garlic, hot pepper, fresh lime and red palm oil and eaten as a side dish.

Flower

The often brilliantly colored flowers are borne in dense heads. Most occur in spikes, and stand like spears in the garden bed. But certain cultivated forms have compact or feathery clusters due to fasciation, the accumulation of genetic malformations that are of huge interest to botanists (Nidavani *et al.*, 2014). Young flowers are regularly used for making soups and stews.

Seeds

The seeds of Celosia are generally black in colour and 1 mm in diameter. Several species of family Amaranthaceae are grown for their grain-like seeds. Although small, these seeds occur in prodigious quantities. In carbohydrate content they equal cereals such as wheat, but have more protein (over 17 % in some strains) and more oil. The grain is ground in flour and used for various food preparation. The seed can be popped and flanked and consumed as snack food.

Fig. 4: Seeds of *Celosia* sps.

Nutritional Attributes

As *Celosia sps* has different edible parts, it can be of significant importance to understand the composition of various edible parts of the plant. Results showed that the nutrients were present in all the parts i.e stem, leaf, root seed and flowers (Prakash *et al.*,1991; Ilodibia *et al.*, 2016; Adegbaju *et al.*, 2019; Sayeed *et al.*, 2020) but in varied proportions. The proximate analysis of different part of *Celosia* sps showed in Table1.

Protein and fat content varied from 3.56 to 16.8 g/100g and 1.10-10.68 g/100g respectively with highest quantity in grain and leaf. The crude fiber content is in the range of 1.10-10.68 g/100g with highest content in grain and leaf . Ash content found to be low in grain i.e 2-3.8 g /100g.

Green leafy vegetables provide humans with adequate amounts of many vitamins and minerals and thus occupy an important place among the food crops. They are valuable in maintaining alkaline reserve of the body because of their nutrient and mineral contents . The perusal of table 1 clearly exposed the importance of Celosia genus as leafy vegetables to our diet.

The high composition of protein and fat in the leaf indicated that it can be used as food supplements especially for malnourished children and more preferably to other parts. Fat is high energy nutrient and does not add to the bulk of the diet. Proteins are used for building and repairing of body tissues, regulation of body processes and formation of enzymes and hormones. The high composition of crude fibre, ash and moisture in the stem indicate the superiority of the stem as a better source of fibre, ash and moisture. Generally, fibre aids and speeds up the excretion of waste and toxins from the body, preventing them from sitting in the intestine or bowel for too long, which could cause a build-up or azotemia.

Analysis of the mineral composition revealed that zinc, phosphorus and iron contents were highest in the leaf 0.823±0.010 percent, 39.768±5.060 percent and 8.192±1.021 percent respectively (Ilodibia *et al.*, 2016). This implies that the leaf will serve as a better source of zinc, phosphorus and iron than other parts of the plant. Iron is an important constituent of haemoglobin found in blood. Iron serves as a carrier of oxygen to the tissues from the lungs by the red blood cell haemoglobin as a transport medium for electrons within cells and as integrated part of important enzyme systems in various tissues. Formation of adenosine triphosphate ATP and synthesis of DNA and RNA while zinc is an essential trace element needed in the body for boosting the immune system and preventing lower respiratory infections (Isong and Essien, 1996).

Table 1: Proximate Analysis of Plant parts *Celosia* sps

Plant parts	Crude Protein (g/100g)	Crude Fiber (g/100g)	Ash (g/100g)	Fat (g/100g)	Moisture (g/100g)	Carbohydrate (g/100g)	Energy (kcal/100g)
Leaf[1]	5.17-10.2	3.53-12.3	15.3-22.43	1.10-1.15	23.84-23.91	45.50-47.6	230-234
Stem	10.20	20.85	23.91	10.09	31.11	4.5-9.04	217-247
Root	9.24	19.34	13.76	5.04	18.50	5.4-10.2	320-340
Grain	11.6-16.8	3-20	2-3.8	6.4-10.9	10-12	60-67	220-250
Flower (fresh weight)	3.56-3.61	2.3-3.2	2.44-2.53	4.58-4.68	79.62-80.75	9.80-11.2	94.74-95.00

Personal elaboration using information from sources: Prakash *et al*, 1991; Ilodibia *et al*, 2016; Adegbaju *et al*, 2019; Sayeed *et al* 2020.

Vitamins A and C contents were highest in the leaf 9.002±0.034 µg/g, and 56.002±3.657 mg/100g respectively (Ilodibia *et al*., 2016). This implies that the leaf of *C. argentea* will serve as a better source of vitamin A and C than the root and stem. Vitamin A enhances vision and healthy skin while vitamin C activates the cell functions. Vitamin C is a powerful antioxidant (Kanu *et al*., 2017). It favors the absorption of iron in the intestine, protects against infections, neutralizes blood toxins and intervenes in the healing of wounds and reduction of blood cholesterol. The study has demonstrated that *Celosia cristata* is a power house of nutrients. Apart from the leaf eaten by our people, other parts of the plant should also be used as food. The result of the study revealed that the leaf has high compositions of crude protein, fat, zinc, phosphorus, iron, vitamin A and vitamin C which indicated that *Celosia cristata* could contribute significantly to human health requirements. Malnourished people could be advised to eat the leaves of *C. cristata* in relatively high amount as food supplement, since it is rich in proteins, minerals and vitamins (A & C). With the versatility of the edible parts of cockscomb it has been widely accepted to be one of the major nutrient sources for the nutrient deficient population combating malnutrition as well as hunger at the same time. Besides macronutrients , Celosia sps are store house of many phyto chemical compounds. The phytochemical constituents of *Celosia* sps are presented in Table.2

Table 2: Phytochemical constituents of *Celosia* species

Metabolite	Phytochemical constituents	Part of the plant
Carbohydrate	Sucrose, disaccharide	Seed
Lipids	Fatty acids, waxes, glycerides, phospholipids, sterols, vitamin A, D, E and K, β-sitosterol 1, palmitic acid and stigmastrol, daucosterol and oleanolic acid	Whole plant
Amino acids, peptides and protein	Bicylic peptides: celogenamide A, B, C, D, E, F, G, H and J, mordin Cyclic peptides: celogenamide A, celogntin K	Seed
Betalain	Betacyanins, amaranthin, isoamaranthin, betalamic acid, miraxanthin V (dopaamine-BX), (S)-tryptophan, (S)-tryptophan-BX, 3-methoxytyramine-BX and yellow betaxanthins	Flower, inflorescence
Phenols and phenolic acid	1-(4-O-β-glucopyranosyl-3- methoxyphenyl) propan-2-ene (citrusin C), 3-O-β-glucopyranosyl-1H-indole (indican), (7E)-6,9-dihydromegastigma-7-ene-3-one-9-O-b-glucopyranoside, (3Z)-hexenyl-1-O-β-(6-O-ß-rhamnopyranosyl-β-glucopyranoside), (3Z)-hexenyl-1-O-β-dglucopyranoside and trans-ferulic acid	Leaves

Metabolite	Phytochemical constituents	Part of the plant
Phenolic glycosides	4-O-β-D-glucopyranosyl-2-hydroxy-6-methoxyacetophenone, 4-O-β-D-apifuranosyl-(1fi 2)-β-dglucopyranosyl-2-hydroxy-6-methoxyacetophenone, eugenyl-O-β-D-glucopyranoside, sucrose, quercetin-3-O-β-D-glucopyranoside, isorhamnetin-3-O-β-D-glucopyranoside, hamnatin-3-O-β-D-glucopyranoside, isorhamnetin-3-O-ß-L-rhamnopyranosyl-(1fi 2)-β-D-glucopyranoside, b-sitosterol, stigmasterol, stigmasterol-3-O-β-D-glucopyranoside	Whole plant
Flavonoid	Isoflavones: 5-Methoxy-6,7-methylenedioxy-2'-hydroxyisoflavone and 2'-methoxy derivative: tlatlancuayin	Aerial parts
Diterpenes steroids	Gibberellic acid	Seed
Seroids	Cristatain, saponins, celosin A, B, C, D and celosin E, celosin F, celosin G	Seed

Source : Adegbaju *et al*., 2019

Carbohydrates

Carbohydrates are the polyhydroxylated aldehydes and ketone, which are formed as result of photosynthesis. Sucrose as a carbohydrate, disaccharide has been accounted along with b-sitosterol (Xue *et al*., 2011) in seeds of *Celocia argentea* which serves as storage of energy.

Lipids

Lipids is an important structural components of cell membranes It acts as permeable barriers to the peripheral environment of cells. In plants, lipids play especially important roles as signaling and energy storage compounds. Plant lipids include triacylglycerols, phospholipids, galactolipids, and sphingolipids. Xue *et al*. (2011) reported following fat components i.e b-sitosterol 1, palmitic acid and stigmasterol 2, daucosterol 3 and oleanolic acid in *Celocia argentea*. The seed of *Celosia argentea* contains a fatty oil known as 'celosia oil' in India.

Lin *et al*. (2002) conducted study to determine the ratio of fatty acid in celosia seed . they used the Soxhlet extraction method to extract the fatty acids using Shimadzu GC-17A and chemical detection with FID. The study showed that the unsaturated fatty acids levels reached up to 79.276 percent and composition in the range of olefinic acid (27.995%) and linoleic acid (44.522%). Celosia Sps also rich in linolenic acid , arachic acid, arachidonic acid, hexadecanoic acid, palmitoleic acid, octadecanoic acid, octadecanoic monoenoic acid (Lin *et al*., 2002 and Zheng *et al*., 2014)

Study conducted by Prakash *et al.* (1991) on 13 lines of *Celosia sps* namely *Celosia argentea, celosia cristata, Celosia plumosa and Celosia whiteii* revealed that oil contents was in the of 56 g kg^{-1} (*C plumosa.* yellow) to 109 g kg^{-1} (*C argentea*). The oils were of the simple palmitic-oleic-linoleic type comprising only these three acids without any great change in the composition. The variation of palmitic acid (1 6 : 0) was from 241 to 353 g kg^{-1}', oleic acid (1 8 : 1) 199 to 259 g kg^{-1} and linoleic acid (18:2) 433 to 524 g kg^{-1}. The total unsaturated acids (oleic and linoleic) showed a variation from 647 g kg^{-1} (C argentea-4) to 759 g kg^{-1} (C plumosa, red) and the saturated/unsaturated acid ratio ranged from 0.32 to 0.55. They also reported that seed oils of Celosia species investigated were found promising with regard to their saturated/ unsaturated fatty acid ratio and fatty acid composition

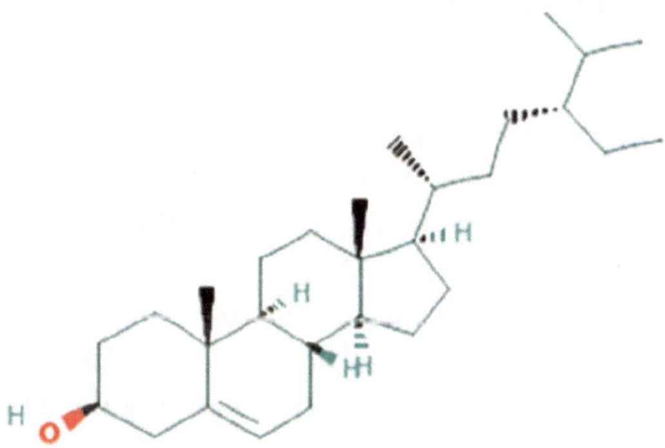

b-sitosterol

Source: https://pubchem.ncbi.nlm.nih.gov/compound/222284#section=2D-Structure

Diémeleou *et al.* (2014 explored possibility of using *Celosia argentea* seed oil for food and pharmaceutical industries application. They carried out physicochemical and nutritive analysis in seed oil. Physicochemical properties of extracted oil were as follow: specific gravity (0.92±0.01), refractive index (1.47±0.00), colour lovibond (20.2±0.00), viscosity at 20°C (53.08±0.28 mPas), cloud point (-2.3±0.00°C) acid value (4.68±1.62 mg KOH/g), peroxide value (10.33±0.58 meq O2/kg), iodine value (124.08±1.22 g I2/100g), saponification value (170.17±1.62 mg KOH/g). Biochemical and nutritive analysis have revealed the following assets: impurities (0.016±0.00%), unsaponifiable matter (1.47±0.21%), phosphorus (0.10±0.00 mg/g), vitamin A (0.52±0.01 mg/g) and vitamin E (0.12±0.01 mg/g). Fatty acids profile of *C. argentea* seed oil highlighted linoleic acids as major fatty acid with amount of 49.94±0.01 percent.

Protien, Peptides and Amino Acid

Amino acids comprising an amine group, a carboxylic acid group are the basic unit of proteins and peptides. They are the building blocks of metabolic processes and mostly water soluble . It is also present as peptide in the plant tissue and known as free amino acid. Pingale *et al.* (2012) reported 18 kinds of amino acids, of which 8 kinds are essential amino acids in Seed of *Celocia argentea* . Kobayashi *et al.* (2001) isolated three novel bicylic peptides, celogentins A 4, B 5, C 6 from the seed of *Celocia argentea* and proved the structures of newly isolated compounds were by NMR and chemical means by comparing with known moroidin. Celogenamide A 7, a cyclic peptide isolated from the seed of *Celocia argentea* and characterized using 2D NMR and MS-MS . Study conducted by Prakash *et al.*(1991) on 13 lines of *Celosia sps* specified that protein content was in the range of 10.1 to 1 16.8 g/100g in seeds.

Dietary fiber

The crude fiber content of grain was in the range of 2.3-3.2 g/100g. Much information about the dietary fibre content of celosia grain is not available. The information pertaining to Celosia leaves is presented here. Celosia leaves *are* abundantly rich in dietary fiber. Adegbaju *et al.* (2019), reported that crude fibre of *C. spicata* (12.3%) has been found to be greater than those of *Arachis hypogea* (3.70%), *Amaranthus hybridus* (8.61%) and *Dioclea reflexa* (9.23%) Crude fibre is very essential for regulating bowel movement, reducing the danger of congenital heart failure, colon cancer, constipation and facilitation of nutrient absorption in the body.

Minerals

The ash content of any food substance gives an knowledge about mineral present in the food . The ash content of Celosia varies from 2-23.91 g/ 100g in different plant parts . Celosia sps is fairly rich in over eighteen minerals (Zheng *et al*., 1996, Markandeya *et al*., 2013). Mineral such as Al, Fe, Ni, Mn, Cu, K, Ti, and Se were far higher in *C. argentea* than that of *C. cristata.* Lin *et al* (2002) reported mineral content of 197, 56, 30, and 160 μg/g for Fe, Mn, Cu, and Zn respectively in *C. argentea.* In another study, Huang and Jiao (2013) estimated seven metal elements such as Zn, Co, Ni, Fe, Cr, Mg, and Cu in the velue of 33.26, 1.39, 0.72, 219.33, 0.35, 283.88 and 5.64 μg/g respectively in *Gannan Semen Celosiae.*

According to the chemical constituent analysis recorded in Dr. Duke's Phytochemical and Ethnobotanical Databases and Adegbaju *et al.* (2019) *Celosia sps* is observed to be a hub of essential minerals and other chemical components. Due to such a broad composition it is stated that the chemicals make the applications of cockscomb broad and wide. Along with its medicinal values it can be also used in dyes and colorants due to the ample amount of pigments present in *Celosia cristata.*

Saponins

Saponins plays a major lowering cancer risk , decrease blood lipids and lower blood glucose response. Saponins exist in the form of oleanane-type triterpenoid in Celosia sps which is considered to be major hepatoprotective compounds. Celosin A (1), Celosin B (2), Celosin C, Celosin D, Celosin E, Celosin F, Celosin G, Celosin I (4), Celosin II(5), Celosin H, Celosin L, Celosin J and Cristain (3) are major saponins isolated from seeds of Celosia Sps. (Sun *et al.*,2010; Xue *et al.*, 2011 and Pang *et al.*, 2014)

Antinutritional Factors

The chemical constituents of any plant and its parts play a significant role in the continuous metabolic process carried in the plants. Some components of plant composition can inhibit some essential components of plants which are crucial for proper body building and nutritional profile build-up of an individual. Antinutrients are the plants secondary metabolites, intended for their protection purpose. In actual, they are polyphenolics that possess antioxidative, anticarcinogenic and antibacterial properties. Due to their complex cage like structure they chelate some of divalent cations with them and also hinder protein digestion as a result of which stunted and malnourished characters appears in consumers. Studies on such anti-nutritional constituents are carried over to understand the inhibitory role of the same. The *Celosia cristata* are having antinutritional factors to some extent in the form of Celogentins and Moroidin are bicyclic peptides isolated from the seeds of Celosia plants and have been which implicated in the inhibition of microtubule formation (Morita *et al.*, 2000), hence which leads to the arrest of mitosis in eukaryotic cells. The ability of celogentins and moroidin in inhibiting the polymerization of tubulin has been reported in cells (Suzuki *et al.*, 2003). This disallows the assemblage of spindle fibres, thereby causing chromosomal aberrations. Nonetheless, the antimitotic potential of the extract of *C. cristata* may favor its uses in the development of drugs to prevent the uncontrolled proliferation of cancer cells of which further investigations is required (Yekeen *et al.*, 2013).

El-Badawy1 *et al.* (2015) reported that spraying Celosia plants with putrescine that is a polyamine synthesized from arginine amino acid increase the concentration of chlorophyll (a & b) and total carotenoids content in leaves. But at the contradictory contest it has been observed that polyamines can affect protein synthesis and nitrogen compounds metabolism at the same time.

Processing Innovations

All parts of celosia plants including leaves, grains, roots and flowers are processed and used in many ways worldwide depending on the traditional wisdom and local requirement of ethnic population. After maturity, the flowers of cockscomb were collected and threshed to remove the tiny seeds. The tiny seeds are dried either by solar drying in an open space or in indoor by supplying external heat source to maintain optimum moisture content. Maintenance of 10-18 percent moisture content is necessary to prevent the insect, fungal and microbial spoilages. Dehulling is not required for celosia grain as it has no hulls unlike common cereals grains. The colour of celosia seeds varies from red to black with size of 0.72-83 mm`in diameter (Kirtikar and Basu, 1935). As that of amaranth, storage tissues are present in the two layers; and the embryo accounting 25 percent of the kernel is present in peripheral part (Sindhu and Khatkar, 2016). Due to extreme small size, normal milling equipment used for cereals and pulses may not suitable for Celosia seed. Modified strong-Scott barley pearler developed by Betschart *et al.* (1981) for making amaranth flour could be effectively used for making celosia flour. Round shaped full starchy perisperm was formed after the complete removal of seed coat and germ parts during the five successive passes through the mill . The fraction having germ portion and seed-coat constituted one fourth of total flour production indicated approximately 75 percent flour extraction rate. Overall concentration of nutrients was 2 to 3 times higher in bran-germ fraction than whole seed flour. Becker *et al.* (1986) efficiently produced amaranth flour fractions using modified mill stone with 0.75 to 0.50 mm gap distance and amplified spindle speed. Like milling of common cereals moisture content of seed plays important role in production of amaranth flour. Flour of celosia can be used alone or in blended form for preparation of various products like biscuits, cookies and pancakes. Whole seeds of celosia are consumed as expanded or puffed cereals and for production of pre-gelatinized flour from puffed grains. Puffed Celosia grains mixed with syrup or jaggery are used to prepare "sweets and savouries" in India which are consumed mostly in fasting periods.

The milling of *Celosia cristata* seeds in parts of tropical Africa is been used as the flour for the production of bread and pancake. Due to increase in the number of chronic disorders generated by glucose intolerance and indigestion as celiac disease, this flour is accepted widely for its nutty flavor bread production as well as the pancakes.

Flaking is an art of preparation of Ready-To-Eat (RTE) foods through moistened seeds; flaking gelatinizes the starch and works to inactivate *lipases*. There are three modes available that are in practice for flaking operation such as edge runner, roller runner and extrusion runner. Flaked products consumed as multi grain food or further process to frying or roasting to snacks either with salt or sweet. They also have been in choice as convenience foods like murmura, kheel and breakfast cereals.

Popping is one of the traditional dry heat (HTST operation) for the preparation of Ready-to-eat category of products. Popped seeds served as evening snacks due to their extraordinary looks acquired by endospermic blast due to moisture and temperature gradient. Sometimes they are further rolled to chikki, bar, laddu and sattu form through sequential steps of seasoning and flavoring.

Health Benefits of Cockscomb

Celosia, also known as wool flowers and cockscomb, is a member of the amaranth family, and originated from the native regions of Asia and Africa. The leaves and the young shoots can be cooked and eaten as a vegetable. Indeed, every part of the celosia plant occupies a valued niche in the world of natural healing. The leaves also have a soft texture and a mild spinach-like taste. They are also pepped up with such things as hot pepper, garlic, fresh lime, and red palm oil and eaten as a side dish. The flowers bring diarrhea under control while the leaves are used as dressings for boils and sores. The seeds ease stresses centered within the chest, and they are also rendered into poultices applied to broken bones (Bakar *et al.*, 2014).

Ethno pharmacology

The grain paste of *Celocia argentea* traditionally used to cure ovarian and uterine diseases in In Aravalli hill region of Rajasthan (Katewa *et al.*, 2004). *Celocia argentea* extensively used in Indian traditional medicine for the treatment of diabetes mellitus (Vetrichelvan *et al.*, 2002). Celosia argentea is also used in traditional medicine for sores, ulcers and skin eruptions (Priya *et al.*, 2004)

Celocia argentea is traditionally used against cold, gastrointestinal, diseases, gynecological diseases, rheumatoid arthritis (Zheng and Xing, 2009) and as fertility regulating agent (Kong *et al.*, 1986) in China . The ethnic Yao tribe of China use the stem, leaf, flower and seed of *Celocia argent*ea for the treatment of hemorrhoids, leucorrhea, profuse uterine bleeding (Li *et al.*, 2006). The grain also used for blurred vision, eye inflammation, eyes bothered by bright light, headache, intestinal worms, painful urination, cough, dysentery, bleeding and diarrhoea.

In south eastern Asian countries such as Indonesia and Vietnam *Celocia argentea* is used traditionally to cure cough, hemostatics and jaundice (Grosvenor *et al.*, 1995). Seeds when in decoction or finely powdered, are considered antidiarrhoeal or aphrodisiac. Whole plant is used for antidote in snake poison while root is used for abdominal colic, gonorrhea and eczema (Stuart,2016).

In USA, midwifery, rural Honduras uses *Celocia argentea* for encouraging lactation and its decoction for hemorrhage (Tictin and Dalle, 2005).

The seeds are prescribed for haemorrhagic conditions including menorrhagia, haematuria, haematemsis, haemoptysis, epistaxis, acute retinal heamorrhage and bleeding haemorrhoids *Celosia argentea* L. has sometimes been useful for treatment of disorders like excessive menstruation and leucorrhea (www.globinmed.com/index.php. 2011)

The seeds of celosia have been extensively used for removing "liver-heat" improving eyesight, clearing wind-heat and as an anti-inflammatory agent, as recorded in Dian Nan's Herbal. seeds were also used as demulcent; for painful micturition and for dysentery (Esmail and -Snafi *et al.*, 2015).

Pharmacology

Anti-Diabetic Activity

Traditionally seeds of celosia used for the treatment of diabetes mellitus. To evaluate anti-diabetic activity, an alcoholic extract of *Celocia argentea* seeds was continually administered alloxan-induced diabetic rats at the rate of 27.8 percent at 250 mg kg^{-1} and 38.8 percent at 500 mg kg^{-1} body weight. The evaluation of anti-diabetic activity was verified by blood glucose and body weight. The study showed significant ($p<0.01$) reduction in the blood glucose in alloxan induced diabetic rats for two weeks with prevention of body weight lost (Vetrichelvan *et al.*, 2002; Barlocco *et al.*, 2002)

Anti-Hepatotoxic Activity

Hepatoprotective effect of water extract of *Celocia argentea* was studied in comparison with *Cassia obtusifolia, Cucurbita moschata and Curcuma aeruginosa* on the CCl4 induced and D galactosamine/lipopolysaccharide, induced liver injury in mice. The study showed that among the other species the water extract of *Celocia argentea* is the most effective (Hase *et al*, 1996). In another experiment conducted by Hase *et al.* (2001), also showed Celosian in seed found to inhibit increase of serum enzymes (GPT, GOT, LTH) and bilirubin levels. The study conducted by Xue *et al.* (2011) also showed significant hepatoprotective effects due to the celosin A 32 and celosin B 33 celosin C 34 and celosin D 35 with oral doses 1.0, 2.0 and 4.0 mg kg $^{-1}$ in CCl4 induced hepatotoxicity in mice.

Anti-inflammatory Activity

Wu *et al.* (2011) studied In *vitro* anti-inflammatory activity using celosin E 36, celosin F 37, celosin G38 and cristatain 31 isolated from the seeds of *Celocia argentea* by keeping Indomethacin as a positive control and lipopolysaccharide (LPS) as a negative control. The celosin F 37 showed the comparable anti-inflammatory activity from its IC50 values. In another study, seed extract of *Celocia argentea* on intraperitoneal administration for 7 days before tumor inoculation significantly arrested liver metastatic caused by intraportal injection of colon 26-L5 carcinoma cell, the same extract also shows immunomodulating properties (Hayakawa *et al.*, 1998).

Anti-miotic (Antitumor) Activity

The celosin compounds E 36, F 37, G 38 and cristatain 31 isolated from the seeds of *Celocia argentea* were tested for their in vitro antitumor activities tin 5 human cancer cell lines (SHG44, HCT116, CEM, MDAMB- 435 and HepG2) by the 3-(4,5- dimethylthiazol-2-yl)-2,5-diphenyltetrazolium bromide (MTT) assay. The cristatain 31 demonstrated better cytotoxic activity among 36, 37, 38 and 31 against Hydroxydaunomycin hydrochloride (DOX) as a positive control (Wu *et al*, 2011).

Antifungal Activity

Anti-fungal activity study using plant seed oil n-hexane extract of *Celosia argentea* was carried against *Aspergillus fumigatus, Candida tropicalis and Trichophyton mentagrophytes*. The seed oil extract showed antifungal activity with minimal inhibitory concentration of 50 percent (Diéméléou *et al.*, 2013).

Antiurolithiatic activity

Antiurolithiatic activity of ethanolic extract of *Celosia argentea* (seed) evaluated in rats and results that groups treated with 250 g and 500 g of Celosia argentea extract showed significant anti urolithiatic activity compared with the standard The result of the study showed and Celosia argentea demonstrated a potent prophylactic effect on formation of kidney stone confirming the folklore about its antiurolithiasis activity (Joshi *et al.*, 2012).

Conclusion

C. argentea, *C. cristata* and *C. isertii* from the family Amaranthaceae has widely found in Asia, China, Africa, America and some parts in India. Because of its high nutrition value it has been considered as an additive in food and pharmaceutical industries. It is a good source of trace elements like zinc, phosphorus and iron which helps in oxygen transport through blood and as a immune booster for the human system. Vitamins like vitamin-A and vitamin-C has found to be present aboundantly in their leaves which helps to improve vision and also to maintain a healthy skin. Studies have proved that phytochemicals present in Celosia possess anti carcinogenic, antioxidant and antibacterial properties. As they are known to have a remarkable of health benifits the flour obtained from celosia seeds were used in the production of pancakes and breads. Further the moistened seeds are flacked to prepare Ready-To-Eat foods on the other hand popped seeds were served as an evening snack. So the above mentioned species from Amaranthaceae family has a notable value for its significant properties. Further studies can also be extended towards the development of functional foods by incorporating the flour, pigment and isolates obtained from celosia at the production of food which leads the way towards "let food be thy medicine, and let medicine be thy food."

References

Abubakar, A., Falusi, A. O., Daudu, O. A. Y., Oluwajobi, A. O., Dangana, M. C., and Abejide, D. R. (2015). Mutagenic effects of sodium azide and fast neutron irradiation on the cytological parameters of M2 Lagos spinach (*Celosia argentea var cristata* L.). World J Agri Res, 3, 107-112.

Adegbaju, Oluwafunmilayo. Dorcas., Gloria, Aderonke, Otunola., and Anthony, Jide. Afolayan., (2019). Potential of Celosia species in alleviating micronutrient deficiencies and prevention of diet-related chronic diseases: a review. AIMS Agriculture and Food, 4(2), 458-484.10.3934/agrfood.2019.2.458

Aladesanw, R. D., Adenawoola, A.R., and Olowolafe, O.G. (2001). Effects of atrazine residue on the growth and development of celosia (Celosia argentea) under screen house conditions in Nigeria. Crop Protection 2001; 20(4), 321-324. https://doi.org/10.1016/S0261-2194(00)00159-9

Badawy, E. M., Kandil, M. M., Mahgoub, M., Shanan, N., and Hegazi, N. (2015). Chemical constituents of Celosia argentea var. cristata L. plants as affected by foliar application of putrescine and alpha-tocopherol. Int. J. Chem. Techol. Res, 8(12), 464-470.

Bakar, D. A., Ahmed, B. A., and Taha, R. M. (2014). *In vitro* callus induction and plant regeneration of Celosia argentea-an important medicinal plant. Brazilian Archives of Biology and Technology, 57(6), 860-866.

Becker, R., Irving, D. W., and Saunders, R. M. (1986). Production of debranned amaranth flour by stone milling. Lebensmittel-Wissenschaft+ Technologie, 19(5), 372-375.

Behera, B. (1974). Cytotaxonomic studies in the family Amaranthaceae. Cytologia, 39(1), 121-131.

Betschart, A.A., Igrving, D.W., Shephard, A.D., and Saunders, R,. (1981). Amaranthus cruentus: milling characteristics, distribution of nutrients within seed components, and the effects of temperature on nutritional quality. Journal of food science 46(4): 1181-87, https://doi.org/10.1111/j.1365-2621.1981.tb03019.x

Cai, Y. Z., Sun, M., and Corke, H. (2005). Characterization and application of betalain pigments from plants of the Amaranthaceae. Trends in Food Science and Technology, 16(9), 370-376.

Celosia cristata, http://www.motherherbs.com/celosia-cristata.html

Dian Nan's Herbal. Shanghai (1994) : Shanghai Bookstore Press.

Diéméléou, C. A., Lessoy, Zoue,, and Niamké, S.L. (2014). Physicochemical and nutritive characterization of linoleic acid-rich oil from seeds of Celosia Argentea, Chiang Mai Journal of Science 41(5):1157-1170

Esmail, Ali., and Al- Snafi.(2015). The chemical constituents and pharmacological importance of celosia cristata – a review. Journal of Pharmaceutical Biology. 5(4):254-261.

Fact Sheet FPS-113, one of a series of the Environmental Horticulture Department, Florida Cooperative Extensio Service, Institute of Food and Agricultural Sciences, University of Florida, Publication date, 1999.

Grosvenor, P. W., Supriono, A., and Gray, D. O. (1995). Medicinal plants from Riau province, Sumatra, Indonesia. Part 2: Antibacterial and antifungal activity. Journal of ethnopharmacology, 45(2), 97–111. https://doi.org/10.1016/0378-8741(94)01210-q

Hamzah, R. U., Lawal, A. R., Madaki, F. M., and Erukainure, O. L. (2018). Methanolic extract of Celosia argentea var. crista leaves modulates glucose homeostasis and abates oxidative hepatic injury in diabetic rats. Comparative Clinical Pathology, 27(4), 1065-1071.

Hase, K., Kadota, S., Basnet, P., Namba, T., and Takahashi, T.(1996) Hepatoprotective Effects of Traditional Medicines. Isolation of the Active Constituent from Seeds of Celosia argentea, Phytother. Res. 10(5). 1996 : 387-392. https://doi.org/10.1002/(SICI)1099-1573(199608)10:5<387::AID-PTR856>3.0.CO;2-V

Hase, K., Xiong, Q., and Kadota, S.(2001) Bioactive Natural Products (Part F), Atta- ur-Rahman, Elsevier 25(6): 459-482.

Hayakawa, Y., Fujii, H., Hase, K., Ohnishi, Y., Sakukawa, R., Kadota, S., Namba, T., and Saiki, I. (1998). Anti-metastatic and immunomodulating properties of the water extract from Celosia argentea seeds. Biological and Pharmaceutical Bulletin, 21(11), 1154–1159. https://doi.org/10.1248/bpb.21.1154

http://gxsti01.vicp.net/ english/ detail/f96f95ff-7125-419b-a265-486842447f7d

http://www. theflowerexpert.com/content/growingflowers/flowersandseasons/cockscomb

https://phytochem.nal.usda.gov/phytochem/plants/show/384?et=

Huang, Z.Q., and Jiao, L.Y., (2013). Determination of seven metal elements in Gannan Semen Celosiae by atomic absorption spectrometry Guangdong Wei Liang Yuan Su Ke Xue, 20: 6-9

Ilodibia, C. V., Chukwuka, C., Chukwuma, U. M., Akachukwu, E. E., Igboabuchi, N. A., and Adimonyemma, R. N. (2016). Anatomical, Proximate, Mineral and Vitamin Studies on Celosia argentea (Linn.). Biotechnology Journal International, 1-7.

Ilodibia1, C.V., Chukwuka1, C., Chukwuma, U. M., Akachukwu, E.E, Igboabuchi, N.A., and Adimonyemma, R. N., (2016).Anatomical, Proximate, Mineral and Vitamin Studies on Celosia argentea (Linn.), British Biotechnology Journal 15(4): 1-7, 10.9734/BBJ/2016/28300

Isong, E. U., and Essien, I. B. (1996). Nutrient and antinutrient composition of three varieties of Piper species. Plant Foods for Human Nutrition, 49(2), 133-137.

Iweala, E. E. J., and Ogidigo, J. O. (2015). Effect of Celosia argentea F. Cristata Schinz. on prostate specific antigen, antioxidant status and hematological parameters in rats induced with benign prostate hyperplasia. Asian J Biochem, 10, 42-51.

Joshi, P.C., Patil,S.A., Sambrekar, S.N.(2012). Evalution of antiurolithiatic activity of ethanolic extract of Celosia argentea (seed) in rats. Universal Journal of Pharmacy. 1(1): 52-60.

Kanu, C. L., Imosemi, I. O., and Malomo, A. O. (2017). A review of the multifaceted usefulness of Celosia argentea Linn. European Journal of Pharmaceutical and Medical Research, 4(10), 72-79.

Katewa, S. S., Chaudhary, B. L., and Jain, A. (2004). Folk herbal medicines from tribal area of Rajasthan, India. Journal of Ethnopharmacology, 92(1), 41–46. https://doi.org/10.1016/j.jep.2004.01.011

Khare, C. P. (2008). Indian medicinal plants: an illustrated dictionary. Springer Science and Business Media.

Kirtikar, K., and Basu, L.,(1935) Indian Medicinal Plants, Vol. III 2nd ed., 2053-2054.

Kobayashi, J., Suzuki, H., Shimbo, K., Takeya, K., and Morita, H. (2001). Celogentins A-C, new antimitotic bicyclic peptides from the seeds of Celosia argentea. The Journal of organic chemistry, 66(20), 6626–6633. https://doi.org/10.1021/jo0103423

Kolade, O. (2017). The Effect of Zn Fertilization on Growth and Nutrition of Celosia Argentea. Texas AandM University-Commerce.

Kong, Y. C., Xie, J. X., and But, P. P. (1986). Fertility regulating agents from traditional Chinese medicines. Journal of Ethnopharmacology, 15(1), 1–44. https://doi.org/10.1016/0378-8741(86)90102-9

Li, S., Long, C., Liu, F., Lee, S., Guo, Q., Li, R., and Liu, Y. (2006). Herbs for medicinal baths among the traditional Yao communities of China. Journal of Ethnopharmacology, 108(1), 59–67. https://doi.org/10.1016/j.jep.2006.04.014

Markandeya, A.G., Firke, N.P., Pingale S.S., and Salunke-Gawali,S.(2013). Quantitative elemental analysis of Celocia argentea leaves by ICP-OES technique using various digestion methods,Int. J. Chem. Anal. Sci., 4 (4): 175-181. 10.1016/j.ijcas.2013.08.003

Mnzava, N. M., Dearing, J. A., Guarino, L., and de Koeijer, H. (1999). Neglected leafy green vegetables crops in Africa: Vol. 2: Bibliography of the genetic resources of traditional African vegetables.

Morita, H., Shimbo, K., Shigemori, H., and Kobayashi, J. I. (2000). Antimitotic activity of moroidin, a bicyclic peptide from the seeds of Celosia argentea. Bioorganic and Medicinal Chemistry Letters, 10(5), 469-471.

Müller, K., and Borsch, T. (2005). Phylogenetics of Amaranthaceae based on matK/trnK sequence data: evidence from parsimony, likelihood, and Bayesian analyses. Annals of the Missouri Botanical Garden, 66-102.

Nath, P., Ohri, D., Jha, S. S., and Pal, M. (1997). Seed protein electrophoresis of wild and cultivated species of Celosia (Amaranthaceae). Genetic Resources and Crop Evolution, 44(3), 241-245.

National Research Council. Lost Crops of Africa. Vegetables. Washington, The National Academies Home, 2006, 93-95.

Nidavani, R. B., Mahalakshmi, A. M., Seema, M., and Krishna, K. L. (2014). Pharmacology Of Celosia Argentea L. Journal of Atoms and Molecules, 4(1), 635.

Olawuyi, O. J., Bamigbegbin, B. J., and Bello, O. B. (2016). Genetic variation of morphological and yield characters of Celosia argentea L. Germplasm. J. Basic Appl. Sci. Int, 13, 160-169.

Pingalea,S.S., Sunita,S.G., Anil,G., Markandeyac., and Narayan, P.F.(2012). Recent Developments in Research of Celocia Argentea: A Review Journal of Pharmacy Research. 5(2):1076-1082

Priya, K. S., Arumugam, G., Rathinam, B., Wells, A., and Babu, M. (2004). Celosia argentea Linn. leaf extract improves wound healing in a rat burn wound model. Wound repair and regeneration : official publication of the Wound Healing Society [and] the European Tissue Repair Society, 12(6), 618–625. https://doi.org/10.1111/j.1067-1927.2004.12603.x

Ramesh, B. N., Mahalakshmi, A. M., and Mallappa, S. H. (2013). Towards a better understanding of an updated ethnopharmacology of Celosia argentea L. Int J Pharm Pharm Sci, 5(3), 54-59.

Rubini, D., Sudhahar, D., and Anandaragopal, K. (2012). Phytochemical investigation and anthelmintic activity of Celosia cristata leaf extract. Int Res J Pharm, 3(5), 335-336.

Sayeed, R., Thakur, M., and Gani, A. (2020). Celosia cristata Linn. flowers as a new source of nutraceuticals- A study on nutritional composition, chemical characterization and in-vitro antioxidant capacity. Heliyon, 6(12), e05792. https://doi.org/10.1016/j.heliyon.2020.e05792

Shanmugam, S., Annadurai, M., and Rajendran, K. (2011). Ethnomedicinal plants used to cure diarrhoea and dysentery in Pachalur hills of Dindigul district in Tamil Nadu, Southern India. Journal of Applied Pharmaceutical Science, 1(8), 94.

Simpson MG. Plant Systematics. 2nd ed. Academic Press, Elsevier, 2010;301–302.

Sindhu, R., Khatkar, B.S. (2016). Physical properties of grains of some pseudocereals. Annals of Agri Bio Research 21(2): 174-78.

Spórna-Kucab, A., Milo, A., Kumorkiewicz, A., and Wybraniec, S. (2018). Studies on polar high-speed counter-current chromatographic systems in separation of amaranthine-type betacyanins from Celosia species. Journal of Chromatography B, 1073, 96-103.

Sun, Z. L., Gao, G. L., Xia, Y. F., Feng, J., and Qiao, Z. Y. (2011). A new hepoprotective saponin from Semen Celosia cristatae. Fitoterapia, 82(4), 591-594.

Suzuki, H., Morita, H., Iwasaki, S., and Kobayashi, J. I. (2003). New antimitotic bicyclic peptides, celogentins D–H, and J, from the seeds of Celosia argentea. Tetrahedron, 59(28), 5307-5315.

Ticktin, T., and Dalle, S. P. (2005). Medicinal plant use in the practice of midwifery in rural Honduras. Journal of Ethnopharmacology, 96(1-2), 233–248. https://doi.org/10.1016/j.jep.2004.09.015

Vetrichelvan, T., Jegadeesan, M., and Devi, B. A. (2002). Anti-diabetic activity of alcoholic extract of Celosia argentea Linn. seeds in rats. Biological and Pharmaceutical Bulletin, 25(4), 526–528. https://doi.org/10.1248/bpb.25.526.

Vetrichelvan, T., Jegadeesan, M., and Devi, B. A. U. (2002). Anti-diabetic Activity of Alcoholic Extract of Celosia argentea L INN. Seeds in Rats. Biological and Pharmaceutical Bulletin, 25(4), 526-528.

Vincken, J. P., Heng, L., de Groot, A., and Gruppen, H. (2007). Saponins, classification and occurrence in the plant kingdom. Phytochemistry, 68(3), 275-297.

Wang, Y., Lou, Z., Wu, Q. B., and Guo, M. L. (2010). A novel hepatoprotective saponin from Celosia cristata L. Fitoterapia, 81(8), 1246–1252. https://doi.org/10.1016/j.fitote.2010.08.011

Wu, Q., Wang, Y., and Guo, M. (2011). Triterpenoid saponins from the seeds of Celosia argentea and their anti-inflammatory and antitumor activities. Chemical and Pharmaceutical Bulletin, 59(5), 666–671. https://doi.org/10.1248/cpb.59.666

Xue, Q., Sun, Z. L., Guo, M. L., Wang, Y., Zhang, G., and Wang, X. K. (2011). Two new compounds from Semen celosiae and their protective effects against CCl4-induced hepatotoxicity. Natural Product Research, 25(8), 772-780.

Xue, Q., Sun, Z. L., Guo, M. L., Wang, Y., Zhang, G., and Wang, X. K. (2011). Two new compounds from Semen celosiae and their protective effects against CCl_4-induced hepatotoxicity. Natural product research, 25(8), 772–780. https://doi.org/10.1080/14786410902833948

Yekeen, T. A., Akintaro, O. I., Akinboro, A., and Azeez, M. A. (2013). Evaluation of cytogenotoxic and nutrient composition of three commonly consumed vegetables in south-western Nigeria. African Journal of Food, Agriculture, Nutrition and Development, 13(2).

Zhang H.C., Zhang, T.T., Du, B.,Cheng, D.Y., and Li, Z.G., (2014). Chemical constituents of Celosia cristata L Chinese Trad. Patent Med., 36 pp. 122-125.

Zheng, X. L., and Xing, F. W. (2009). Ethnobotanical study on medicinal plants around Mt.Yinggeling, Hainan Island, China. Journal of Ethnopharmacology, 124(2), 197–210. https://doi.org/10.1016/j.jep.2009.04.042

20

Biotechnological Interventions in Millets and Pseudo Cereals: A Review

Janifer Raj Xavier, Pal Murugan Muthaiah, Gopal Kumar Sharma and Anil Dutt Semwal

Defence Food Research Laboratory, Siddarthanagar, Mysore-570011, India

Introduction

Small size grains namely millets are produced by extraneous grasses and the word millet is derived from "mille' a French word synonymous to thousand indicating handful of millet grains in a handful measure contains thousand grains (Taylor and Emmambux, 2008). Small millets and pseudo cereals belong to Gramineae family and were major ancient foods of Asian and African continents while Western countries used them as only bird feed. Scientific interventions revealed extra ordinary traits such as lower glycemic index, gluten-free protein, abundant minerals, vitamins and antioxidant phytocomponents, of these lesser known crops. Traditional food uses of millets include fermented and non-fermented breads, porridges and snacks and are divergent in society. Millets are gaining importance in Western nations as well, as components of multigrain and gluten free products due to their specific health benefits (Chandrasekara *et al.*, 2012). Sensory acceptability and anti-nutritional factors are current limitations of millets and pseudo cereal which could be overcome by, scientific interventions of processing parameters and products such as composite flours in combination with cereals and pulses. Industrial applications of scientific findings and development of newer procedures and novel food products from millets and pseudo cereals have been reported.

Biotechnological interventions and scientific techniques for modification, improvement and increasing value of microbes, plants, animals and foods for economic interest has gained popularity by latest technological advances to study at molecular level collectively termed as 'omics'. They present newer strategies and opportunities for development of small millets and pseudo cereals based processing conditions. Microbial processes such as fermentation has been used since ancient times by human kind for nutritional benefits and cultural heritage. Even today, fermentation is widely used as a preservation technology and one third of processed foods belong to the category of fermented foods. Biotechnological methods used in processes involved in production of fermented foods, aid in improving nutritional quality and sensory score, ensure viable starter cultures free of toxin producing microorganisms and are capable of producing antimicrobial substances to curb growth of pathogenic microbes along with synthesis of vitamins and important amino acids. Furthermore, fermentation technology also produces additives such as enzymes, flavouring compounds, preservatives for use in food processing applications as well improves digestibility of complex food ingredients. However, utmost care and attention has to be given when introducing modern technological advances in food processing sector and stringent risk assessment protocols need to be followed.

Metagenomics approaches also support in identification of unknown and non-culturable microorganisms involved in traditional fermented foods, thereby aid in defining the starter cultures for production of region and community specific fermented foods.

Omics approaches using latest genomic tools for millets and pseudo cereals would augment research on these crops to achieve nutritional security. Molecular tools for improvement of millets and pseudo cereals are gaining popularity as they are source of valuable traits such as abiotic and biotic stresses as well as nutrient quality. Recent advances in next generation sequencing technologies help in rapid whole genome sequencing and trait characterisation of lesser known crops. Marker assisted breeding through identification of genes responsible for specific traits are upcoming trends. Genome editing tools have larger scope in editing the genes accountable for anti-nutrient factors and production of crops with beneficial attributes (Figure. 1). Recent advances in development of functional foods, marker assisted molecular breeding, genetic engineering using recombinant DNA technology, gene cloning, *in vitro* plant regeneration and fermentation processes pertaining to using small millets and pseudo cereals have been discussed in this chapter.

Fig. 1: Schematic representation of locations worldwide showing millet cultivation for food use

Functional Foods Based on Millets and Pseudo Cereals

Traditional fermented foods based on millets are specific to geographical regions and traditional folklore. Lactic acid producing microorganisms, yeasts and molds form the predominant genera of organism responsible for fermentation. Back sloping is the most common starter method in traditional practices of fermentation, while instances of natural fermentation by wild cultures are also allowed to obtain the desirable product. Basic steps of cleaning, steeping, milling, sieving, malting, mashing, fermentation, filtration, sedimentation, cooking and extraction are involved in production processes of fermented foods. *Ben-saalga, Boza, Burukutu, Bushera, Doro, Fura, Jandh, Kodoko jaanr, Koko, Koozh, Kununzaki, Madua Apong, Malwa/Ajon, Mangisi, Ogi, Oshikundu, Oti-oka, Shakparo Ale, Togwa, Uji, Zoomkoom* and *Xiao mi jiao* are examples of fermented millet-based products available globally. Fermentation and malting of millets and pseudo cereals improve nutrient availability, reduces anti-nutritional factors and dietary bulk (Coulibaly *et al.,*2012). Microorganisms involved in natural fermentation is dominated by lactic acid producers, while spoilage organisms such as yeast, fungi and other genera of bacteria were also reported (Galati *et al.,*2014). Modern methods of fermentation using biotechnological approaches add on beneficial components through bioprocessing of millets using scientific interventions at various stages of fermentation process. Reduction in anti-nutrient factors, imparting peculiar sensory parameters and taste, improvement of protein digestibility and enhanced bioavailability of calcium and other minerals after fermentation of millets and pseudo cereals make them more attractive as starter material for development of fermented foods.

Technological advances such as efficient starter cultures, modern fermentation processes and equipment have made an impact on development of newer fermented products using millets and pseudo cereals. Methods and procedures used for development of cheese starter cultures are being used for cereal and millet-based foods. Bulk production of starter cultures in tailormade medium and subsequent addition of lesser quantity of starter inoculum into fermentation vats is practised. Optimisation of media ingredients for specific strains of microorganisms and development of bioreactors for controlled growth is essential step in development and production of starter cultures in large scale. Application of basic scientific knowledge towards a product resulting from microbial technology, is an essential step in transformation of lab level experiments into technologies of high, intermediate and lower levels. Process development methods and cost of establishment of scale-up protocols are major factors in translation of basic research into applied science. Traditional fermentative processes involve use of mixed cultures based on substrates, availability of starter material and environmental conditions. High value fermented products with added nutritional value and sensorial properties can be developed using microorganisms suitable for industrial upscaling under controlled co-culture and fermentation conditions (Ghoshal, 2018). Role of genomics in deciphering information on microorganisms responsible for millet fermentation is growing rapidly through characterisation of microorganisms growing in specialised ecological niches as well ensure quality and uniformity of fermented beverages based on millets and pseudo cereals. Denaturing gradient gel electrophoresis (DGGE) and Amplified ribosomal DNA restriction analysis (ARDRA) are tools which are important to study taxonomy of unculturable microorganisms of traditional fermented foods. Second generation whole genome sequencing techniques such as Illumina and pyro sequencing can be used to sequence and identify unculturable strains used in preparation of fermented foods (Bader *et al.,*2018). Commercialisation of millet-based foods on a larger scale would require scale-up of fermentative processes such as starter cultures, containers, rooms with controlled temperatures and support of network of researchers.

Finger millet (*Elusine corocana*), has been used since ancient time as fermented foods such as pancake, porridge and beverages (Gomez, 1993). Finger millet substituted for rice and used along with black gram can be made into pancakes such as *dosa* and *idly* in Ethiopia (Gebrekidan *et al.,*1982). Antinutrient factors namely trypsin inhibitor activity, tannins and phytates reduced by 32, 52 and 20 percent, respectively on fermentation by microflora naturally present in finger millet grains. Improved protein and starch digestibility, amino acid profile and availability of minerals such as calcium, zinc, phosphorus and iron were reported in fermented finger millet-based foods in comparison to non-fermented grains (Singh and Raghuvanshi, 2012). Proteins present in grains are acted upon by microbial enzymes while starch digestibility improved due to weakening of bonds during soaking and fermentation. Varma and Patel (2013) reported use of fermented finger millet products such as *dosa* and *idly* with improved taste, fiber, calcium and protein content and reduced anti nutrient factors. Fermented finger millet porridge, *koozh* is consumed widely in state of Tamil Nadu. Finger millet is ground into flour, mixed with water and the slurry is left for natural fermentation for 16 to 18 hours, and cooked with rice (20% w/v) with excess water and made into a porridge with thick consistency. It is further fermented for 24 hours to obtain a semi solid porridge called *kali* and consumed by mixing water (1:6 w/v) and required amount of salt using hands along with onions and green chilies (Thirumangaimannan and Gurumurthy, 2013). Subash *et al.* (2015) reported enhanced bioavailability of B group of vitamins in *Lactobacillus salivaricus* fermented finger millet. Finger millet-based gruel like product based on skimmed milk powder was formulated using lactic acid bacteria at fermentation temperature of 45°C (Mugocha *et al.,* 2000). Composite bread prepared using fermented finger millet flour using *Saccharomyces cerevisiae* and lactic acid bacteria had longer shelf life and increased vitamin B content with reduced gluten content in comparison to white bread (Mythrayee and Pavithra,2017). Fermentation of foxtail millet (*Setaria italica*) using *L. paracasei* Fn032 derived trypsin resistant peptides namely FFMp4 and FFMp6 was reported. The derived peptides were antimicrobial against *Escherichia coli* ATCC 8099 and had significant antioxidant activity (Amadou *et al.,* 2013). Buckwheat incorporation up to 50 percent improved overall amino acid profile, flavonoid content, antioxidant activity and higher over all acceptability of *idly* prepared using composite batter containing parboiled rice and common buckwheat (*Fagopyrum esculentum* Möench) (Kumari *et al.,* 2020). Kang *et al* (2017) reported solid state fermentation of buckwheat using three strains of of *Agaricus* such as *Agaricus blazei* (Murrill SH26), *A. bisporus* (AS2796) and *A. bisporus* (G1). Enhanced phenolic content and antioxidant activity were reported for fermented grains in comparison to control without fermentation. Fermentation of buckwheat flour using *L. plantarum* and incorporation of the same during production of gluten free muffins was reported. Increased poly phenolics content, furosine, lysine, antioxidant activity and browning index were observed (Zieliński *et al.,*2017). Lactic acid bacterial fermentation of buckwheat was carried out using Fresco DVS 1010 and *L. rhamnosus* GG. Probiotic products with shelf stability of 14 and 7 days were obtained using mixed and single culture fermentation, respectively (Liptáková *et al.,* 2017). Fermented buckwheat sprouts using *Saccharomyces cerevisiae* and *L. rhamnosus* showed enhanced antioxidant activity and restored antioxidant levels in liver of hyperlipidemic mice in comparison to unfermented buckwheat sprouts. Rutin, isoorientin, orientin, isovitexin, vitexin and polyphenolic content increased content on fermentation (Chen *et al* ., 2020).

Prebiotics are non-digestible carbohydrate molecules which aid in proliferation of beneficial microbiota in human gut. Millets are source of prebiotics, barnyard millet reported the highest number of prebiotics around 14.7g/100g followed by foxtail millet (8.0g/100g), little millet (6.4-7.0), kodo millet (6.4-7.0g/100g), finger millet (3.6g/100g), pearl millet (1.2 - 2.3g/100g) and proso millet (0.7-2.2 g/100g) (Rao *et al.,* 2017). β-glucans derived from kodo millet and pearl millet, resistant starches, and hemicellulose component of millet brans have been reported for their prebiotic properties (Avashthi *et al.,*2018). Production of xylooligosaccharides (XOS), a heat resistant prebiotic molecule by enzymatic hydrolysis using endo-xylanase derived from *Thermomyces lanuginosus* on finger millet (CO 9) seed coat substrate has been reported. Water soluble xylan from finger millet seed coat yielded 72% of XOS. Acid hydrolysis for production of prebiotics results in toxic residues while enzymatic methods are considered as greener approach (Palaniappan *et al.,*2017). Furthermore, prebiotic nature of millets is being explored recently for production of functional symbiotic beverages using probiotic cultures (Stefano *et al* 2017). Probiotic beverage developed using finger millet, oat and bovine milk along with *L.acidophilus* NCDC14 (1%), *L. casei* NCDC297 (5%), *L. rhamnosus* NCDC347 (5%) in pure or co-cultures was reported. Highest sensory score was obtained for *L. acidophilus* based functional beverage. Thus, millets are promising components of futuristic symbiotic foods (Kumar *et al.,* 2019). Shaikh *et al* (2017) reported a functional probiotic beverage prepared from malted finger millet using bacterial strains, *L. helveticus* MTCC 5463 and *Streptococcus thermophiles* MTCC 5460. Fermentation of finger millet grains using *L. casei* 431® for production of probiotic beverage using cocoa and bovine milk was reported (Fasreen *et al.,* 2017). Millet based fermented products were used as probiotic treatment for children suffering from diarrhoea (Lei *et al.,* 2006). Solid state fermentation of finger millet using the GRAS fungus *Rhizopus oryzae* was reported for mobilisation of bound phenolics up to 18.64 mg gallic acid equivalent/g with enhanced antioxidant activity while enzymatic treatment resulted in leaching out of phenolics during processing operations (Yadav *et al* 2013). Lactic acid bacteria (14 strains) and fungi (*R. oligosporus*) based submerged fermentation of raw and roasted buckwheat flour resulted in increase in polyphenlics and rutin content and results were culture specific. Advanced glycation end-products were reduced on fermentation (Zieliński *et al.,* 2019).

Crop Improvement

(i) Molecular marker based crop breeding

Identifiable DNA sequences associated with inheritance of a trait or linked gene, found at specific locations of the genome are called as molecular markers. Marker-assisted breeding, conserving genetic resources and genotype verification makes molecular markers important for modern breeding programmes. Genetic diversity studies using molecular markers and marker assisted selection associated with specific traits are essential components of molecular breeding of crops. Conventional crop improvement methods used selection as a tool method for breeding millets and pseudo cereals. Novel germplasm of millets with valuable traits such as nutrients, biotic and abiotic stress tolerance are source for genes essential for millet improvement. Crop improvement of millets and pseudo cereals may target further enhancement of nutritional quality and reduction in anti-nutrient factors. Identification and isolation of genes responsible for specific traits in millet and pseudo cereal using germplasm collections has been reported. Potential nature of DNA markers in hastening the processes of traditional breeding through marker assisted selection is established in recent years. Germplasm resources of millets and pseudo cereals are essential for crop improvement using various molecular techniques such as allele mining, quantitative trait loci (QTL) and fine mapping, gene tagging, marker trait associated genome wide selection and through background and forward selective breeding approaches (Collard *et al.,* 2008).

Various organizations at national and international levels are involved in preserving millet and pseudo cereal germplasm. Putative genes or gene candidates for biotic and abiotic stresses has been identified and deposited in open source databases developed by NIPGR (www.nipgr.res.in/foxtail.html) such as transposable element based marker database (FmTEMDb), mi-RNA markers(Yadav *et al.,* 2014a; Yadav *et al.,* 2014b) and intron length polymorphism markers (Muthamilarasan *et al.,* 2014) for use by plant breeders. Plant genetic resources are essential for regular supply of genetic material for crop improvement programmes through traditional and modern breeding strategies. Conservation of germplasm, characterization in terms of specific traits and their sustainable utilization in breeding programmes for biofortification would pave way for nutritional enhancement of millets and pseudo cereals (Muthamilarasan and Prasad, 2015). Varietal development for higher yields and disease resistance was the aim behind germplasm characterisation and maintenance during 1980s while nutritional quality gained importance only during 21st century. Superior lines with farmer friendly qualities such as early flowering, increased grain yield and adaptation to local conditions are screened for desirable nutritional traits thus accelerating breeding programmes for nutrition. Diverse genotypes are hybridized to generate populations to map genes responsible for nutrient rich grains through marker assisted tagging of genomic regions. Evaluation of core germplasm collection for specific nutritional traits essential to incorporate diverse phenotypes into breeding experiments to produce recombinant inbred lines useful for biofortification.

International Crop Research Institute for Semi-Arid Tropics (ICRISAT) holds the largest millet germplasm in their gene bank. Core and mini core collections of finger millet and core collections for other small millets. 6084 accessions of *E. corocana* and *E. africana* based on inflorescence type, 1542 accessions of foxtail millet from 23 countries grouped into three races such as *italica, maxima* and *moharia* as well ten sub races, 749 accessions of barnyard millet belonging to India and Japan, 665 accessions of kodo millet and 473 accessions of little millet arc available in ICRISAT gene bank (Gupta *et al.,* 2017). Recent biotechnological advances can give a boost to major areas of millet and pseudo cereal improvement programmes by quickening breeding experiments for elimination of genes responsible for anti-nutrient factors and enhance resistance to fungal infections. Molecular markers identified in millets showed transferability across genera up to 85 percent success rate. Characterisation of germplasm resources based on trait specificity through genetic diversity analysis using various dominant and co-dominant molecular markers were reported. Identification of desirable genotypes and comparisons with available genotypes of land races is a basic necessity for choosing trait specific genes. Genetic diversity analysis of 386 accessions of prosomillet (*Panicum miliaceum* L.) indicated significantly higher Shannon index, a measure of characterizing species diversity in a community for quantitative traits such as fat content, 1000-seed weight, top 3 leaves area, main stem diameter, and plant height in the range 2.092 to 2.052 in comparison to qualitative traits. Quantitative and qualitative traits showed correlation between each other emphasising important of core collections as valuable genetic resources for breeding purposes for traits (Zhang *et al.,* 2012) Vetriventhana *et al.* (2016) studied 200 accessions of proso millet belonging to 30 different countries in terms of phenotypic diversity to identify high grain yielding and nutritious accessions. Highly diverse race was patentissimum while ovatum and compactum were less diverse in nature. Large seed size, higher grain yield, mineral and protein rich accessions were identified and information generated would be useful for developing varieties which are adaptable for specific regions. Foxtail millet (*Setaria italica*) germplasm accessions (741 nos.) were evaluated for high yielding varieties with characteristics such as tiller number, panicle length and lesser flowering time (Nirmalakumari and Vetriventhan,2010). *Panicum miliaceum* L. accessions from China across 12 provinces were assesses to facilitate breeding of especially of millet cultivars free of amylose. Genes responsible for waxy endosperm (*Wx-S*) were characterised by gene sequencing and marker (M5-R11) linked to waxy endosperm trait of common millet accessions was identified (Wang *et al* 2018). Buckwheat (*Fagopyrum* sp.) cultivated worldwide belongs to two major species. *Fagopyrum esculentum* Moench (common buckwheat) is consumed widely in comparison to *F. tataricum* Gaertn. (tartary buckwheat). *F. tataricum* is rich in vitamins of B group, protein, fiber and rutin, a flavanoid. *F. tartaricum* accessions (190 nos.) were DNA fingerprinted using Amplified Fragment Length Polymorphism (AFLP) markers for rutin content and most accessions showed wide variation of rutin content from 10 to16 μg/mg of rutin, 14 percent of tested accessions contained 17 μg/mg to 30 μg/mg of rutin while 5 percent accessions had low rutin content of ≤ 10 μg/mg (Gupta *et al.,* 2012).

Inter Simple Sequence Repeats (ISSR) and Random Amplified Polymorphic DNA (RAPD) markers discriminated finger millet accessions based on seed coat colour (Gupta *et al.,* 2010). Katsu *et al* .(2019) developed a DNA based markers with RNA polymorphism to discriminate tartary buckwheat and products developed such as flour and dried noodle using '*Manten-Kirari*', a cultivar with trace-rutinosidase content. Buckwheat is rich in rutin content and identification of cultivars with rutinosidase content is useful to prevent rutinosidase contamination. '*Manten-Kirari*' was differentiated from major tartary buckwheat varieties (03 markers), common buckwheat varieties (02 markers) and rutin rich cultivar '*Hokkai T8*' (08 markers) for the first time. Manimekalai *et al.* (2018) assessed genetic diversity among 61 barnyard millet (*Echinochola frumentacea*) accessions of Southern India using phenotypic characterisation and Simple Sequence Repeat markers (SSRs) markers. 12 clusters were found using Jaccard's coefficient and enormous genetic diversity was observed for important agronomic and nutritional quality. Tiwari *et al.* (2018) reported genetic diversity analysis of Indian little millet (*Panicum sumatrense*) genotypes using RAPD markers. The polymorphism information content ranged between 0.102 to 0.517. Krishna *et al.* (2018) reported cross transferability of 33 and 2 simple sequence repeat markers SSRs belonging to finger millet and foxtail millet, respectively in eight other millets. The identified markers were capable of explaining diversity using polymorphic information content (PIC) values, principal component analysis and unweighted pair group method with arithmetic mean (UPGMA) clustering method among millet accessions.

Millets are source of starches for local communities that depend on millet-based foods. Amylose and amylopectin form the major portion of grain starches which for the waxy and non-waxy grain types. Waxy genotypes of cereals and millets, presently in use were selected by conventional breeding techniques and recent advances in molecular marker assisted breeding aids in selection of waxy mutants in one or more alleles from diverse millet germplasm. Waxy phenotype is controlled by a dominant allele *Wx*, while loss of function of recessive allele *wx* is responsible for waxy phenotype by substitution of cysteine in place of tyrosine amino acid (Fukunaga *et al.*, 2002). Genome editing tools induce site specific mutations and convert non-waxy types to waxy grain type. Waxy grains with 100% amylopectin and 0% amylose content are glutinous, easily digestible and form an important component of infant weaning foods. Derived cleaved amplified polymorphic sequences (dCAPS) identified waxy landraces of proso millet from germplasm of Japan and other countries (Araki *et al.,* 2012). Foxtail draft genome of two foxtail millet cultivars, helped in rapid identification of waxy elite (Zhang *et al.,* 2012). SSRs and single nucleotide polymorphisms (SNPs) identified using molecular characterisation of *Opm* gene identified QTLs for amino acid content. High tryptophan genotypes of finger millet and expressed sequence tags (ESTs) could lead to improved seed protein quality Setarins, genes responsible for prolamine content were identified in foxtail millet and quantitative real-time Polymerase Chain Reaction (PCR) analysis confirmed their role in protein quality of millets on over expression setarins at spikes (Goron and Raizada, 2015).

Whole genome sequences have to be deciphered for millets and pseudo cereals as they are need of the hour for quick identification of SSRs, SNPs and Insertion-deletion mutations (InDels). Recent biotechnological advances can give a boost to major areas of millet and pseudo cereal improvement programmes by quickening breeding experiments for elimination of genes responsible for anti-nutrient factors and enhance resistance to fungal infections. Millets have anti-nutrient components such as tannins, saponins, phytic acid and bound phenolics which make them unsuitable for bio fortification. RNA interference technology and other gene editing tools such as transcription activator like effector nucleases (TALENs), zinc finger nucleases (ZFNs) and clustered regularly interspaced short palindromic repeats (CRISPR) are possible technological interventions to edit or eliminate genes responsible for anti-nutrient factors (Vinoth *et al.,* 2017). Biosynthetic pathway for phytic acid production in cereals has been characterised and enzymes such as myo-inositol-3-phosphate synthase (MIPs), myo-inositol-3-phosphate 5/6-kinase (MIK) and Inositol 1,3,4,5,6- pentakisphosphate 2-kinase (IPK1) were reported to control phytate content in crops (Kuwano *et al.,* 2009). Gene silencing of *IPK1* using RNAi technology reduced phytate content in rice (Sweetman *et al.,* 2006) while maize with low phytate content was developed by site specific insertional mutation of zinc finger nucleases (ZFNs) (Shukla *et al.,* 2009). Furthermore, precise genetic engineering methods are potential to develop millets and pseudo cereals biofortified with nutrients.

Bai *et al.* (2013) reported application of whole genome sequence available for foxtail millet landrace using Solexa sequencing technology through Genome Analyzer ii (GA ii) for creation of rich tag library for future breeding experiments. Genes responsible for important agronomic qualities are controlled by action of many genes present in different loci, termed as QTL. Identification and selection of QTLs form a major milestone in pooling them to bring out the desired trait, genotyping technologies involving high-throughput methods quickly identify QTLs and are cost effective in comparison to traditional selection methods. Next generation sequencing using Illumina and Roche 454, identified SNPs and SSR markers in finger millet (Gimode *et al.,* 2016). Hittalmani *et al.* (2017) reported finger millet ML-365 whole genome sequence using Illumina platform and SOLiD technology. A genetic map of high density, five markers of co-dominant nature and identification of QTLs of eleven important traits of foxtail millet using RAD-Seq technology was reported. Foxtail millet whole genome sequence indicated presence of 2824 drought responsive genes.

Granule bound starch synthase I (*GBSSI*) gene was characterized using genomic sequence of buckwheat. It composed of 3947 nucleotides, 14 exons and 13 introns. Resemblance in the range of 73.4%–77.8% and 67.6%–70.4% was observed for starch sysnthesis genes of dicots and monocots, respectively (Wang *et al.,* 2014). Plants are capable of recognizing various environmental signals and substantial changes occur in primary and secondary metabolism thereby adaptation to environmental stresses are met out successfully. Comparative analysis of transcriptome using illumina sequencing approach of common buckwheat accessions cultivated in simulated saline alkali conditions using 100 mM sodium chloride revealed regulatory genes controlling salt tolerance and resistance. 26,672 unigenes annotated in four databases included 47 upregulated and 07 down regulated genes during salt stress. The derived information is an important tool for breeding of common buckwheat for salt tolerance (Lu *et al.,* 2018). Jeon *et al.* (2018) reported cold temperature induced changes in transcripts of phenylpropanoid biosynthetic pathway (anthocyanins and proanthocyanidins) and metabolism of carbon and nitrogen response in tartary buckwheat cultivars. Up regulation of genes responsible for production of phenyl propanoids, increase in sugars and reduction of amino acids were observed in metabolomic profiles using GC-TOFMS. Tricarboxylic acid cycle derived organic acids, phenylpropanoid biosynthetic transcripts and its products, anthocyanins and proanthocyanidins, carbon-nitrogen metabolism increased on exposure to cold stress in comparison to plants grown at 25°C. Adaption to environmental stresses through primary secondary metabolic pathways were reported by transcriptomic and metabolic profile of tartary buckwheat (Jeon *et al.,* 2018). Kotrade *et al.* (2020) studied genes responsible for drought response among *Paspalum* species such as *P. bisulcatum*, *P. laetum* and *P. turgidum*. RNA sequencing of transcripts from plants exposed to severe drought like situations were studied. 150 genes were expressed in *P. laetum* and *P. turgidum* and indicated their more drought tolerant nature in comparison to *P. bisulcatum* using comparative transcriptional analysis. Discoveries pertaining to specific genes for important nutritional traits would be useful for crop improvement programmes of related cereals and millets (Muthamilarasan and Prasad, 2015). Zinc transporters engineered finger millets using overexpression of *OsZIP1,* transgenics showed zinc content of 10 to 15 mg/kg in first generation followed by 20 mg/kg in second generation. Manganese levels also improved in the transgenics engineered for enhancement of zinc content of grains (Ramegowda *et al.,* 2013).

(ii) Gene transformations and *in vitro* plant regeneration

Plant tissue culture or micro propagation or in vitro plant regeneration involves regeneration of a whole plant from small sections of plant tissue, or entire structures such as buds, and culturing them using specific plant growth media under artificial conditions. Genetic transformation experiments on crops require protocols for growing plants *in vitro* under artificial conditions after gene transformation using tissues or whole plant parts. Genetic engineering of crops using genes for desirable traits can be obtained using direct methods such as gene gun and in direct methods through *Agrobacterium*-mediated transformation (AMT) by infection of the host part with desired gene carried by *Agrobacterium* (Shrawat and Lo¨rz , 2006). Genetic modification can be obtained by transferring genes between organisms belonging to different kingdoms termed as distant transfer, genes transferred from one species to other names as close transfer and modification of genes already present in the host to change the expression (Kothari *et al.,* 2005; O'Kennedy *et al.,* 2006). Cereal gene transformations were obtained by *Agrobacterium* such as LBA4404, EHA10, EHA105, EHA101, AGL0 and AGL1 (Cheng *et al* 2004) while millet genetic transformations reported LBA4404 for foxtail millet (Fang *et al.,* 2007) and millet transformation is still in infancy. Finger millet gene transformations resulted in transgenic finger millet plants with resistance to abiotic stresses (Anjaneyulu *et al.,* 2014) such as drought and salinity as well disease resistance (Latha *et al.,* 2005). Biolistic method of gene transformation using callus explants for RbcS/Ft with promoters such as cauliflower mosaic virus CaMV35S,rice actin gene (ActI) and maize ubiquitin gene (UqI) was reported (Gupta *et al.,* 2001) while disease resistant plants were obtained by transformation of embryonic callus (Latha *et al* 2005). Fang *et al.* (2007) reported AMT transformation of callus using *Zm13a* gene in finger millet. Barnyard millet gene transformation with rice ribulose 1,5 biphosphate carboxylase gene *RbcS* (small subunit gene promoter) Ft5 promoter (C4 isoform of phosphoenolpyruate carboxylase gene from *Flaveria trinervia*) (Gupta *et al.,* 2001). Transgenic pearl millet (*Pennisetum glaucum* (L.) R. Br.) plants resistant to downy mildew were reported (Latha *et al.,* 2006).

In vitro regeneration of crops act as starting material or explant for genetic transformation and regeneration into whole plants after transformation, screening for genes for localized insertion are essential steps after gene transformation to make the whole process successful. Rangan (1976) reported successful *in vitro* plant regeneration of cereal and millets during 1970s using specific media ingredients to grow monocots. Yemets *et al.* (2013) reported an efficient regeneration and microplast isolation protocol for finger millet (*E. coracana* (L.) Gaertn.) from callus using N6 medium. Chromosomal transfer between alien crops and finger millet may be possible using the developed protoplast. Genetic engineering and successful plant regeneration of foxtail millet engineered for down regulated phosphate transfer genes was reported. *Agrobacterium*-mediated transformation (AMT) produced foxtail millet plants down regulated phosphate transporter family genes (PHT1) such as *SiPHT1:2 SiPHT1:3* and *SiPHT1:4* and increased phosphate availability in roots (Ceasar *et al.,* 2013). Samantaray *et al.* (2001) standardised protocols for *in vitro* plant regeneration of chromium and nickel tolerant *Echinochloa colona* L. using callus cultures derived from mesocotyly explants, leaf tip and base. Murashige and Skoog (1962) medium (mg/l) supplemented with 0.5 benzyl adenine (BA), 3.0 2,4-D, 1.5 chromium and 2.0 nickel supported the growth of calli while somatic embryos were established using MS medium supplemented with 1.0-1.5 mg/L BA, 0.5-1.5 kinetin and 0.25-0.5 naphthalene acetic acid. *Paspalum notatum* Flugge. (Bahia grass) subjected to genetic transformation using biolistic gene gun method was reported (Altpeter and James, 2005).

Unpark (2018) summarised information regarding biotechnological approaches reported for buckwheat improvement through plant tissue culture, hairy root culture and gene transformations which are essential for future studies on the crop. Akulov *et al.* (2018) reported total phenolic content and phenolic composition in long-term (3–10 years) of tartary buckwheat calli with different types of cell differentiation and distinct morphogenic ability.

In non-morphogenic calli, the content of phenolic compounds was 2.9–7.5 times less compared to morphogenic calli. Rutin content was found in traces in non-morphogenic calli while 0.87% rutin was reported in morphogenic calli. Higher rutin content may be attributed to synthesis in the proembryonal cell complexes during embryonic calli development. Phenolic acids, catechin, epicatechin, quercetin, and quercitrin were also found in morphogenic calli. Transgenic lines developed contains transgenes and have to be subjected to stringent testing as they contain introduced genes from inter and intra species to confer desired traits. Harmful effects and allergenicity are to be tested as a proof for safe food use. Detailed protocols to establish safety of transgenics has been provided by Food and Agricultural Organisation (FAO) and World Health Organisation (WHO) (Halsberger, 2003)

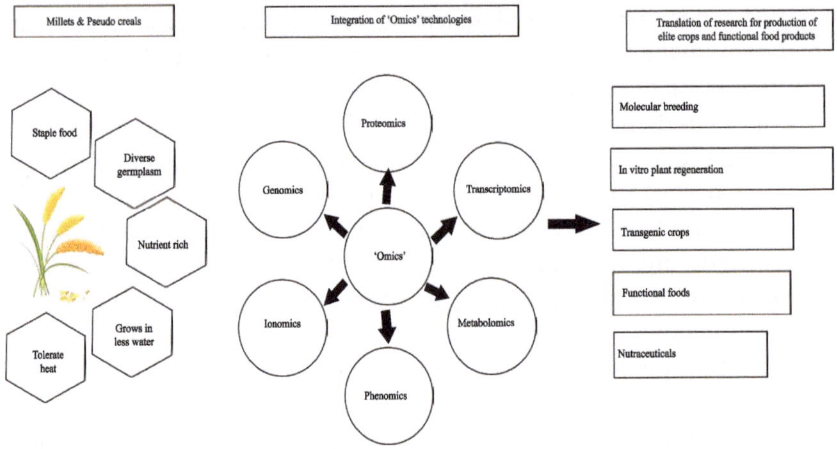

Fig. 2: Schematic representation of biotechnological approaches for development of millets and pseudo cereals

Conclusions and Future Prospects

Biotechnology is an advanced technology which is an amalgamation of various streams of science and could bring revolutionary and desired changes for crop improvement and development of processed foods. Modern biotechnological tools has been successfully used in improving yield as well as for reduction of anti-nutritional factors, enhancing nutritive value and aid in improvement of functional value of processed foods as desired by customer/users for use as nutraceutical and functional food. Multifaceted biotechnological tools and their application in field of food fermentation, microbial intervention in food processing for production of food additives as well as molecular diagnostics for food safety and hygiene are well known. Furthermore, biotechnology as a tool would enhance developments pertaining to millet and pseudo cereal based crops and food processing in a larger scale. Commercialisation of millet and pseudo cereals based foods and their importance in international and national feeding programmes would definitely make farmers develop interest to grow these lesser known crops in larger scale. This would also regenerate newer opportunities for farming community and larger use of millet-based foods would pave way for nutritious and health society, by eradication of malnutrition and thereby deficiency diseases.

References

Akulov Anton N, Elena A. Gumerova and Natalya I.Rumyantseva (2018). Cell Cultures of Fagopyrum tataricum as a Source of Biologically Active Phenolic Compounds In: Buckwheat Germplasm in the World pp 259-270 https://doi.org/10.1016/B978-0-12-811006-5.00025-2

Kang Man, Fei-Hong Zhai, Xin-Xin Li, Jin-Liang Cao and Jian-Rong Han (2017). Total phenolic contents and antioxidant properties of buckwheat fermented by three strains of Agaricus Journal of Cereal Science 73: 138-142 https://doi.org/10.1016/j.jcs.2016.12.012

Altpeter F, James VA (2005) Genetic transformation of turftype bahiagrass (Paspalum notatum Flugge) by biolistic gene transfer. Int Turfgrass Soc Res J 10:1–5

Amadou Issoufou Guo-WeiLe Tidjani AmzaJin and SunYong-HuiShi (2013). Purification and characterization of foxtail millet-derived peptides with antioxidant and antimicrobial activities. Food Research International 51(1): 422-428

Anjaneyulu, E., Reddy, P. S., Sunita, M. S., Kishor, P. B., and Meriga, B. (2014). Salt tolerance and activity of antioxidative enzymes of transgenic finger millet overexpressing a vacuolar HC-pyrophosphatase gene (SbVPPase) from Sorghum bicolor. J. Plant Physiol. 171, 789–798. doi: 10.1016/j.jplph.2014.02.001

Araki, M.,Numaoka,A.,Kawase,M.,andFukunaga,K.(2012). Origin of waxy common millet, *Panicum miliaceum* L. in Japan. Genet. Res. Crop Evol. 59, 1303–1308. doi: 10.1007/s10722-011-9755-9

Avashthi H, Pathak RK, Pandey N, Arora S, Mishra AK, Gupta VK, Kumar A. 2018. Transcriptome-wide identification of genes involved in Ascorbate–Glutathione cycle (Halliwell–Asada pathway) and related pathway for elucidating its role in antioxidative potential in finger millet (*Eleusine coracana* (L.)). 3 Biotech. 1-18.

Bader Johannes, Christopher Brigham, Milan K. Popović (2018) Chapter 5 - Development of Controlled Co-cultivations for Reproducible Results in Fermentation Processes in Food Biotechnology In: Advances in Biotechnology for Food Industry A volume in Handbook of Food Bioengineering. Edited by: Alina Maria Holban and Alexandru Mihai Grumezescu pp 135-165

Bai,H.,Cao,Y.,Quan, J., Dong, L., Li, Z., Zhu, Y., et al. (2013). Identifying the genome-wide sequence variations and developing new molecular markers for genetics research by re- sequencing a land race cultivar of foxtail millet. PLoS ONE 8:e73514. doi: 10.1371/journal.pone.0073514

Ceasar, S. A., Hodge, A., Baker, A., & Baldwin, S. A. (2014). Phosphate concentration and arbuscular mycorrhizal colonisation influence the growth, yield and expression of twelve PHT1 family phosphate transporters in foxtail millet (*Setaria italica*). PLoS One, 9(9), e108459.

Chandrasekara A, Shahidi F. (2012). Bioaccessibility and Antioxidant Potential of Millet Grain Phenolics as Affected by Simulated in Vitro Digestion and Microbial Fermentation. J Funct Foods. 4: 226–237.

Chen Tiejun, Meizi Piao,Syed Md.Ehsanur Rahman, Lehong Zhang and Yang Deng (2020). Influence of fermentation on antioxidant and hypolipidemic properties of maifanite mineral water-cultured common buckwheat sprouts Food Chemistry 321: 126741 https://doi.org/10.1016/j.foodchem.2020.126741

Cheng M, Lowe BA, Spencer TM, Ye X, Armstrong CL (2004). Factors influencing Agrobacterium-mediated transformation of monocotyledonous species. *In Vitro* Cell Dev Biol Plant 40:31–45

Collard Bertrand C. Y. and David J. Mackill (2008). Marker-assisted selection: an approach for precision plant breeding in the twenty-first century. Phil. Trans. R. Soc. B (2008) 363, 557–572 doi:10.1098/rstb.2007.2170

Coulibaly, A., Kouakou, B., and Chen, J. (2012). Extruded adult breakfast based on milled and soybean: Nutritional and functional qualities, source of glycemic low food. J. Nutr. Food Sci. 2(7):1–9. doi.:10.1016/ j.plaphy.2018.02.001

Fang FQ, Qian Z, Guang MA, Jing JY (2007). Co-suppression of Si401, a maize pollen speciWc Zm401 homologous gene, results in aberrant anther development in foxtail millet. Euphytica 163:103–111

Fasreen MMF, Perera ODAN, Weerahewa HLD. (2017). Development of finger millet based probiotic beverage using lactobacillus casei431®. OUSL. J. 128-138.

Fukunaga K, Kawase M, Kato K. 2002. Structural variation in the Waxy gene and differentiation in foxtail millet [*Setaria italica* (L.) P. Beauv.]: implications for multiple origins of the waxy phenotype.Mol Genet Genomics 268: 214–222.

Galati, A., Oguntotinbo, F. A., Moschetti, G., Crescimanno, M., and Settanni, L. (2014). The cereal market and the role of fermentation in cereal-based food production in Africa. Food Rev. Int. 30:317–337.

Gebrekidan, B., 1982. The Ethiopia Sorghum Improvement Project. Terminal Report. IDRC, Ottawa.

Ghoshal Gargi (2018) Biotechnology in Food Processing and Preservation: An Overview In: Advances in Biotechnology for Food Industry A volume in Handbook of Food Bioengineering Edited by: Alina Maria Holban and Alexandru Mihai Grumezescu pp 27-54

Gimode D, Odeny DA, de Villiers EP, Wanyonyi S, Dida MM, Mneney EE, et al . (2016). Identification of SNP and SSR Markers in Finger Millet Using Next Generation Sequencing Technologies. PLoS ONE 11(7): 0159437. https://doi.org/10.1371/journal.pone.0159437

Gomez MI. Preliminary studies on grain quality evaluation for finger millet as a food and beverage use in the southern African region. In: Advances in small millets. Riley KW., Gupta SC, Seetharam A, Mushong, JN New Delhi, India, Oxford and IBH Publishing, 1993, 289-296.

Goron, T.L.,andRaizada,M.N.(2015). Genetic diversity and genomic resources available for the small millet crops to accelerate a New Green Revolution. Front. Plant. Sci. 6:157. doi: 10.3389/fpls.2015.00157

Gupta Nidhi , Sunil K.Sharma, Jai C.Rana and Rajinder S.Chauhan (2012). AFLP fingerprinting of tartary buckwheat accessions (*Fagopyrum tataricum*) displaying rutin content variation Fitoterapia 83(6): 1131-1137 doi.:10.1016/j.fitote.2012.04.015

Gupta P, Raghuvanshi S, Tyagi AK (2001) Assessment of the efficiency of various gene promoters via biolistics in leaf and regenerating seed callus of millets, *Eleusine coracana* and *Echinochloa crusgalli.* Plant Biotechnol 18: 275–282

Gupta, R., Verma, K., Joshi, D. C., Yadav, D., and Singh, M. (2010). Assessment of genetic relatedness among three varieties of finger millet with variable seed coat color using RAPD and ISSR markers. Genet. Eng. Biotechnol. J. 2, 1–9.

Gupta, S. M., Arora, S., Mirza, N., Pande, A., Lata, C., Puranik, S., et al . (2017). Finger Millet: a "certain" crop for an "uncertain" future and a solution to food insecurity and hidden hunger under stressful environments. Front. Plant Sci. 8:643. doi: 10.3389/fpls.2017.00643

Halsberger, A.G., 2003. GM Food: The risk assessment of immune hypersensitivity reactions covers more than allergenicity. Food, Agriculture and Environment 1, 42–45 /http://www.biotech-info.net/hypersensitivity.htmlS.

Hittalmani, S., Mahesh, H.B., Shirke, M.D. et al . (2017). Genome and Transcriptome sequence of Finger millet (*Eleusine coracana* (L.) Gaertn.) provides insights into drought tolerance and nutraceutical properties. BMC Genomics 18, 465 https://doi.org/10.1186/s12864-017-3850-z

Jeon Jin, Jae Kwang Kim, Qi Wu and Sang UnPark (2018). Effects of cold stress on transcripts and metabolites in tartary buckwheat (*Fagopyrum tataricum*) Environmental and Experimental Botany (155): 488-496 https://doi.org/10.1016/j.envexpbot.2018.07.027

Katsu Kenjiro , Tatsuro Suzuki, Kaien Fujino, Toshikazu Morishita and Takahiro Noda (2019). Development of a DNA marker for variety discrimination specific to 'Manten-Kirari' based on an NGS-RNA sequence in Tartary buckwheat (*Fagopyrum tataricum*)495: 51-57 https://doi.org/10.1016/j.foodchem.2019.05.087

Kothari SL, Kumar S, Vishnoi RK, Kothari SL, Watanabe KN (2005). Applications of biotechnology for improvement of millet crops: review of progress and future prospects. Plant Biotechnol 22:81–88

Kotrade Peter, Anahita Werner, Ingo Ebersberger and Wolfgang Brüggemann (2020) Comparative transcriptomics reveal insights into the drought response of the three *Panicum species P. bisulcatum, P. laetum* and *P. turgidum*.Plant Gene 23: 100232 https://doi.org/10.1016/j.plgene.2020.100232

Krishna Ajeesh T.P., T.Maharajan, R. Host Antony David, M.Ramakrishnan S. Antony Ceasar, V.Duraipandiyan G. Victor Roch and S.Ignacimuthu (2018) Microsatellite markers of finger millet (*Eleusine coracana* (L.) Gaertn) and foxtail millet (*Setaria italica* (L.) Beauv) provide resources for cross-genome transferability and genetic diversity analyses in other millets Biocatalysis and Agricultural Biotechnology 16: 493-501 doi.10.1016/j.bcab.2018.09.009

Kumar A, Kaur A, Tomer V, Rasane P, Gupta K. 2019. Development of nutricereals and milk based beverage: Process optimization and validation of improved nutritional properties. J. Food. Process. Eng. 1-9

Kumari Supriya, Seerat Bhinder, Balwinder Singh, Amritpal Kaur and Narpinder Singh (2020) Effect of buckwheat incorporation on batter fermentation, rheology, phenolic, amino acid composition and textural properties of idli J Food Sci Technol 54(6):1425–1432 DOI 10.1007/s13197-017-2561-4

Kuwano, M.,Mimura,T.,Takaiwa,F.,andYoshida,K.T.(2009).Generation of stable 'low phytic acid' transgenic rice through antisense repression of the1 D–myoinositol 3– phosphate synthase gene using the18-kDa oleosin promoter. Plant Biotechnol.J. 7, 96–105. doi: 10.1111/j.1467-7652.2008.00375.x

Latha MA, Rao KV, Reddy TP, Reddy VD (2006) Development of transgenic pearl millet (*Pennisetum glaucum* (L.) R. Br.) plants resistant to downy mildew. Plant Cell Rep 25:927–935

Latha MA, Venkateswara Rao K, Dashavantha Reddy V (2005) Production of transgenic plants resistant to leaf blast disease in finger millet (*Eleusine coracana* (L.) Gaertn.). Plant Sci 169:657–667

Lei Vicki, Henrik Friis and Kim Fleischer Michaelsen (2006) Spontaneously fermented millet product as a natural probiotic treatment for diarrhoea in young children: An intervention study in Northern Ghana.International Journal of Food Microbiology, 295 (3): 246-253

Liptáková Denisa, Zuzana Matejčeková and Ľubomír Valík (2017) Lactic Acid Bacteria and Fermentation of Cereals and Pseudocereals In: Fermentation Processes pp 221-253 DOI: 10.5772/65459

Lu Qi-Huan, Ya-Qi Wang, Jin-Nan Song and Hong-Bing Yang (2018) Transcriptomic identification of salt-related genes and de novo assembly in common buckwheat (*F. esculentum*) Plant Physiology and Biochemistry 127: 299-309

Manimekalai M., M. Dhasarathan , A. Karthikeyanb , J. Murukarthick , V.G. Renganathan , K. Thangaraj, S. Vellaikumar , C. Vanniarajan and N. Senthil (2018)Genetic diversity in the barnyard millet (*Echinochola frumentacea*) germplasms revealed by morphological traits and simple sequence repeat markers Current Plant Biology, 14: 71-78

Mugocha PT, Taylor JRN, Bester BH. Fermentation of a composite finger millet dairy beverage. World Journal of Microbiology and biotechnology. 2000; 16:341-344.

Murashige T, Skoog F (1962) A revised medium for rapid growth and bioassays with tobacco tissue cultures. Physiol Plant 15: 473–497

Muthamilarasan, M., Venkata Suresh, B., Pandey, G., Kumari, K., Parida, S. K., and Prasad,M. (2014). Development of 5123 intron-length polymorphic markers for large-scale genotyping applications in foxtail millet. DNA Res. 21, 41–52. doi: 10.1093/dnares/dst039

Muthamilarasan,M., and Prasad,M.(2015).Advances in Setaria genomics for genetic improvement of cereals and bioenergy grasses. Theor.Appl.Genet. 128, 1–14. doi: 10.1007/s00122-014-2399-3

Muthamilarasan, M., Dhaka, A.,Yadav, R., and Prasad, M. (2016). Exploration of millet models for developing nutrient rich graminaceous crops. Plant Sci. 242, 89–97. doi: 10.1016/j.plantsci.2015.08.023

Mythrayee R, Pavithra A. Comparative Study on Nutritive Content of finger millet-wheat composite bread fermented with lactic acid bacilli and yeast. Journal of Biotechnology and Biochemistry. 2017; 3:15-21.

Nirmalakumari A, Vetriventhan M. 2010. Characterization of foxtail millet germplasm collections for yield contributing traits. Elect J Plant Breed 1:140–147.

O'Kennedy MM, Grootboom A, Shewry PR (2006) Harnessing sorghum and millet biotechnology for food and health. J Cereal Sci 44:224–235

Palaniappan Ayyappan, V. Geetha Balasubramaniam & Usha Antony (2017) Prebiotic potential of xylooligosaccharides derived from finger millet seed coat, Food Biotechnology, 31:4, 264-280, DOI: 10.1080/08905436.2017.1369433

Ramegowda, Y.,Venkategowda, R., Jagadish, P., Govind, G., Hanumanthareddy, R. R., Makarla, U (2013). Expression of a rice Zn transporter, OsZIP1, increases Zn concentration in tobacco and finger millet transgenic plants. Plant Biotechnol.Rep. 7, 309–319. doi: 10.1007/s11816-012-0264-x

Rangan TS (1976) Growth and plantlet regeneration in tissue cultures of some Indian millets: *Paspalum scrobiculatum* L., *Eleusine coracana* Gaertn. and *Pennisetum typhoideum* Pers. Z Pflanzenphysiol 78: 208–216

Rao, B. D., Bhaskarachary, K., Christina, G. D. A., Devi, G. S., & Tonapi, V. (2017). Nutritional and health benefits of millets. ICAR Indian Institute of Millets Research (IIMR): Rajendranagar, Hydrabad, PP. 112.

Samantaray Sanghamitra , Gyana Ranjan Rout and Premananda Das (2001) Induction, selection and characterization of Cr and Ni-tolerant cell lines of *Echinochloa colona* (L.) Link in vitro.Journal of Plant Physiology 158(10): 1281-1290 https://doi.org/10.1078/0176-1617-00533

Shaikh AM, Sreeja V, Desai RR. 2017. Effect of malted and unmalted finger millet flour and its rates of incorporation on quality attributes of finger millet enriched probiotic fermented milk product. Int. J. Curr. Microbiol. App. Sci. 2258-2266.

Shrawat AK and Lo¨rz H (2006) Agrobacterium-mediated transformation of cereals: a promising approach crossing barriers. Plant Biotechnol J 4:575–603

Shukla, V.K., Doyon,Y., Miller, J.C., DeKelver, R.C., Moehle, E.A.,Worden, S.E., et al . (2009). Precise genome modification in the crop species Zeamays using zinc-finger nucleases. Nat. Lett. 459, 437–443. doi: 10.1038/nature07992

Singh P, Raghuvanshi RS. Finger millet for food and nutritional security. African Journal of Food Science.2012; 6:77-84.

Stefano Elisa Di, Jessica White, Shannon Seney, Sharareh Hekmat and Tim McDowell Mark Sumarah 4 and Gregor Reid (2017) A novel millet-based probiotic fermented food for the developing world nutrients 2017, 9, 529; doi:10.3390/nu9050529

Subhash B, Kakade, Hathan BS. Finger Millet Processing: Review. International Journal of Agriculture Innovations and Research. 2015; 2319-1473.

Sweetman, D.,Johnson,S.,Caddick,S.E.,Hanke,D.E., and Brearley,C.A. (2006). Characterization of an Arabidopsis inositol 1,3,4,5,6-pentakisphosphate 2-kinase(AtIPK1). Biochem.J. 394, 95–103. doi: 10.1042/BJ20051331

Taylor, J.R.N., and Emmambux,M.N.(2008).“Gluten: free foods and beverages from millets,”in Gluten-Free Cereal Products and Beverages (Food Science and Technology), edsE.K. Arendt and F. Dal Bello (Cambridge, MA: Academic Press), 119–148.

Thirumangaimannan Geetha & Kalaichelvan Gurumurthy (2013) A study on the fermentation pattern of common millets in Koozh preparation – a traditional South Indian food Indian Journal of Traditional Knowledge 12(3):512-517

Tiwari Neelam, Sharad Tiwari and Niraj Tripathi (2018) Genetic characterization of Indian little millet (*Panicum sumatrense*) genotypes using random amplified polymorphic DNA markers Agriculture and Natural Resources 52(4): 347-353

Upadhyaya, H., Gowda, C., and Reddy, V. G. (2007). Morphological diversity in finger millet germplasm introduced from Southern and Eastern Africa. J. SAT Agric. Res. 3, 1–3.

Varma V and Patel S. Value added products from nutria cereals: Finger millet. Emiates Journal of Food and Agriculture. 2013; 25:169-176

Vetriventhan, M., Upadhyaya, H.D., Dwivedi, S.L., Pattanashetti, S.K.,and Singh, S. K. (2016). “Finger and foxtail millets,” in Genetic and Genomic Resources for Grain Cereals Improvement, eds M.Singh and H.D. Upadhyaya (Cambridge, MA: Academic Press),291–319.

Vinoth A and Ravindhran R (2017) Biofortification in Millets: A Sustainable Approach for Nutritional Security. Front. Plant Sci. 8:29. doi: 10.3389/fpls.2017.00029

Wang J., Wang Z., Du X., Yang H., Han F., Han Y., et al ., (2017). A high-density genetic map and QTL analysis of agronomic traits in foxtail millet [*Setaria italica* (L.) P. Beauv.] using RAD-seq. PLoS ONE 12 (6): e0179717. https://doi.org/10.1371/journal. pone.0179717

Wang Ruiyun, Wang Haigang, Xiaohuan Liua , Xu Jia , Ling Chenb , Ping Luc , Minxuan Liuc, Bin Tengd and Zhijun Qiao (2018) Waxy allelic diversity in common millet (*Panicum miliaceum* L.) in China. The crop journal 377-385 https://doi.org/10.1016/j.cj.2018.02.004 2214-5141

Wang Xun, Bo Feng, Zhibin Xu, Francesco Sestili, Guojun Zhao, Chao Xiang, Domenico Lafiandra and Tao Wang (2014) Identification and characterization of granule bound starch synthase I (GBSSI) gene of tartary buckwheat (*Fagopyrum tataricum* Gaertn.) Gene 534 (2): 229-235 https://doi.org/10.1016/j.gene.2013.10.053

Yadav Chandra Bhan, Venkata Suresh Bonthala, Mehanathan Muthamilarasan, Garima Pandey, Yusuf Khan and Manoj Prasad (2014) Genome-wide development of transposable elements-based markers in foxtail millet and construction of an integrated database, DNA Research, 2014, 1–12 doi: 10.1093/dnares/dsu039

Yadav Geetanjali, Anshu Singh, Patrali Bhattacharya, Jude Yuvraj and Rintu Banerjee (2013) Comparative analysis of solid-state bioprocessing and enzymatic treatment of finger millet for mobilization of bound phenolics. Bioprocess and Biosystems Engineering. 36: 1563–1569

Yadav, C.B., Muthamilarasan, M., Pandey, G. Yusuf Khan and Manoj Prasad (2014). Development of novel microRNA-based genetic markers in foxtail millet for genotyping applications in related grass species. Mol Breeding 34: 2219–2224 https://doi.org/10.1007/s11032-014-0137-9

Yemets AI, Klimkina LA, Tarassenko LB, Blume YB (2003) Efficient callus formation and plant formation of goosegrass [*Eleusine indica* (L.) Gaertn.]. Plant Cell Rep 21: 503–510

Zhang,G., Liu,X., Quan,Z., Cheng,S., Xu,X.,Pan,S (2012).Genome sequence foxtail millet (*Setaria italica*) provides insights into grass evolution and biofuel potential. Nat. Biotechnol. 30, 549–554. doi: 10.1038/nbt.2195

Zielin´ski Henryk, Zuzana Ciesarova´, Kristı´na Kukurova´, Danuta Zielinska, Dorota Szawara-Nowak, Małgorzata Starowicz and Małgorzata Wronkowska (2017) Effect of fermented and unfermented buckwheat flour on functional properties of gluten-free muffins LWT - Food Science and Technology 81: 35-41https://doi.org/10.1016/j.lwt.2020.109042

Zieliński Henryk, Dorota Szawara-Nowak, Natalia Bączek and Małgorzata Wronkowska (2019) Effect of liquid-state fermentation on the antioxidant and functional properties of raw and roasted buckwheat flours. Food Chemistry 271: 291-297